CHEMICAL ENGINEERING METHODS AND TECHNOLOGY

INDIUM

PROPERTIES, TECHNOLOGICAL APPLICATIONS AND HEALTH ISSUES

Chemical Engineering Methods and Technology

Additional books in this series can be found on Nova's website under the Series tab.

Additional E-books in this series can be found on Nova's website under the E-book tab.

Materials Science and Technologies

Additional books in this series can be found on Nova's website under the Series tab.

Additional E-books in this series can be found on Nova's website under the E-book tab.

CHEMICAL ENGINEERING METHODS AND TECHNOLOGY

INDIUM

PROPERTIES, TECHNOLOGICAL APPLICATIONS AND HEALTH ISSUES

HSAIO G. WOO
AND
HUANG TSAI CHOI
EDITORS

nova publishers
New York

For permission to use material from this book please contact us:
Telephone 631-231-7269; Fax 631-231-8175
Web Site: http://www.novapublishers.com

LIBRARY OF CONGRESS CATALOGING-IN-PUBLICATION DATA

Indium : properties, technological applications and health issues / editors, Hsaio G. Woo and Huang Tsai Choi.
p. cm.
Includes index.
ISBN 978-1-62257-696-8 (hardcover)
1. Indium--Research. I. Woo, Hsaio G. II. Tsai Choi, Huang.
QD181.I5I53 2011
546'.677--dc23 2012030288

Published by Nova Science Publishers, Inc. † New York

CONTENTS

PREFACE

In this book, the authors gather and present current research from across the globe in the study of the properties, technological applications and health issues relating to indium. Topics discussed include the properties of indium, gallium and Ga-In alloys in confined geometry; indium as a co-catalyst in Novel Cu or Ni catalysts for reduction of biomass derived fatty acids to alcohols; reliability of indium tin oxide thin films; indium reagents for chemical transformations; indium atoms using electron and optical spectroscopy; indium sulfide-manganese ore; rare earth ions doping on the photoluminescent properties of indium hydroxide; and hydrothermal synthesis of indium tin oxide nanoparticles without chlorine contamination.

Chapter 1 - The review of the authors' recent studies is focused on properties of nanocomposites consisted of opal and porous glass matrices which pores are filled with gallium, indium, and their alloys. The atomic mobility in confined melts, melting and freezing in metallic particles within pores, and changes in the Knight shift of NMR lines in the nanocomposites will be discussed. The experimental results were obtained by NMR, acoustic and resistance measurements, and X-ray powder diffraction. Atomic diffusion in confined melts was found to slow down drastically compared to the bulk counterparts. Additional slowing down in atomic mobility was observed for the confined metals in the supercooled state. The Knight shift of the NMR lines which is proportional to the electron susceptibility was shown to decrease in liquid confined metals and alloys. The melting and crystallization for metallic components of nanocomposites were noticeably moved to lower temperatures and strongly diffused. New crystalline phases of gallium were found to emerge under nanoconfinement.

Chapter 2 - Supported copper or especially nickel catalysts are very important for hydrogenation of various organic compounds in the practice. Activity, selectivity and stability can be greatly improved using co-catalysts: e.g. chromium compounds with Cu (Adkins catalyst for fatty alcohol production) or molybdenum compounds with Ni (for hydrodesulfurization [HDS], hydrodeoxygenation [HDO], hydrodenitrogenation [HDN], etc. catalysts in petroleum refining.

Recently, the authors discovered novel highly efficient bimetallic supported catalysts which were exceedingly active and selective in the hydrodeoxygenation (HDO) of octanoic acid to octanol at moderate pressure and temperature in addition very effective in reduction of acetic acid to ethanol. The catalysts contained indium as co-catalyst and a metal of high hydrogenation activity, such as, copper or nickel, on various support. A significant synergism can be observed when indium with another metal of hydrogenation activity is applied. Appearance of Cu_2In or Ni_2In phases, intermetallic compounds on the surface of metal particles results in significant increase of desired hydrogenation activity, i.e., in stepwise hydrogenation of carboxylic acids stopped at alcohol formation and in inhibiting hydrogenations when hydrodecarbonylation of carboxylic acids (loss of one carbon atom) and hydrogenolysis of hydrocarbon products (loss of further carbon atoms) proceed. Moreover, in presence of mobile indium atoms or of indium containing nano clusters mono- or bimolecular alcohol dehydration capability of the support is selectively poisoned.

In order to have more information about the role of indium, experiments are carried out using quite different hydrogenation metals, Cu and Ni. Hydrogenation experiments are carried out with long chain fatty acid like octanoic acid for production of octyl alcohol and short chain acetic acid (produced in large amounts from biomass) for production of ethyl alcohol, ethyl acetate or acetaldehyde. The main question is how the co-catalyst second metal of moderate hydrogenation activity as gallium, cadmium, tin, but first of all indium, modify the activity of the supported hydrogenation metal and the surface properties of the support.

Chapter 3 - Indium antimonide is known as one of the best studied A^3B^5 compounds both theoretically and experimentally. Well established band structure and developed technology have made InSb an attractive material in experimental physics and practical applications. Manganese doped InSb is a unique material, which has provided us a chance to study metal-insulator transition (MIT) simultaneously under a variety of experimental conditions (impurity concentration variations, magnetic field, uniaxial and hydrostatic compression). The results of the experiments with manganese doped InSb single crystals have stimulated us to explore the nature of Colossal Magnetoresistance (CMR) and Excitonic Insulator (EI) phase-effects which interpretation has remained the matter of discussions for decades. Comparative study of magnetotransport effects in p-InSb(Mn) and uniaxially stressed p-InSb(Ge) crystals provided in the CHAPTER have shown that specific p-InSb(Mn) properties could rather be related to the influence of Jahn-Teller distortion of crystal lattice caused by Mn^{+2} ions than to the interaction between manganese ions as it usually takes place in magnetic materials – DMS, MS and Manganite Perovskites. Coexistence of electrons and holes at temperatures below 4K at manganese impurity concentration $\sim 10^{17} cm^{-3}$ around the metal-insulator transition in p-InSb(Mn) crystals gives the ground to explain the MIT in the framework of excitonic insulator phase model. An external magnetic field causes destruction of permanent excitons in InSb(Mn) and so causes the increase in charge carrier concentration, i.e. magnetic field induces CMR and drives IMT transition. The existence of CMR in the nonmagnetic semiconductor has led us up to the concept of the same physical CMR nature origin in DS, DMS, MS and Manganite Perovskites, i.e. Colossal Magnetoresistance can be described by excitons behavior in magnetic field. An existence of the gap energy Δ in InSb(Mn) in the millikelvin range of temperature could be used in creation of advanced cryogenic devices - thermometers, bolometers and selective sensors for millimeter and radio-frequency wavelength.

Chapter 4 - Indium-mediated organometallic reactions have elicited considerable attention from the discovery of the remarkable reactivity of the metal in organic or aqueous media. Organometallic species may be postulated when In(0) or In (I) is inserted into a carbon-halogen bond. Organoindium reagents may also be prepared by transmetalation. This chapter is specifically devoted to alkynyl, alkenyl and aryl indium reagents and their use in organic synthesis.

Chapter 5 - One of the primary applications of indium has been in the form of tin doped indium oxide (ITO). ITO is the preferred transparent conducting layer in many applications, such as liquid crystal displays, plasma displays, touchscreens, and photovoltaics. The high electrical conductivity, optical transmittance and moisture resistance make ITO a dominant transparent conductor. Recently, there has been a tremendous rise in production of portable electronic devices and a necessity of flexible devices is increasing more than ever. To produce flexible devices, flexible substrates should replace conventional rigid substrates. Although ITO has a low sheet resistance and very good transparency, ITO thin film deposited on a flexible substrate produces cracks at low tensile strains. This failure is due to the brittleness of ITO. Therefore, the mechanical stability of ITO on flexible substrates should be thoroughly considered in the design and manufacturing of flexible devices. Furthermore, the thermal expansion coefficient mismatch between the ITO film and the substrate creates cyclic internal stress on the film when the device is subjected to fluctuating temperatures. This could create micro-cracks and consequently damage the film as the micro-cracks propagate throughout the film over a prolonged exposure to the fluctuating temperatures. In this chapter, the effects of mechanical loads such as bending and tensile loads are investigated and discussed. In addition, the effect of mechanical and electrical properties of the ITO films exposed to thermal and isothermal conditions are discussed. The film-response to the mechanical and/or thermal loads is detailed in terms of different measures, namely electrical resistance, surface morphology, film composition, film crystallinity, transmittance and failure modes, such as cracks and thermal damage.

Chapter 6 - Recently, the utility of chiral indium(III) complexes or indium(III) salts as Lewis acids in asymmetric organic synthesis has received a great deal of interest due to their highly catalytic activity and excellent stereoselectivity, along with relatively low toxicity, good stability in air and water, and recyclability. There are a great number of examples involving indium(III) reagents in organic synthesis appeared in the last few years. A highlight of current developments on the use of chiral indium(III) complexes and indium(III) salts in various functional group transformations such as stereselective Friedel-Crafts alkylation, allylation reaction, Diels-Alder cycloaddition, Mukaiyama aldol condensation, carbonyl ene reaction, alkynylation reaction, Mannich-type reaction, Prins cyclization, ring opening of *meso*-epoxides, Petasis reaction, cycloaddition, *etc* is presented here, with particular emphasis on reaction diastetreoselectivity and enantioselectivity.

Chapter 7 - Indium (In) is classified as one of the post-transition metals and is located in group 13, similar to boron and aluminum, on the periodic table of the elements. Interest in the use of indium and indium-related reagents is growing. Many recent investigations in synthetic chemistry have focused on indium or indium-related reagents, and have shown distinct progress. A variety of applications have been explored, and there have been reports detailing organic reactions that utilized indium or indium-related reagents to realize chemical transformations, bond-formations, and bond-extensions. This chapter covers the valence of indium metal, highlights the chemistry of indium or indium-related reagents, such as chemoselective protection and deprotection methodologies, and includes applications of indium chemistry. These applicable examples demonstrate further prospects and suggest an expansive potential for the field.

Chapter 8 - In this chapter the authors overview important characteristics of indium and present the authors' experimental results regarding indium lines obtained by electron spectroscopy and laser induced breakdown spectroscopy (LIBS). In the first series of experiments high purity indium granules were melted in the Knudsen type oven in order to obtain the effusive atomic beam which was perpendicularly crossed by monoenergetic electron beam. In the second series of experiments investigated were both high purity indium granules, to get reference signals, and suspected indium traces within basis plate of liquid crystal display. The surface of sample was hit by laser beam in order to obtain the plasma plume that emits the spectral lines which are the fingerprints of the present atomic and ionic species. Optical emission spectrum has been acquired using streak camera. The most intense In (I) lines of the spectrum at 303.9 nm, 325.6 nm, 410 nm, 451 nm are detectible in both series of experiments, however with different resolution. The optical diagnostics of the indium lines could be useful for detecting indium in electric waste, because the detecting device could be made portable. In this chapter the authors also show similarities and differences between the measurements obtained by the electron spectroscopy and laser induced breakdown spectroscopy applied on indium as the target.

Chapter 9 - Indium is traditionally extracted from lead-zinc, tin-sulfide, copper- and copper-zinc-sulfide, skarn sulfide-magnetite deposits as an trace element at bulk content of some dozens ppm. The discovery in Russia in 1988 of sulfide-manganese ores with 30 to 300 ppm content of indium that had a considerable quantity of indium sulfides revealed the natural existence of a potentially industrial type of indium ore unknown before.

Chapter 10 - Tin-doped indium oxide (ITO) thin films were fabricated by the sol-gel spin-coating method with different indium sol-gel solutions synthesized from different indium precursors: $In(NO_3)_3$ or $InCl_3$ (denoted as N-ITO and Cl-ITO, respectively). For both N-ITO and Cl-ITO thin films, the increase of mobility and the reduction of carrier concentration/resistivity with increasing annealing temperatures from 400-700 °C are related to the increase of crystallization/densification and the annihilation of oxygen vacancies. The refractive index (1.84 at $\lambda = 550$ nm), packing density (0.83), resistivity (4.27×10^{-3} Ω-cm), and optical band gap (3.95 eV) of N-ITO thin films are higher than that of Cl-ITO thin films, which can be attributed to the higher densification, lower crystallinity, and more free charge carriers of N-ITO thin films. These properties make the indium nitrate-derived ITO thin films have better potential applications for some commercial products.

Chapter 11 - Crystalline rare earth ions -doped indium hydroxide ($In(OH)_3:RE^{3+}$) nanostructures were prepared by a rapid and efficient Microwave-Assisted Hydrothermal (MAH) method. The influence of doping rare earth (RE) ions (Eu, Tb and Tm) on the luminescence of $In(OH)_3$ was studied. Nanostructures were obtained at a low temperature. FE-SEM images confirm that these samples are composed of 3D nanostructures. XRD, optical diffuse reflectance and photoluminescence (PL) measurements were used to characterize the products. Emission spectra of RE-doped indium hydroxide samples under excitation (350.7 nm) presented broad band emission referent to the indium hydroxide ($In(OH)_3$) matrix. Relative intensities of the RE^{3+} emissions increased as the concentration of this ion increased from 0, 1, 2, 4 and 8 mol %, of Eu^{3+}, but the luminescence is drastically quenched for the $In(OH)_3$ matrix.

Chapter 12 - Indium tin oxide ($In_2Sn_{1-x}O_{5-y}$) nanoparticles were synthesized by hydrothermal method from stable indium tin acetylacetone complexes and postannealed at 600 °C. The absence of chlorine ions shortens the synthesis process, decreases the particle agglomeration and improves the particle purity. The introduced complexing ligand acetylacetone decreases the obtained nanoparticle size. The improved powder properties accelerates the sintering of the $In_2Sn_{1-x}O_{5-y}$ nanoparticles and reaches a relative density of 96.4% of theoretical density (TD) when pressureless sintered at 1400 °C.

In: Indium
Editors: Hsaio G. Woo and Huang Tsai Choi

ISBN: 978-1-62257-696-8

Chapter 1

PROPERTIES OF INDIUM, GALLIUM, AND GA-IN ALLOYS IN CONFINED GEOMETRY

E. V. Charnaya[1,2], Cheng Tien[1], M. K. Lee[1] and Yu. A. Kumzerov[3]
[1]Department of Physics, National Cheng Kung University, Taiwan
[2]Institute of Physics, St Petersburg State University, St Petersburg, Russia
[3]A. F. Ioffe Physico-Technical Institute RAS, St Petersburg, Russia

ABSTRACT

The review of our recent studies is focused on properties of nanocomposites consisted of opal and porous glass matrices which pores are filled with gallium, indium, and their alloys. The atomic mobility in confined melts, melting and freezing in metallic particles within pores, and changes in the Knight shift of NMR lines in the nanocomposites will be discussed. The experimental results were obtained by NMR, acoustic and resistance measurements, and X-ray powder diffraction. Atomic diffusion in confined melts was found to slow down drastically compared to the bulk counterparts. Additional slowing down in atomic mobility was observed for the confined metals in the supercooled state. The Knight shift of the NMR lines which is proportional to the electron susceptibility was shown to decrease in liquid confined metals and alloys. The melting and crystallization for metallic components of nanocomposites were noticeably moved to lower temperatures and strongly diffused. New crystalline phases of gallium were found to emerge under nanoconfinement.

1. INTRODUCTION

New functionalities which were suggested recently in microelectronics, computer science, biology, medicine, and engineering have triggered profound interest in detailed understanding of properties of various nanostructured materials. The particular interest is focused on metals which role in practical applications could be hardly overestimated. Studies of physical properties of nanostructured metals are also of great importance for fundamental physics, in particular, for such areas as nanoscience, superconductivity, physics of phase transitions, crystallography.

Many efforts were recently exerted to improve the understanding of the influence of size reduction on behavior and physical properties of various metals in solid and in liquid state. Metallic nanoparticles, thin films, and nanostructures of different geometry as well as nanocomposites with metallic components were investigated. Many of them have already found numerous applications and are intensively tested for new promising practical uses. Among the most promising are metallic nanostructures obtained by embedding metals into mesoporous matrices. Such nanocomposites have many advantages related to their particular configuration. The nanoconfinement also prevents materials from harmful influence of the environment and ensures the stability of their properties. Depending on pore size, shape, and filling, confined nanostructured metals can be considered as ensembles of weakly linked nanoparticles, wires, or extended percolated system and they manifest the relevant features.

Here we present a review of recent studies of metallic nanocomposites consisted of mesoporous matrices filled with Ga, In, and Ga-In alloys. The chapter is structured as follows. In Section 2 we will provide a short description of porous matrices which were used for nanocomposites with metallic inclusions. In Section 3 we will discuss the peculiarities of the melting and crystallization phase transitions for confined metals. It will be shown that the melting and freezing processes of nanostructural metals are strongly altered compared to those in bulk because of size reduction, peculiarities of heterogeneous crystallization within nanopores, and also due to formation of different crystalline phases. Such studies were very important also for understanding freezing and melting of metallic thin films and isolated fine metallic particles.

In Section 4 we will consider the slowdown of atomic mobility in confined liquid metals and alloys. It will be shown that the effect of nanoconfinement on atomic self-diffusion depends on pore size and is heightening with decreasing pore sizes. The experimental data on atomic mobility were obtained by NMR. The theoretical model which allows us to evaluate the correlation time of atomic mobility from spin relaxation measurements will be discussed.

In Section 5 we will show that the Knight shift of the NMR lines decreases for confined metals and alloys. This decrease has the clear size dependence.

2. Porous Matrices

2.1. Opal Matrices

Opal matrices or photonic crystals consist of amorphous silica spheres which are close packed (Figure 2.1). The ideal cubic packing has octahedral and tetrahedral voids between touched spheres. To make the opal structure strong, it is sintered. The sintering leads to a decrease between centers of the adjacent spheres and therefore to the reduction of the octahedral and tetrahedral pores. In the studies of metals under nanoconfinement, the opal matrices with diameters D of silica spheres of about 100, 210, 240, 250, and 260 were used. For ideal cubic close packing the tetrahedral and octahedral pores have sizes d=0.225 D and d=0.414 D, respectively. Here d is the diameter of a sphere that would just fit into voids. Every constituting sphere is associated to 1 octahedral and 2 tetrahedral voids.

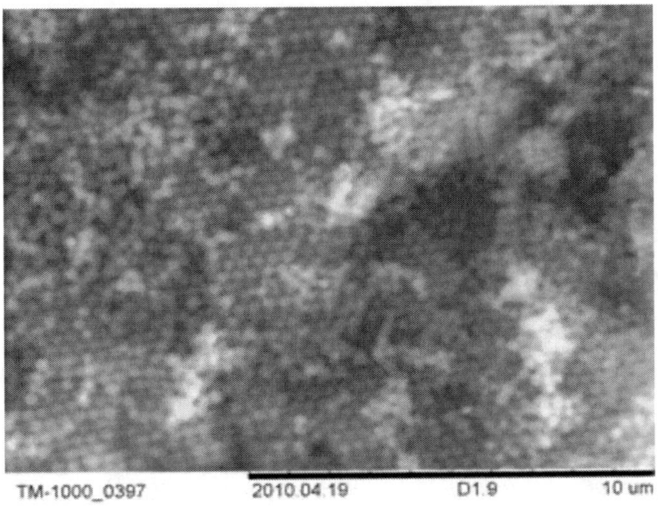

Figure 2.1. Structure of the opal matrix.

Real size of pores within opal photonic crystals is somewhat smaller. It can be measured using mercury porosymetry, small angle X-ray diffraction, nitrogen adsorption-disorption isotherms, electron microscopy. For instance, mercury intrusion porosimetry for the opal sample with D=210 nm revealed a pore size distribution between 19 and 35 nm with distinct maxima at two radii of 30 and 20 nm.

2.2. Porous Glasses

Porous glasses used in studies of confined metals were prepared from phase-separated soda borosilicate glasses with pore structure produced by acid leaching. The leaching leads to the formation of interconnected pore network. Some of porous glasses were laboratory made. The size of their pores varied from 2 to 200 nm. Also Vycor glasses with pores of 6 nm were used as mesoporous matrices. The pore size for the glasses under study was determined using mercury intrusion porosimetry and electron microscopy.

2.3. Pore Filling

The liquid metals or alloys were embedded into the porous matrices under high pressure up to 10 kbar. The filling of the total pore volume was from 30 to 95%.

3. Melting and Freezing of Metals under Nanoconfinement

Different experimental techniques can be used in investigations of melting and freezing in confined metals. Some methods can provide only indirect information on the melting and freezing processes.

For those methods, the phase transitions are registered upon alterations which they produce in properties of the sample, for instance, in ultrasound velocity or resistance. Nevertheless, the indirect methods make it possible to study special phenomena such as reversible and irreversible behavior, percolation and relaxation.

Other techniques, such as NMR and x-ray diffraction, can give immediate information on the amount of liquid and frozen fractions.

NMR provides direct information on the melting and freezing processes in confined metals because the intensity of the appropriate NMR signals are proportional to the total amount of metals in liquid or solid states. For metals with non-cubic crystalline lattice, the anisotropy of the Knight shift in the solid state leads to extremely broad resonance line from a number of particles with chaotic orientations of the crystallographic axes. Therefore, the only signals from metals in liquid state can be observed. Their integral intensity as a function of temperature gives the amount of unfrozen confined material through the melting and crystallization phase transitions.

When only a single crystalline phase is formed upon cooling, one can expect the behavior of the NMR signal intensity upon freezing and melting shown in Figure 3.1.

However, when some particular metal can be frozen into different crystalline modifications, the simple hysteresis loop as in Figure 3.1, transforms into a composite loop with several steps corresponding to formation of different structures at the appropriate temperatures. The latter is the case of gallium under nanoconfinement. Measurements of the intensity of NMR signals corresponded to the liquid state then give the temperatures of the onset and end of the freezing and melting and demonstrate visually how these processes diffuse under particular nanoconfinement. Additionally, one can study the irreversible or reversible behavior through the melting and freezing phase transitions using partial thermal cycles when cooling (warming) changes to warming (cooling) before the freezing (melting) is finished. The x-ray powder diffraction technique seems to be very suitable for studying the melting and freezing phase transitions in porous media since it shows directly the amount of frozen fractions as well as the structure of confined materials and the size of nanoparticles formed within pores during cooling.

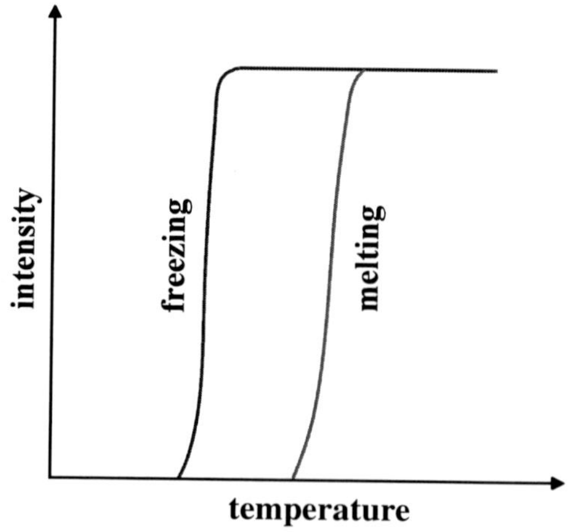

Figure 3.1. General temperature dependence of the intensity of the NMR signal from liquid phase upon cooling (freezing) and warming (melting).

Combining the x-ray diffraction with some other direct or indirect methods seems to be also fruitful for further studies of the melting and freezing phase transitions and for comparing results obtained using different experimental techniques.

Measurements of temperature dependences of ultrasound velocity and resistivity provide indirect information about the melting and freezing phase transitions because both of them noticeably change when confined metals or alloy change their states from liquid to solid and vice versa.

The acoustic methods are especially helpful since they allow obtaining the results in a rather simple and quick way.

3.1. Ga In Opal Matrices and Porous Glasses

Since the first studies of the melting and freezing phase transitions in confined gallium metal [1-4] it was clear that nanostructured gallium can be frozen into various crystalline modifications. X-ray powder diffraction measurements showed this fact in a most direct way. NMR, acoustic, and resistivity measurements also evidenced the multiphase structure of confined gallium.

The particular structures which emerge upon cooling as a result of crystallization processes depend strongly on pore size and pore geometry. Some of the crystalline modifications within pores coincide with known different bulk structures (metallic gallium shows a pronounced tendency to polymorphism already in bulk at ambient pressure [5-14]). However, some of gallium structures which were observed for this metal under nanoconfinement, were completely different from those in bulk gallium.

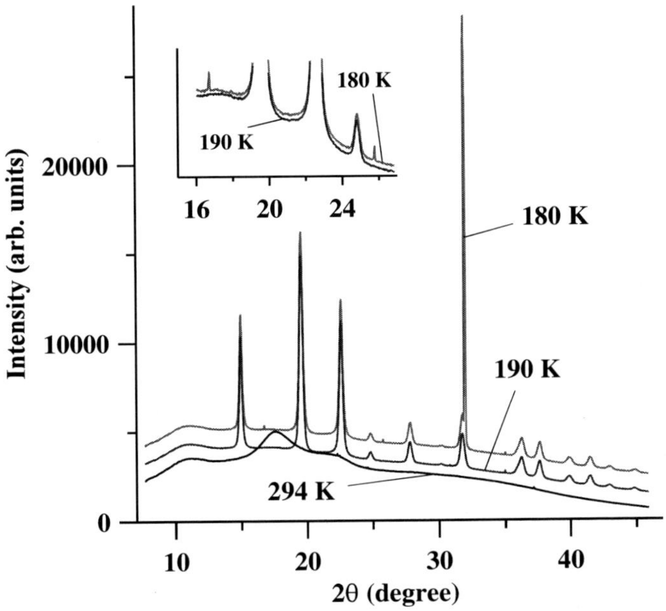

Figure 3.2. X-ray patterns of the gallium loaded porous glass with pore size 7 nm at 190 K (blue) and 180 K (red) obtained upon cooling and at 294 K (black) obtained upon successive warming. The patterns are shifted vertically for better visibility. The inset shows scaled patterns at 190 and 180 K.

Detailed X-ray studies of two samples of porous glasses with mean pore diameters of 7 and 3.5 nm revealed two new gallium crystalline modifications which were unknown until now [15,16]. The evolution of gallium structure within pores of 7 nm in diameter upon cooling from 320 K down to 180 K and consecutive warming is illustrated in Figure 3.2. Gallium remains in liquid state till 260 K with a typical sharp-peak-less pattern (Figure 3.2 shows a pattern at 294 K as an example). A crystalline structure with peaks corresponding to those shown in Figure 3.2 for 190 K appears first upon cooling at 255 K. This set of peaks corresponds to the most pronounced peaks of the x-ray pattern for α-Ga [17] except for the missed one at $2\theta \cong 11.5$ degrees. Therefore, the peaks can be ascribed to disordered α-Ga which freezing and melting are affected by confinement. Similar disorder under nanoconfinement was found in solid argon, nitrogen, and some other substances [18-20] and was ascribed to numerous stacking faults. As can be seen from Figure 3.2, the peaks belonging to this set are broadened. The sizes d of confined gallium crystallites were evaluated using the Scherrer's equation $d = k\lambda/(B\cos\theta)$ with the shape correction constant $k = 0.95$ for spherical particles and full width at half maximum B of the related Bragg peaks. To estimate the size of gallium crystallites formed within pores the peak at $2\theta \cong 19.6$ degrees was chosen since it is a singlet. The Scherrer equation yields a size of 140 Å which is about twice as large as the pore diameter. Thus, the front of crystallization into α-Ga covers only few neighbor pores.

At 180 K another set of peaks appears upon cooling. One of them (at $2\theta \cong 31.95$ degrees) is very intensive, two others (at $2\theta \cong 25.77$ and $2\theta \cong 16.73$ degrees) being very weak (see inset in Figure 3.2). This set of peaks does not correspond to patterns of any known gallium crystalline structures including high pressure ones. Because of the remarkable difference in the diffraction angles, these peaks cannot be explained by disorder in bulk structures. Such set of peaks was not found previously under nanoconfinement, either. Continuing the list of the reported gallium crystalline structures (from α- to θ-Ga [21]), this phase was named ι-Ga [15,16]. The size of confined ι-Ga crystallites evaluated with the Scherrer equation is noticeably bigger than pore size (about 280 Å). It means that ι-Ga crystallites should have a dendrite-like shape following the tortuous pore network.

The temperature dependences of the amount of frozen ι and α gallium crystalline modifications are shown in Figure 3.3.

The melting temperature of confined α-Ga defined as a point where the derivative of intensity with temperature is maximal equals $T_{m,\alpha}$=275 K. The melting is complete at 285 K (the melting point for bulk α-Ga is 303 K). The new ι phase melts completely at 260 K with the mean melting temperature $T_{m,\iota}$=250 K. For both phases the freezing process is shifted to lower temperature compared to melting.

The crystalline gallium structure within pores of smaller sizes is still more complex. Upon cooling from 320 K confined gallium remains in the supercooled state at least till 150 K. During the first cooling-warming cycle gallium was not yet frozen at 140 K (Figure 3.4). Then upon further cooling, a single peak at 2θ=12.05 degrees appears. At lower temperatures another set of broad peaks emerges combined with narrower peaks. Such patterns are seen below 130 K down to the lowest temperature achieved and upon sequential warming up to about 175 K. An example of a pattern at 170 K obtained upon warming is shown in Figure 3.4. Then, structures responsible for narrow peaks melt and the only set of broad peaks survives up to about 220 K.

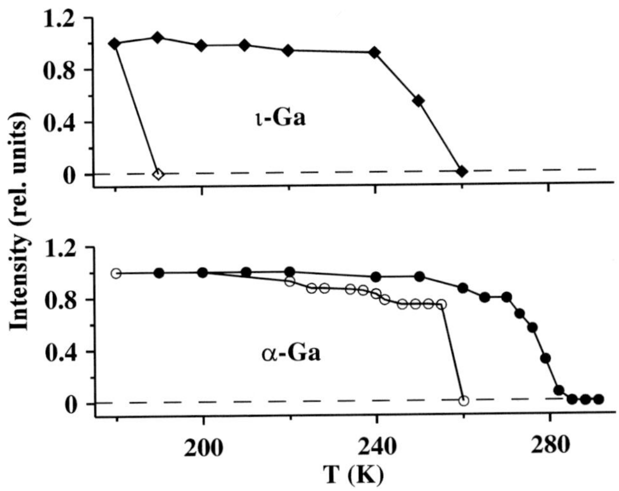

Figure 3.3. Temperature dependences of the intensity of x-ray peaks at 2θ=31.95 (ι-Ga, diamonds) and 19.63 (α-Ga, circles) degrees for the gallium loaded porous glass with pore size 7 nm. Open and closed symbols correspond to cooling and warming, respectively. Solid lines are guides for the eye.

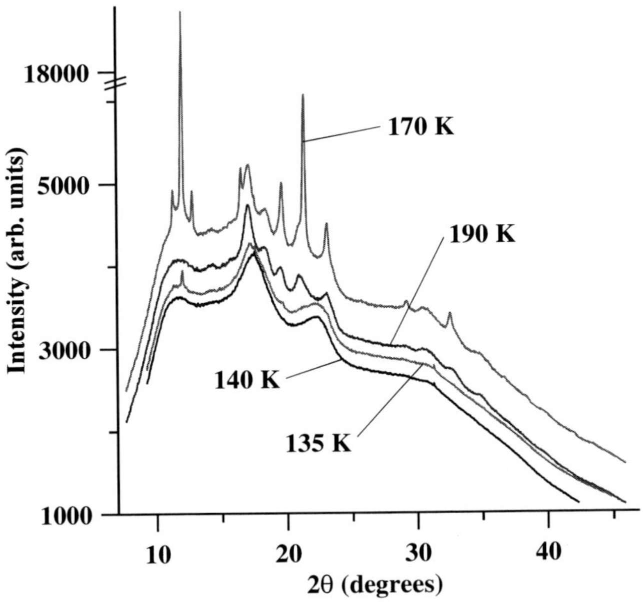

Figure 3.4. X-ray patterns of the gallium loaded porous glass with pore size 3.5 nm at 140 K (black) and 135 K (green) obtained upon cooling and at 170 K (red) and 190 K (blue) obtained upon successive warming. The patterns are shifted vertically for better visibility.

All broad peaks coincide well with strong peaks in the δ-Ga x-ray powder diffraction pattern [17]. The temperature dependence of the amount of δ-Ga is shown in Figure 3.5 for two cooling-warming cycles. There is a pronounced thermal hysteresis between freezing and melting of δ-Ga, the temperature ranges of crystallization and melting being quite reproducible. The mean melting temperature is $T_{m,\delta}$=215 K and confined δ-Ga melts completely up to 230 K, while bulk δ-Ga melting point is 253.7 K [5]. The Scherrer equation gives for confined δ-Ga particles a size of 60 Å, about twice as large as the pore diameter. Confined δ-Ga was stable after solidification and till melting while it is metastable in bulk. Thus, nanoconfinement stabilizes δ-Ga.

A set of narrower peaks at 2θ=12.05, 11.35, and 12.85 degrees (Figure 3.4) does not belong to any known crystalline modifications of gallium as in the case of the ι-Ga modification. This modification was named κ-Ga. Several other narrower peaks appear at about the same temperatures as the three low-angle peaks of κ-Ga. These peaks could also belong to κ-Ga. However, two peaks at 2θ=21.4 and 23.15 degrees could belong as well to β-Ga [5] formed within pores, other peaks being suppressed by lattice disorder.

Three peaks at 2θ=16.57, 19.7, and 32.6 degrees coincide with intensive lines belonging to a confined gallium phase previously found in porous glasses with pore sizes 4 nm and opal photonic crystals [3,4]. That modification was identified with a bc tetragonal structure with lattice constants a=b=0.325 nm and c=0.495 nm [22]. It differs from the high-pressure GaIII phase which is also bc tetragonal [11]. Continuing the list of the observed gallium modifications, this tetragonal structure was labeled as λ-Ga in [15,16]. Thus, one can conclude that narrower peaks in Figure 3.4 could belong to κ-Ga, only, or to a combination of κ-Ga with β- and λ-Ga. The temperature dependence of the amount of confined κ-Ga is shown in Figure 3.5. The κ-Ga freezing and melting processes are quite reproducible. The mean melting temperature is $T_{m,\kappa}$=180 K. While κ- and δ-Ga crystallize at near temperatures, $T_{m,\kappa}$ is much lower than $T_{m,\delta}$. Peaks which certainly belong to κ-Ga have the same width. The Scherrer equation gives then for the size of κ-Ga crystallites an estimate of 350 Å. It shows that the crystallization front went through several neighbor pores and the κ-Ga crystallites should have a dendrite-like shape similar to those of the ι-Ga phase.

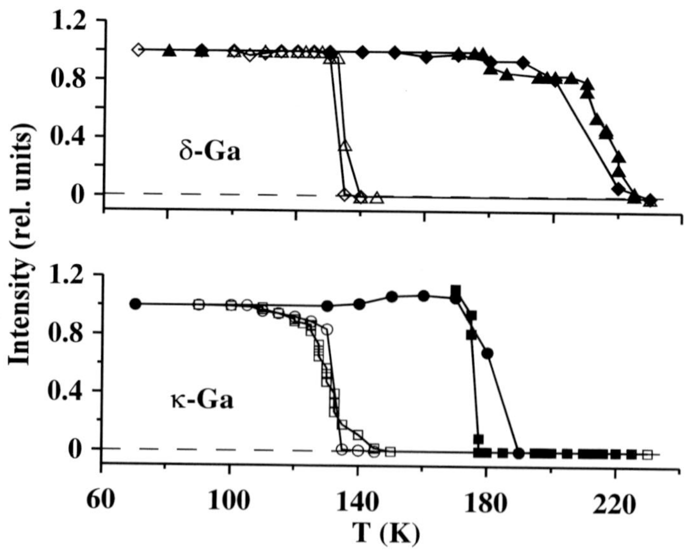

Figure 3.5. Temperature dependences of the intensity of x-ray peaks at 2θ=12.05 (κ-Ga) and 17.20 (δ-Ga) degrees for the gallium loaded porous glass with pore size 3.5 nm. Open and closed symbols correspond to cooling and warming, respectively. Circles and diamonds mark results obtained during the first cooling-warming thermocycle, squares and triangles – during the second thermocycle. Solid lines are guides for the eye.

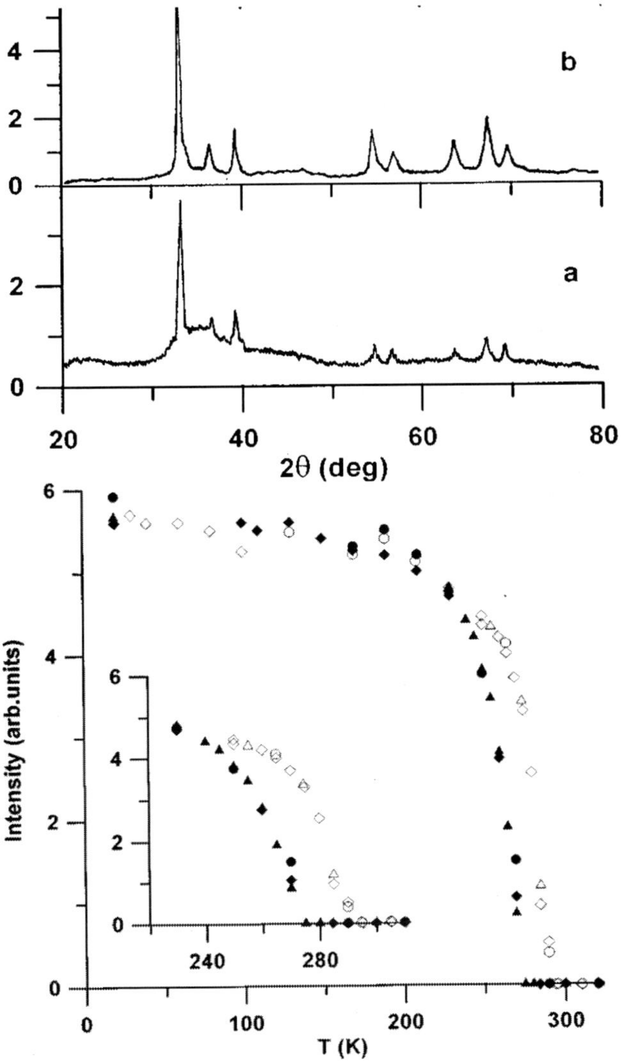

Figure 3.6. Left panel: the x-ray patterns at 250 K (a) and 100 K (b) in the gallium loaded porous glass with pores of 4 nm. Right panel: The temperature dependence of the relative intensity of the x-ray-diffraction peak at 2θ=33°. Solid symbols, cooling; open symbols, warming. Different symbols correspond to three consecutive temperature cycles. The inset shows the scaled hysteresis.

Note, that the ι, κ, and λ modifications are also responsible for unusual temperatures of the superconducting transition in confined gallium, discussed in [16].

During x-ray diffraction experiments, upon cooling as well as upon warming, the samples were kept at several temperatures for more than one hour and repetitive measurements were carried out. No changes in the diffraction patterns were noticed which confirms that the crystalline structures observed corresponded to equilibrium.

For confined α- and δ-Ga the obtained data show remarkable reduction of the melting temperature compared to their bulk counterparts, for about 26 K and 39 K, respectively. Such reduction of melting is known to be caused by size effects and was observed for different materials embedded into nanopores. The theoretical models described in Section 3.4 predict for small particles a shift ΔT of the melting to low temperatures which is inversely proportional to particle size due to the increase of the surface-to-volume ratio.

Since this ratio in the case of confinement depends on pore size d, one can assume the relationship $\Delta T/T_c=k/d$, where T_c is the bulk melting point and k is a constant specific for the substance under study. k has a meaning of the critical size, corresponding to the melting reduction till absolute zero. Then one can obtain for α- and δ-Ga $k_\alpha=0.53$ nm and $k_\delta=0.40$ nm, respectively. The result obtained for α-Ga will be compared with data for other pore sizes in Section 3.4. The emergence of a single λ-Ga structure was observed for the porous glass with 4 nm pores [4]. The X-ray patterns corresponded to this modification of the confined gallium and the temperature variation of the amount of the crystalline phase are shown in Figure 3.6. The x-ray results in Figures 3.2-3.6 show the most distinct features of the crystallization and melting phase transitions for metallic gallium under nanoconfinement:

a) The melting temperature for particular gallium crystalline structures is reduced compared to the relevant bulk melting points;
b) There is a pronounced and reproducible hysteresis between melting and freezing;
c) The temperature ranges of melting and freezing are broadened;
d) Gallium can form different crystalline modifications under nanoconfinement, some of them being not known for bulk gallium.

The issues (a)-(c) are common for other materials embedded into mesoporous matrices [19]. The issue (a) will be discussed in more details in Section 3.4. The quite reproducible thermal hysteresis between freezing and melting is probably related to heterogeneous crystallization within pores which becomes inconvertible at a temperature when the crystallization nucleus equals a pore size. The broadening of the melting and freezing temperature ranges are probably due to pore size distribution as was suggested for other confined substances [23,24]. The results obtained by other methods and for other porous matrices agree with the above conclusions (a)-(d). Meanwhile, some peculiarities were also observed for some particular samples. Figures 3.7 and 3.8 show the variations with temperature of the integral intensity of liquid metal NMR signal from the ^{71}Ga isotope in porous glasses with 2 and 16 nm pores [25]. First, the samples were warmed up to 330 K well above the bulk gallium melting point (303 K) to be assured that the total amount of gallium in the pores was melted. Then, the samples were cooled down to 155 K and warmed up again. Under such conditions, the temperature dependences of the integral intensity were quite reproducible. Gallium within the porous glass with 16 nm pore size remained melted until about 265 K where the abrupt complete freezing occurred. Solidification of gallium within a glass with 2 nm pore size was diffused. It started at about 185 K and was still incomplete till 155 K. Melting was smeared in both samples under study. One can see simple crystallization-melting hysteresis loops in both samples.

However, the extremely strong reduction of the melting temperature in the sample with 2 nm pores evidences that gallium within pores of this sample was frozen into a structure which differs from α-Ga while for the sample with a pore size of 16 nm the shift of the melting range to low temperatures is not so remarkable that indicates the formation of the bulk-like α-Ga. The average value of the melting temperature (293 K) will be used in Sec.3.4 to plot the size dependence of the melting shift.

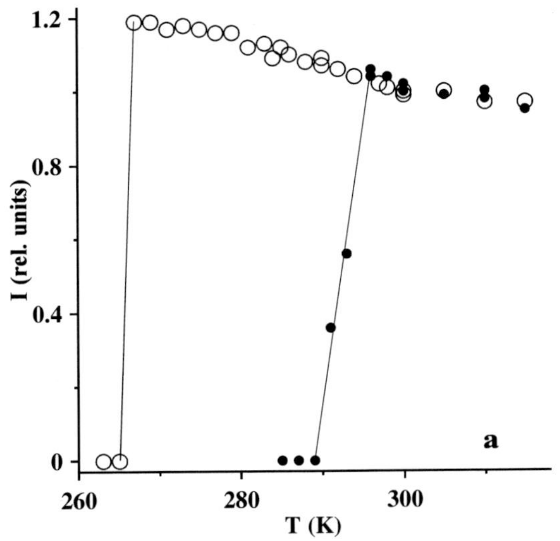

Figure 3.7. Temperature dependences of the NMR signal integral intensity I in a porous glasse with 16 nm pores. Open symbols – cooling, closed symbols – warming. Solid lines are guides for the eye.

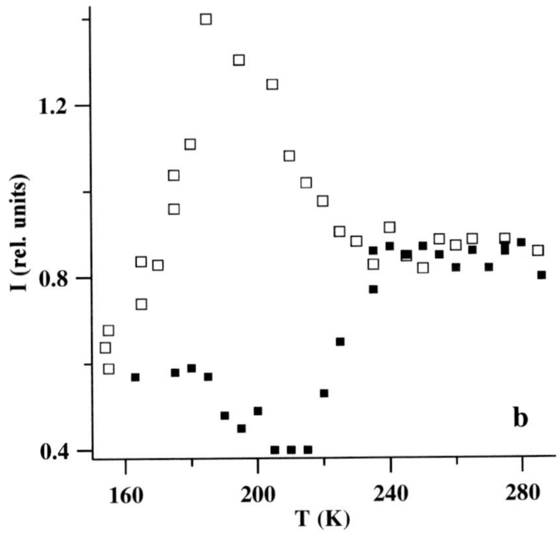

Figure 3.8. Temperature dependences of the NMR signal integral intensity I in a porous glasses with 2 nm pores. Open symbols – cooling, closed symbols – warming. Solid lines are guides for the eye.

Acoustic and NMR studies carried out for the sample of porous glass with pores of 8 nm filled with gallium [26] showed that the freezing and melting phase transitions and the formation of different crystalline structures depend drastically of the thermal regimes, that is on the rate of the temperature changes and limit temperatures achieved upon cooling and warming.

The variations with temperature of the NMR signal intensity and alterations in the relative changes of the ultrasound velocity correspond to the full freezing-melting cycle (Figure 3.9). One can see two superimposed hysteresis loops which demonstrate the occurrence of two different crystalline modifications, at least.

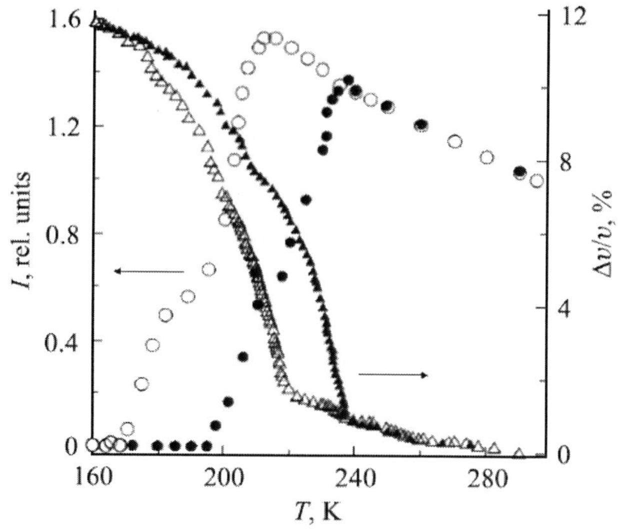

Figure 3.9. Temperature dependences of the NMR signal intensity I (circles) and relative changes in the ultrasound velocity Δv/v (triangles) for the full freezing-melting thermal cycle upon cooling (open symbols) and warming (closed symbols) in the porous glass with 8 nm pores filled with gallium.

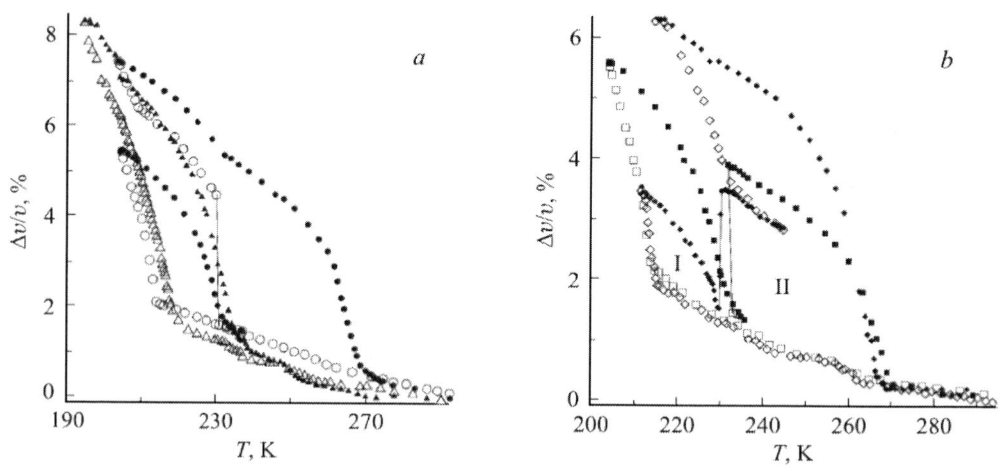

Figure 3.10. Temperature dependences of the relative changes in ultrasound velocity Δv/v upon cooling (open symbols) and warming (closed symbols) in the porous glass with 8 nm pores filled with gallium for two thermal cycles (a and b) started at room temperature. a) Triangles: cooling down to 194 K, warming up to room temperature; circles: cooling down to 405 K, warming to 237 K, cooling to 204 K, and worming to room temperature. b) Diamonds: cooling down to 204 K, warming up to 235.5 K, cooling to 232 K, and warming to room temperatures. Squares: cooling to 212 K, warming to 245 K, cooling to 215 K, and warming up to room temperature.

The x-ray patterns monitored at 211 K after cooling from room temperature evidenced that the crystalline structure which forms first upon cooling is β-Ga [26]. However, the melting and freezing processes are strongly affected by alterations in thermal cycling (Figure 3.10). Other gallium structure can emerge within pores upon consecutive thermal cycles. This is manifested by the appearance of additional hysteresis loops and changes in the melting temperature. X-ray patterns confirm that α- and λ-Ga were also formed within pores. The average temperature of the upper melting process (265 K) in Figure 3.10 (a) can be ascribed to the melting of confined α-Ga. Figure 3.10 evidences that some of the crystalline structures of confined gallium are stable only within particular temperature ranges and can appear from liquid even upon warming (Figure 3.10 (b)).

A very clear picture which revealed the formation of different gallium modifications within pores of an opal matrix was obtained in [3,27]. The opal consisted of silica spheres with diameter 250 nm. Figure 3.11 shows two quite separated hysteresis loops obtained by ultrasonic measurements. Figure 3.12 shows x-ray data for the same sample [3]. The obtained X-ray powder diffraction results showed that three different gallium modifications emerge within pores of the opal matrix: α-Ga, β-Ga, and a tetragonal structure which later was named λ-Ga. Upon cooling the first structure which appears is λ-Ga, then at slightly lower temperature α-Ga emerges. At lower temperatures β-Ga also appears. By comparing x-ray and acoustic data one can see that the two hysteresis loops in ultrasound velocity correspond to the freezing and melting of α-Ga (the upper temperature loop) and β-Ga (the low temperature loop). However, the ultrasound velocity was insensitive to the formation of the tetragonal λ-Ga. It shows that the use of indirect methods to detect the melting and solidification phase transitions can be somewhat limited.

The freezing transition for α-Ga during the second thermal cycle after heating till 320 K was observed at much lower temperature than during the first cycle when the sample was heated only till 303 K. Such behavior was completely reproducible. The difference in the freezing of α-Ga can be treated as an evidence of some residual crystallization centers which remain within pores after warming till 303 K and disappear after warming till 320 K.

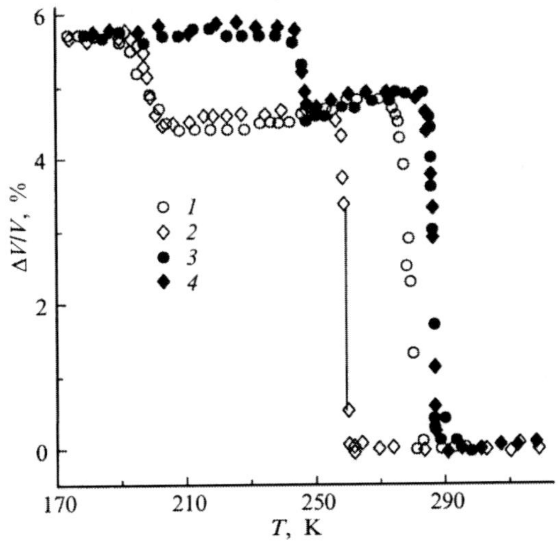

Figure 3.11. Temperature dependences of the relative changes in the ultrasound velocity for the opal matrix with silica spheres of 250 nm which pores were filled with gallium upon cooling (1,2) and warming (3,4). Data were monitored during two thermal cycles after heating the sample up to 303 K (1,3) and 320 K (2,4).

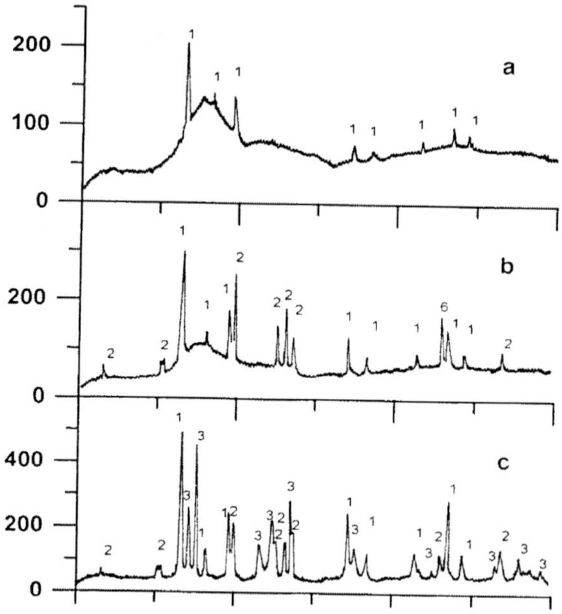

Figure 3.12. The x-ray patterns of gallium confined within the opal at 270 (a), 240 (b), and 100 (c) K. Peaks corresponding to the various solid gallium modifications are marked by numbers: 1-λ-Ga, 2-α-Ga, 3-βGa..

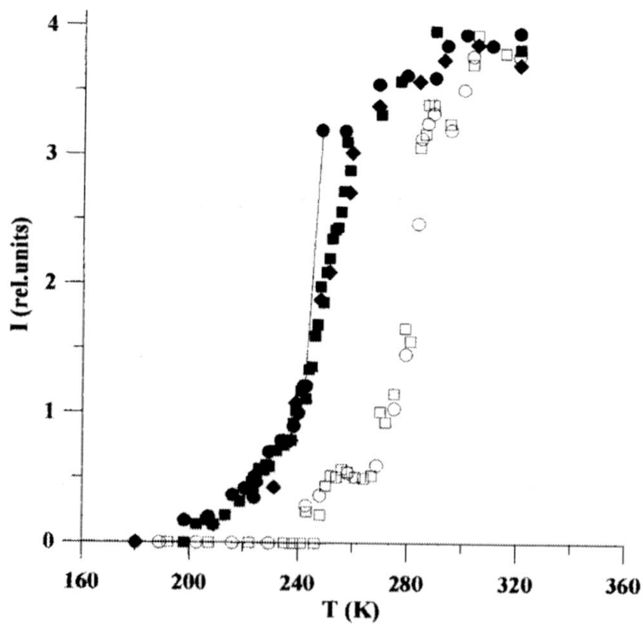

Figure 3.13. Intensity I of the ^{71}Ga NMR signal against temperature on cooling (solid symbols) and on warming (open symbols) in the Vycor glass filled with gallium. Circles—cooling down from 320 K, diamonds and squares—cooling down from 300 K. The straight line provides a guide for the eye.

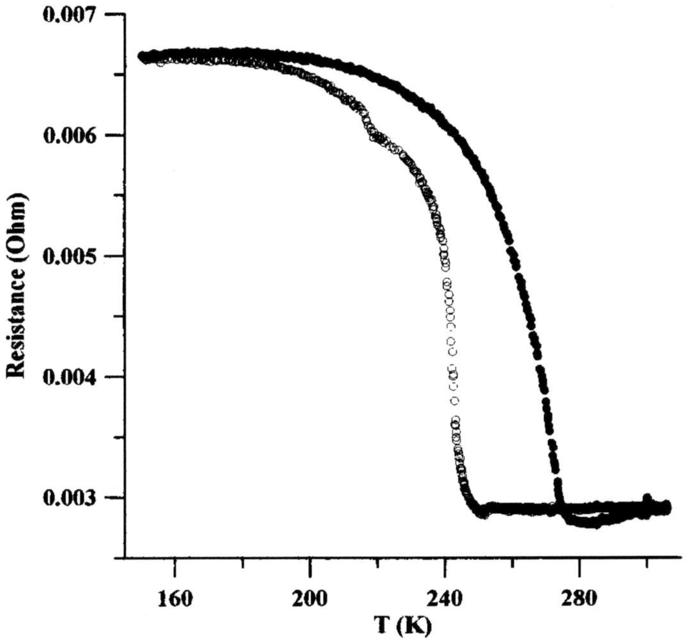

Figure 3.14. The temperature dependence of resistance obtained during a full temperature cycle in the Vycor glass filled with gallium. Open circles—cooling, solid circles—warming.

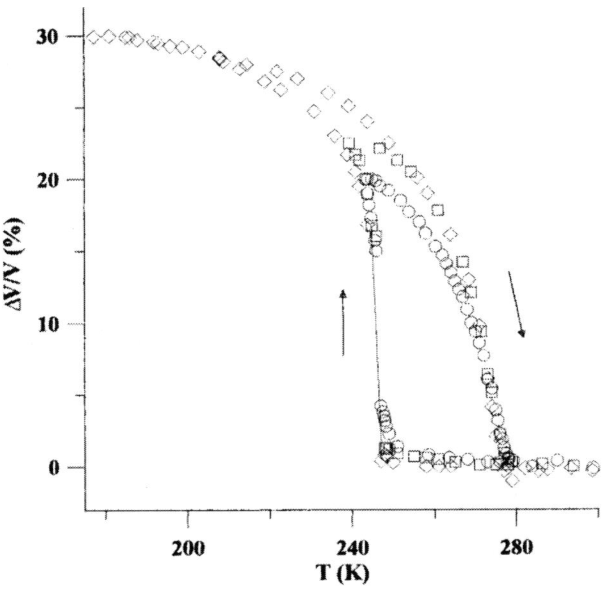

Figure 3.15. Relative changes of the longitudinal ultrasound velocity versus temperature upon cooling and warming cycles in the Vycor glass filled with gallium. Arrows indicate the direction of changing temperature.

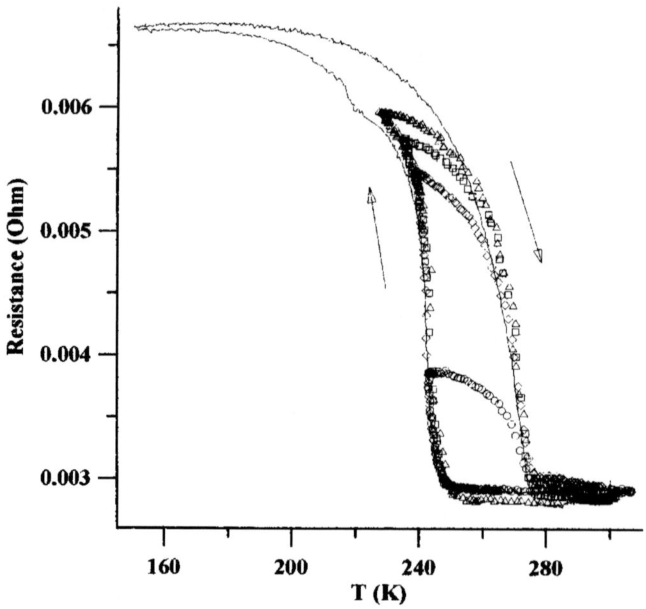

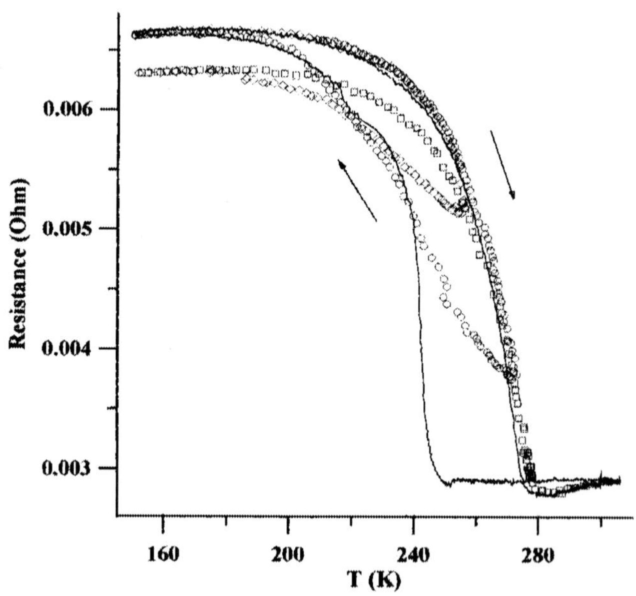

Figure 3.16. Changes in resistance versus temperature upon partial cooling and warming cycles in the Vycor glass filled with gallium. Arrows indicate the direction of changing temperature. The solid line shows changes in resistance during the full temperature cycle from Figure 3.13.

Such residual centers induce crystallization of α-Ga and reduce the temperature range where the supercooled gallium survives.

Combined NMR, resistance and acoustic studies of the melting–freezing phase transitions were carried out for gallium embedded in Vycor glass [2]. The results are shown in Figures.3.13-3.15. NMR data revealed the emergences of three gallium structures within pores which are the same as in the opal (see Figure 3.12) according to their melting temperatures. Meanwhile, acoustic velocity and resistance were sensitive only to the phase transitions in confined α-Ga and β-Ga in full agreement with the results obtained for gallium within the opal matrix. Additional measurements of resistance and ultrasound velocity during incomplete thermal cycles (Figures.3.15 and 3.16) clear showed that the freezing and melting processes in confined gallium are irreversible. These results contradict the reversible behavior of mercury in a porous glass during most of the temperature range of melting [28]. One can suppose that the melting phase transition of the confined α-Ga gallium is broadened due to distribution of the pore sizes and also maybe is affected by the melting of λ-Ga while the melting of mercury was treated within the framework of the model of liquid skin on a solid core (see Section 3.4).

It follows also from Figure 3.16 that in spite of the irreversibility, the solidification starts immediately when cooling begins after incomplete melting. Nevertheless, the temperatures of complete freezing remain equal to those obtained for full temperature cycles in contrast to the noticeable narrowing of the hysteresis loops observed for gallium confined within an opal [27].

3.2. In in Opal Matrices

Indium is a very important metal for many various applications. Therefore, the melting and freezing phase transitions in indium nanostructures were repeatedly investigated. The melting of thin indium films on germanium substrates [29], discontinuous indium films on amorphous silicon nitrite substrates [30], indium threads and nanoparticles, indium in an aluminum matrix (see references in [31,32]), indium nanoparticles in porous glass [33], and indium nanoparticles on the surface of quartz glass [32], on the surface of graphite [34], and on the surface of WSe_2 [35] were studied. For zero-, one-, and two-dimensional nanostructures of indium with a free surface a decrease in the melting temperature was revealed.

For the particles with nonfree surfaces an increase in the melting temperature was observed, which was attributed to the effect of the interaction of an indium particle with a matrix. In all cases solid confined indium had the bulk crystalline structure.

However, only few papers are devoted to the peculiarities of the melting and freezing transitions for indium embedded into mesoporous matrices. In [33] the crystallization and melting of indium nanoparticles embedded into Vycor glass were studied by differential scanning calorimetry. The pronounced lowering of the temperature of melting was found along with its broadening.

The freezing was found to be shifted to low temperature compared to melting and also broadened. Additionally, it was emphasized that the calorimetric measurements confirmed the heterogeneous crystallization of indium under nanoconfinement.

NMR and acoustic studies carried out in [36,37] for the opal photonic crystal filled with indium showed that the temperature range of melting moves remarkably to low temperatures with decreasing the pore filling. The hysteresis loops for the ultrasound velocity near the freezing and melting transitions of confined indium can be seen in Figure 3.17. The mean melting temperature decreases with decreasing the amount of indium confined to mesopores (Figure 3.18).

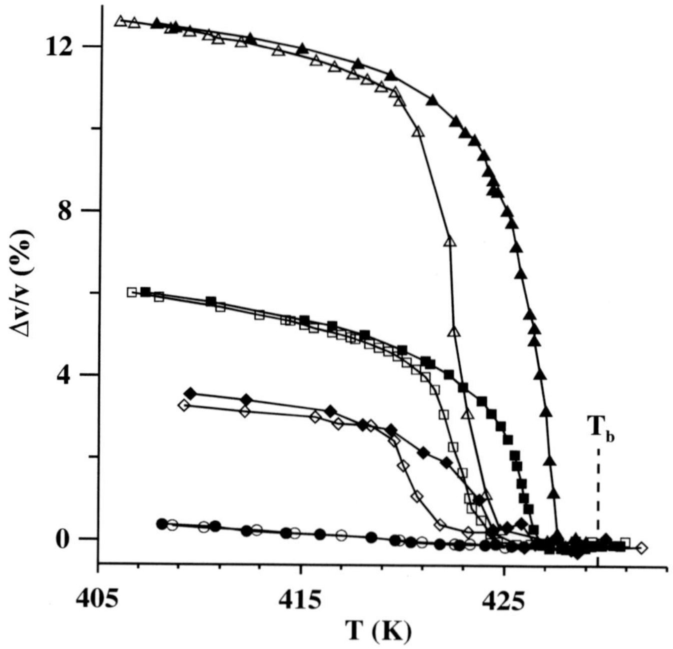

Figure 3.17. Temperature dependences of relative velocity of longitudinal ultrasonic waves during heating (filled symbols) and cooling (empty symbols) for the sample of indium loaded opal photonic crystal. Triangles, squares, diamonds and circles correspond to filling factors Z = 93, 79, 51 and 35%. Solid lines connect experimental points.

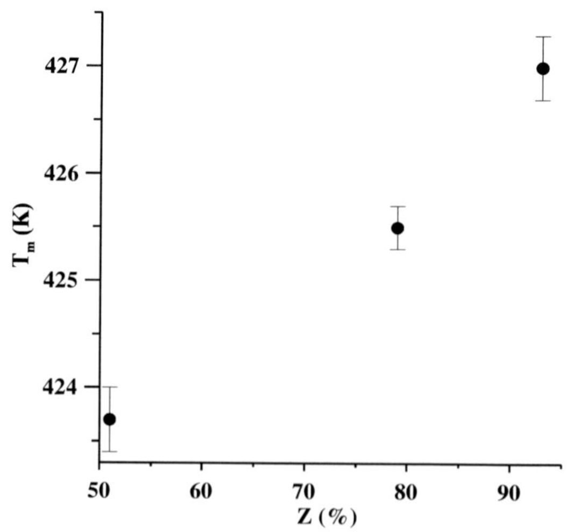

Figure 3.18. Dependence of mean melting temperature on the pore filling factor for the sample of indium loaded opal photonic crystal. The data were extracted from acoustic measurements shown in Figure 3.16.

Dependence of the melting temperature on the filling factor Z can be explained by indium redistribution inside the opal pore network in the process of Z variations. A decrease in Z took place when the sample was kept at temperatures above the melting temperature of bulk indium and indium in the pores was in a melted state.

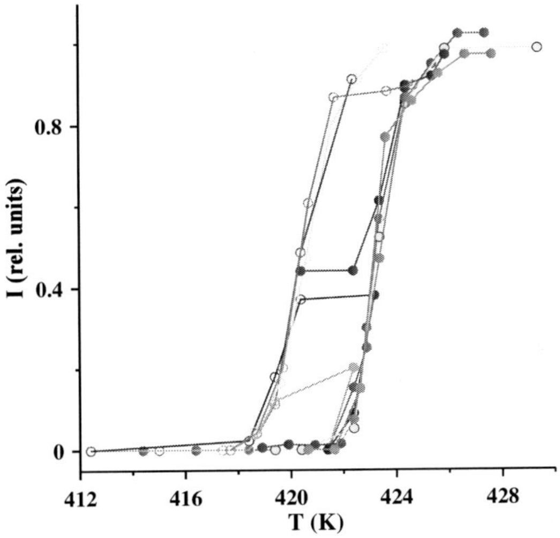

Figure 3.19. Temperature dependence of the integral intensity I of the NMR signal from ^{115}In in the sample of the opal photonic crystal loaded with indium for several thermal cycles marked with different colors. The pore filling was 51%.

Since liquid indium partially wets the walls of a silicate matrix (the wetting angle at the contact surface of liquid indium with quartz glass being 40°), the melt tends, first of all, to fill the pores of small sizes. Thus, with a decrease in Z indium frees large pores which yield a decrease in the particle sizes within pores. As the temperature of particle melting is related to particle sizes the process of indium redistribution in pores leads to the lowering of the mean melting temperature with decreasing Z. NMR and acoustic studies also confirmed the completely reproducible and irreversible character of the melting and freezing phase transitions at the same filling factor for full and partial thermal cycles. An example of NMR data is shown in Figure 3.19. Contrary to the case of gallium embedded into mesoporous matrices, the freezing process in indium was not shifted to high temperatures and the hysteresis loop was not narrowed when cooling was started at a temperature below the complete melting of confined indium.

Therefore, the general features of the melting and freezing phase transitions for confined indium can be summarized as follows:

(a) The melting and freezing transitions move to lower temperatures compared to the melting point of bulk indium;
(b) The hysteresis between the melting and freezing processes is completely reproducible even for partial thermal cycles;
(c) The melting ad freezing processes were completely irreversible and the freezing (or melting) during partial thermal cycles started only at temperatures corresponding to these processes at full cycles melting-freezing.

3.3. Ga-in Alloy under Nanoconfinement

Gallium-indium eutectic alloys have extensive applications from mercury free thermometers and flexible electronic components to light emitting diodes.

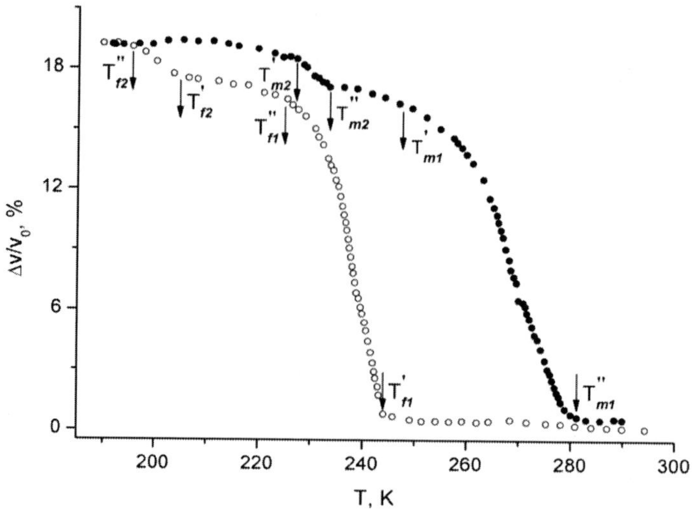

Figure 3.20. Temperature dependence of the relative change of the velocity of the longitudinal ultrasonic wave in the porous glass with pore size 8 nm filled with Ga-In alloy for the full cooling–heating thermal cycle. Open symbols - cooling and closed symbols - heating. Temperatures in the panel mark the onset and end of the freezing and melting processes for two hysteresis loops.

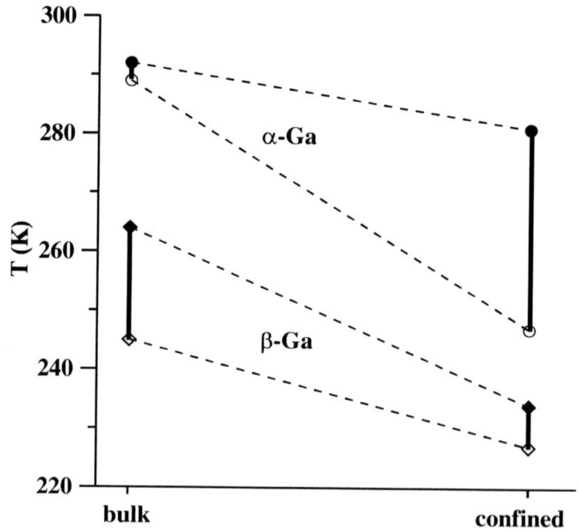

Figure 3.21. Melting intervals for mixtures with α-Ga and β-Ga for the bulk and nanostructured alloy extracted from Figure 3.19. Closed and open symbols show the liquidus and solidus temperatures for the bulk alloy and temperature of the completion and beginning of melting for a nanostructured alloy, respectively. Solid bold lines show the regions of the coexistence of solid and liquid phases.

With further reactions the Ga-In alloys are used in printable semiconductors and solar cells.

The expansion of potential applications of the Ga-In alloys over nanodevices requires better understanding the effect of size reduction on their physical properties. However, until recently the melting and crystallization processes in confined geometry were studied only for pure metals and the first paper devoted to this problem in an alloy was appeared in 2011 [38]. The studies were carried out using ultrasound techniques.

A nanocomposite consisted of porous glass matrices filled with a gallium–indium eutectic alloy was used for studies. The average pore diameter was 8 nm. The gallium–indium alloy had a composition of 90 at.% Ga and 10 at.% In close to the eutectic point.

The temperature dependence of the relative velocity of the longitudinal acoustic wave in the sample for the complete cooling–heating thermal cycle is presented in Figure 3.20. The anomalous changes in the temperature dependence of the velocity are due to the phase transitions (melting and crystallization) in the alloy within pores. The beginning and completion of the phase transitions were determined as boundaries of the regions, where the anomalies of the velocity at heating and cooling of the sample are observed. It is seen that melting and crystallization occur in two stages. The appearance of the double hysteresis loop shows that different crystalline modifications with different liquidus and solidus temperatures are formed upon solidification of indium–gallium alloy in mesopores. It is known (see [39] and references therein) that the bulk gallium and indium alloy has a phase diagram of the eutectic type with a solidus temperature of 289 K and the eutectic point at 14.2 at.% In. The alloy of the 90 at.% Ga and 10 at.% In composition has a liquidus temperature of approximately 292 K. A mixture of crystalline gallium of the α-modification and crystalline solid solution, which consists mainly of indium with a small admixture of gallium and has the structure of pure indium, is formed below the solidus. However, upon overcooling of the bulk alloy, formation of the metastable β-gallium phase is also possible (see [39] and references therein). The metastable phase diagram in the case of the formation of β-Ga also belongs to the eutectic type. The solidus temperature for it decreases strongly and is equal to 245 K. The concentration of 6.2 at.% In corresponds to the eutectic point. The results obtained (Figure 3.20) allow us to assume that the two-stage crystallization and melting process of the gallium–indium alloy in porous glass is due to the formation of two different mixtures in pores, in which gallium crystallizes in the α- and β-Ga structures. Figure 3.21 shows that the melting region in nanopores for both mixtures is shifted to low temperatures. For the mixture with α-Ga, the melting region under confined geometry is strongly broadened compared to the region of the coexistence of the liquid and crystalline phases in the bulk alloy, which probably is due to the dispersion of the pore sizes existing in porous glass. At the same time, for the mixture with β-Ga, the melting region in nanopores, on the contrary, is noticeably narrowed.

Temperature dependences of the velocity of the ultrasound at partial cycles are presented in Figure 3.22 under conditions, when cooling or heating of the sample begins before the completion of the melting within pores or completion of the alloy crystallization, respectively. The reproducibility of the experimental results and their correspondence to data obtained for the full cycle shown in Figure 3.20 is obvious. Figure 3.22 also demonstrates the irreversibility of the melting or crystallization under nanoconfinement. The irreversible behavior noticeably differed for cooling and warming. The velocity curves obtained during cooling came to the cooling branch for the full cycle only at the end of the first stage of the crystallization, whereas, during heating, the velocity curves came to the heating curve for the full cycle long before the completion of melting. Such behavior resembles the irreversibility obtained for pure gallium.

It should be emphasized that the reproducibility of the hysteresis loops in Figure 3.22 indicates the stable character of the formation of the mixture with β-Ga in nanopores, unlike the bulk alloy. Stabilization of the metastable metal phase in nanopores was also shown for pure gallium (see Section 3.1).

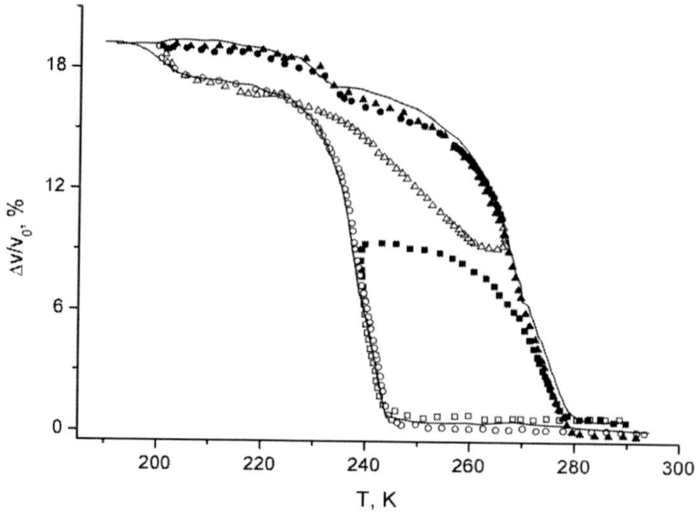

Figure 3.22. Temperature dependence of the relative change of the velocity of the longitudinal ultrasonic wave in the porous glass with pore size 8 nm filled with Ga-In alloy for several partial cooling–heating thermal cycles. Open symbols - cooling and closed symbols - heating. The straight line shows the velocity changes during the full cycle from Figure 3.19.

Therefore, one can conclude that the Ga-In alloy embedded into the particular mesoporous matrix has the following properties:

(a) Contrary to the case of bulk eutectic Ga-In alloys, the confined alloys forms mixtures where Ga crystallizes into to stable structures: α- and β-Ga;
(b) Temperatures of liquidus and solidus are reduced for both gallium modifications compared to bulk.

3.4. Theoretical Models

The results obtained for the melting and freezing phase transitions for metals under nanoconfinement (as well as for other substances) are usually treated on the basis of models developed for free particles.

Thermodynamic models of small particle melting predict a reduction of the melting temperature when the melt wets its own solid surface. The first idea of size-effects on melting of small metallic particles was advanced by Pawlow [40]. He suggested that the melting point can be considered as a triple point where the solid, melted and vapor states are in equilibrium and included in the thermodynamic potential the surface energy which is important because of the increased surface to volume ratio in small particles. The decrease of the melting temperature ΔT for a spherical particle with radius R is then given by

$$\frac{\Delta T}{T_b} = \frac{2(\gamma_{sv} - \gamma_{lv})}{L\rho R}, \qquad (3.1)$$

where $\Delta T=T_b-T_m$, T_b is the melting point of the relevant bulk and T_m is temperature of particle melting, γ_{sv} and γ_{lv} are the surface tensions for the solid-vapor and liquid-vapor boundaries, respectively, L is the latent heat of fusion, ρ is the density. In (3.1) the difference between the density of liquid and solid was neglected. Often the Gibbs-Thompson equation is used to treat the melting temperature reduction in small particles

$$\frac{\Delta T}{T_b}=\frac{2\gamma_{sl}}{L\rho R}, \tag{3.2}$$

where γ_{sl} is the surface tension at the boundary between solid and its melt.

It was also suggested (see [41], for instance) to write the free energy balance for the whole particle before and after the melting or freezing phase transition which yields

$$\frac{\Delta T}{T_b}=\frac{3(\gamma_{sv}-\gamma_{lv})}{L\rho R}. \tag{3.3}$$

More complicated models account for the formation of a liquid skin on the surface of the solid core. First such an idea was advanced in [42,43]. The necessary condition for the thin surface layer melting is

$$\gamma_{sv}-\gamma_{sl}-\gamma_{lv}>0. \tag{3.4}$$

Later it was shown that the model predicts a stable liquid skin on the solid surface in a temperature range below the accomplishment of melting only when one takes into account the influence of the solid core on the structure of the surface liquid layer [44]. Since the surface tension on the free surface of the liquid layer should approach γ_{sv} when the layer thickness $\delta\rightarrow 0$, the free energy of the liquid skin assumes the form

$$F_\delta=-N'\mu_s+N'\mu_l+4\pi R^2\gamma_{lv}+4\pi r^2\gamma_{sl}+(4\pi R^2\gamma_{sv}-4\pi R^2\gamma_{lv}-4\pi r^2\gamma_{sl})f(\delta), \tag{3.5}$$

where $f(\delta)=\exp(-\delta/\xi)$ for the metallic surface, ξ is a parameter of the order of several angstrems which can be named the extrapolation length. In (3.5) N′ is the number of particles in the layer, r is the radius of the solid core, and μ_s and μ_l are chemical potentials of solid and liquid, respectively. The free energy extremum is achieved when

$$L\rho\frac{\Delta T}{T_b}=\frac{2\gamma_{sl}}{R-\delta}(1-e^{-\frac{\delta}{\xi}})+(\gamma_{sv}-\gamma_{lv}-\frac{(R-\delta)^2}{R^2}\gamma_{sl})\frac{R^2}{(R-\delta)^2}\frac{e^{-\frac{\delta}{\xi}}}{\xi}. \tag{3.6}$$

This relationship was obtained in [44]. Using (3.6) one can find the condition for the equilibrium state with a solid core surrounded by the liquid skin. In particular, the melted skin appears at a temperature corresponding to δ=0

$$L\rho\frac{\Delta T}{T_b}=(\gamma_{sv}-\gamma_{lv}-\gamma_{sl})\frac{1}{\xi} \quad (3.7)$$

and achieves its maximal thickness with increasing temperature. Then the solid core melts sharply.

The liquid skin model predicts the reversible melting until the solid core melts. This model was used to treat the experimental results obtained for confined mercury and tin [28,45]. Since the melting process is reversible as for gallium, indium, and gallium-indium alloy within pores, it means that the pore size distribution dominates the broadening of the melting transition under nanoconfinement and the role of the possible liquid skin formation is not important. Then it is reasonable to use the relations like (3.1)-(3.3) which predict a linear dependence of the melting temperature reduction on the inverse particle size. Moreover, the reduction of the melting temperature arises actually due to the increased influence of the surface and the surface to volume ratio for confined particles is defined by the size of pores. This justifies the use in the relation which follows from (3.1)-(3.3)

$$\Delta T = kT_b / d \quad (3.8)$$

(k is a coefficient of proportionality) the pore size as d.

The validity of (3.8) can be checked for gallium within porous glasses. For α-Ga the mean melting temperature was plotted as a function of 1/d in Figure 3.22. One can see that the linear dependence is satisfied approximately. Assuming that this trend remains valid till very small pores, one can find the critical size corresponding to zero melting temperature: d_c=0.64 nm.

4. Atomic Mobility in Confined Metals

Recent studies of atomic mobility, relaxation and flowing in liquids embedded into various porous matrices have evidenced remarkable alterations compared to bulk (see for instance [46-49] and references therein). Alterations induced by confinement depend on whether liquids wet the inner surface of matrices, on pore sizes and tortuosity of the pore network, on length of molecular chains for polymers, and on other factors. Since NMR is known to provide immediate information about dynamics in bulk liquids through nuclear spin relaxation measurements, the NMR methods were applied successfully to study mobility in liquids within pores [47-52]. Nevertheless, in spite of numerous studies of liquids within porous matrices, nothing was known about peculiarities in spin relaxation and dynamics in confined metallic melts until the first paper devoted to this matter appeared [53].

4.1. Nuclear Spin Relaxation in Liquid Metals

Studies of nuclear spin-lattice relaxation for the bulk melted gallium and indium (see [54] and references therein) have shown that it occurs via coupling of nuclear magnetic moments with conduction electrons and of nuclear quadrupole moments with dynamic gradients of electric fields produced by atomic motion, the main contribution to relaxation is magnetic. Since atomic mobility in bulk liquid metals is fast enough, total relaxation is described by a single exponential (the extreme narrowing limit) with the field-independent relaxation times.

The most convenient way to separate the two contributions which can be used for bulk rubidium, antimony, and gallium melts (see [54,55] and references therein) is to measure the spin relaxation rate for two isotopes with different magnetic dipole moments γ and electric quadrupole moments Q. The spin relaxation rate due to the magnetic contribution is proportional to the squared gyromagnetic ratio while the relaxation rate due to the quadrupole contribution (in the extreme narrowing limit) is proportional to the squared quadrupole moment. Using these relationships and the experimental relaxation times for two isotopes, one can evaluate the relaxations times corresponding to the different nuclear couplings. Gallium has two isotopes, ^{69}Ga and ^{71}Ga, with quite distinct γ and Q and similar natural abundances. Therefore, the separation of two contributions to relaxation is rather accurate [56]. However, while indium has two naturally occurring isotopes, ^{115}In and ^{113}In, they have almost identical nuclear gyromagnetic ratios γ and quadrupole moments Q. It prevents the separation of magnetic and quadrupole relaxation from a comparison of the two isotopes. In addition, the natural abundance of the ^{113}In isotope is very low and relaxation measurements for it need too much time. The quadrupole contribution for bulk indium was estimated from theoretical calculations.

The extreme narrowing limit remains valid when the product $\omega_0\tau_c$ is much less than unit. Here ω_0 is the Larmor frequency and τ_c is the appropriate correlation time. Magnetic relaxation is caused by conduction electrons and the correlation time for them is always much shorter than $1/\omega_0$. However, as it will be shown in Section 4.2, the correlation time of atomic mobility, which is of the order of $10^{-11}\div10^{-12}$ s in bulk melts in agreement with the self-diffusion rate in liquid bulk metals [57], can become longer under nanoconfinement. Therefore, to treat the spin relaxation data for confined metals, one must use the general relationships which are valid for any $\omega_0\tau_c$. Such analytical general relationships for quadrupole relaxation can be derived only for spin $I\leq3/2$. This is just the case of gallium isotopes ($I=3/2$). But the ^{115}In isotope has spin $I=9/2$. Fortunately, the experimental data for liquid indium in confined geometry were obtained only for the case when the extreme narrowing approximation was still correct (Section 4.2.3).

The longitudinal quadrupole relaxation in liquids for the spin 3/2 is given by the [58]:

$$\frac{M(t)}{M_0}=1-b\left[\frac{4}{5}\exp\left(-2\left(\frac{eQ}{\hbar}\right)^2 J_{-22}(2\omega_0)t\right)+\frac{1}{5}\exp\left(2\left(\frac{eQ}{\hbar}\right)^2 J_{-11}(\omega_0)t\right)\right], \tag{4.1}$$

where M(t) and M_0 are time-dependent and equilibrium magnetizations, respectively, 1-b is the starting relative magnetization, e is the electron charge, $J_{-ii}(\omega)$ are the spectral densities of the electric field gradient correlation function at the nuclear site.

In the extreme narrowing limit the exponents in (4.1) are reduced to $-C\tau_c t$ [58], where C is a constant, which is proportional to Q^2, and τ_c is the correlation time of low-frequency electric field gradient fluctuations produced by atomic motion. The constant C includes ion contribution corrected for the antishielding factor.

The product $C\tau_c$ has the meaning of the inverse time of quadrupole relaxation. In the general case one can assume that the correlation function of atomic mobility is represented by $\exp(-t/\tau_c)$, thus, the exponents in (4.1) are reduced to $-C\tau_c t/(1+j^2\omega_0^2\tau_c^2)$, where j=1,2.

The relaxation time T_{1m} corresponding to magnetic contribution to the total relaxation process in liquid metals which is caused by coupling with conduction electrons, is related to the Knight shift by the Korringa relation:

$$T_{1m}TK^2 = const/(\gamma\alpha)\,, \tag{4.2}$$

where T is the temperature, K is the Knight shift, α is the correction factor which accounts for the effects of electron correlation and exchange.

Total relaxation for spin I=3/2 can be described by the following relationship

$$\frac{M(t)}{M_0} = 1 - b\left[\frac{4}{5}\exp\left(-\frac{C\tau_c t}{1+4\omega_0^2\tau_c^2}\right) + \frac{1}{5}\exp\left(-\frac{C\tau_c t}{1+\omega_0^2\tau_c^2}\right)\right]\exp\left(-\frac{t}{T_{1m}}\right). \tag{4.3}$$

4.2. Experimental Results

4.2.1. Ga

As it was mentioned above, the two gallium isotopes have slightly different abundance, spin equal to 3/2, and distinct gyromagnetic ratios γ and quadrupole moments Q (Q=0.171 and 0.107 barn and $\gamma_n=6.44\cdot10^7$ and $8.18\cdot10^7$ rad$\cdot$T$^{-1}\cdot$s^{-1} for ^{69}Ga and ^{71}Ga, respectively). This allows separating magnetic and electric spin interactions especially when NMR measurements are carried out at various external magnetic fields.

Measurements of longitudinal spin relaxation in liquid confined gallium were carried out for several samples of gallium loaded porous glasses with pore size 2, 8, and 16 nm and opal consisted of silica spheres with size 250 nm [25,53,59] at room and at low temperatures. Let us consider first the results obtained at room temperature.

The freezing transition for gallium within nanopores is lowered compared to that in bulk (Section 3.1) and confined gallium still remains liquid at room temperature. While the melting point of bulk metallic gallium is 303 K it can be supercooled very easy below room temperature and remains liquid at room temperature for long time. Thus, spin-lattice relaxation for liquid gallium embedded into porous matrices as well as for bulk liquid gallium can be measured at ambient conditions.

Most of measurements were carried out at four different magnetic fields (about 7, 9, 11, and 17 T) using Bruker MSL300, Avance400, MSL500, and Avance750 NMR pulse spectrometers. The inversion recovery procedure was applied to observe the restoration of longitudinal nuclear magnetization.

The longitudinal magnetization recovery curves obtained at room temperature for gallium within the porous glasses with 16 and 8 nm pores are depicted in Figures.4.1 and 4.2, respectively, along with the results for bulk liquid gallium.

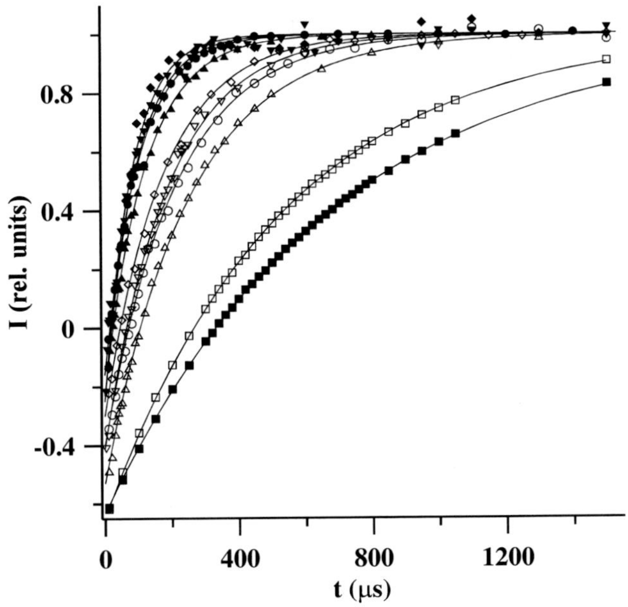

Figure 4.1. NMR signal integral intensity I versus time after a 180° pulse for the gallium loaded porous glass with 16 nm pores and in bulk (squares). Closed and open symbols correspond to ^{69}Ga and ^{71}Ga, respectively. Diamonds, inversed triangles, circles, and triangles correspond to measurements at magnetic field of 7, 9, 11, and 17 T, respectively. Solid lines are theoretical calculations.

Since atomic mobility in bulk liquid gallium is fast enough, total relaxation is described by a single exponential (the extreme narrowing limit) with the field-independent relaxation times. The relaxation time for ^{71}Ga is shorter than for ^{69}Ga since magnetic interaction with conduction electrons dominates relaxation.

In our measurements T_1^{69} =678 μs and T_1^{71} =534 μs at room temperature in perfect agreement with previous data. Figures.4.1 and 4.2 show drastic enhancement in the relaxation rate for both gallium isotopes in confined melts compared to bulk liquid gallium. In contrast to relaxation in bulk, relaxation for confined gallium at the same field and in the same sample is noticeably faster for the ^{69}Ga isotope with greater quadrupole moment and smaller gyromagnetic ratio.

Moreover, the relaxation rate visibly depends on magnetic field for gallium within the porous glasses.

The fact that relaxation is faster for the isotope with larger quadrupole moment shows that the quadrupolar contribution dominates longitudinal relaxation in confined geometry contrary to the bulk case.

The remarkable enhancement of the quadrupole relaxation rate under nanoconfinement as well as its dependence on magnetic field show that the spectral densities of the electric field gradient correlation function which determine quadrupole spin relaxation in confined gallium are strikingly increased and the extreme narrowing approximation is no longer valid.

The procedure which allows extracting the information about atomic mobility from data for spin relaxation was suggested in [25,53,59]. The method is based on the separation of magnetic and quadrupole contributions to relaxation first for bulk liquid gallium and then for confined gallium using the general relationship (4.3).

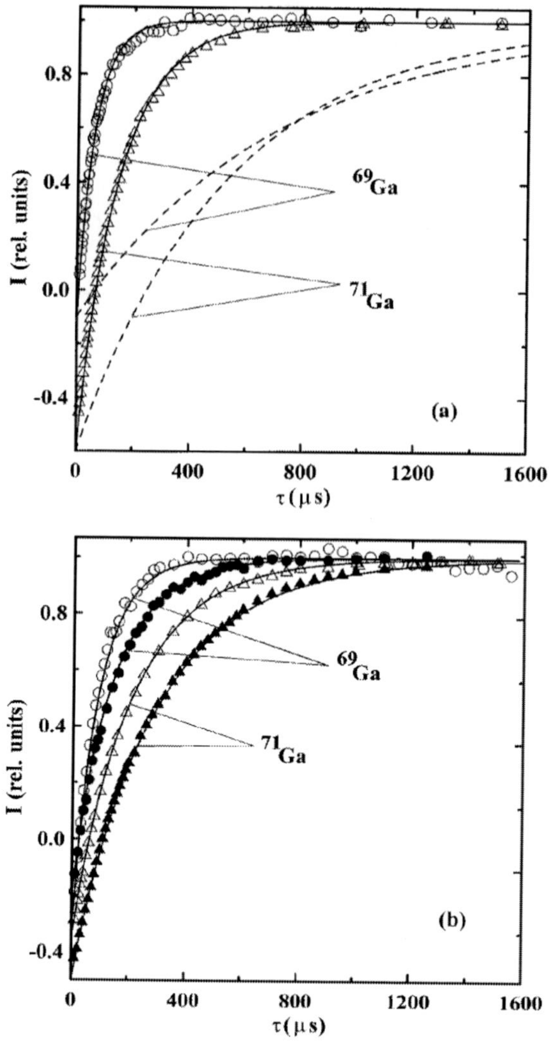

Figure 4.2. Magnetization recovery curves (t is the time between first and second pulses) for liquid gallium embedded into the porous glass with 8 nm pores. Circles and triangles mark experimental results for ^{69}Ga and ^{71}Ga, respectively. Solid lines are theoretical calculations. (a) Measurements in magnetic field 7 T; dashed lines show restoration of magnetization in bulk liquid gallium. (b) Measurements in magnetic field 11 T (open symbols) and 14 T (closed symbols).

The magnetic and quadrupolar contributions to spin-lattice relaxation in bulk liquid gallium can be separated [56] for both gallium isotopes using their gyromagnetic ratios and quadrupole moments and $T_{1m}^{71} = \frac{\gamma_{n69}^2}{\gamma_{n71}^2} T_{1m}^{69}$. At room temperature it yields T_{1m}^{71} =585 μs, T_{1m}^{69} =944 μs, T_{1Q}^{71}=6130 μs, and T_{1Q}^{69} =2400 μs in agreement with previous theoretical calculations and experimental data [54].

The measurements of the Knight shift in liquid gallium showed that it was influenced by confinement but very slightly (see [25,53,59] and Section 5). Since the time T_{1m} must satisfy the Korringa relation (4.2) it can be assumed to keep its bulk value for further estimates for confined gallium.

Table 4.1. Correlation times of electric field gradient fluctuations at room temperature for liquid gallium embedded into mesoporous matrices and in bulk

sample	Porous glass 2 nm	Porous glass 8 nm	Porous glass 16 nm	Opal	Bulk
τ_c (10^{-11} s)	70	80	37	20	1.4

Then one can fit the full set of recovery curves for confined gallium within porous matrices at room temperature using only two fitting parameters, C_{69} and τ_c, for each sample, taking into account that $C_{71}=\frac{Q_{71}^2}{Q_{69}^2}C_{69}$. The best fit for the porous glass with 16 nm pore size was obtained when C_{69} and τ_c were assumed to be $3\cdot10^{13}$ s^{-2} and $3.7\cdot10^{-10}$ s, respectively. The fitting is shown in Figure 4.1 together with the experimental results.

The value of C_{69} coincides with the relevant value $3.2\cdot10^{13}$ s^{-2} found for spin relaxation in liquid gallium confined within a synthetic opal and porous glass with 8 nm pores. It also agrees well with theoretical calculations of the quadrupole relaxation rate [54] in bulk liquid gallium. This value of C_{69} corresponds to the correlation time in bulk liquid gallium $\tau_c \cong 1.4\cdot10^{-11}$ s.

The spectral density at low frequencies, which determines the value of C for both isotopes, depends on structure of melts. Thus, it seems well founded that C keeps the same value from bulk liquid gallium to pore size of about 8 nm at least. However, the best fit for the porous glass with 2 nm pore size was obtained when $C_{69}=1.9\cdot10^{14}$ s^{-2} and $\tau_c=7\cdot10^{-10}$ s. Moreover, the experimental relaxation curves could not be fitted for the same quadrupole parameters C as for bulk. The alterations in C for gallium within so fine pores indicate some changes in structure or chemical bonds for confined gallium melt which influence electric field gradients at gallium nuclei.

The result obtained shows that the correlation time for liquid gallium in the porous glass considerably increased compared to bulk. The values of the correlation time τ_c found are listed in Table 4.1.

On can see from Table 4.1 that when C remains unchanged, there is a monotonic increase in the correlation time with decreasing pore size. And for the finest pores the correlation time is about the same as for the 8 nm porous glass. Note that weak monotonic alterations with decreasing pore size were also seen for the Knight shift in confined liquid gallium (see Section 5).

The increase in the correlation time for liquid gallium embedded into nanoporous matrices evidences drastic slowing down the atomic diffusion at room temperature under confinement, the self-diffusion coefficient being inversely-proportional to τ_c [60]. Some trend in slowing down the mobility in liquid gallium thin films was also reported in [61] according to quasielastic neutron scattering study where it was found that the self-diffusion coefficient was too low to be resolved, at least less than 10^{-5} cm^2/s compared to $3\cdot10^{-5}$ cm^2/s in bulk.

Note that results obtained in [59] showed that slowdown of self-diffusion becomes noticeable already for small gallium droplets of 5 microns in size.

It is also of great interest to find which is the temperature dependence of the atomic mobility in confined liquid gallium, whether it differs or not from that in bulk liquid gallium and how it is influenced by partial freezing of gallium within nanopores.

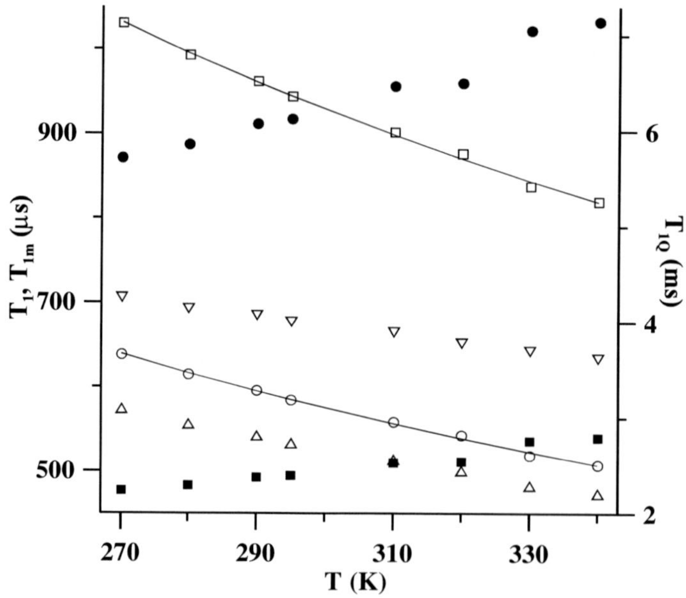

Figure 4.3. Temperature dependences of relaxation times for the gallium isotopes in bulk melt. Open and closed symbols relate to left and right axes, respectively. Triangles and inversed triangles are and T_1^{69} . Open and closed circles are and T_{1Q}^{71} . Open and closed squares are and T_{1Q}^{69} . Solid lines represent 1/T dependences.

Assuming that the values of C do not vary with temperature, one can use the relationship (4.3) for evaluating temperature dependences of the correlation time in confined liquid gallium from relaxation measurements at different temperatures. First it is necessary to evaluate and in bulk at various temperatures. For this the temperature dependences of spin-lattice relaxation in bulk melt must be carried out. The experimental data and results of calculations for relaxation in bulk gallium are shown in Figure 4.3. The temperature dependences of the magnetic contribution were found to satisfy perfectly the Korringa relation, in agreement with previous studies. This justifies the use of the 1/T law to evaluate the magnetic relaxation times, and T_{1m}^{71} , below the bulk gallium freezing point.

The calculated correlation times at different temperatures relative to their values at room temperature for liquid gallium in the porous glasses with 16 and 2 nm pores and for bulk melt are shown in Figure 4.4.

While for bulk liquid gallium the temperature dependence of τ_c is rather weak, it becomes much stronger under confinement. Above and within the temperature range of freezing, the correlation time increases by more than a factor of 3 for both porous matrices. It should be also noted that for gallium within pores of 2 nm in size the correlation times obtained upon warming could differ from those obtained upon cooling (see Figure 4.4). Besides, magnetization restoration curves and, therefore, correlation times calculated could be slightly different for different measurements.

This effect may be caused by redistribution of gallium within pore volume upon changing temperature.

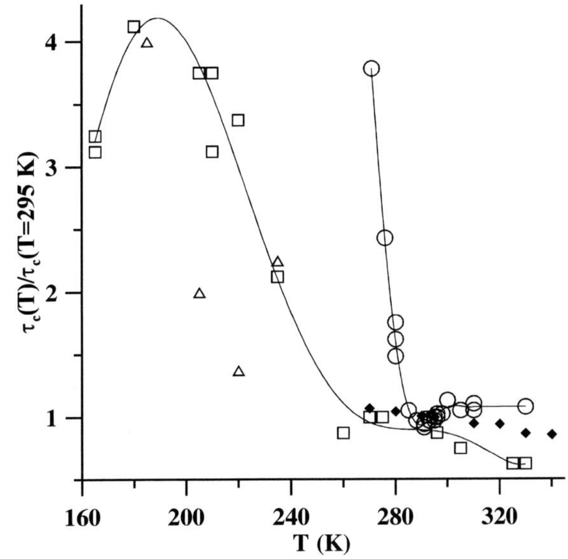

Figure 4.4. Correlation times versus temperature relative their magnitudes at room temperature for bulk gallium melt (solid diamonds), porous glass with 16 nm pores (circles), and porous glass with 2 nm pores (squares – cooling, triangles – warming). Solid lines are guides for the eye clarifying temperature dependences of the correlation time for the porous glass with 16 nm pores and for the porous glass with 2 nm pores upon cooling.

The increase in the correlation time in confined geometry also leads to shortening the transverse spin relaxation time as was found in [25,53,59]. It appears, in particular, in the broadening of NMR lines. However, the experimental lineshapes might be also influenced by inhomogeneity of the Knight shift and therefore are less suitable for evaluating the correlation time of electric field gradient fluctuations on nuclear sites.

One can conclude that:

(a) data on nuclear spin relaxation in liquid gallium under nanoconfinement can be used to evaluate the correlation time of atomic mobility;
(b) atomic mobility in confined liquid gallium at room temperature slows down compared to the bulk gallium melt and the slowdown becomes stronger with decreasing the pore size;
(c) additional slowdown of atomic mobility is observed at low temperature, especially in the range of partial freezing.

4.2.2. In

Indium isotopes also have quadrupole moments and therefore the atomic mobility in the indium melt should affect nuclear spin relaxation. The scenario which was used for liquid confined gallium can be applied to liquid indium under nanoconfinement. The obvious problem is that, as was already mentioned, it is impossible to separate magnetic and quadrupole contributions to spin relaxation in bulk melted indium on the basis of experimental data for only ^{115}In. Instead, one can use theoretical estimates for the qudrupole contribution. Measurements of nuclear spin-lattice relaxation were carried out for the isotope ^{115}In in liquid metallic indium embedded into artificial opals and a porous glass [62].

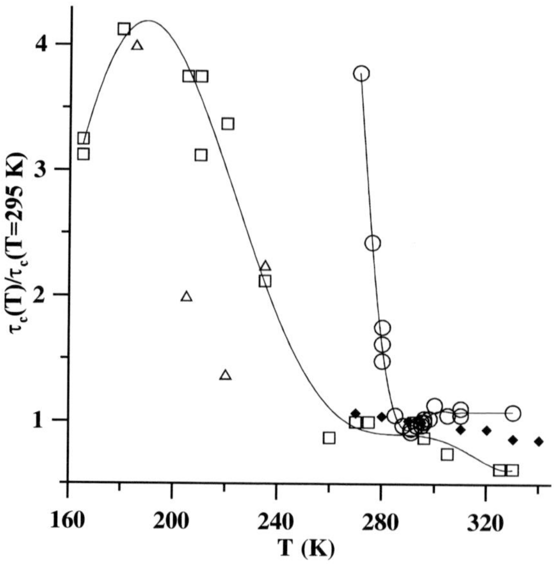

Figure 4.5. ^{115}In longitudinal magnetization recovery curves (I is the integral intensity of NMR signals; τ is the time between first and second pulses) for melted indium embedded into the porous glass (closed diamonds), Opal-1 (open circles), and Opal-2 (open squares), for bulk liquid indium (closed circles), and for melted indium on the surface of Opal-1 (closed squares). The inset shows full recovery curves for bulk and surface liquid indium. Solid lines are single exponentials.

The opals under study consisted of close-packed silica spheres with diameters of about 240 and 210 nm. The porous glass had pores of about 7 nm. Relaxation measurements were performed at magnetic field of 9.4 T. The inversion recovery procedure was used to obtain the longitudinal magnetization versus time dependences. The Knight shift and NMR lineshape were also measured for ^{115}In as well as for ^{113}In. Similar measurements were performed also for bulk liquid indium.

The melting and freezing temperatures for confined indium are reduced compared to the bulk melting point (see Section 3.2). Moreover, liquid indium within pores can be supercooled to lower temperatures.

NMR relaxation measurements for confined indium were carried out first at 415 K when indium on the sample surfaces is completely frozen to prevent the influence of signals from the surface melt. The relevant NMR characteristics for bulk liquid indium were measured at 430 K. The ^{115}In longitudinal magnetization restoration curves obtained are shown in Figure 4.5. The remarkable enhancement of spin-lattice relaxation for confined indium compared to relaxation in bulk was found. For the opals the relaxation rate increased more than by a factor of four. The relaxation acceleration is even more pronounced for indium within the porous glass. This drastic acceleration of spin-lattice relaxation in confined liquid indium can arise because of increasing either the magnetic or the quadrupole contribution.

One can estimate the possible alteration of the magnetic contribution for the opals and porous glass compared to bulk using the Korringa relation (4.2) as it was done for gallium. Measurements of NMR line positions for bulk and confined liquid indium showed that the Knight shift in the opals coincided with that in bulk within the experimental accuracy limits. The Knight shift in the porous glass was decreased compared to bulk by about 100 ppm, that is, by about 1 % of the Knight shift in the bulk indium melt. The fact that the Knight shift remained almost unchanged indicates clearly that the magnetic contribution to nuclear spin relaxation for confined liquid indium remained unchanged also. Thus, we should ascribe the relaxation acceleration to the enhancement of the quadrupole contribution.

As it will be proved below, the extreme narrowing approximation remains valid for liquid indium within the opal and porous glass samples thus simplifying the treatment of the quadrupole relaxation for confined indium. Within the framework of the extreme narrowing limit the total longitudinal magnetization restoration process should be described by a single exponential with the spin-lattice relaxation time T_1 given by

$$T_1 = (T_{1m}^{-1} + C\tau_c)^{-1}. \qquad (4.4)$$

Implying T_{1m} not to be influenced by confined geometry in agreement with the Korringa relation, we must suggest that the acceleration of magnetization restoration is caused by enhancement of the $C\tau_c$ term.

The quadrupole and magnetic contributions to spin relaxation of the ^{115}In isotope for bulk liquid indium were separated in [63,64] based upon temperature dependences of the relaxation rate. The quadrupole contribution for ^{115}In was also estimated theoretically [63,65]. Theoretical models predicted for the quadrupole relaxation rate $C\tau_c$ in bulk a value of about 3200 s^{-1} [65] and 1900 s^{-1} [63] near the melting point, while from experiment $C\tau_c$<1900 s^{-1} [54,63]. In the estimates for the quadrupole relaxation rate in confined liquid indium the both above values of $C\tau_c$ were used [62]. The total spin-lattice relaxation rate measured for bulk liquid indium just near the melting point was 5320 s^{-1} (T_1=188 μs). It gives two possible estimates for the magnetic contribution: T_{1m}^{-1}=2120 s^{-1} for $C\tau_c$ = 3200 s^{-1} and T_{1m}^{-1}=3420 s^{-1} for $C\tau_c$ = 1900 s^{-1}. This allows calculating the quadrupole relaxation rates for the opals and porous glass. All $C\tau_c$ values obtained are listed in Table 4.2. It follows from Table 4.2 that the quadrupole relaxation rate increased more than by a factor of six in both opal samples and by a factor of 20 in the porous glass, this conclusion keeps valid independently of assumptions made for quadrupole relaxation in bulk indium. Moreover, quadrupole relaxation dominates for liquid indium within the porous matrices. One can assume that alterations in relaxation arise mainly due to the increase in the correlation time of the atomic movements. The correlation time in bulk can be estimated from the known value for the self-diffusion as τ_c~12·10^{-12} s in agreement with [63]. The values of τ_c for confined liquid indium calculated under the assumption of constant C are listed in Table 4.2. It is easy to see that the product $\omega_o\tau_c$ remains much less than unity, which validates the extreme narrowing approximation assumed. The increase of the correlation times of atomic motion for liquid indium in the opals and porous glass estimated from the spin relaxation measurements evidences a drastic slowdown of the translational diffusion in confined indium melt. Such slowdown has clear size dependence similarly to the case of confined gallium. The increase in the correlation time for indium within pores should also influence the quadrupole contribution to transverse spin relaxation. Within the framework of the extreme narrowing approximation, the longitudinal and transverse relaxation times are equal to each other, thus T_2 for liquid indium should be given by (4.4). On the other hand, the acceleration of transverse spin relaxation in liquids leads to the NMR line broadening. As is seen from Table 4.2, the experimental resonance lines for confined melts are much broader than in bulk.

Table 4.2. ^{115}In spin-lattice relaxation rate T_1^{-1} (measured at 415 K for confined indium and at 430 K for bulk and surface indium), rate of quadrupole relaxation T_{1q}^{-1}, correlation time of atomic motion τ_c, and linewidth at half-height δ for indium confined to opals and porous glass, bulk indium and indium on the surface of the opal matrix with 240 nm size of silica spheres. T_{1q}^{-1} and τ_c are calculated starting from two values of T_{1q}^{-1} in bulk liquid indium, 3200 s^{-1} (upper lines) and 1900 s^{-1} (lower lines)

	Bulk In	In in opal (240 nm)	In in opal (210 nm)	In in porous glass	In on surface
T_1^{-1} (s^{-1})	5320±50	21300±800	25700±1200	77000±15000	7340±90
T_{1q}^{-1} (s^{-1})	3200	19200	23600	75000	5220
	1900	17900	22300	73000	3920
τ_c ($\mu s \cdot 10^{-6}$)	12	71	87	280	19.3
		113	140	460	24.5
δ (ppm)	19.4	166	235	510	26.5

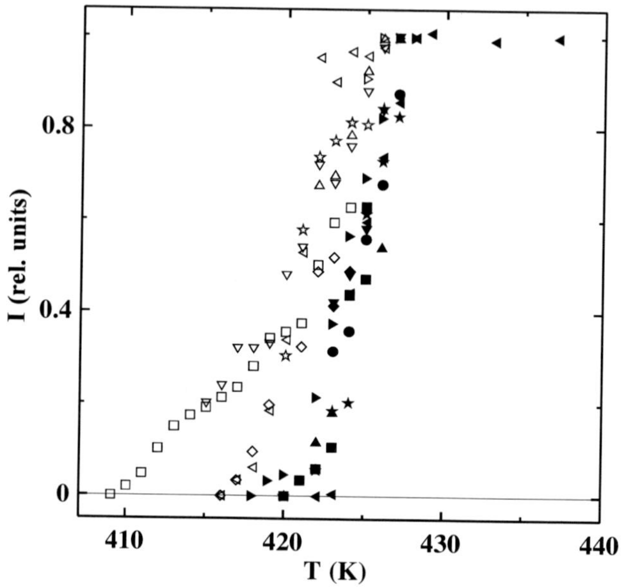

Figure 4.6. Temperature variations of the integral intensity I of the ^{115}In NMR signal from liquid indium confined to the opal matrix with the silica sphere size of 240 nm upon warming (closed symbols) and cooling (open symbols). Different symbols correspond to different thermal cycles.

However, the NMR lines for confined indium calculated using the equality T_2=T_1 are narrower than the experimental ones, which evidences the influence of some additional mechanism of broadening. Similar excess NMR line broadening was observed for liquid gallium in confined geometry. One can suggest that such additional broadening is caused by difference of the Knight shift in pores of different sizes. The ^{115}In spin-lattice relaxation acceleration and line broadening were also observed for a thin film of melted indium on the surface of opal consisted of spheres with 240 nm size. The film thickness was not quite homogeneous, but its average value can be estimated as 5 microns. The results for this film are included into Table 4.2. They allow us to suggest that the slowdown of the atomic diffusion in liquid metallic indium is noticeable even for dimensions of 5 microns.

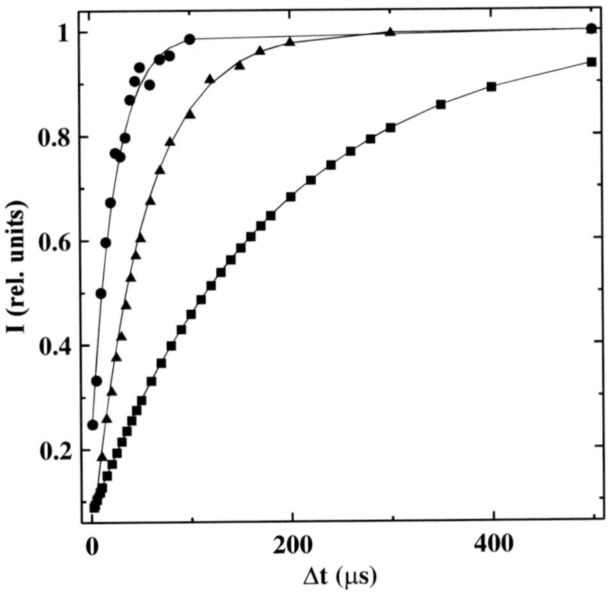

Figure 4.7. ^{115}In nuclear magnetization restoration curves for liquid indium confined to the opal matrix with the silica sphere size of 240 nm obtained at 427 K (triangles) and 419 K (circles) and for bulk melted indium (squares) obtained at 430 K.

Additional slowdown of self-diffusion was found by NMR in liquid metallic indium confined within nanopores of an opal photonic crystal with size of silica spheres 240 nm within temperature ranges of indium diffused freezing and melting [37]. To control the melting and solidification processes and to get information about relative amount of unfrozen indium, the integral intensity of the NMR signal from liquid indium was measured upon warming and cooling.

Both melting and freezing for confined indium were found to be diffused and the melting and freezing temperatures were pronouncedly reduced compared to the bulk melting point (Figure 4.6).

The freezing process could occur in a different way depending on the highest temperature achieved upon warming, thus providing opportunity of relaxation measurements at the same temperature for various full amounts of unfrozen indium. The melting process was complete till 428 K.

The ^{115}In longitudinal magnetization restoration curves obtained at 427 and 419 K are shown in Figure 4.7.

They correspond to totally melted indium and to partly frozen indium (I=0.03), respectively. One can see a remarkable enhancement of spin-lattice relaxation for totally melted confined indium compared to relaxation in bulk and even stronger enhancement of spin restoration rate at the lower temperature and reduced relative intensity corresponding to coexistence of liquid and frozen indium within pores.

Measurements of T_1 at various temperatures and various relative intensities of the NMR signal revealed clear correlations between the relaxation time and the relative intensity of the NMR signal, that is the portion of unfrozen confined indium (Figure 4.8). In contrast, no evident correlations were found between the relaxation time and temperature (see the inset in Figure 4.8).

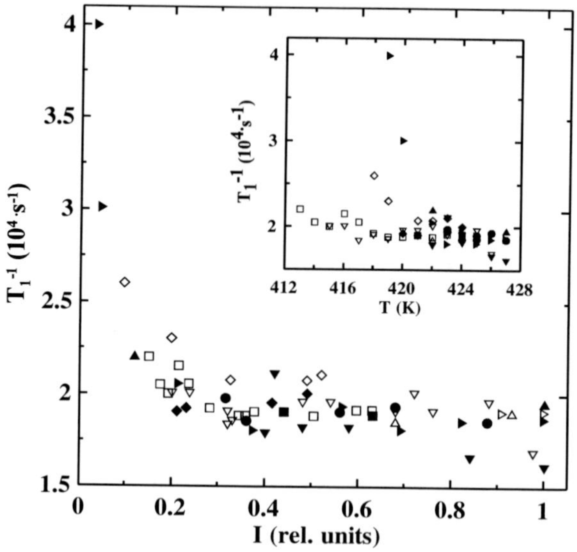

Figure 4.8. The [115]In spin-lattice relaxation rate versus the integral intensity I of the NMR signal for liquid indium confined to the opal matrix with the silica sphere size of 240 nm upon warming (closed symbols) and cooling (open symbols). Different symbols correspond to different thermal cycles. The inset shows the temperature dependences of T_1^{-1}.

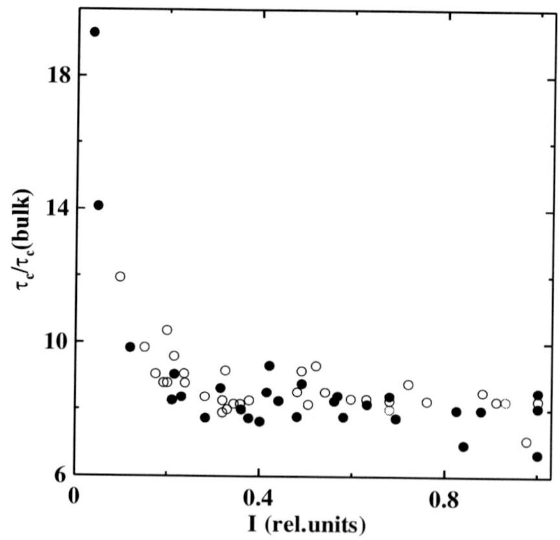

Figure 4.9. The correlation time of atomic mobility τ_c in liquid indium confined to the opal matrix with the silica sphere size of 240 nm relative to that in bulk versus the integral intensity I of the [115]In NMR signal through the melting and freezing phase transitions. Here was assumed to be 3420 s^{-1} at 430 K. Closed symbols correspond to warming, open symbols – to cooling.

One can conclude that the additional acceleration of nuclear spin-lattice relaxation in confined liquid indium is driven by the amount of unfrozen indium within pores rather than by temperature itself.

The dependence of the correlation time on the relative intensity of NMR signal is shown in Figure 4.9 when T_{1m}^{-1}=3420 s^{-1} was assumed for bulk. A similar curve can be plotted for another estimate of T_{1m}^{-1}. One can see that the correlation time increases by 3 times upon decreasing the relative amount of melted indium from 1 to 0.03. Note, that the product $\omega_o\tau_c$ still remains much less than unity. The results obtained manifest the drastic slowdown of self-diffusion in confined indium within the temperature range of the freezing and melting phase transitions in addition to the general slowdown of confined indium self-diffusion compared to bulk. The decrease of the self-diffusion coefficient under partial indium crystallization follows the portion of unfrozen indium within pores rather than temperature.

4.2.3. Ga-In alloy

The potential applications of the Ga-In alloys in nanodevices depends on atomic mobility in confined melts which is directly related to the processes of atomic diffusion and viscosity. Following data discussed in Sections 4.2.1 and 4.2.2, a slowdown of atomic mobility can be also expected for nanostructured liquid gallium-indium eutectic alloys. NMR studies of atomic mobility in a gallium-indium alloy with composition 90 at.% Ga and 10 at.% In embedded into porous glasses were carried out in [66,67]. Nuclear spin relaxation in the Ga-In alloy was confined to two different porous glasses with pore size of 200 and 5 nm. Data were obtained at two magnetic fields, 9.4 and 17.6 T, using Bruker Avance400 and Avance750 pulse spectrometers.

The use of different magnetic fields allowed more accurate evaluations of the correlation times and alterations in electric field gradients on nuclei. Bulk Ga-In alloy of the same composition was also studied to make a comparison. The bulk Ga-In alloy of the composition under study is liquid at room temperature (295 K) as its liquidus temperature is near 20 °C [68].

The alloy within pores is also completely melted at room temperature according to [38,66]. Therefore, studies of mobility in confined and bulk alloy can be carried out at 295 K. NMR spectra for the isotopes ^{71}Ga, ^{69}Ga, and ^{115}In showed that the Knigh shift for two gallium isotopes and ^{115}In for alloy confined to porous glasses decreases only slightly compared to that in bulk.

Table 4.3. The Knight shift (*K*), linewidth at half height (*Δ*), total longitudinal relaxation time (T_1), quadrupole constant (*C*), and times corresponding to magnetic (T_{1m}) and quadrupole (T_{1Q}) contributions to longitudinal spin relaxation for the Ga-In alloy. Values of the Knight shift for ^{69}Ga are the same as for ^{71}Ga

isotope	^{71}Ga				^{69}Ga				^{115}In	
alloy	bulk	200 nm	5 nm		bulk	200 nm	5 nm		bulk	200 nm
			17.6 T	9.4 T			17.6 T	9.4 T		
K (ppm)	4256±1	4255±1	4070±5	4035±5					8504±2	8498±3
Δ (ppm)	6.0±0.5	10±1	80±5	130±8	6.5±0.5	11±1	140±10	230±15	18±1	30±2
T_1 (μs)	519±3	509±10	88	59	617±5	577±30	32	24	220±8	201±15
T_{1m} (μs)	586±9				946±15				410±40	
T_{1Q} (μs)	4500±800	3800±600	103	66	1800±400	1500±300	33	25	490±40	400±50
C (μs^{-2})	15.7	15.7	51		40.2	40.2	130		146	146

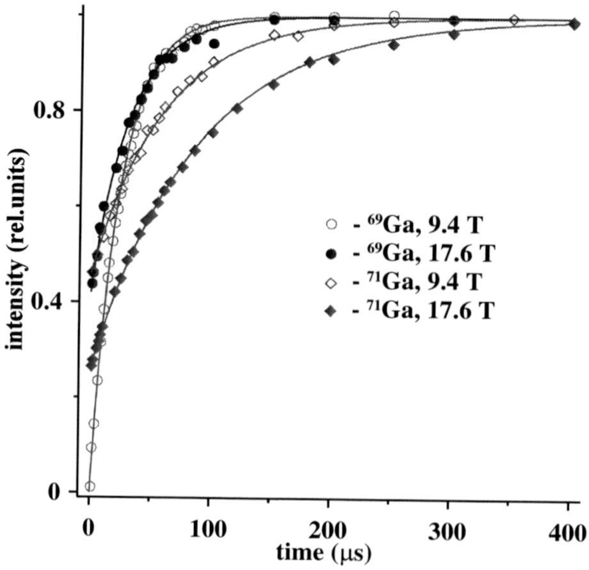

Figure 4.10. Longitudinal magnetization recovery curves for the ^{71}Ga (diamonds) and ^{69}Ga (circles) isotopes at field 9.4 T (open symbols) and 17.6 T (closed symbols) for Ga-In alloy confined to porous glass with 5 nm pores. Solid lines show theoretical calculations.

The recovery of the longitudinal nuclear magnetization for gallium and indium isotopes in bulk alloy was single exponential similarly to other bulk liquid metals and alloys and the relaxation rate was the same within experimental accuracy at both magnetic field. Single exponential and field-independent relaxation corresponds to the extreme narrowing limit. The extreme narrowing limit remains valid for the alloy embedded into 200 nm pores where relaxation was only slightly faster and lines slightly broader than in bulk alloy.

Therefore, longitudinal relaxation in the alloy within 200 nm pores as well as in bulk can be defined by relaxation times. The times averaged over several measurements are shown in Table 4.3.

For the liquid Ga-In alloy confined to 5 nm pores, longitudinal relaxation for both gallium isotopes was much faster compared to that in bulk.

The experimental magnetization recovery curves for ^{71}Ga and ^{69}Ga for the alloy within 5 nm pores are shown in Figure 4.10. One can see that recovery of longitudinal magnetization is noticeably faster at magnetic field 9.4 T for both isotopes. The field dependence of the relaxation rate evidences that the extreme narrowing limit is no longer valid. In this case the spin magnetization recovery is not single exponential and in the particular case of spin 3/2 is described by (4.3). However, the recovery curves remain still quite similar to single exponentials and can be approximately characterized by relaxation times. This facilitates the comparison with relaxation in bulk alloy and alloy within bigger pores. The times evaluated from the curves in Figure 4.10 are also listed in Table 4.3. Note that the relaxation times for the alloy confined to 5 nm pores are shorter for the ^{69}Ga isotope with larger quadrupole moment while spin-lattice relaxation in bulk and within bigger pores is faster for the ^{71}Ga isotope with higher gyromagnetic ratio.

The gallium and indium quadrupole relaxation rates and correlation time of atomic mobility in the liquid Ga-In alloy can be evaluated using measurements of spin relaxation at two different magnetic fields and under two different nanoconfinements.

Table 4.4. Evaluated correlation times and their ratio to the value in bulk for the Ga-In alloy

alloy	bulk	200 nm	5 nm
$\tau\,(10^{-6}\ \mu s)$	14	17	370
τ/τ_b	1	1.2	26

The magnetic and quadrupole contributions to spin-lattice relaxation for two gallium isotopes can be separated using the same approach as was used for pure bulk gallium. The obtained times are listed in Table 4.3.

The correlation time in bulk liquid alloy τ_b was estimated in [66] from the diffusion coefficient D to be equal to τ_b=1.4 10^{-11} s. Then, one can find and for bulk (see Table 4.3).

The same approach can be applied to the alloy within 200 nm pores as the extreme narrowing approximation is still valid for this pore size. On the other hand, from the Korringa relation taking into account that the Knight shift in liquid alloy within 200 nm pores was almost the same as in bulk one can conclude that the magnetic relaxation for the alloy within 200 nm pores did not change compared to bulk, either. Assuming the invariance of the times of magnetic relaxation for both gallium isotopes one can evaluate the times of quadrupole relaxation (see Table 4.3). One can easily get that both approach give self-consistent estimates.

The times of quadrupole relaxation for the alloy in the glass with 200 nm pores are shorter than in bulk. It gives τ_{200} =1.7 10^{-11} s and τ_{200}/τ_b=1.2. Note that the ratio between the correlation times does not depend on the assumed magnitude of τ_b. The obtained value for the correlation time in the alloy within 200 nm pores can be then used to estimate the quadrupole and magnetic contributions to spin relaxation for indium nuclei. Magnetic relaxation time for ^{115}In in 200 nm pores being the same as in bulk because of very small alterations in the Knight shift, the indium relaxation times in bulk and 200 nm pores satisfy the equations

$$(T_{1b}^{In})^{-1} = C_b^{In}\tau_b + (T_{1mb}^{In})^{-1},$$

$$(T_{1,200}^{In})^{-1} = C_b^{In}\tau_{200} + (T_{1mb}^{In})^{-1}, \qquad (4.5)$$

where the quadrupole constant was also implied to be the same in bulk and within 200 nm pores. The equations (5) allow evaluating the values of and T_{1mb}^{In}. They are shown in Table 4.3.

Note that the obtained quadrupole relaxation rate in bulk alloy $(T_{1Qb}^{In})^{-1} = C_b^{In}\tau_b$ =2044 s^{-1} is just between two theoretical estimates reported for liquid indium: 3200 s^{-1} and 1900 s^{-1}.

For the alloy confined to porous glass with 5 nm pores, nuclear spin-lattice relaxation for two gallium isotopes is accelerated and depends on magnetic field. Similarly to bulk pure gallium and indium the acceleration of relaxation can be ascribed to the remarkable increase in the quadrupole contribution.

The field dependence of relaxation evidences that the extreme narrowing limit is no more correct. The results of fitting the experimental magnetization recovery curves with theoretical relationships are shown in Figure 4.10. The fitting gives τ_5 =3.7 10^{-10} s and quadrupole constants listed in Table 4.3. The good fitting for the four experimental curves with only two adjusting parameters confirms the validity of the theoretical model used.

One can see that the correlation time of electric field gradients for the Ga-In alloy within 5 nm pores is much longer than in bulk or in the alloy within 200 nm pores which evidences a remarkable slowdown of atomic mobility under such nanoconfinement.

One should emphasize that the quadrupole constant is also remarkably changed for the alloy in fine pores. This shows strong alterations in electric field gradients driven by structural changes in the melt confined to 5 nm pores. To make the increase in the quadrupole contribution for the alloy within 5 nm pores more evident in comparison to bulk and 200 nm pores, the approximate times of quadrupole relaxation calculated using the equation are also listed in Table 4.3. These values show that quadrupole relaxation within 5 nm pores is about 40 times as fast as in bulk. The obtained correlation times for the alloy within porous glasses are collected in Table 4.4.

One can conclude that nanoconfinement leads to drastic slowdown of atomic mobility in liquid metallic gallium and indium as well as in Ga-In alloy. The slowdown shows clear size dependence.

5. Knight Shift under Nanoconfinement

5.1. Size Dependence of the Knight Shift

The Knight shift in liquid metals with closed inner electron subshells originates mainly from the Fermi contact interaction of the spin polarized conduction electrons with nuclei [69,70]:

$$K_S = \frac{8}{3}\pi\chi_S\Omega\left\langle|\Psi_F(0)|^2\right\rangle, \tag{5.1}$$

where χ_s is the electron spin susceptibility per unit volume, Ω is the atomic volume, is the probability density at the nucleus of s-like electrons on the Fermi surface. The contributions from the orbital hyperfine interaction and core polarization are small. Measurements on bulk liquid metals were consistent with theoretical examination and with spin-lattice relaxation studies. According to (5.1), the Knight shift caused by the contact interaction is directly proportional to the electron spin susceptibility.

Until recently, influence of particle size reduction on the Knight shift was studied mostly for small isolated or supported metallic particles and thin layers in solid state (see [71-74] and references therein). At low temperatures theories predict alterations in the Knight shift due to quantum size effects rising from increasing the conduction electron level separation (see [75] and references therein). However, at higher temperatures the Knight shift could be affected chiefly by surface induced spatial variations in the density of electrons on the Fermi level which influence the total electron spin susceptibility because of increasing the surface to volume ratio (see [76] and references therein). According to theoretical calculations carried out using the local density functional formalism the spin susceptibility for simple metals with closed inner electron subshells should show damped Friedel oscillations near the metal surface; the integral surface magnetic susceptibility increases compared to bulk. This model was applied to treat the strong broadening of NMR lines due to broad distribution in the Knight shift for small solid particles of such metals as Pb, Cu, or Ag [73,77]. The drastic resonance line broadening prevented clear observation of size dependence of the Knight shift.

Contrary to the case of solid metals, the NMR lines for liquid nanoparticles should be narrowed due to atomic mobility. Thus, the Knight shift measured for confined liquid tin nanoparticles should correspond to whole particles.

Studies of the Knight shift for liquid metals under nanoconfinement were carried out for gallium, indium, tin, mercury and for gallium-indium alloy [25,37,45,53,59,62,67,78,79].

For gallium the Knight shift was measured for both gallium isotopes, ^{69}Ga and ^{71}Ga, with spin I=3/2. For indium the Knight shift was measured also for two isotopes, ^{115}In and ^{113}In, with spin I=9/2. Tin has three naturally occurring isotopes which nuclei have non-zero magnetic moments. All three isotopes have spin I=1/2. The best NMR signals are seen for ^{119}Sn and the Knight shift was measured for this isotope. Mercury has two naturally occurring isotopes with non-zero magnetic moments, however only ^{199}Hg with spin I=1/2 provides good NMR signals and the Knight shift was measured for this isotope.

For all pure liquid metals and Ga-In alloy in confined geometry the Knight shift was found to decrease compared to the relevant bulk melts. The alterations in the value of the Knight shift

$$\Delta K = K - K_b \tag{5.2}$$

(K_b is the Knight shift in bulk) differed for different metals. To reveal some regularities in the Knight shift alterations the plot of ΔK versus atomic number for liquid gallium, indium, tin, and mercury embedded into porous glass with 4 nm pore radius are shown in Figure 5.1. The Knight shift for gallium and mercury were obtained at room temperature. Data for tin and mercury were obtained near the relevant melting temperatures. (The point for confined mercury was taken from the dependence of the Knight shift on pore size shown in Figure 5.2). One can see from Figure 5.1 that the Knight shift change increases monotonically with increasing atomic number. However, since the Knight shift for bulk liquid metals, K_b, itself varies with atomic number, the fractional changes in the Knight shift under confinement, $\Delta K/K_b$, are more indicative. They are listed in Table 5.1. The fraction $\Delta K/K_b$ is very similar for four melts and seems to be practically independent of atomic number. This result supports the suggestion that the confinement induced decrease in the Knight shift for liquid metals is mainly related with alterations in the electron spin susceptibility (in accordance with (5.1)) which in bulk depends only on the number of conduction electrons per unit volume within the framework of the free electron model. Table 5.2 shows the values of the Knight shift for liquid gallium under various nanoconfinement. One can see a general tendency for the decrease in the Knight shift with decreasing size of pores, while the dependence of the Knight shift on the pore size is not quite monotonous.

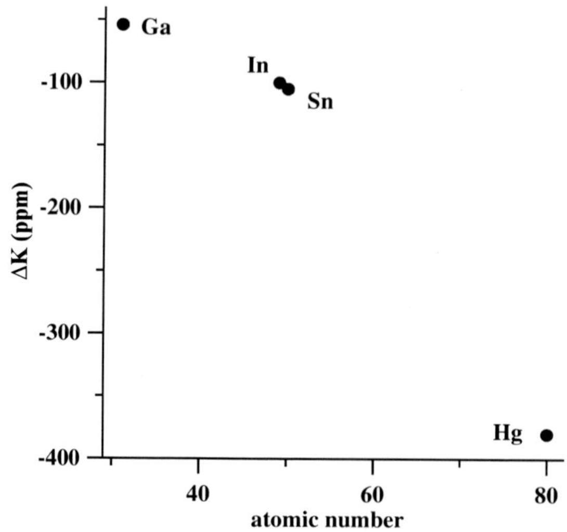

Figure 5.1. The change ΔK of the Knight shift for liquid tin, mercury, indium, and gallium embedded into porous glass with 4 nm radius versus their atomic number.

Table 5.1. Fractional change in the Knight shift $\Delta K/K_b$ in liquid metals confined within porous glass with 4 nm pores and Knight shift K_b in bulk melts

metal	Sn	Hg	In	Ga
$\Delta K/K_b$ (%)	-1.43	-1.38	-1.26	-1.19
K_b (%)	0.73	2.74	0.79	0.45

Table 5.2. Difference ΔK between the Knight shift confined liquid gallium and in bulk at room temperature. The asterisk * marks size of silica spheres

sample	bulk	porous glasses						Vycor glass	opals	
Pore size, nm		3.5	4	5	8	200 (8 nm)	200 (200 nm)	6	100*	250*
-ΔK, ppm	0	192	137	74	54	74	17	143	71	14

Nevertheless, if to select the porous matrices with similar pore geometry, namely, porous glasses produced by the same technology or the opals, the size dependence for the selected samples becomes monotonous. The Knight shift at room temperature versus inverse pore diameter is shown in Figure 5.2 for the glasses with 3.5, 4, 5, and 8 nm pores. The size dependence can be rather well fitted by the empirical relation

$$K=K_b\{1-\alpha\exp(-d/d_o)\}, \quad (5.2)$$

where d is the pore diameter, α and d_o are the fitting parameters. The fitting curve is also shown in Figure 5.2. The fitting parameters for these porous glasses are: α=0.14, d_o=2.6 nm.

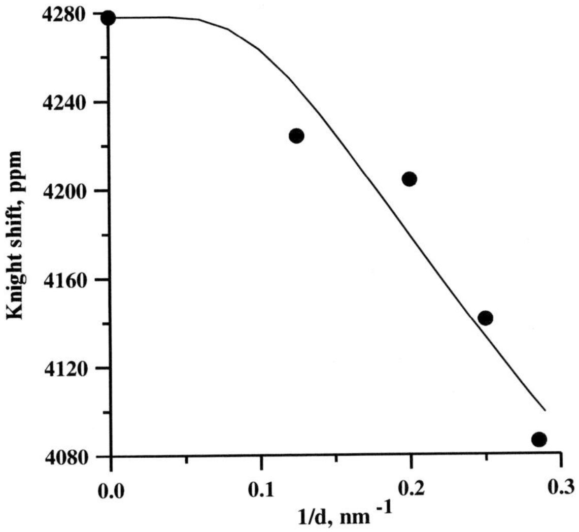

Figure 5.2. Dependence of the Knight shift for liquid gallium in porous glasses on the inverse pore diameter at room temperature. The solid curve is an exponential approximation (5.2).

For other porous matrices the empirical relation (5.2) with the above fitting parameters is not longer valid, this might evidence the pronounced role of pore geometry in the dependence of the Knight shift on pore sizes.

For liquid indium embedded into opal matrices theKnight shift was almost not affected. However, for indium within porous glass with 8 nm pores the Knight shift was decreased by about 100 ppm.

Changes in the gallium and indium Knight shift for the confined Ga-In alloy with composition 90 at.% Ga and 10 at.% In was found in [66,67]. The values of the Knight shift are presented in Table 4.3. The Knigh shift for two gallium isotopes obtained at fields 9.4 and 17.6 T for bulk alloy and alloy confined to porous glass with 200 nm pores coincided with each other within experimental accuracy as is expected. For the glass with 5 nm pores, the Knight shift was different at different magnetic fields while it coincided within experimental accuracy for both isotopes at the same field. The general trend of decreasing the Knight shift with decreasing the pore size remains valid for the alloy as it was for the pure metals. Signals from ^{115}In were observed only for bulk alloy and alloy within bigger pores.

Since the Knight shift measured for confined liquid metallic nanoparticles should correspond to whole particles, the decrease in the Knight shift observed reflects a size dependence of averaged values of the shift. According to (5.1) one should expect that the enhancement predicted of the integral surface electron susceptibility could lead to a total increase in the Knight shift for liquid nanoparticles of non-transition metals. However, the decreasing of the shift observed for confined liquid metals and alloy contradicts these theoretical predictions. The decrease of the Knight shift cannot either be explained with the additional Laplace pressure which increases with decreasing sizes of small particles since, for instance, the Knight shift in bulk liquid mercury is known to increase with decreasing volume or increasing pressure. Therefore, the reduction of the Knight shift under nanoconfinement as well as field dependence of the shift for gallium within small pores need in more studies and theoretical treatment.

Table 5.3. The coefficient of linear regression dK/dT calculated for the temperature range 220 to 310 K for liquid gallium under different nanoconfinement and for bulk melted gallium

sample	bulk	porous glasses								Vycor glass	opals	
Pore size, nm		2	3.5	4	5	8	16	200 (8 nm)	200 (200 nm)	6	100*	250*
dK/dT, ppm/K	-0.29	0.28**	-0.58	-0.77	0**	-0.32	-0.34	-0.76	-0.66**	-0.22	-0.46	-0.31

The asterisk * marks size of silica spheres; double asterisks ** mark coefficients calculated for the temperature range 260 to 320 K

5.2. Variations of the Knight Shift under Nanoconfinement with Temperature

Variations in the Knight shift for metals under nanoconfinement with temperature were studied mostly for liquid gallium. It was obtained that the temperature dependences for confined gallium were quite different from those in bulk melted gallium [25,78]. Some examples are presented in Figures 5.3-5.5. The temperature dependence of the Knight shift can be characterized by a coefficient dK/dT.

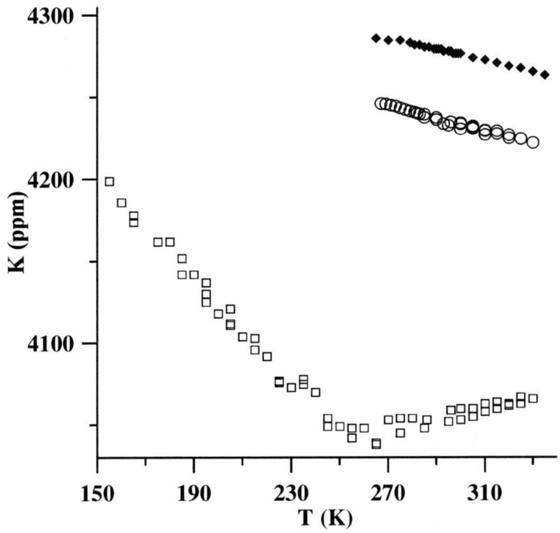

Figure 5.3. Temperature dependences of the liquid gallium Knight shift K relative to GaAs for the porous glasses with 2 nm pores (squares) and 16 nm pores (circles) and for bulk (diamonds).

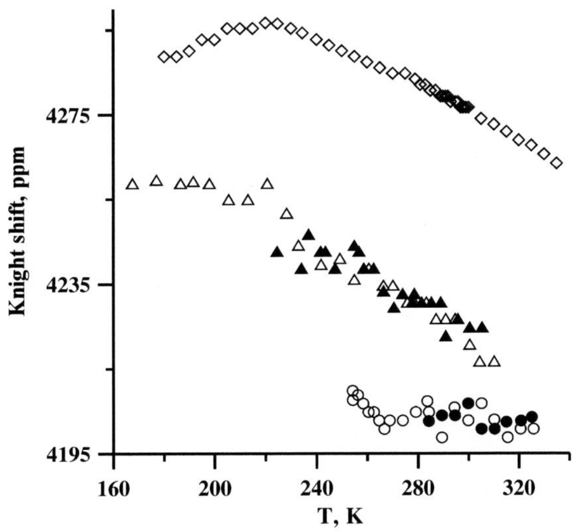

Figure 5.4. Temperature dependence of the Knight shift for liquid gallium in the porous glasses with pore size of 5 nm (circles) and of 8 nm (triangles) and for bulk liquid gallium (diamonds). Open circles—cooling; closed circles—warming.

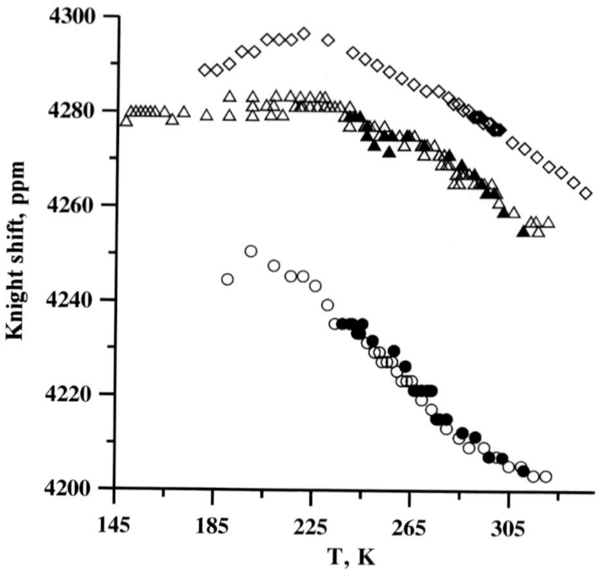

Figure 5.5. Temperature dependence of the Knight shift for liquid gallium in the opals with silica sphere diameters of 100 nm (circles) and of 250 nm (triangles) and for bulk liquid gallium (diamonds). Open circles—cooling; closed circles—warming.

Such coefficients are collected in Table 5.3 for gallium within different nanocomposites studied. The coefficient dK/dT remains negative while changed for most of porous matrices. However, for the smallest pore size of 2 nm the coefficient dK/dT becomes positive in some temperature range covering room temperature (Figure 5.4).

Quite unusual situation with the splitting of the gallium NMR line and bifurcation of the Knight shift versus temperature dependence was observed in [80] for the gallium loaded opal matrix consisted of silica spheres. Diameter of silica spheres was 250 nm. The sequence of NMR spectra obtained for ^{71}Ga upon cooling from 330 K shows a surprising splitting in the temperature range between 260 and 220 K (Figure 5.6). The positions of the two components of the NMR line at different temperatures are shown in Figure 5.7.

The bifurcation of the Knight shift dependences on temperature was treated in [80] as an evidence for the phase transition in confined liquid gallium which is seen near the gallium freezing. Later a theoretical model [81] was suggested to treat the phase transition in melted gallium.

Therefore, we can conclude that:

- The Knight shift of NMR lines in confined liquid mercury, tin, gallium, and indium metals and in Ga-In eutectic alloy decreases compared to bulk;
- Size effects on the Knight shift becomes stronger for smaller pores. This tendency is monotonous for the porous matrices of the same pore geometry but different pore size;
- Temperature dependence of the Knight shift for liquid gallium is noticeably affected by nanoconfinement;
- Anomalies of the line shape and Knight shift for gallium within an opal suggested a phase transition in the liquid structure near the freezing temperatures.

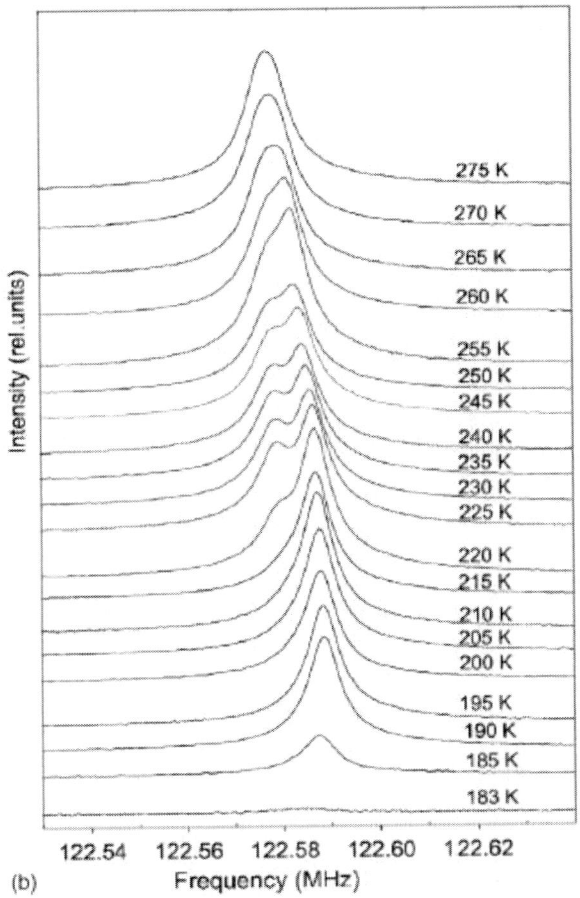

Figure 5.6. NMR spectra of ^{71}Ga in the opal with 250 nm silica spheres. Between 260 and 220 K the spectrum clearly exhibits a splitting.

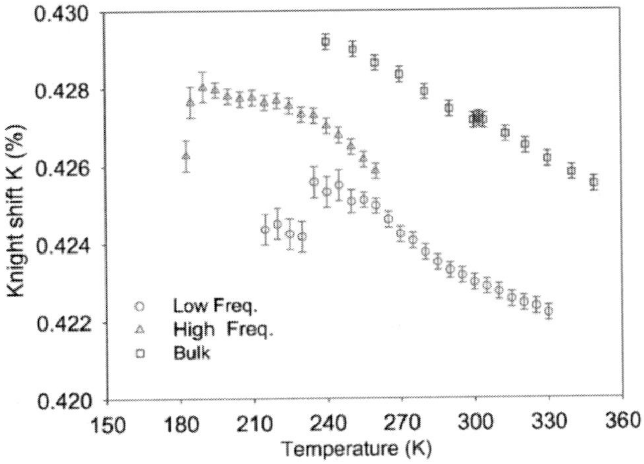

Figure 5.7. The temperature dependence of the Knight shifts of low-frequency (LF) and high-frequency (HF) ^{71}Ga NMR signals in opal with 250 nm silica spheres. The squares are the Knight shift of Ga in bulk liquid gallium for comparison.

Conclusion

Experimental studies which were reviewed in Sections 3-5 revealed pronounced alterations in properties of liquid gallium, indium, and their alloy under nanoconfinement. Most of the findings could be treated using predictions of modern theoretical models while some of them still need in more studies.

The shift to low temperatures of the melting and freezing transitions is a result of the influence of size reduction due to the increase in the role of surface. Even when the size of confined particle is larger than the pore diameter which is possible when the particle shape is dendrite, the surface to volume ratio is still defined by the pore size. It gives an explanation for the clear dependence of the melting temperature on the pore diameter. The melting diffusion for gallium, indium, and their alloy can be attributed to pore size distribution as it is commonly believed for other confined substances. In addition, the melting and freezing in confined gallium is very often affected by formation of different crystalline structures from the supercooled melt, some of them were not known for bulk gallium. The crystallization under nanoconfinement is likely induced by heterogeneous nucleation and this explains the reproducible character of the melting-freezing hysteresis loops. Sharp or diffused crystallization is probably defined by connectivity of the network of confined particles and, therefore, sharp freezing can be observed under high pore filling.

The remarkable enhancement of nuclear spin-lattice relaxation for gallium, indium, and their alloy was shown to arise because of the increase in the quadrupole contribution which depends on the atomic mobility in melts. Theoretical models allow us to evaluate the correlation time of atomic mobility using measurements of spin relaxation. It was found that the correlation time which has the physical meaning of the time between two consecutive hops is increased for melts within pores more than by an order of magnitude. Additional changes in spin relaxation under nanoconfinement upon decreasing temperature evidence the additional slowdown of atomic mobility.

The reduction of the Knight shift for liquid metals embedded into mesopores compared to bulk was proved not only for gallium, indium, and Ga-In alloy, but also for mercury and tin. It was suggested that such a reduction follows the decrease in the electron susceptibility. However, there is no theoretical interpretation of this effect. Nevertheless, a phenomenon of the Knight shift bifurcation in a particular porous matrix loaded with gallium was understood and treated as a result of a phase transitions in confined melted gallium.

References

[1] B.F. Borisov, E.V. Charnaya, Yu.A. Kumzerov, A.K. Radzhabov, and A.V. Shelyapin, *Solid. State Commun.*, 92, 531 (1994).

[2] B.F. Borisov, E.V. Charnaya, T. Loeser, D. Michel, C. Tien, C.S. Wur, and Yu.A. Kumzerov, *J. Phys.*, Condens. Matter 11, 10 259 (1999).

[3] E.V. Charnaya, C. Tien, K.J. Lin, and Yu.A. Kumzerov, *J. Phys., Condens. Matter*, 10, 7273 (1998).

[4] E.V. Charnaya, C. Tien, K.J. Lin, and Yu.A. Kumzerov, *Phys. Rev.*, B 58, 11 089 (1998).

[5] L. Bosio, H. Curien, M. Dupont, and A. Rimsky, *Acta Crystallogr.,* B 29, 367 (1973).
[6] L. Bosio, A. Rimsky, M. Dupont, and H. Curien, *Acta Crystallogr.,* B 28, 1974 (1972).
[7] L. Bosio, A. Defrain, H. Curien, and A. Rimsky, *Acta Crystallogr.,* B 25, 995 (1969).
[8] H. Curien, A. Rimsky, and A. Defrain, Bull. *Soc. Fr. Mineral. Cristallogr.,* 84, 260 (1961).
[9] J. Donohue, The Structure of the Elements (Wiley, New York, 1974).
[10] R.D. Heyding, W. Keeney, and S.L. Segel, *J. Phys. Chem.,* Solids 34, 133 (1973).
[11] L. Bosio, J. Chem. Phys. 68, 1221 (1978).
[12] O. Degtyareva, M.I. McMahon, D.R. Allan, and R.J. Nelmes, *Phys. Rev. Lett.,* 93, 205502 (2004).
[13] E.L. Gromnitskaya, O.F. Yagafarov, O.V. Stalgorova, V.V. Brazhkin, and A.G. Lyapin, *Phys. Rev. Lett.,* 98, 165503 (2007).
[14] O. Schulte and W.B. Holzapfel, *Phys. Rev.,* B 55, 8122 (1997).
[15] M.K. Lee, C. Tien, E.V. Charnaya, H.-S. Sheu, and Yu.A. Kumzerov, *Phys. Lett.,* A 374, 1570 (2010).
[16] E.V. Charnaya, C. Tien, M.K. Lee, and Yu.A. Kumzerov, *J. Phys., Conden. Matter,* 21, 455304 (2009).
[17] Powder Diffraction Files (PDF-4+, 2008), International Center for Diffraction Data, 12 Campus Boulevard, Newtown Square, PA, U.S.A. 19073-3273.
[18] P. Huber, D. Wallacher, and K. Knorr, *Phys. Rev.,* B 60, 12666 (1999).
[19] H. K. Christenson, *J. Phys., Condens. Matter,* 13, R95 (2001).
[20] P. Huber and K. Knorr, Phys. Rev. B 60, 12657 (1999).
[21] D. Teske and J.E. Drumheller, *J. Phys., Condens. Matter,* 11, 4935 (1999).
[22] I.G. Sorina, E.V. Charnaya, L.A. Smirnov, Yu.A. Kumzerov, and C. Tien, *Phys. Solid. State,* 40, 1407 (1998).
[23] J.H. Strange, M. Rahman, and E.G. Smith, *Phys. Rev. Lett.,* 71, 3589 (1993).
[24] O.V. Petrov and I. Furó, NMR cryoporometry: Principles, applications and potential. Progr. in Nuclear Magnetic Resonance Spectroscopy, 54, 97–122 (2009).
[25] E.V. Charnaya, C. Tien, W. Wang, M.K. Lee, D. Michel, D. Yaskov, S.Y. Sun, and Yu.A. Kumzerov, *Phys. Rev.,* B 72, 035406 (2005).
[26] B.F. Borisov, E.V. Charnaya, A.V. Gartvik, C. Tien, Yu.A. Kumzerov, and V.K. Lavrentev, *Phys. Solid State,* 46, 2286 (2004).
[27] J.M. Dereppe, B.F. Borisov, E.V. Charnaya, A.V. Shelyapin, M.M. Nassar, and Yu.A. Kumzerov, *Phys. Solid State,* 42, 193 (2000).
[28] B.F. Borisov, E.V. Charnaya, P.G. Plotnikov, W.-D. Hoffmann, D. Michel, Yu.A. Kumzerov, C. Tien, and C.S. Wur, *Phys. Rev.,* B 58, 5329 (1998).
[29] G. Krausch, T. Detzel, H. Bielefeldt, et al., *Appl. Phys.,* A 53, 324 (1991).
[30] M. Zhang, M.Yu. Efremov, F. Schiettekatte, et al., *Phys. Rev.,* B 62, 10548 (2000).
[31] D. Xie, M.P. Wang, W.H. Qi, and L.F. Cao, Mater. *Chem. Phys.,* 96, 418 (2006).
[32] M.A. Tagliente, G. Mattei, L. Tapfer, M. Vittori Antisari, and P. Mazzoldi, *Phys. Rev.,* B 70, 075418 (2004).
[33] K.M. Unruh, T.E. Huber, and C.A. Huber, *Phys. Rev.,* B 48, 9021 (1993).
[34] M.K. Zayed, M.S. Hegazy, and H.E. Elsayed-Ali, *Thin Solid Films,* 449, 254 (2004).
[35] M. Dippel, A. Maier, V. Gimple, et al., *Phys. Rev. Lett.,* 87, 095505 (2001).
[36] B.F. Borisov, A.V. Gartvik, E.V. Charnaya, and Yu.A. Kumzerov, *Acoustical Physics,* 55, 816 (2009).

[37] E.V. Charnaya, C. Tien, M.K. Lee, and Yu.A. Kumzerov, *Phys. Rev.,* B 75, 212202 (2007).

[38] A.L. Pirozerskii, E.V. Charnaya, E.N. Latysheva, A.I. Nedbai, Yu.A. Kumzerov, and A.S. Bugaev, *Acoustical Physics,* 57, 637 (2011).

[39] T.J. Anderson and I. Ansara, *J. Phase Equilib.,* 12, 64 (1991).

[40] P. Pawlow, Z. *Phys. Chem.,* 65, 545 (1909).

[41] E. Molz, A.P.Y. Wong, M.H.W. Chan, and J.R. Beamish, *Phys. Rev.,* B 48, 5741 (1993).

[42] C.R.M. Wronski, *Br. J. Appl. Phys.,* 18, 1731 (1967).

[43] C.J. Coombes, *J. Phys.,* 2, 441 (1972).

[44] R. Kofman, P. Cheyssac, A. Aouaj, Y. Lereah, G. Deutscher, T. Ben-David, J. M. Penisson, and A. Bourret, *Surf. Sci.,* 303, 231 (1994).

[45] E.V. Charnaya, C. Tien, M.K. Lee, Yu.A. Kumzerov, *Phys. Rev.,* B 75, 144101 (2007).

[46] Dynamics in Small Confining Systems V, edited by J.M. Drake, J. Klafter, P.E. Levitz, R.M. Overney, and M. Urbakh (MRS, Warrendale, PA, 2001).

[47] P. Levitz, J.P. Korb, and D. Petit, *Europ. Phys. J.,* E 12, 29 (2003).

[48] M. Weber, A. Klemm, and R. Kimmich, *Phys. Rev. Lett.,* 86, 4302 (2001).

[49] S. Granick, *Science,* 253, 1374 (1991).

[50] S. Stapf, R. Kimmich, and R.-O. Seitter, *Phys. Rev. Lett.,* 75, 2855 (1995).

[51] S.H. Anastasiadis, K. Karatasos, G. Vlachos, E. Manias, and E.P. Giannelis, *Phys. Rev. Lett.,* 84, 915 (2000).

[52] J.-P. Korb, M. Whaley-Hodges, and R.G. Bryant, *Phys. Rev.,* E 56, 1934 (1997).

[53] E.V. Charnaya, T. Loeser, D. Michel, C. Tien, D. Yaskov, and Yu.A. Kumzerov, *Phys. Rev. Lett.,* 88, 097602 (2002).

[54] J.M. Titman, *Physics Reports,* 33, 1 (1977).

[55] F.A. Rossini and W.D. Knight, *Phys. Rev.,* 178, 641 (1969).

[56] A.L. Kerlin and W.G. Clark, *Phys. Rev.,* B 12, 3533 (1975).

[57] J. Bosse, D. Quitmann, and C. Wetzel, Phys. Rev. A 28, 2459 (1983).

[58] P.S. Hubbard, *J. Chem. Phys.,* 53, 985 (1970).

[59] C. Tien, E.V. Charnaya, P. Sedykh, and Yu.A. Kumzerov, *Phys. Solid State*, 45, 2352 (2003).

[60] A.Abragam, Principles of nuclear magnetism (Clarendon Press, Oxford, 1989).

[61] H. Konrad, C. Karmonik, J. Weissmüller, H. Gleiter, R. Birringer, and R. Hempelmann, *Physica,* B 234-236, 173 (1997).

[62] E.V. Charnaya, C. Tien, Yu.A. Kumzerov, and A.V. Fokin, *Phys. Rev*., B 70, 052201 (2004).

[63] F.A. Rossini, E. Geissler, E.M. Dickson, and W.D. Knight, *Advan. Phys.,* 16, 287 (1967).

[64] W.W. Warren and W.G. Clark, Bull. Am. Phys. Soc. 11, 916 (1966).

[65] C.A. Sholl, *Proc. Phys. Soc.,* (London) 91, 130 (1967).

[66] E.V. Charnaya, C. Tien, M.K. Lee, and Yu.A. Kumzerov, *J. Phys., Condens. Matter,* 22, 195108 (2010).

[67] D.Yu. Podorozhkin, C. Tien, E.V. Charnaya, M.K. Lee, L.J. Chang, D. Michel, J. Haase, Yu.A. Kumzerov, *Physica,* B 407, 2063 (2012).

[68] C.E.T. White and H. Okamoto, Phase Diagrams of Indium Alloys and their Engineering Applications, Monograph Series on Alloy Phase Diagrams, ASM International, Materials Park, OH, 1992.

[69] W.D. Knight and A.G Berger, Ann. Phys., 8, 173 (1959).

[70] G.C. Carter, L.H. Bennett, and D.J. Kahan, Metallic shifts in NMR (Pergamon Press, Oxford, 1977).

[71] C.P. Slichter, *Phil. Mag.,* B 79, 1253 (1999).

[72] E.L. Nagaev, *Phys. Rep.,* 222, 199 (1992).

[73] J.J. van der Klink and H.B. Brom, *Prog. Nucl. Magn. Res. Spectr.*, 36, 89 (2000).

[74] P.-A. Vuissoz, J.P. Ansermet, and A. Wieckowski, *Phys. Rev. Lett.,* 83, 2457 (1999).

[75] K. Kimura, Phys. Rev., B 42, 6939 (1990).

[76] M.C. Desjonqueres and D. Spanjaard, Concepts in Surface Physics (Springer, Berlin, Heidelberg, New York, 1998).

[77] M.J. Williams, P.P. Edwards, and D.P. Tunstall, Discuss. *Faraday Soc.,* 92, 199 (1991).

[78] E.V. Charnaya, D. Michel, C. Tien, Yu.A. Kumzerov, and D. Yaskov, *J. Phys., Condens. Matter, 15*, 5469 (2003).

[79] C. Tien, E.V. Charnaya, M.K. Lee, and Yu.A. Kumzerov, *J. Phys., Condens. Matter,* 19, 106217 (2007).

[80] C. Tien, E.V. Charnaya, W. Wang, Yu.A. Kumzerov, and D. Michel, *Phys. Rev.,* B 74, 024116 (2006).

[81] D.A.C. Jara, M.F. Michelon, A. Antonelli, and M. de Koning. *J. Chem. Phys.,* 130, 221101 (2009).

In: Indium
Editors: Hsaio G. Woo and Huang Tsai Choi
ISBN: 978-1-62257-696-8

Chapter 2

Indium: An Efficient Co-catalyst in Novel Cu or Ni Catalysts for Selective Reduction of Biomass Derived Fatty Acids to Alcohols

György Onyestyák, Szabolcs Harnos* and Dénes Kalló

Institute of Materials and Environmental Chemistry, Research Centre for Natural Sciences, Hungarian Academy of Sciences, Budapest, Hungary

Abstract

Supported copper or especially nickel catalysts are very important for hydrogenation of various organic compounds in the practice. Activity, selectivity and stability can be greatly improved using co-catalysts: e.g. chromium compounds with Cu (Adkins catalyst for fatty alcohol production) or molybdenum compounds with Ni (for hydrodesulfurization [HDS], hydrodeoxygenation [HDO], hydrodenitrogenation [HDN], etc. catalysts in petroleum refining.

Recently, we discovered novel highly efficient bimetallic supported catalysts which were exceedingly active and selective in the hydrodeoxygenation (HDO) of octanoic acid to octanol at moderate pressure and temperature in addition very effective in reduction of acetic acid to ethanol. The catalysts contained indium as co-catalyst and a metal of high hydrogenation activity, such as, copper or nickel, on various support. A significant synergism can be observed when indium with another metal of hydrogenation activity is applied. Appearance of Cu_2In or Ni_2In phases, intermetallic compounds on the surface of metal particles results in significant increase of desired hydrogenation activity, i.e., in stepwise hydrogenation of carboxylic acids stopped at alcohol formation and in inhibiting hydrogenations when hydrodecarbonylation of carboxylic acids (loss of one carbon atom) and hydrogenolysis of hydrocarbon products (loss of further carbon atoms) proceed. Moreover, in presence of mobile indium atoms or of indium containing nano clusters mono- or bimolecular alcohol dehydration capability of the support is selectively poisoned.

In order to have more information about the role of indium, experiments are carried out using quite different hydrogenation metals, Cu and Ni. Hydrogenation experiments are carried out with long chain fatty acid like octanoic acid for production of octyl alcohol and short chain acetic acid (produced in large amounts from biomass) for production of ethyl alcohol, ethyl acetate or acetaldehyde. The main question is how the co-catalyst second metal of moderate hydrogenation activity as gallium, cadmium, tin, but first of all indium, modify the activity of the supported hydrogenation metal and the surface properties of the support.

1. Introduction

Addition of a second component to metal catalysts is widely used in order to enhance activity and/or selectivity. The removal of heteroatoms (S, N, O) from petroleum fractions is one basic step in refining to produce transportation fuels and chemicals of desired quality. Using petroleum fractions the heteroatoms should be removed at a low concentration level of contaminating compounds. Nowadays, biomass materials coming to the front as renewable resources contain oxygen in very high concentration. So, novel, more efficient catalysts are needed.

Fatty alcohol is one of the first chemicals of natural origin (produced from fatty acid content of fats or vegetable oils) and used in high volume. In 1931, Adkins suggested to use copper chromite catalysts for selective hydrogenation of fatty acid methyl esters to fatty alcohols [1]. The use of chromium is limited by its toxicity. There are worldwide efforts to find alternative catalyst systems.

The major industrial use of supported metal sulfides is the removal of sulfur, nitrogen, oxygen and metals from the most hydrocarbon streams in refineries by reduction in hydrotreating processes. Hence this is the catalytic technology comprising the largest material processing per year. Hydrotreating catalysts contain molybdenum and cobalt or nickel, supported on γ-Al_2O_3. Molybdenum sulfide is traditionally considered as the catalyst and Co or Ni is considered to promote the Mo activity [2]. Removal of oxygen from oxygen-containing molecules was especially important for the treatment of coal-derived liquids. Biomass-derived feedstocks have been recently introduced into refineries posing new challenges for refining processes and catalysis. Due to high oxygen content of biomass feedstocks, deoxygenation processes play an important role in upgrading of these raw materials. The economics seems to be favorable for conversion of edible biomass (vegetable oils) to fuel containing carboxyl groups only which can be efficiently converted. Investigations of conventional sulfided HDS catalysts, such as CoMo, NiMo, NiW supported on alumina, have revealed that complete conversion of triglycerides to hydrocarbons can be attained over these contacts [3]. The studies have shown that two reaction pathways, i.e. hydrodeoxygenation and hydrodecarboxylation, are responsible for hydrocarbon formation. The extent of these different reaction routes depends on the catalyst choice and reaction conditions [4, 5].

When HDS catalysts are applied addition of sulfur containing compounds into the reaction zone is required to maintain the catalyst activity. If non-sulfided catalysts are used the inconvenient sulfurization step can be omitted. Thus, the goal of this work is to develop non-sulfided catalysts for upgraded bio-fuel production.

Recently, we discovered novel highly efficient bimetallic alumina supported catalysts which are highly active in the selective reduction of octanoic acid to octanol [6]. The catalysts contained zerovalent indium as co-catalyst and a metal of hydrogenation activity, such as copper or nickel.

Indium is a rare post transition metal, the 61st most abundant element of the Earth core, however, plays important role in various applications. Consequently, its price is high and increases with the increasing demands. In the middle of 1980s, the development of indium phosphide semiconductors and indium tin oxide films for liquid crystal displays (LCD) aroused much interest. By 1992, the thin film application had become the largest end use. Catalytic applications appeared in various forms nowadays, but this role seems to be rather promising.

A number of studies showed that In-zeolites have significant activity in the selective catalytic reduction of NO to N_2 with light hydrocarbons in the presence of excess oxygen (NO-SCR) [7], which reaction has great practical importance especially in the abatement of NO_x emission of boilers and engines fuelled by natural gas. Alumina supported In_2O_3 catalyst demonstrates also good NO_x reduction performance [8]. Supported In_2O_3 catalyst was proven to be a novel efficient catalyst for dehydrogenation of propane with CO_2 [9] or N_2O [10]. Pure In_2O_3 was revealed to act as a surprisingly efficient and promising, highly active and selective catalyst for steam reforming of methanol to $CO_2 + H_2$ [11]. Although increasing attention has lately been paid to indium containing catalysts for methanol steam reforming, first Y Men et. al. assumed that using alumina supported Pd in intimate contact with In_2O_3, the active phase presumably is in form of a PdIn alloy [12].

In the trimetallic SnInPt/alumina gasoline-reforming catalyst In is present in zerovalent unalloyed form or in Pt_xIn alloys not changing the electronic properties of Pt [13]. Indium was not found to be a good promoter instead of copper for the Fe-based Fischer-Tropsch synthesis (FTS) catalyst because of the melting of indium during the FTS reaction [14]. Platinum catalysts supported on corundum promoted by indium and tin were successfully applied for the selective dehydrogenation to propylene [15]. Although the catalysts were reduced in hydrogen flow at 500 °C in situ the reactor, surprisingly the structure of the bimetallic active components on corundum could be described as relatively highly dispersed metal-ionic system. Tin and indium addition to $MgAl_2O_4$ supported Pt catalyst lead to an excellent performance in the butane dehydrogenation reaction [16].

Nowadays, bimetallic In containing catalysts are coming to the front in reductions, too. Technologies for removal of nitrate from drinking water and waste water will be required in the near future, and catalytic nitrate reduction is one of the most promising methods. So far, the understanding of nitrate reducing catalysts has been very poor. InPt/alumina catalysts were characterized by EXAFS and XANES [17]. The results confirm that for the nitrate hydrogenation, an intimate contact between the two metals is necessary in order to carry out the reaction efficiently. During the reaction, the amount of bimetallic particles decreased and unalloyed Pt considerably increased, which led to a partial deactivation of the catalyst. Selective hydrogenation of acrolein to allyl alcohol with molecular hydrogen is extremely difficult. Silver catalyst is used typically for oxidation reactions rather than for hydrogenation, however an important improvement of selectivity to allyl alcohol is possible, when bimetallic $AgIn/SiO_2$ catalysts are used [18]. Even in a strongly reducing atmosphere In remains to some extent positively charged. Concerning the silver centre, comparable redox behavior could not be observed.

Biomass, not appropriate for human and animal feeding, became a desired renewable raw material and energy source. The processing is focused to economical production of more valuable substances, chemicals, energy carriers such as hydrogen, methane or liquid fuel or to produce easily transportable electric energy. The alcohols are important chemicals and lower alcohols, first of all ethanol, can be directly used as a fuel. The anaerobic fermentation of sugar is a well known process for ethanol production. However, the use of sugar or starch for fuel production decreases the food output of the agriculture, raises food prices and is economically, environmentally and socially disadvantageous. Present research aims to establish scientific bases of a new technological and material platform for alcohol production.

Chemicals and fuels can be produced from biomass by thermochemical, biochemical and catalytic technologies, and combination thereof. With exception of the sugar route most of the important technological platforms provide fatty acids or volatile fatty acids (VFAs) [19]. The increasing demand for biofuel initiates research to efficiently convert VFAs (first of all acetic acid) to alcohols instead of deeply reduced, gaseous alkanes or biogas. The traditional technology for obtaining fatty alcohols involves esterification of the acid and catalytic hydrogenolysis of the ester at elevated temperature and high pressure to two alcohols over old fashioned Adkins catalyst. The technology of fatty alcohol production cannot be directly adopted for the partial hydrodeoxygenation (HDO) of VFAs to get lower alcohols. Recently we discovered novel highly efficient bimetallic supported catalysts, which were active and selective in the HDO of octanoic acid to octanol at moderate pressure and temperature [6]. The catalysts contained indium and another metal of hydrogenation activity, such as, copper or nickel, on oxide support [20, 6]. Research is needed to study in details these novel catalyst compositions and catalytic process.

2. Preparation and Testing of Catalysts

Copper (CuZ) and nickel (NiZ) zeolites A, P, X and Y were prepared from NaA, NaP, NaX and NaY commercial products by conventional aqueous ion-exchange using Cu $(acetate)_2$ and Ni $(acetate)_2$. The obtained samples had a metal content of about 7–16 wt. %. Copper or nickel zeolite preparations e.g. with 10 wt. % metal content are marked as 10CuZ or 10NiZ where Z denotes A, P, X, Y zeolite.

Alumina-supported Cu and Ni catalysts were prepared on γ-Al_2O_3 using impregnation with NH_4OH solution of Cu $(acetate)_2$ and Ni $(acetate)_2$, then dried and calcined at 550 °C for 3 hours to decompose the acetate to oxide and $9Cu/Al_2O_3$ and $9Ni/Al_2O_3$ were obtained, i.e. preparations containing 9 wt. % copper and 9 wt. % nickel.

Composite catalysts were prepared by adding indium (III) oxide powder to the CuZ, NiZ, Cu/Al_2O_3 and Ni/Al_2O_3 samples and grinding the mixture in agate mortar.

A commercial, conventionally used Adkins catalyst (consisting of 72 wt. % $CuCr_2O_4$ and 28 wt. % CuO) was tested for comparison with the novel catalyst preparations.

X-ray powder diffractograms (XRD) were recorded using XRD cell for measurements at room temperature and for *in situ* measurements in gas flow (hydrogen) at elevated temperatures. The mean crystallite size of the copper particles was estimated by the Scherrer equation.

Catalytic hydroconversion of acetic acid (CH_3COOH), octanoic acid ($C_7H_{13}COOH$), octanal ($C_7H_{13}CHO$) /as intermediate compound/, and of octanol ($C_8H_{13}OH$) as product were investigated. Catalyst was generally pretreated in hydrogen flow in the high-pressure fixed bed flow-through catalytic reactor at 450 °C and 21 bar. The reactions were carried out at 21 bar total pressure in a flow of acetic acid or octanoic acid/H_2/He mixtures. The reaction temperature and the H_2 partial pressure were varied in the ranges of 200-380 °C and 3.6-21.0 bar, respectively.

The reaction was allowed to run for one hour at each temperature to attain steady state and for a second hour to collect reactor effluent for analysis. The reactor effluent collected in the second hour of catalytic run was depressurized, and cooled to room temperature. The product mixtures were analyzed by gas chromatograph (GC) equipped with a capillary column and a flame ionization detector. The non-condensing products (CO, CO_2, CH_4 and light hydrocarbons) were analyzed using an on-line GC with capillary column and thermal conductivity detector.

Product compositions are expressed in weight percents and plotted in stacked area graphs where the distance between two neighboring curves represents the concentration of a given component at a given temperature.

3. Results and Discussion

A major 21st century goal for academia, government and industry should be the efficient and economical utilization of biomass resources principally for production of transportation fuels and chemicals. Some of our present works focused on the hydroconversion of sunflower oil [21, 22] or octanoic acid [5] to aliphatic paraffins (C_{17} or C_{18} and C_7 or C_8). The reaction proceeded in consecutive steps of fast hydrogenolysis to carboxylic acids and propane, and hydrodeoxygenation (HDO) of the carboxylic acid intermediates. Furthermore, the increasing demand for biofuel initiates research to efficiently convert short chain carboxylic acid produced from biomass by biological destruction (first of all acetic acid) to alcohol preserving the original chain length. Such a process however needs novel, well engineered, tailor-made, bi- or multi-component catalysts similarly to the conventional, old-fashioned Adkins or HDS catalysts.

The copper catalyses selective addition of hydrogen to carbon-oxygen bonds but scarcely the hydrogenolysis of the carbon-carbon bonds in alcohols produced by hydrogenation of triglycerides. From the early 30s up to now the chemical industry transforms triglycerides, primarily vegetable oil, by catalytic hydrogenation to fatty alcohols [1]. For fatty alcohol production chromium-containing copper (Adkins) catalysts are used. The use of toxic chromium involves environmental and health hazards, therefore, research was continued to develop new, environmental benign hydrogenation catalysts for the process. An obvious approach of the alcohol selectivity problem is to use low-activity supported copper catalyst, containing copper in high dispersion as possible. Furthermore, in order to increase the conversion and improve selectivity at moderate pressure the reaction temperature must be increased and/or a catalyst modifier must be used that is less hazardous than chromium.

The regularly structured crystalline microporous aluminosilicates like the zeolites were supposed to be ideal supports for the stabilization of highly dispersed copper clusters or a perfect dispersion of Cu-atoms [23, 24]. However, the conversion of octanoic acid (OA) over more silicious and consequently hydrothermally stable, reduced Cu-zeolites (e.g. mordenite and ZSM-5 having low copper contents because of their low aluminum content) was low, i.e.

only 6-7 wt. % at 380 °C and 21 bar. Zeolites A, X and P having the possible lowest Si/Al ratio (around 1), consequently containing the highest number of charge compensating cations as possible, fully ion-exchanged with copper cations already show much higher activities in the reduction of octanoic acid (see Table 1.). However, it is well-known that Al-rich zeolites, shown above, do not preserve their structure in the H-form inevitably generated when the Cu^{2+} cations are reduced with H_2. The 14CuP, 15CuX and 16CuA preparations have similar chemical composition and show similar structural instability under the conditions of activation and of OA hydroconversion [20, 25]. In addition, H-zeolites formed during the reduction of Cu^{2+} cations having high Brönsted acidity which is very efficient in mono- or bimolecular alcohol dehydration producing alkenes or ethers as undesirable by-products.

Table 1. The yields of main products of octanoic acid hydroconversion (weight %) at 380 °C, WHSV=2 h^{-1} and 21 bar over various zeolite base catalysts

Catalyst	Octane	Octenes	Octanal	Octanol	Conversion
7CuY	1.3	20.8	0.9	2.2	41.7
7CuY + 10 % In_2O_3	**3.3**	**49.6**	**1.3**	**2.1**	**84.4**
15CuX	3.9	35.3	1.5	7.3	62.0
15CuX + 10 % In_2O_3	**–**	**32.1**	**4.5**	**31.1**	**87.8**
16CuA	–	4.6	5.3	42.0	62.7
16CuA + 10 % In_2O_3	**–**	**2.7**	**6.6**	**63.0**	**85.4**
14CuP	–	16.4	3.7	25.1	58.7
14CuP + 10 % In_2O_3	**–**	**8.4**	**5.3**	**61.1**	**88.67**

The H-zeolite is a solid acid. If the zeolite structure contains protons as charge compensating cations in high concentration the solid acid hydrolyses the zeolite structure. Attempts to substitute the mobile (active) protons with mono- or bivalent metals added in form of salt or vapor were not successful.

In presence of In_2O_3 copper zeolite could be reduced with H_2 without formation of H-zeolite but formation of In^+Z^- [26, 27]:

$$In_2O_3 + 3H_2 + Cu^{2+}Z^- \leftrightarrow 2In^+Z^- + Cu^0 + 3H_2O \qquad (1)$$

where Z^- represents a negatively charged lattice site.

Nevertheless the integrity of the structure of zeolites A, X or P could not be maintained, only small domains of microporous zeolite structures could be preserved upon dehydration/reduction. Nevertheless, the obtained In,Cu composite catalysts embedded in destructed zeolite structure showed much higher activity and octanol selectivity than the simple supported Cu catalysts (see Table 1 and Figure 1) [20, 25]. This observation gave the base of this research.

Cu-P, -A, -X and -Y-zeolites were pretreated in hydrogen flow at 450 °C before octanoic acid hydroconversion. During pretreatment reduction/dehydration proceed and partly amorphous aluminosilicate supported copper catalysts were formed, which contain copper nanoparticles in sufficiently high dispersion. The diffraction lines, assigned to metallic copper phase were evaluated between 200 and 350 °C (e.g. see diffractograms of 14CuP in Figure 2a, b). Thus, metal clusters formed when the zeolite structure has been deeply damaged. The size of copper clusters has important relevance. The widths of the diffraction lines assigned to Cu^0 particles are strikingly different for the hollow CuX and the more compact CuA or CuP. The average copper particle sizes in CuX, CuA and in CuP are 26, 7 and 17 nm, respectively. In CuY, of higher Si/Al ratio than CuX (both of the same faujasite structure), in the stable, faultless microporous faujasite framework with much less lattice aluminum atoms copper cations have free access for the reducing agent hydrogen and the formed copper atoms can migrate to the surface of the zeolite crystals without barrier, so the average copper particle size is much larger, 55 nm. P-, A- and X-zeolite structures collapsed loosing most of the Brönsted acidic sites, whereas Y can preserve its microporous framework containing the charge compensating protons in high concentration.

The activity and selectivity of the catalysts studied can be characterized by the product distributions as function of reaction temperature or time on stream represented in stacked area graphs (e.g. in Figure 1 for CuP). For a simpler comparison of the catalytic behaviors on the various zeolite based copper catalysts are summarized in Table 1 involving only the main products. The conversion on CuY of the lowest copper content and of the lowest metal surface are a bit lower than on other Cu-zeolites but the selectivities are highly different. CuY catalyst remained a real microporous H-zeolite without any structural damage loaded copper particles on the outer surface of zeolite crystallites. So, having high Brönsted acid site concentration being equivalent with the formed copper atoms, it is very active in dehydration of the alcohol obtained through the step-by-step reduction of the fatty acid (see Table 1). Contrary, the highly damaged, destructed zeolites of low Si/Al (X, A, P) lost their microporous structure together with the charge compensating Brönsted acidic protons. Consequently, the alcohol intermedier molecules are preserved in much higher concentration depending on the remaining Brönsted acidity characteristic of the parent zeolite structure (zeolite X lattice was the less damaged resulting in high octene yield) (see Table 1).

Adding 10 wt. % In_2O_3 to all four zeolite Cu-catalysts the main effect is that all activities increase (Table 1). Over destructed Cu-zeolites (X, A, P) indium doping is still more efficient, octanol yields increase strikingly while octene yields significantly decrease. For CuY catalyst having high faultless Brönsted acidity the influence is reflected only in the higher activity, but it remained almost completely selective for octene yield. The great affect of indium admission as co-catalyst seems to be complex. Under pretreatment and reaction conditions active components, copper and indium exist in zerovalent form. Presumably copper-indium intermetallic compounds (most probably Cu_2In) (Figure 2) are responsible for the activity increase since it has been found that the activity of reduced In_2O_3 (In^0) is negligible under these conditions and the activity of copper is essentially lower. Further question is how and why is suppressed the remaining low dehydration activity as a consequence of the appearance of metallic indium atoms on the surface of support and/or metal particles. Activity of destructed zeolites in the octanol dehydration reflected by octene yields is different (Table 1) depending presumably on remaining Brönsted acidity. Modification with a large amount of In_2O_3 (added plus 10 wt. % In_2O_3 of the mass of Cu-zeolite catalyst) cocatalyst strikingly increases the activity (cf. e.g. Figures 1a and A, b and B) preserving it at a permanent level and the high selectivity by eliminating dehydration as a consecutive reaction which is followed by double bond hydrogenation:

$$\text{octanoic acid} \xrightarrow[-H_2O]{+H_2} \text{octanal} \xrightarrow{+H_2} \text{octanol} \xrightarrow{-H_2O} \text{octenes} \xrightarrow{+H_2} \text{octane} \quad (2)$$

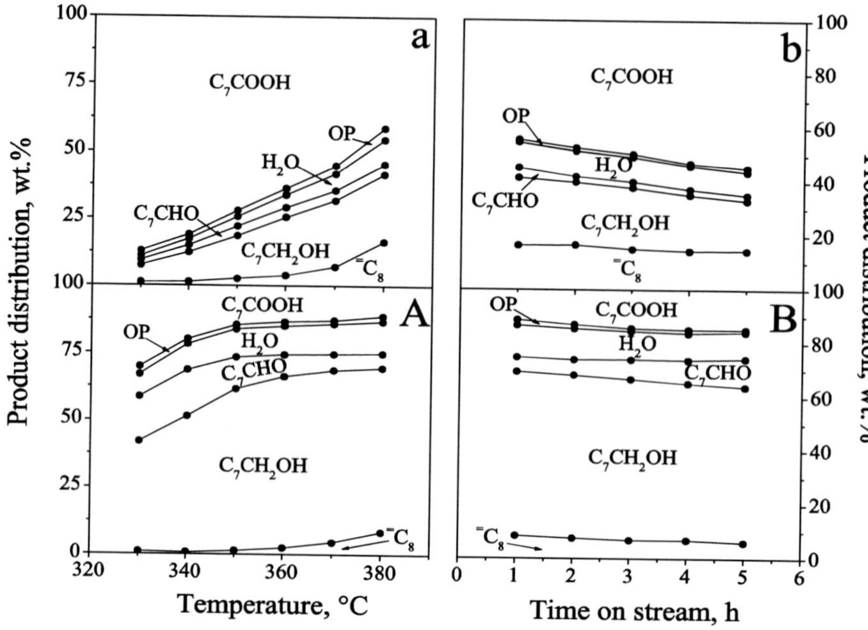

Figure 1. Octanoic acid hydroconversion over 14CuP (a; b) and doped with + 10 wt. % In_2O_3 (A; B) catalysts are characterized by distributions of main products as a function of reaction temperature (a; b) and the time-on-stream (A; B) at 21 bar total pressure. The temperature was kept at 380 °C in the case of time-on-stream measurement. The WHSV of OA was 2.0 h^{-1}. Legends: $^{=}C_8$ octenes; C_7CH_2OH octanol; C_7CHO octanal; H_2O water; C_7COOH octanoic adid; OP other products.

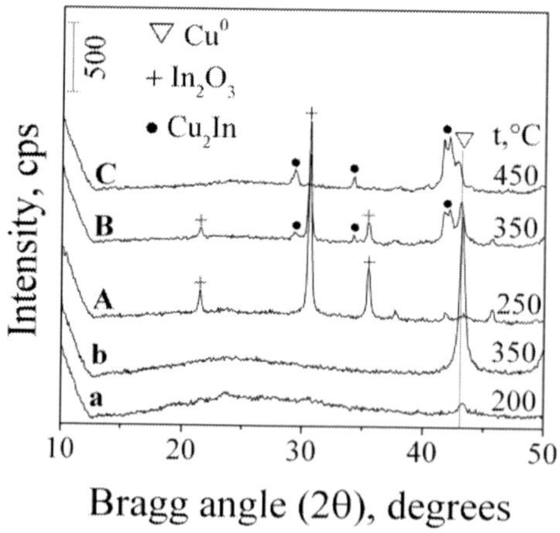

Figure 2. XRD patterns of 15CuX (a, b) and doped with 36 wt. % In_2O_3 (A, B, C) samples recorded at the indicated temperatures after 30 min treatment at each temperature in H_2 flow. The temperature was increased in 100 or 50 °C segments.

When product distribution is plotted against the reaction temperature the efficiency of indium-doping seems to be more significant at lower temperature (Figure 1a, A) and the consecutive nature of this reaction is reflected in Figure 1A showing the concentration change of the octanal intermedier compound.

The moderate decay of activity can be also eliminated by In admission (compare Figure 1b and 1B).

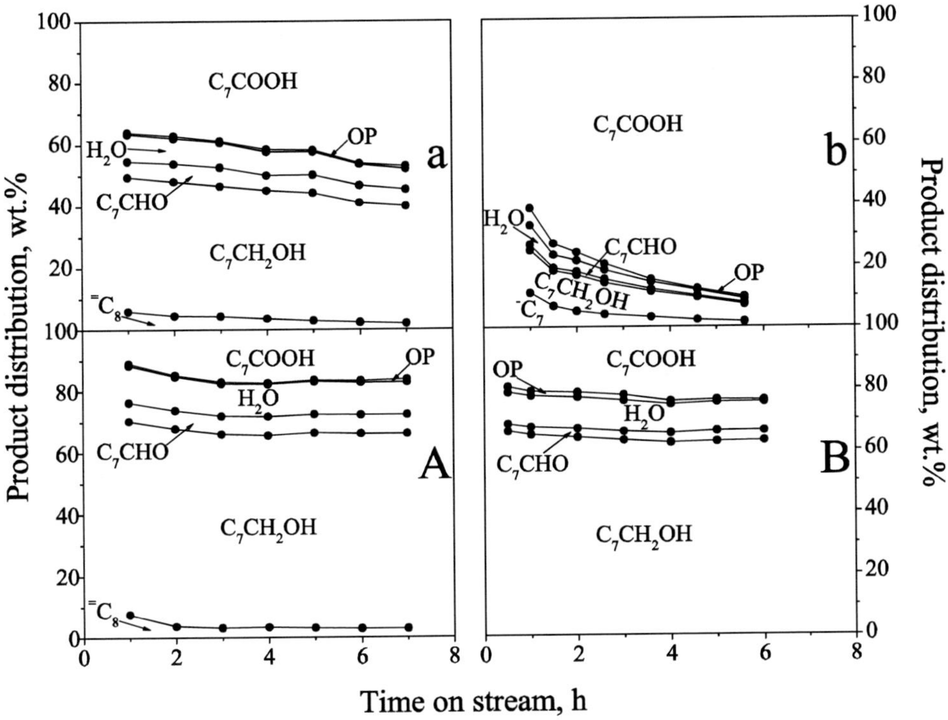

Figure 3. Octanoic acid hydroconversion over 16CuA (a) and doped with + 10 wt. % In_2O_3 (A) catalysts at 380 °C and over NiA (b) and doped with +10 wt. % In_2O_3 (B) catalysts at 300 °C are charactrized by distributions of main products as a function of time–on–stream at 21 bar total pressure. The WHSV of OA was 2.0 h^{-1}. Legends as in Figure 1. In addition: $^{-}C_7$ heptane.

Summing up, well dispersed copper particles can be generated on aluminosilicate supports formed by destructive H_2 reduction/dehydration of zeolite CuP, CuA and CuX. Such amorphised type catalysts show high activity and selectivity in the catalytic hydroconversion of carboxylic acids to alcohols. The selectivity for alcohol depends on the original zeolite structure. Zeolite CuA (small pore) seems to be a better catalyst precursor than CuX (large pore), having the same composition but different frameworks consisting of the same building units (linked truncated octahedras). The activity and the alcohol yield of the supported Cu catalysts can effectively be increased by In_2O_3 doping. Appearance of indium metal in the active metallic copper containing phase highly increases the activity and selectivity. It seems likely that indium effectively eliminates the dehydration of octanol to octenes for Cu-zeolite catalysts of low Brönsted acid site concentration.

The catalytic behaviour of zeolite based amorphised catalyst samples significantly depends on the original zeolite structure. Although the destructed zeolite CuA sample has the smallest specific surface area (only 8 m^2/g), its activity is similar to that of CuX, CuP but the octanol selectivity is the highest without indium doping (Table 1) and In admission enhances only the conversion (cf. Figures 3a and A).

Non-sulfided Ni-catalysts for sunflower oil hydroconversion to green Diesel [21, 22] were proven to be too active and catalysed effectively the side reactions besides desired alkane formation. Indium doping offers itself to attain better selectivity and more moderate reaction conditions over the too active nickel catalysts. In Figure 3 the octanoic acid hydroconversion over CuA (at 380 °C) and NiA (at 300 °C) catalysts with and without indium doping are compared. The activity differences are reflected by the high reaction temperature difference for attaining comparable activities. On the NiA catalyst fast activity decay can be observed (Figure 3b) contrary to the other three catalysts (in Figs 3a, A, B). This sample can split the C-C bonds resulting in the hydrodecarbonylation reaction producing chain shortened alkane (heptane) and CO which is quickly hydrogenated to methane (CO and CH_4 are plotted together with the other products (OP)). Admission of 10 wt. % In_2O_3 results a drastic or rather breath-taking improvement of catalytic behaviour reflected by permanent conversion at high level together with high octanol selectivity. Compared with the Cu-zeolites it is important to mention, that Ni^{2+} cations in the negatively charged zeolite lattice can be reduced only under drastic conditions, at much higher temperature than Cu^{2+}, i.e. above 350 °C. Consequently at low level of nickel cluster formation after partial reduction the crystallinity of the zeolite can be preserved in higher extent. HT-XRD diffractograms confirm this assumption. Such effect can be enhanced by using partially exchanged Ni,Na-A zeolite which contains nickel cations exchanged only to 20 % of the originally present sodium cations. However the catalytic behaviors were very similar for such highly different samples. Consequently, although observing nice effects using the indium co-catalyst loaded Cu- or Ni-zeolites, it seems to be practical to apply more simple and more conventional supports (Al_2O_3, SiO_2, TiO_2, etc.) than amorphous aluminosilicates of zeolite origin which are too complex although very interesting for detailed researches.

Alumina Supported InNi and InCu Composite Catalysts

Alumina supported Ni-catalysts prepared in two different ways (one of them is a non-pyrophoric Raney type nickel catalyst) were found to catalyze highly selectively the hydrodecarbonylation of fatty acids formed as intermediates of triglyceride hydrodeoxygenation [22]. In order to make clearer the scheme of transformations octanoic acid was used as reactant. Catalysts showed high activity and sufficient decarbonylation selectivity but CO methanation and paraffin hydrogenolysis side reactions producing methane in high concentration are very disadvantageous. Non-pyrophoric Raney nickel catalyst was found to be more active than conventional Ni/Al_2O_3 for the side reactions, too. Although the Ni contents are nearly the same, in non-pyrophoric Raney-Ni the nickel phase can be found mainly in metallic state in contrary to supported Ni-catalysts, where the oxide phase is dominant in spite of the same reduction treatment. Poisoning with Sn or Pb in order to eliminate the activities for the hydrogen consuming side reactions was not successful. However, applying the experiences obtained with the indium doped destructed zeolite supported copper catalysts addition of large amount of In_2O_3 (36 wt. %) was found to be also completely effective in suppressing side reactions over the non-pyrophoric Raney nickel catalyst (see in Table 2). It governed the process again to a new deoxygenating pathway, instead of the chain-shortening hydrodecarbonylation namely into the step-by-step reduction route of carboxylic acid stopped at alcohol formation. Decreasing the amount of In_2O_3 additive the original reaction pathway is gradually restored producing mainly heptane beside octanol whereas paraffin hydrogenolysis is still effectively eliminated (see low hexane yields obtained over indium doped samples in Table 2). It is a great disadvantage that indium (III) oxide is needed in a very high amount to attain these results. Decreasing the amount of indium oxide the selectivity for the main products (heptane or octanol) is changing inversely, even with 18 wt. % of In_2O_3 55 % heptane and only 15 % octanol is obtained (see Table 2). It is interesting to note, that 4 wt. % indium oxide already increases the alcohol production by two orders of magnitude. The methanation of primary product CO is complete, but hydrogenolysis of heptane to hexane, etc. is highly restricted.

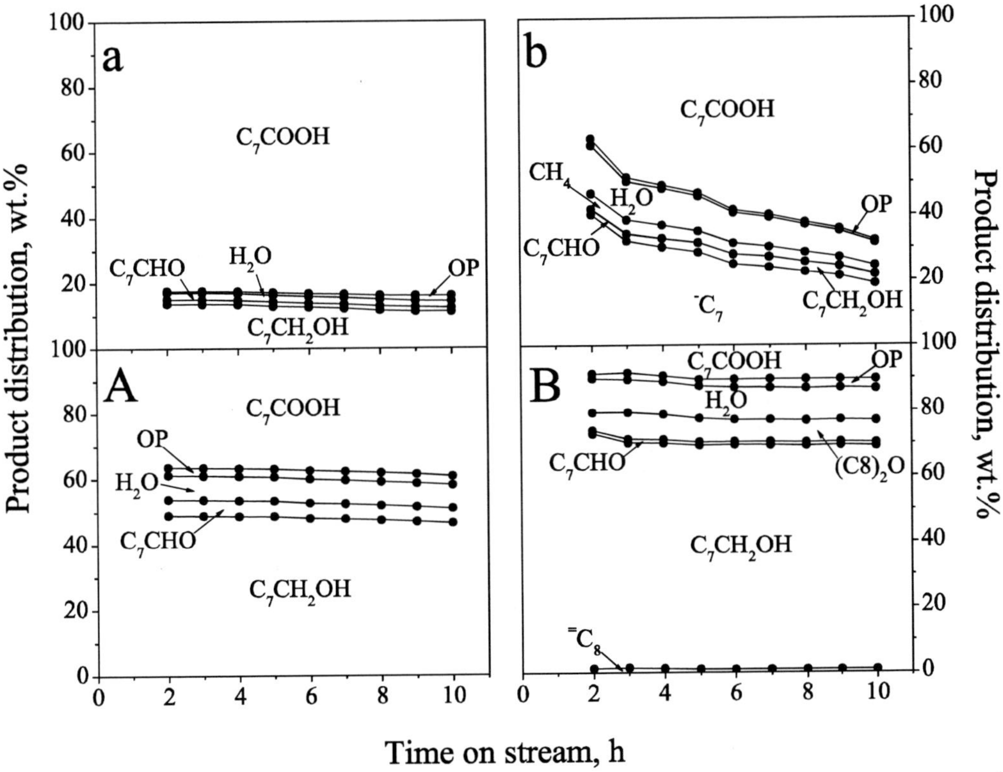

Figure 4. Octanoic acid hydroconversion over $9Cu/Al_2O_3$ (a), $9Cu/Al_2O_3$ + 10 wt. % In_2O_3 (A), $9Ni/Al_2O_3$ (b) and $9Ni/Al_2O_3$ + 10 wt. % In_2O_3 (B) are characterized by main products as a function of time–on–stream at 300 °C and 21 bar total pressure. The WHSV of OA was 2.0 h^{-1}. Legends as in Figure 1. In addition: $(C_8)_2O$ dioctyl ether.

Considering only the mass of active phases, the non-pyrophoric Raney type sample contains nickel mainly in metallic form whereas in the alumina supported sample the oxide phase is dominant. However this condition does not strictly influence the ratio of active surfaces. Another important fact is that starting with a 50-50 wt. % nickel-aluminum alloy an aluminum oxy-hydroxide supported, finely dispersed NiAl intermetallic phase can also be formed. By heating up to 450 °C during the pretreatment the support is transformed to γ-

Al_2O_3 being similar support as in the case of impregnated sample used for comparison [22]. The structure of the non-pyrophoric Raney-Ni catalyst is not well determined like that of destructed zeolites, so detailed investigations regarding the interactions of indium co-catalyst with the main reducing metal, the nickel is worth to be done only with the simplest alumina supported nickel sample, i.e., with the $9Ni/Al_2O_3$ catalyst.

Octanoic acid hydroconversion on monometallic Ni/Al_2O_3 catalyst shows selective hydrodecarbonylation resulting in heptane formation (Figure 4b). The activity of Ni/Al_2O_3 depends on Ni content. CO formed beside water in hydrodecarbonylation is completely hydrogenated to methane and some excess methane can be also formed in hydrogenolysis of heptane as indicated by formation of lower hydrocarbons (e.g. hexane) similarly for NiA (Figure 3b). Modification with co-catalyst strikingly increases the activity, maintaining it at a constant level and drastically changing the selectivity (Figure 4B) by eliminating hydrodecarbonylation as a main reaction and H_2 consuming side-reactions thus governing the process to a different consecutive deoxygenating reaction pathway, i.e., a step-by-step reduction with H_2 and stopping this hydrogenation reaction series at alcohol formation before dehydration to alkenes or dioctylether and saturation of the formed olefines [6]. The results are well comparable at the same reaction temperature (already at 300 °C) obtained mainly over the indium doped composite bimetallic catalysts (Figure 4A, B). Influences of indium doping for both base metal catalysts seem to be very similar only with some shift on the reaction temperature scale.

Table 2. The yields of main products of octanoic acid hydroconversion (wt. %) at 300 °C, WHSV=3 h^{-1} and 21 bar in 5 h time-on-stream over a non-pyrophoric Raney nickel type catalyst (NPR27Ni) modified with various amounts of In_2O_3

Catalyst	Hexane	Heptane	CH_4	Octanal	Octanol	Conversion
NPR27Ni + 0 % In_2O_3	4.2	32.7	9.6	-	0.1	68.2
+ 4 % In_2O_3	0.9	57.6	8.3	0.8	10.6	90.1
+ 18 % In_2O_3	0.3	55.1	7.6	0.5	14.9	89.6
+ 36 % In_2O_3	0.02	10.0	1.2	1.6	57.9	83.2

While the carboxylic acid hydroconversion to alkanes increases with the Ni-content of Ni/Al_2O_3 samples, for In_2O_3 doped samples at constant indium content the conversion can be lower when the Ni-content is above an optimum value (see conversion curves obtained over 3, 9 and $18Ni/Al_2O_3$ samples in [6]). Decrease of nickel loading below 9 wt. % is also disadvantageous resulting in the downfall of conversion and octanol yield followed by an increase of octanal selectivity, because only a lower reduction state can be attained over less active catalyst [6]. Furthermore, using Ni-free alumina sample containing only the indium co-catalyst significant conversion can be observed at much higher reaction temperature (above 350 °C) and the main product is octanal. Without any metal loading the alumina support can produce only traces of octanal and octanol between 350 and 380 °C.

At a fixed Ni-loading (9 wt. %) decreasing the amount of added indium oxide in a wide range hardly affects the selectivity of the catalyst composites [6]. Hence it seems likely to apply in low concentration the expensive, but strongly efficient In_2O_3 co-catalyst. It can be estimated that both active metal components can be used in about 10 wt. % to attain high carboxylic acid conversion with high and stable octanol yield.

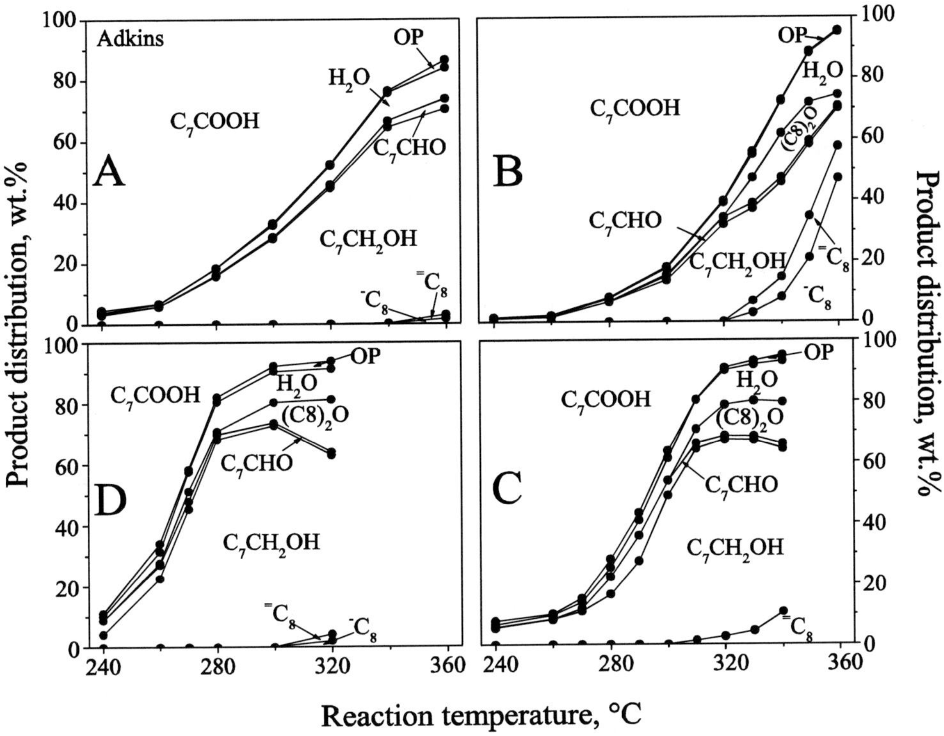

Figure 5. Octanoic acid hydroconversion over commercial Adkins (A), $9Cu/Al_2O_3$ (B), $9Cu/Al_2O_3$ + 10 wt. % In_2O_3 (C) and $9Ni/Al_2O_3$ + 10 wt. % In_2O_3 (D) catalysts characterized by distributions of main products between 240 and 360 °C at 21 bar total pressure. The WHSV of OA was 2.0 h^{-1}. Legends as in Figure 1. In addition: $^-C_8$ octane, $(C_8)_2O$ dioctyl ether.

In Figure 5A and D the catalytic properties of a commercial, conventionally used Adkins catalyst (consisting of 72 wt. % $CuCr_2O_4$ and 28 wt. % CuO) and our bimetallic alumina supported composite ($NiIn/Al_2O_3$) are compared along the increasing reaction temperature under identical conditions. Although both samples are able to produce alcohol from fatty acid with high selectivity, the novel catalyst seems, however, to be more active: 80 % overall conversion of OA on $9Ni/Al_2O_3$ catalyst dopped with 10 wt. % In_2O_3 is already attained at 280 °C while on copper chromite only at 360 °C. Indium free $9Cu/Al_2O_3$ sample shows much poorer catalytic properties (Figure 5B) than conventional Adkins catalyst (Figure 5A), however after doping of the highly efficient indium the activity exceeds that of commercial Adkins type copper catalyst (Figure 5C).

Dioctyl ether formation in bimolecular octanol dehydration on $9Ni/Al_2O_3$ dopped 10 wt. % In_2O_3 samples becomes important only at high octanol yield above 280 °C (Figure 5D). Monomolecular octanol dehydration appears only above 300 °C. Both these acid catalyzed side reactions may be attributed to alumina. However, these undesirable reactions can be eliminated on highly active and stable $9Ni/Al_2O_3$ + 10 wt. % In_2O_3 catalyst composite applying lower reaction temperature and higher space velocity.

Beside the main metal phase active in hydrogenation, the applied support and its interactions with the indium co-catalyst may be of importance to ensure the advantageous catalytic properties. These interactions may be negligible correlated to the hydrogenation of reactant carboxylic acids under moderate conditions however they can become significant at much higher reaction temperature. Product compositions are displayed in Figures 6 after contacting intermediate product aldehyde (Figure 6a, b, c) and desired end product alcohol (Figure 6A, B, C) with alumina support, alumina and In_2O_3, $9Cu/Al_2O_3$ and In_2O_3. Octanal is extremely reactive in aldol type condensation on pure alumina support leading to dimmer formation (Figure 6a). Active sites for aldol condensation can be effectively deactivated by metallic indium admission while octyl aldehyde is reduced to octyl alcohol and octanol is hardly dehydrated (Figure 6b).

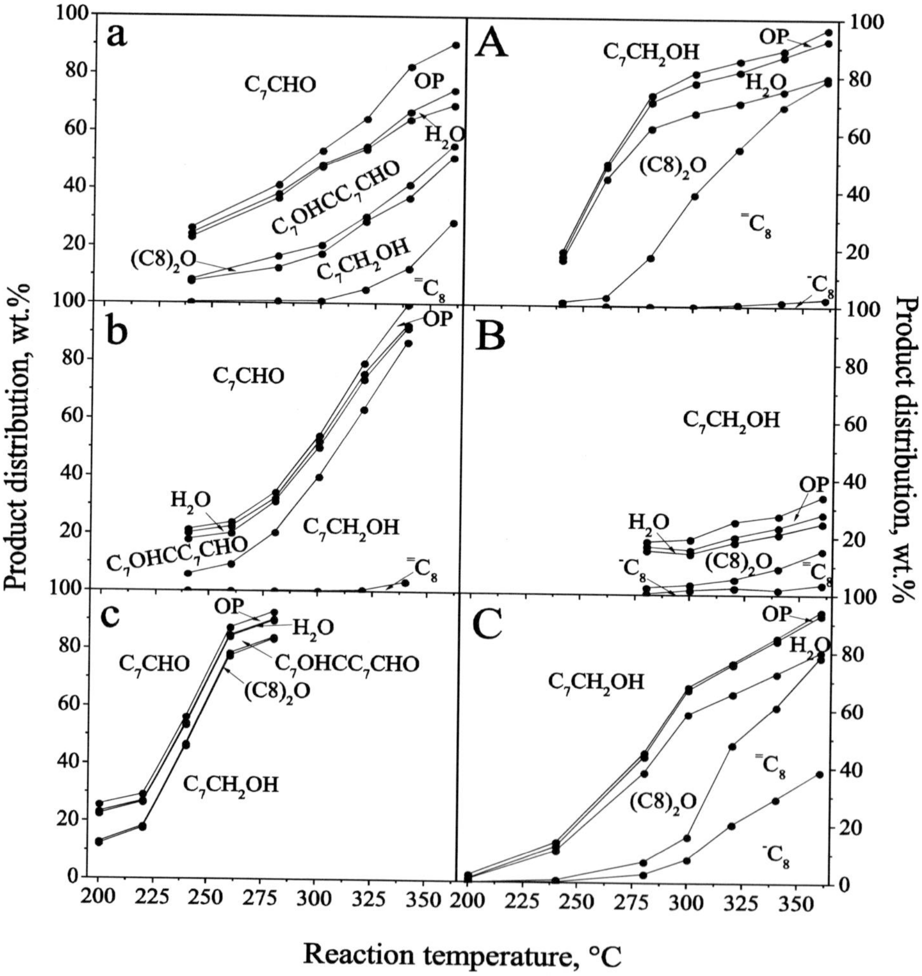

Figure 6. Octanal (a; b; c) and octanol (A; B; C) hydroconversion over Al_2O_3 (a; A), Al_2O_3 + 10 wt. % In_2O_3 (b; B) and $9Cu/Al_2O_3$ + 10 wt. % In_2O_3 (c; C) catalysts characterized by distributions of main products between 200 and 360 °C at 21 bar total pressure. The WHSV of octanal and octanol was 2.0 h^{-1} similarly to octanoic acid reactant. Legends as in Figure 1. In addition: C_7OHCC_7CHO 2–hexyldec 2–enal (the product of octanal aldol condensation), $^{-}C_8$ octane, $(C_8)_2O$ dioctyl ether.

Octanol can easily be dehydrated to dioctyl ether and octenes over the alumina support (Figure 6A). At lower reaction temperature the bimolecular dehydration forming dioctyl ether is dominant. On increasing the temperature the monomolecular dehydration resulting in olefin formation becomes the main reaction pathway. Probably this change of dehydration mechanism is essentially affected by the decreasing alcohol coverage of the surface at increasing temperature. Indium on the surface of alumina effectively suppresses alcohol dehydrations (Figure 6B).

Using octanal as reactant dioctyl ether yield is very low because of the competing aldol condensation alcohol formation is suppressed on the alumina surface (Figure 6a). Comparison of figures in the first column of Figure 6 gives some insight into the role of indium: i., aldol condensation of octanal is greatly suppressed by indium (Figure 6b), ii., the dehydrations of octanol are suppressed as well by indium suggesting that indium blocks the active sites of the alumina support (Figure 6b, Figure 6B) and iii., indium seems to be active in aldehyde reduction, at higher temperature than the main hydrogenating metal, copper (Figure 6c) or nickel. Correlation of Figure 6C and Figure 5C indicate that the presence of the reactant fatty acid can play an important role to poison the active centers of dehydration together with indium, because octanol can be highly dehydrated over CuIn/Al_2O_3 catalyst, i.e., in presence of copper and indium on alumina, but in absence of the organic acid reactant. Figure 6c demonstrates that copper can promote the reduction with hydrogen of the aldehyde at much lower temperature than indium (Figure 6b) being more active hydrogenating metal, but the distribution of various components, types of sorption sites and the coverage of organic compounds on each those over the catalyst surface are very important. Summarized it can be stated that indium can deactivate dehydrating active sites, when carbonyl group is present in the reactant (acid or aldehyde adsorption is necessary).

NiIn/Al_2O_3 catalyst is more active in octanoic acid hydroconversion than CuIn/Al_2O_3 since the conversion and octanol yield curves are shifted by 20 – 30 °C to higher temperatures for Cu,In/Al_2O_3 (Figure 5C, D). However, the product distributions are the same on both bimetallic catalysts. Hence similarity of catalytically active phases seems probable.

For 9Ni/Al_2O_3 catalyst at the routine pretreatment temperature, at 450 °C only a small fraction of NiO is reduced and complete reduction can be attained only at 650 °C (Figure 7b) [6]. While in 9Cu/Al_2O_3 catalyst copper oxide is already fully reduced at 200 °C and copper particles of ~20 nm average diameter (∇ in Figure 7a) are formed and these particles do not change up to 650 °C (and not only up to 450 °C as shown in Figure 7a). Reduction of admixed In_2O_3 can be observed at much higher temperature, i.e., between 350 and 450 °C (see Figure 7A and Figure 7B) detected by disappearence of In_2O_3 diffraction lines (+). In_2O_3 reduction is completed up to 450 °C on all 9Cu/Al_2O_3 and 9Ni/Al_2O_3 samples when indium is in liquid form (see Figure 7). The generated zerovalent indium can already form copper-indium alloy (•) above 350 °C (see Figure 7A). In the case of this composition copper particles, i.e., the pure metal phase can not be detected below the reduction temperature of In_2O_3 indicating either the formation of too small Cu^0 nanoparticles not detectable by XRD method or more probably the alloying of the low concentration Cu^0 with small amount of reduced In_2O_3 as shown by the appearance of a new phase (marked with • in Figure 7A) which can be assigned to some kind of CuIn alloy. Mainly Cu_2In phase, but Cu_4In and Cu_9In_4 phases may also be present in lower concentration. These phases were determined based on ICDD database [27], the corresponding ICDD numbers are Cu_2In: 42-1475, Cu_4In: 42-1477 and Cu_9In_4: 42-1476. After treatment at 450 °C the 9Cu/Al_2O_3 sample mixed with 20 wt. %

In_2O_3 (which amount is more than necessary to form the Cu_2In intermetallic compound), cooling down the composite sample to R.T. the presence of excess solid metallic indium phase (*) can be also detected (see Figure 7A). Frozen indium particles with average sizes of ~130 nm are larger than copper particles indicating higher mobility of indium atoms above the melting point (156.4 °C) than that of copper atoms. $9Cu/Al_2O_3$ + 10 wt. % In_2O_3 composite contains Cu and In nearly in 2:1 atomic ratio, so all amounts of two metals are consumed in formation of the alloy phase (Cu_2In) having an average size of ~10 nm [29].

When less In_2O_3 is added Cu and Cu_2In phases are present simultaneously [29], when indium is in excess it appears as a distinct pure metal phase which can only be detected below its melting point 156.4 °C. Formation of Cu and/or Cu_2In alloy nanoparticles, and appearance of zerovalent indium on the support is evident on the basis of HT-XRD results but the clearing of catalytic properties, i.e., complete understanding of the multiple behaviors of such a complex ensemble of active sites requires further comprehensive investigations.

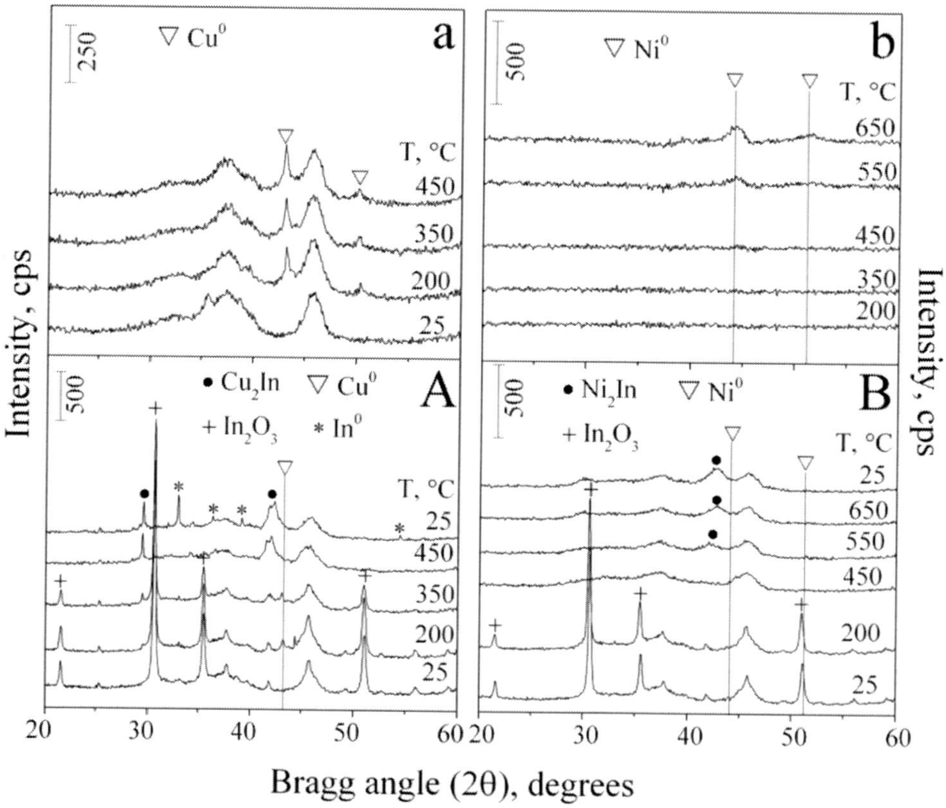

Figure 7. XRD patterns of $9Cu/Al_2O_3$ (a) and doped with 10 wt. % In_2O_3 (A), $9Ni/Al_2O_3$ (b) and doped with 10 wt. % In_2O_3 (B) catalysts. The diffractograms were recorded at the indicated temperatures after 30 min treatment at each temperature in H_2 flow. The temperature was increased in 100 or 50 °C segments.

Regarding the pretreatment conditions of the novel, highly efficient NiIn/alumina composite catalyst, influence of the pretreatment temperature on activity and selectivity in OA hydroconversion in a wide range of 350 and 550 °C can not be observed [6]. In_2O_3 is completely reduced while Ni^{2+} only partly, as shown by HT-XRD patterns in Figure 7b, B.

Melting point of In being 156.4 °C (boiling point is >2000 °C) thus the formed metal should be in liquid phase which may interact with the environment: can penetrate in the surface of the alumina support and/or can form alloy phase(s) immediately with the appearance of nickel clusters arising later in increasing amounts along increasing reduction temperature. At the routine pretreatment temperature, at 450 °C only a small part NiO can be reduced [6, 21] and a complete reduction can only be attained at 650 °C (Figure 7b). After a similar procedure when the treatment is carried out at 650 °C instead of 450 °C (see Figure 7B) and cooling to R.T. In^0 diffraction lines can not be anymore detected and a new line /marked with asterisk (*)/ at 43.2° appears in the diffractogram which can be assigned to some kind of InNi alloy, mainly to $InNi_2$ phase, but $InNi_3$ and $InNi_4$ phases can be also present in lower concentration similarly to the behaviour of Cu-catalysts [30]. It can be concluded: if the catalyst contains enough reduced metallic nickel clusters and the total mass of indium is in zerovalent state, InNi alloy is formed and In^0 is not detectable. Furthermore, it was proven that the appearance of zerovalent indium is necessary to octanol production. This requirement is already satisfied by a pretreatment temperature higher than 350 °C.

Metallic nickel phase can be detected in partly reduced monometallic Ni/Al_2O_3 catalysts (Figure 7b) [21]. Ni^0 particles formed are real nanoclusters having average size in the range 5-8 nm which do not change when the temperature is elevated. The formation of NiIn alloy on the active surface of catalysts can be ascertained on the basis of HT-XRD results (Figure 7B). The catalytic behavior of $InNi/Al_2O_3$ catalyst in a wide pretreatment temperature range does not change. Presumably the active InNi intermetallic surface can be established with small masses of indium and nickel and further growing of the bulk phase hardly affects the active surface.

Outstanding Efficiency of Indium as Co-catalyst Compared to its Neighbours in the Periodic Table

Indium is a very soft, malleable and easily fusible post–transition metal chemically similar to gallium and thallium, and shows the intermediate properties between the two metals. Melting point and boiling point trends are just opposite, where indium has a middle position. Beside chemical properties these physical characteristics may be important in evolution of copper or nickel co-catalysts of outstanding efficiency, e.g., mobility of reduced metal. Appearance of various alloys or intermetallic compounds (some are shown in Figures 2 and 7) can play important role in formation of the catalytically active centres on the surface. HT-XRD method was proven to be very useful to observe the phase changes of the main mass of indium however the catalytically active sites are in very low concentration. The first results of HT-XRD investigations are shown in Figure 8 for 10 wt. % In_2O_3, Ga_2O_3, Tl_2O_3, CdO or SnO doped $9Cu/Al_2O_3$.

In_2O_3 is reduced to metal between 350 and 450 °C (Figure 8d) contrary to Ga_2O_3 which can be reduced only at much higher temperature, between 550 and 650 °C (Figure 8c). It is important to emphasize that reduction of copper and indium oxides does not take place parallel, which can give better opportunity for alloy or intermetallic phase formation (copper particles are formed at lower temperature than metallic indium). Gallium oxide reduction starts at much higher temperature when the formation of copper particles is accomplished. Both post-transition metals can form some kind of alloys or intermetallic phases with copper (e.g. Cu_2In) indicated by the diffraction line (●) appearing at lower Bragg angle than the more intensive diffraction line of copper (∇). Cu_2In phase (Figure 8d) seems to be stable, the line is preserved up to 650 °C similarly to Ni_2In (Figure 7B). In the case of the Ga doped sample the similar signal shows maximum intensity between 550 and 600 °C (Figure 8c). Above 600 °C the intensity of this line significantly decreases parallel with the formation of pure Cu phase. This observation suggests that gallium should be more volatile than indium although its boiling point is higher, but its melting point is lower. These HT-XRD results suggest that gallium may be similarly efficient co-catalyst as indium, but the pretreatment temperature is limited by 550 °C (Figure 8c).

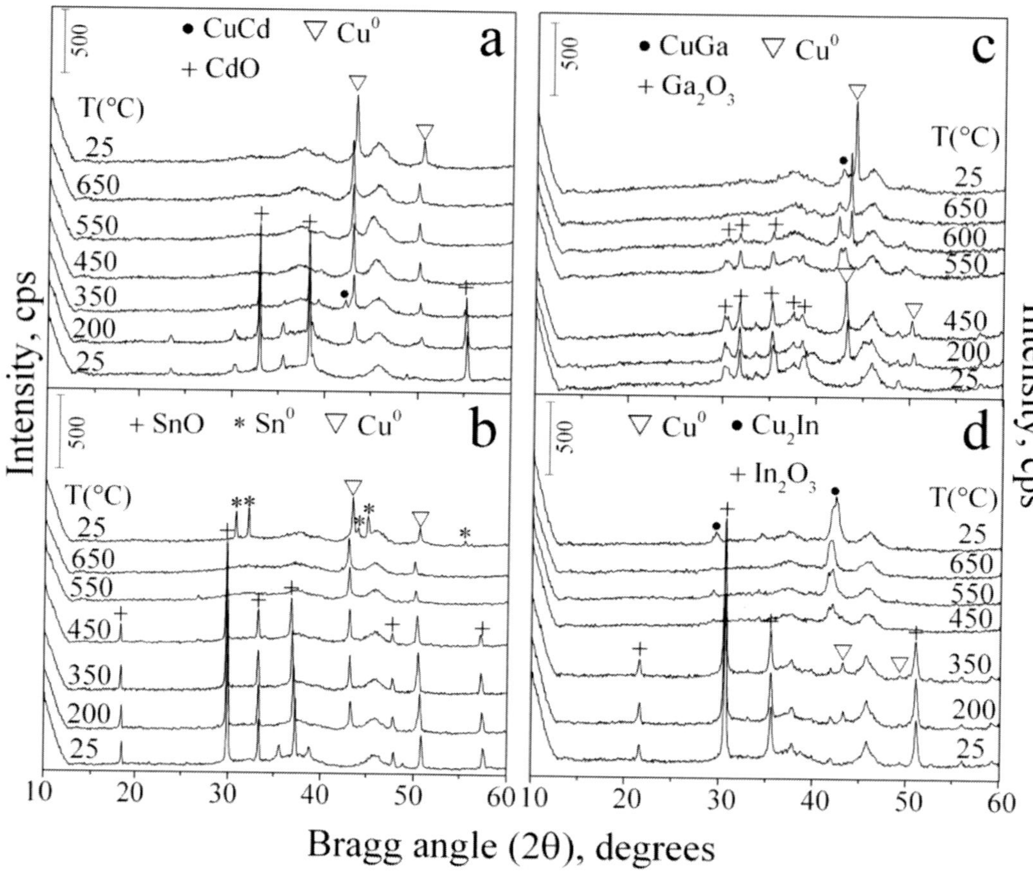

Figure 8. XRD patterns of $9Cu/Al_2O_3$ catalysts doped with 10 wt.% of CdO (a), SnO (b), Ga_2O_3 (c) and In_2O_3 (d). The diffractograms were recorded at the indicated temperatures after 30 min treatment at each temperature in H_2 flow. The temperature was increased stepwise by 100 or 50 °C.

Cadmium, the left hand side neighbour of indium in periodic system can produce similarly intermetallic compounds like indium or gallium, but the signal can be detected only at 350 °C with very low intensity (Figure 8a). However this line disappears when the temperature is increased, because cadmium is volatile, as found in solid state ion-exchange experiments [31]. Consequently cadmium doped sample can be reduced and used only up to 350 °C.

The HT-XRD diffractograms of tin doped sample shown in (Figure 8b) demonstrate that although SnO can be reduced completely to metallic tin up to 550 °C with similar physical properties as indium (e.g. melting point is 232 °C), surprisingly no any formation of alloys or intermetallic compounds can be observed in a wide range of pretreatment temperature: copper and tin can be detected in separate phases in this composite. Thallium (III) oxide showed similar behaviour during the HT-XRD measurement without any trace of formation alloys or intermetallic compounds.

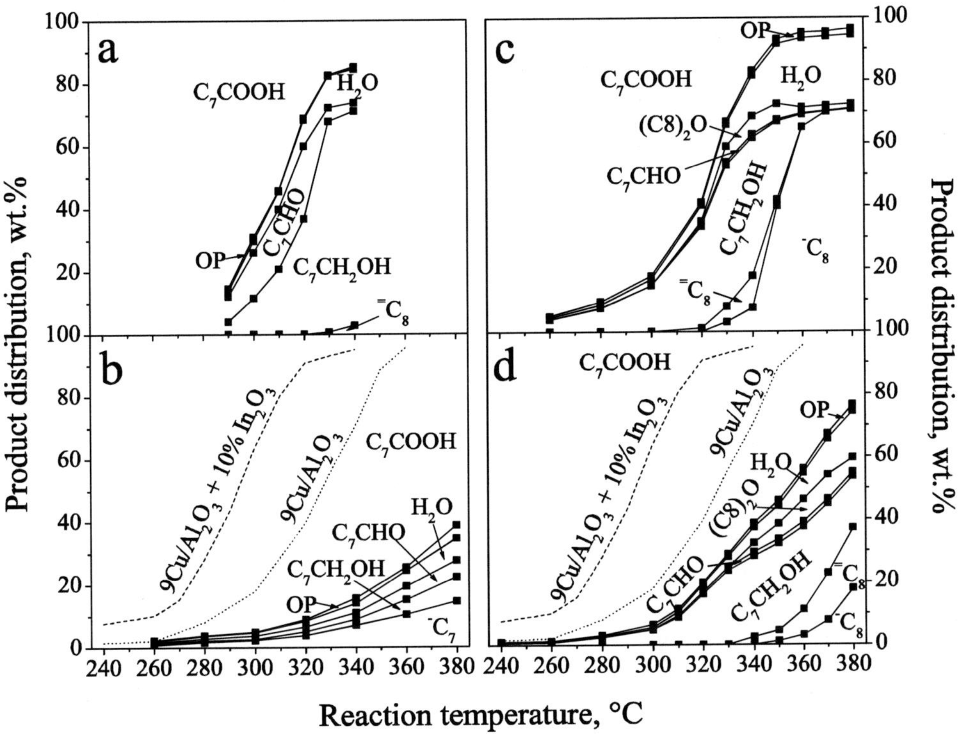

Figure 9. Octanoic acid hydroconversion over $9Cu/Al_2O_3$ + 10 % CdO pretreated at 350 °C (a), $9Cu/Al_2O_3$+ 10 % SnO pretreated at 450 °C (b), $9Cu/Al_2O_3$+ 10 % Ga_2O_3 pretreated at 450 °C (c) and 550 °C (d) catalysts characterized by distributions of the main products between 240 and 360 °C at 21 bar total pressure using stacked area graphs. WHSV of OA was 2.0 h^{-1}. Figure 9b, d also shows the activity of $9Cu/Al_2O_3$ (dotted line) and $9Cu/Al_2O_3$ + 10 % In2O3 (dashed line) as a function of reaction temperature.

Catalytic test results obtained for the doped copper loaded alumina catalysts are summarized in Figure 9 representing the changes of octanoic acid hydroconversion depending on the reaction temperature. Conversion curves plotted with dashed and with dotted lines relate to $9Cu/Al_2O_3$ and $9Cu/Al_2O_3$ +10 % In_2O_3 catalysts in Figure 9b and 9d as reference points (full figures can be find in Figure 5B and C). Figure 9c perfectly overlaps Figure 5B which means that the admission of Ga_2O_3 has no any influence to the catalytic behavior after pretreatment at 450 °C. This observation is accordance with the fact that Ga_2O_3 can be reduced only at higher temperature. However reducing at 550 °C gallium should be present totally in metallic form (Figure 8c), but instead of improvement of catalytic properties, essentially lower conversions were measured (Figure 9d). Gallium atoms deactivate partly the surface of copper particles and the bimetallic particles detected in the diffractograms of Figure 8c do not improve the catalytic behavior.

Cadmium can also form bimetallic phase in a small extent, and cadmium doping can increase the activity and improve the alcohol selectivity (Figure 9a). The problem is here, that it is easy to loose the cadmium co-catalyst due to its high volatility. SnO can be completely reduced above 450 °C, but SnCu alloy formation can not be detected by XRD method. Presence of tin atoms causes an effective poisoning together with unfavourable change in the selectivity producing chain shortened heptane instead of C_8 compounds. Probably copper surface is fully poisoned, perhaps covered with tin atoms and the observed conversion is catalysed only by tin. Thallium showed similar changes in the XRD cell as tin. This element and most of its compounds are highly poisonous, hazardous materials, so the catalytic tests were omitted with thallium doped composite.

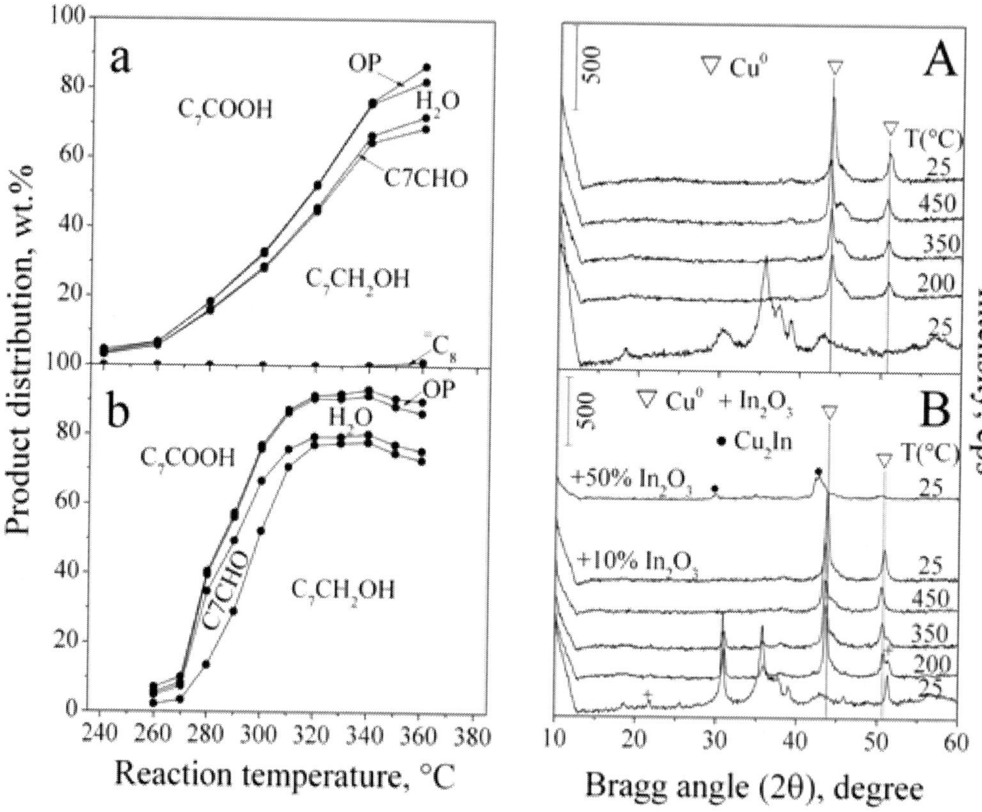

Figure 10. Octanoic acid hydroconversion over commercial Adkins catalysts without (a) and with 10 wt. % of co-catalyst (b) characterized by distributions of the main products between 240 and 360 °C at 21 bar total pressure using stacked area graphs. WHSV of OA was 2.0 h^{-1}. XRD patterns of commercial Adkins catalyst (A) and doped with 10 and 50 % In_2O_3 (B) samples recorded at the indicated temperatures after 30 min treatment at each temperature in H_2 flow. The temperature was increased in 100 or 50 °C segments.

It can be concluded that indium as co-catalyst has unique quality which results in creating highly advantageous CuIn or NiIn bimetallic composites. Appearance of bimetallic phases seems to be essential that an improvement in qualities of copper or nickel catalyst could be attained (with In, or Cd). However the formation of some alloys or intermetallic compounds is not enough to come to the desired catalyst (see gallium). It can be supposed that the highly active species are presumably present in very low concentration, i.e., in form of very thin layers or small clusters on the active metal surface and their identification seems rather difficult.

Commercial Adkins catalyst has very high copper content which is present under the reaction conditions in metallic form (see Figure 10A). Adding 10 wt. % In_2O_3 the catalytic behaviour drastically changes (c.f. Figs. 10A and b) becoming totally similar to that of $InCu/Al_2O_3$ supported composite catalyst. In the HT-XRD patterns influence of 10 wt. % In_2O_3 admission can not be detected (Figure 10B). Admission of 50 wt. % In_2O_3 results in a radical change of the diffractograms. Only Cu_2In phase can be detected alone which points to the formation of this characteristic phase the presence of which is coupler with outstanding catalytic behaviour.

HYDROCONVERSION OF ACETIC ACID TO BIOETANOL AND ETHYLACETATE

Numerous biomass conversion pathways, biomass degradation platforms generate carboxylic acids. Regarding the need of transportation and the industrial sectors for chemicals, the selective catalytic hydroconversion of bio-acids to alcohols, ethers, and esters is a process of high commercial interest. Up to 150 °C the distillate of the bio-oil, obtained from lignocellulose pyrolysis contains lower carboxylic acids and aldehydes in relatively high concentration. Fuel alcohol can be obtained by selective reduction of these components. Instead of the less advantageous thermochemical routes, a novel and favorable microbiological destruction /MixAlco/ process is elaborated [19]. In principle, the biogas process is suggested to be stopped after the rapid anaerobic acidogenic digestion stage to produce volatile fatty acids (VFA), mainly acetic acid [32-33], not followed by the slow methanogenesis step. The overall chemical reaction without any loss of biomass conducted by species of anaerobic bacteria, including members of the genus Clostridium may be represented as:

$$C_6H_{12}O_6 \rightarrow 3\ CH_3COOH \tag{3}$$

The efficiently selective hydrogenation of such volatile fatty acids (VFAs, e.g. acetic acid) to alcohols in a continuous flow system working in vapor phase under mild conditions seems to be still a problematic step.

As it has been shown indium doping of supported Ni or Cu catalyst proved to be highly efficient for reduction of the long chain fatty acids to aliphatic alcohols opening a novel route for catalyst development. The same catalyst systems seem promising for the selective reduction of VFAs, short chain carboxylic acid homologues to alcohols. Acetic acid (AA), the most easily available platform material obtainable from degradation of non edible biomass, as a model reactant is of great importance.

Highly dispersed copper particles can be produced on amorphous aluminosilicate by dehydration/H_2-reduction of low-silica Cu-zeolites. The obtained Cu/aluminosilicate preparations have high activity and selectivity also in the hydroconversion of biomass-derived carboxylic acids (e.g. acetic acid) to alcohols, aldehydes and esters. The catalytic properties depended on the original zeolite structure as it has been found for octanoic acid. Hydrogenation activity and selectivity of the supported Cu/aluminosilicate catalysts can be significantly modified by In addition. The catalytic activity and selectivity of different catalyst compositions are reflected by the product distribution as a function of temperature (see Figure 11). Only traces of gaseous products (ethylene, CH_4, CO and CO_2) were detected on Adkins and indium doped catalysts. Main products formed from AA over all In-free Cu-zeolite catalysts were ethanol and ethyl acetate (see e.g. on CuA in Figure 11a). The acid is reduced first to acetaldehyde and then to ethanol. Depending on the reaction conditions, fractions of the reactant acid and the product alcohol formed ethyl acetate according to the Fischer esterification mechanism. In addition, ethyl acetate can be also formed directly from two acetaldehyde molecules according to the Tishchenko reaction.

With exception of 16CuA all the other zeolite based catalysts were significantly more active than commercial chromium-containing Adkins catalyst and demonstrate different selectivity being the main product ethyl acetate unlike ethanol. Such a direct reaction route to the valuable ester product can be economically more attractive and environmentally benign [34]. When fatty acids of longer chains, e. g. OA, were reduced over the same catalysts, the addition of In_2O_3 to the catalyst drastically increased the conversion and octanol yield. Using AA as reactant, the In_2O_3 addition changed the selectivity and the conversion as shown for 16CuIn/A (Figure 11b). Acetaldehyde appeared in high concentration at lower conversions and the ethyl acetate yield was low. The product distribution could be hardly influenced by increasing the amount of the In_2O_3 modifying agent between 5 and 36 wt. %. The greatest change takes place at low loading, below 2 wt. % suggesting that indium is effective even in low concentration.

The catalytic activity and selectivity of the alumina supported catalyst preparations are characterized by the product distribution as a function of reaction temperature and are shown in Figures 11d, e and f. Main valuable products formed from acetic acid (AA) are ethanol and ethyl acetate. As by-products acetaldehyde, acetone, ethane, methane, carbon dioxide, carbon monoxide and water were determined in various amounts depending on the composition of the catalysts. Carboxylic acid is converted first to aldehyde then to alcohol, fractions of the reactant acid and the products can form ethyl acetate according to the Fischer and/or Tishchenko esterification mechanism. From acetic acid after AA-anhydride formation decarboxylation takes place resulting in formation of acetone and CO_2. Ethanol can be easily dehydrated in a monomolecular way to ethylene or to diethyl-ether. Similarly to long chain fatty acids two hydrodeoxygenation routes [10] can be distinguished: (i) consecutive, step-by-step reduction of carboxyl group or (ii) one step hydrodecarbonylation resulting in carbon monoxide and water.

X-ray diffraction patterns recorded and evaluated proves that in the novel $InNi/Al_2O_3$ composite catalysts, activated at 450 °C in H_2 flow before the reaction, In_2O_3 is reduced and indium metal far above its melting point (156.4 °C) is present under the reaction conditions and finely dispersed bimetallic $InNi_2$ particles are formed (Figures 7b and B). Consequently, the appearances of these two characteristic components in the samples containing initially admixed In_2O_3 are responsible for the drastic changes of catalytic properties (cf. Figures 11e and f). In general, the modification by indium changes the hydrogenation activity of the nickel catalyst for the C−O or C=O bonds and controls the extent of side reactions.

Unlike octanoic acid for fatty acids of short carbon chain, e.g. acetic acid (AA) as reactant contacted with Ni/Al_2O_3 or $InNi/Al_2O_3$ catalyst and (Figures 11e and f) beside alcohol the ester is a typical product, as well. However, neither ether nor alkene formation can be detected contrarily to octanoic acid. Alkyl-chain lengths of reactants affect the adsorption capabilities which determine the coverage of the reactant or product compounds on the same active surface and thus results in significantly different catalytic activities and product distributions.

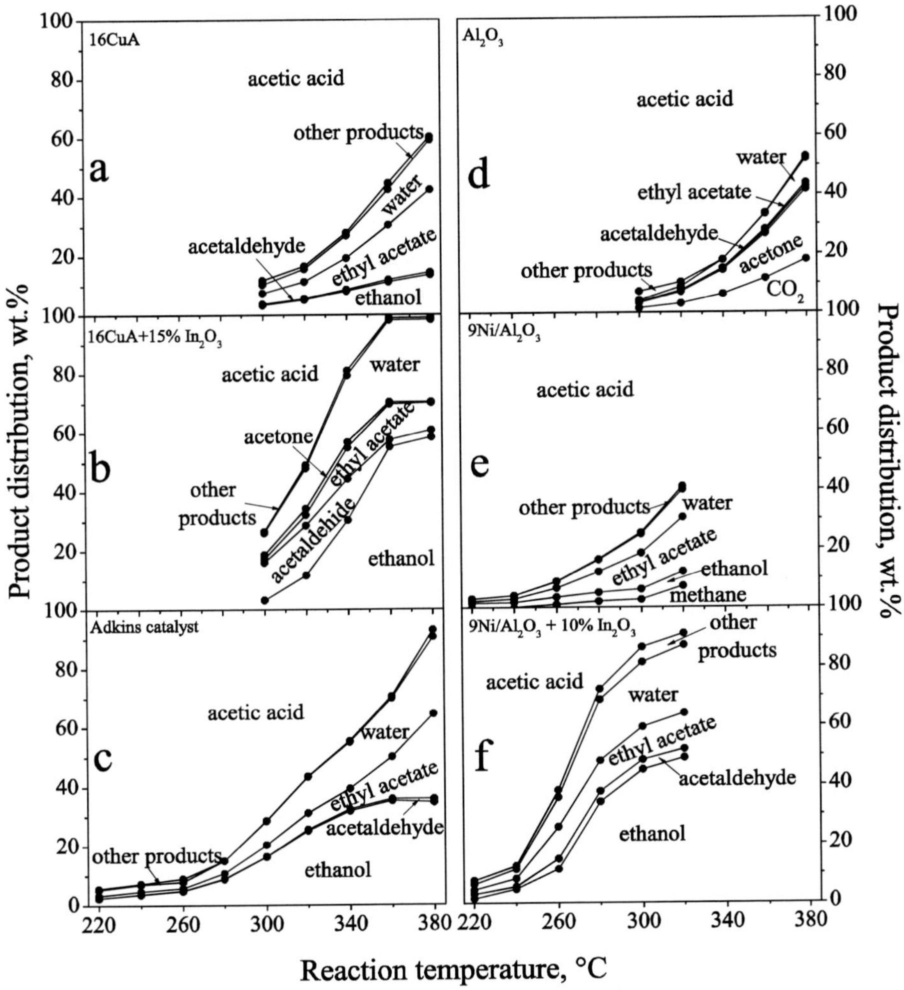

Figure 11. Acetic acid hydroconversion over 16CuA (a), 16CuA + 15 wt. % In_2O_3 (b), commercial Adkins (c), Al_2O_3 (d), $9Ni/Al_2O_3$ (e) and $9Ni/Al_2O_3$ + 10 wt. % In_2O_3 (f) catalysts characterized by distributions of the main products between 220 and 380 °C at 21 bar total represented by stacked area graphs. WHSV of acetic acid was 1.0 h^{-1}.

$9Ni/Al_2O_3$ catalyst (Figure 11e) was similarly active as the commercial chromium-containing Adkins catalyst (Figure 11c) but shows different selectivity being the main product ethyl acetate unlike ethanol. Octanoic acid hydroconversion on monometallic Ni/Al_2O_3 catalyst shows selective hydrodecarbonylation resulting in heptane formation as main product (Figure 4b). Using acetic acid as reactant similar transformation can proceed also, but in much lower extent as shown in Figure 11e by methane formation at increasing reaction temperature when from one acetic acid molecule one methane can form by hydrogenation of carbon monoxide evolved in hydrodecarbonylation and a second one from the remaining methyl group. However the other reaction path the step-by-step reduction of acetic acid to ethanol seems to be dominant, which can come to end immediately with a fast esterification of reactant acid and the product alcohol resulting low ethanol yield as shown in Figure 11e. At the same time a quite different parallel reaction path for the considerable ethyl acetate production can be easily conceivable from the reactive intermedier, acetaldehyde (Tishchenko reaction as a special case of the Cannizzaro reaction). Alumina support can play a determining role in ester formation either Ni or $InNi_2$ particles are present. Here is an exciting room for future research for understanding the reaction mechanism of ethyl acetate formation.

Modification by adding 10 wt. % In_2O_3 co-catalyst to Ni/Al_2O_3 catalyst highly increases the activity and strikingly changes the selectivity (c.f. Figs. 11 e and f). Hydrodecarbonylation is completely eliminated (methane is not formed) governing thus the process clearly to the consecutive deoxygenating reaction pathway, i.e., to a step-by-step reduction with H_2 and stopping this hydrogenation reaction series at ethanol formation allowing esterification to a limited extent, however, ester can be formed from intermediate aldehyde too, as mentioned before. The boiling points of these two main products are nearly the same thus being their separation rather expensive. Using the product mixture as fuel the separation is not necessary. Compared with the commercial, conventionally used Cu-chromite (Adkins catalyst) catalyst (Figure 11c) and the bimetallic alumina supported composite ($InNi/Al_2O_3$) the novel catalyst system (Figure 11f) seems to be essentially more active and selective for producing ethanol.

Beside the metal phase being active in hydrogenation reactions, the support applied and its interactions with the indium co-catalyst may be of great importance to ensure the advantageous catalytic properties. Furthermore Ni and/or In free alumina can catalyze undesirable side reactions. Figure 11d demonstrates the catalytic behavior of γ-alumina support with reactant AA. It has some activity at higher reaction temperatures than the $InNi/Al_2O_3$ composite catalyst on which complete hydroconversion of AA can be attained in this temperature range. Above 300 °C the support can selectively catalyze acetone formation proceeding via bimolecular dehydration of AA followed by decarboxylation. No products of the desired consecutive AA reduction can be detected. Consequently all the products shown in Figure 11e are formed only over nickel particles. Contrary the liquid indium metal (working far above its melting point) presumably well dispersed on alumina support shows significant hydrogenating activity producing all the compounds observable in consecutive AA reduction steps as main products but by 80-100 °C higher reaction temperature than over the $InNi/Al_2O_3$ composite (not shown in Figure 11). Besides this, partition of the original activity of the alumina support is still preserved in presence of indium since significant amounts of acetone and CO_2 are produced. It means that indium atoms at this concentration can deactivate only a fraction of such active sites all over the surface of alumina support.

On comparing hydroconversion of octanoic or ethanoic acids over $InNi/Al_2O_3$ composite catalyst (cf. Figures 4B and 11f at 300 °C) similar conversions can be observed at the same mass feed rate of acids suggesting that the *molecular* reduction rate of the carboxylic group on the bimetallic surface (perhaps only to a formyl group) is higher for lighter carboxylic acid than for heavier one. In addition the product compositions are different: in hydroconversion of octanoic acid either aldehyde or ester can not be detected but some alcohol dehydration takes place, while for acetic acid both ester formation is considerable and aldehyde formation can be observed but no dehydroxylation.

The activity dependence on the reactant partial pressures indicates the rate-controlling Langmuir-Hinshelwood kinetics (not shown).

Combining relatively simple pyrolytic (or biochemical) technologies resulting in production of acetic acid with the catalytic transformations shown, bioethanol and chemicals, such as ethyl acetate, can be obtained from biomass, e.g. from lignocellulose.

Regarding the continuation of this research the main question is how the second metal, the indium modifies the activity of the supported hydrogenation metal and the surface properties of the oxide support? It is essential to gain basic knowledge of the conversion kinetics and the mechanisms at molecular level for engineering effective catalyst composition and catalytic process. What are the elementary reactions of the HDO process and what is the kinetics of the reaction? What are the rate determining step and the most abundant surface intermediate of the consecutive HDO reaction? What is the molecular mechanism of the elementary reaction steps? How to prepare well tailored catalyst of required selectivity? What is the structure of the active site? What factors determine the activity, selectivity and catalyst stability?

Conclusion

Indium doping of supported Cu or Ni catalysts has been discovered to be highly efficient in reduction of long or short chain fatty acids with hydrogen opening a novel route for catalyst development. Hydrogenation of carboxylic acid resulting in selectively produced alcohol can similarly be increased by In_2O_3 doping over Cu as well as over Ni supported catalysts. Presence of metallic indium or rather alloying copper and nickel particles can effectively stop the step-by-step catalytic reduction after alcohol formation and hinders the consecutive mono- or bimolecular dehydration of product. High conversion with high alcohol selectivity can be attained in a wide loading range of Cu or Ni and In on various supports. During the pretreatment in hydrogen flow copper oxide is reduced to metallic nanoparticles at lower temperature than In_2O_3, but equally below the pretreatment temperature (450 °C). These differences are not reflected by any change of catalytic behaviour in a wide range of Cu and In content. Contrary, nickel cations can be reduced only at higher temperature than In_2O_3, but the catalytic behaviour is also independent of the pretreatment temperature i.e., on degree of reductions. Minute amounts of the active metals are participating in catalytic transformations.

Appearance of Cu_2In and Ni_2In alloy phases on the surface of metal particles results in significant increase of hydrogenation activity, however, the presence well–dispersed indium atoms or clusters in liquid state on the surface of support seems to be determinative for the selective poisoning the alcohol dehydration capability of the support. The beneficial properties of indium acting as co-catalyst seems to be unique compared to its neighbors in the periodic table (Ga, Tl, Cd and Sn) similarly to its outstanding role in electronic industry.

The activity dependence on the reactant partial pressures reveals Langmuir-Hinshelwood kinetics with rate controlling surface reaction over the composite catalyst.

ACKNOWLEDGMENT

The authors wish to express their appreciation to Mrs. Ágnes Farkas Wellisch for her technical assistance. Thanks is due to the Hungary-Slovakia Cross-border Co-operation Programme (Project registration number: HUSK/1101/1.2.1/0318) for supporting this research.

REFERENCES

[1] H. Adkins, R. Connor, *J. Am. Chem. Soc.* 1931, 53, 1091-1095.
[2] R. Prins, V.H.J. de Beer, G.A. Somorjai, *Catal. Rev. Sci. Eng.* 1989, 31, 1-41.
[3] P. Simacek, D. Kubicka, G. Sebor, M. *Pospisil, Fuel* 2009, 88, 456-460.
[4] D. Kubicka, L. Kaluza, *Appl. Catal.* A 2010, 372, 199-208.
[5] L. Boda, Gy. Onyestyák, H. Solt, F. Lónyi, J. Valyon, A. Thernesz, *Appl. Catal.* A 2010, 374, 158-169.
[6] Gy. Onyestyák, Sz. Harnos, D. Kalló, *Catal. Com.* 2011, 16, 184-188.
[7] F. Lónyi, H. Solt, J. Valyon, A. Boix, L.B. Gutierrez, *J. Mol. Catal.* A 2011, 345, 75-80.
[8] J.A.Perdigon-Melon, A. Gervasini, A. Auroux, *J. Catal.* 2005, 234, 421-430.
[9] M. Chen, J. Xu, Y.M. Liu, Y. Cao, H.Y. He, J.H. Zhuang, *Appl. Catal.* A 2010, 377, 35-41.
[10] M. Chen, J.L. Wu, Y.M. Liu, Y. Cao, K.N. Fan, *Catal. Com.* 2011, 12, 1063-1066.
[11] H. Lorenz, W. Jochum, B. Klötzer, M. Stöger-Pollach, S. Schwarz, K. Pfaller, S. Penner, *Appl. Catal.* A 2008, 347, 34-42.
[12] Y. Men, G. Kolb, R. Zapf, M. O'Connell, A. Ziogas, *Appl. Catal.* A 2010, 380, 15-20.
[13] A. Jahel, P. Avenier, S. Lacombe, J. Olivier-Fourcade, J.C. Jumas, J. *Catal.* 2010, 272, 275-286.
[14] W.M. Hexana, N.J. Coville, *Appl. Catal.* A 2010, 377, 150-157.
[15] S.B. Kogan, M. Herskowitz, *Catal. Com.* 2001, 2, 179-185.
[16] S.A. Bocanegra, A.A. Castro, O.A. Scelza, S.R. de Miguel, *Appl. Catal.* A 2007, 333, 49-56.
[17] F.A. Marchesini, C.A. Querini, E.E. Miró, F.G. Requejo, J.M. Ramallo-López, *Catal. Com.* 2008, 10, 355-358.
[18] F. Haass, M. Bron, H. Fuess, P. Claus, *Appl. Catal.* A 2007, 318, 9-16.
[19] H.N. Chang, N.J. Kim, J. Kang, C.M. Jeong, *Biotechn. Bioproc. Eng.* 2010, 15, 1-10.
[20] Sz. Harnos, Gy. Onyestyák, J. Valyon, M. Hegedűs, Z. Károly In Proceedings of 10th Pannonian International Symposium on Catalysis; M. Derewinski, B. Sulikowski, A. Wegrzynowicz; Ed; *Polish Zeolite Association and Institute of Catalysis and Surface Chemistry*, PAS: Kraków, 2010; pp. 334-341.
[21] Sz. Harnos, Gy. Onyestyák, D. Kalló, *Reac. Kinet. Mech. Cat.* 2012, DOI: 10.1007/s11144-012-0424-6.
[22] Sz. Harnos, Gy. Onyestyák, Á. Szegedi, D. Kalló, Fuel 2012, (in print).
[23] Beyer, H., Jacobs, P.A., Uytterhoeven, J. B. *J. Chem. Soc. Faraday Trans.* 1 1976, 72, 674-685.

[24] R.G. Herman, J.H. Lunsford, H. Beyer, P.A. Jacobs, J.B.J. Uytterhoeven, *J. Phys. Chem.* 1975, 79, 2388-2394.

[25] Sz. Harnos, Gy. Onyestyák, D. Kalló, *Micropor Mesopor Mat.* 2012, DOI: 10.1016/j.micromeso.2012.03.011.

[26] H.G. Karge, In: Progress in zeolite and Microporous Materials (H. Chon, S.K. Ihm, Y.S. Uh, Eds.) *Stud. Surf. Sci. Cat.* 105C (1997) 1901.

[27] H.K. Beyer, R.M. Mihályi, Ch. Minchev, Y. Neinska, V. Kanazirev *Micropor. Mater.* 1996, 7, 333-341.

[28] B.D. Cullity In *Elements of X-ray Diffraction* 2nd edition, Addison-Wesley Publ Co.: Reading, MA, 1978, pp

[29] Gy. Onyestyák, Sz. Harnos, D. Kalló, *Catal. Com.* 2012, (in print).

[30] Z. Bahari, E. Dichi, B. Legendre, J. Dugue, *Thermochim. Acta* 2003, 401, 131-138.

[31] Gy. Onyestyák, F. Lónyi, J. Valyon, *J. Therm. Anal. Calor.* 2005, 79, 561-565.

[32] M.T. Holtzapple, C.B. Granda, *Appl. Biochem. Biotechnol.* 2009, 156, 525-536.

[33] V. Pham, M.T. Holtzapple, M. El-Halwagi, *J. Ind. Microbiol. Biotechnol.* 2010, 37 1157-1168.

[34] Atkins, M.P., *Top. Catal.* 2003, 24, 185-186.

In: Indium
Editors: Hsaio G. Woo and Huang Tsai Choi

ISBN: 978-1-62257-696-8

Chapter 3

METAL-INSULATOR TRANSITION IN INSB INDUCED BY DOPING, MAGNETIC FIELD, UNIAXIAL STRESS AND HYDROSTATIC PRESSURE

Sergey A. Obukhov
The IOFFE Physical Technical Institute of the Russian Academy of Sciences, St. Petersburg, Russia

ABSTRACT

Indium antimonide is known as one of the best studied A^3B^5 compounds both theoretically and experimentally. Well established band structure and developed technology have made InSb an attractive material in experimental physics and practical applications. Manganese doped InSb is a unique material, which has provided us a chance to study metal-insulator transition (MIT) simultaneously under a variety of experimental conditions (impurity concentration variations, magnetic field, uniaxial and hydrostatic compression). The results of the experiments with manganese doped InSb single crystals have stimulated us to explore the nature of Colossal Magnetoresistance (CMR) and Excitonic Insulator (EI) phase- effects which interpretation has remained the matter of discussions for decades. Comparative study of magnetotransport effects in p-InSb(Mn) and uniaxially stressed p-InSb(Ge) crystals provided in the CHAPTER have shown that specific p-InSb(Mn) properties could rather be related to the influence of Jahn-Teller distortion of crystal lattice caused by Mn^{+2} ions than to the interaction between manganese ions as it usually takes place in magnetic materials – DMS, MS and Manganite Perovskites. Coexistence of electrons and holes at temperatures below 4K at manganese impurity concentration ~$10^{17}cm^{-3}$ around the metal-insulator transition in p-InSb(Mn) crystals gives the ground to explain the MIT in the framework of excitonic insulator phase model. An external magnetic field causes destruction of permanent excitons in InSb(Mn) and so causes the increase in charge carrier concentration, i.e. magnetic field induces CMR and drives IMT transition. The existence of CMR in the nonmagnetic semiconductor has led us up to the concept of the same physical CMR nature origin in DS, DMS, MS and Manganite Perovskites, i.e. Colossal Magnetoresistance can be described by excitons behavior in magnetic field. An existence of the gap energy Δ in InSb(Mn) in the millikelvin range of temperature could be used in creation of advanced cryogenic devices - thermometers, bolometers and selective sensors for millimeter and radio-frequency wavelength.

1. INTRODUCTION

The CHAPTER is devoted to one of the challenging problems in modern solid state physics – Metal-Insulator Transition (MIT) and aims to discuss both fundamental physics and MIT applications within the wide range of solid state materials - Doped Semiconductors (DS), Strongly Correlated Electron Systems (SCES), Diluted Magnetic Semiconductors (DMS), Magnetic Semiconductors (MS), High Temperature Superconductors (HTS) and Colossal Magnetoresistive Perovskites (CMR). Phenomenologically, MIT is known as a physical effect observed in solid state materials with the change of the material resistivity from metal conductivity type to semiconductor or turns into insulator as a result of technological manipulations or under external forces. It should be mentioned that in experiments we can observe both - transition from metal to insulator (MIT) and insulator to metal transition (IMT) depending on the material type and external forces. Compound variation or material doping with impurities can be related to the technological methods, which induce MIT. Another type of methods, when MIT is driven by external forces, belongs to temperature and magnetic field variations, uniaxial stress and hydrostatic pressure. In this CHAPTER the accent is put on the transport effects around MIT in InSb crystals doped with nonmagnetic and magnetic impurities.

1.1. Experimental Observations of MIT

Experimental MIT studies started in 1950-s with the progress in DS technology. Resistivity-temperature dependence experiments in semiconductor single crystals revealed MIT induced by temperature decrease as it is shown in Fig.1 for p-Ge(Ga) [ref.1], n- Ge(As) [ref.2] and p-InSb(Mn) [ref. 3]. It was experimentally demonstrated that variation in impurity concentration can also bring about transition in DS from the metal type of conductivity to semiconductor type. This effect was observed in Ge, Si and other DS. The example of resistivity dependence on impurity concentration for p-Ge(Ga) [ref.4] and for n-Si(P) [ref.5] is given in Fig. 2. Besides MIT induced in DS by temperature and impurity concentration variations it was also demonstrated experimentally that similar transition can be observed if we put semiconductors in the magnetic field. For example n-Ge(Sb) [ref.6], n-InSb(Te) [ref.7] and p-InSb(Ge) [ref.8] single crystals which had a metallic type of conductivity in the absence of magnetic field demonstrated noticeable magnetic field driven increase in specific resistivity, named Giant Positive Magnetoresistance (GPM), (Fig. 3). Magnetic field can induce in semiconductors not only GPM but also dramatic fall of specific resistivity named Giant Negative Magnetoresistance (GNM). In DMS and MS resistance goes down when magnetic field of several Tesla is applied as it is demonstrated in Fig.4 for p-$Hg_{1-x}Mn_xTe$ [ref.9] and n-$Gd_{1-x}v_xS_4$ [ref.10]. These changes resulted in semiconductors transition from semiconductor conductivity type to metal conductivity, namely, insulator-metal transition (IMT). Historically GNM effect has been related to the magnetization of semiconductor compounds containing transition elements, e.g manganese, or rare earth magnetic ions - Gd,

Eu and etc. Surprisingly, magnetoresistance studies of p-InSb crystals doped with Mn at manganese concentration N_{Mn}~$10^{17}cm^{-3}$ [ref.11] demonstrated that GNM can be observed also in non-magnetic doped semiconductor, p-InSb(Mn), as it is depicted in Fig.4.

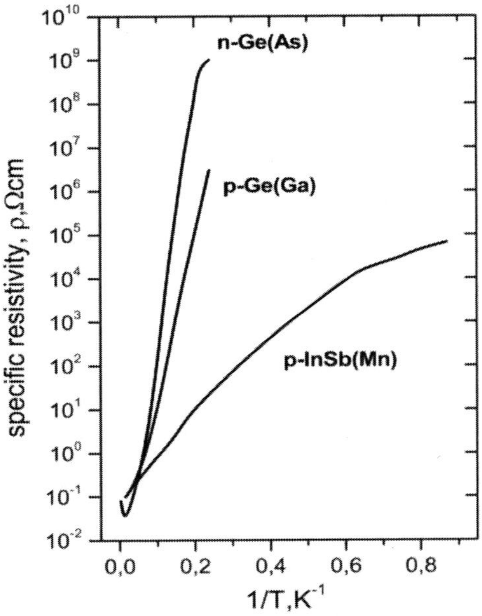

Figure 1. Metal-insulator transition in n-Ge(As) at n=1,32×$10^{15}cm^{-3}$, p-Ge(Ga) at p=2,4×$10^{15}cm^{-3}$ and p-InSb(Mn) at p=5,4 ×$10^{16}cm^{-3}$, data adapted from [1,2,3] respectively.

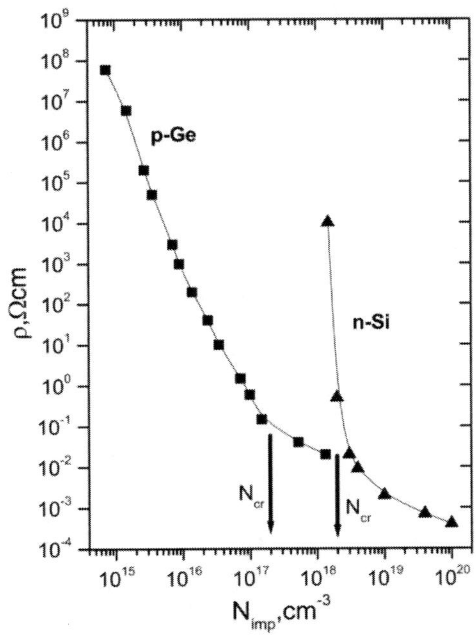

Figure 2. p-Ge(Ga) and n-Si(P) resistivity plotted against impurity concentration at T=4,2K, data adapted from [4,5]; N_{cr} is a critical concentration of MIT.

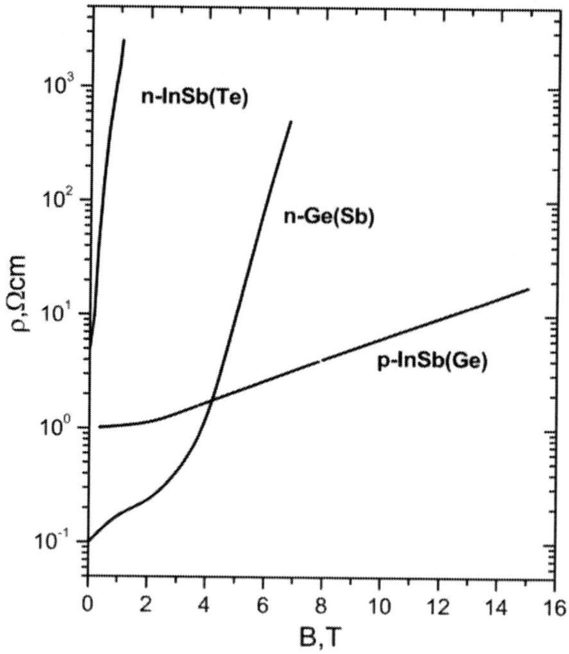

Figure 3. Giant Positive Magnetoresistance in n-Ge(Sb) at T=30mK, in n-InSb(Te) at 1,6K and p-InSb(Ge) at 1,84K, data adapted from [6÷8] respectively.

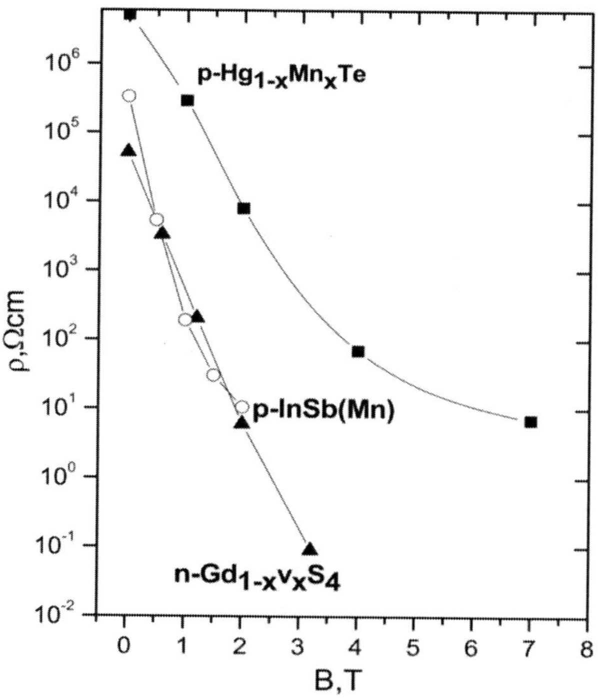

Figure 4. Giant Negative Magnetoresistance in p-$Hg_{1-x}Mn_xTe$ at T=1,66K, n-$Gd_{1-x}v_xS_4$ at 4,2K and p-InSb(Mn) at 0,4K, data adapted from [9÷11] respectively.

Another type of MIT in semiconductors can be driven by uniaxial and hydrostatic pressure. Transition from metallic conductivity to semiconductor conductivity type under uniaxial compression was observed in p-Si(B)[ref.12], p-Ge(Ga) [ref.13], p-InSb(Mn) [ref.14] and p-InSb(Ge) [ref.8] as the result of Giant Piezoresistance in these semiconductors as it is displayed in Fig.5. It can be seen from Fig.5 that under uniaxial compression at T=1,72K the specific resistivity increases from 10 to 10^6 depending on semiconductor type.

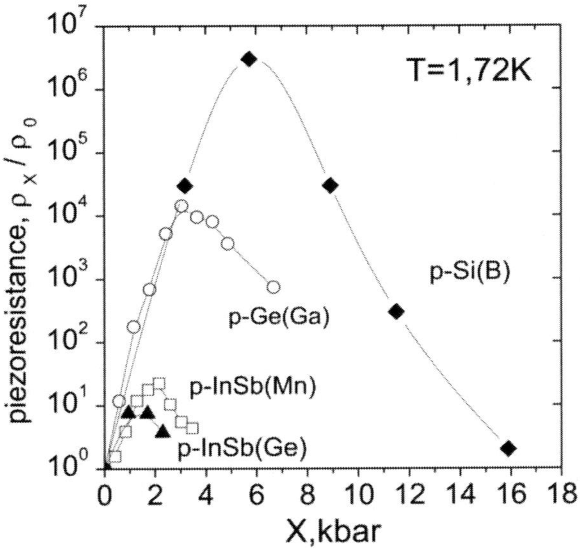

Figure 5. Longitudinal piezoresistance ρ_X/ρ_0 in p-Si(B) at p=2,7×10^{18}cm^{-3}; p-Ge(Ga) at p=1×10^{17}cm^{-3}; p-nSb(Mn) at p=2×10^{17}cm^{-3}; p-InSb(Ge) at p=2×10^{17}cm^{-3}; data adapted from [12÷14,8], respectively; ρ_X is the specific resistivity under uniaxial stress X and ρ_0 is the resistivity in the absence of stress.

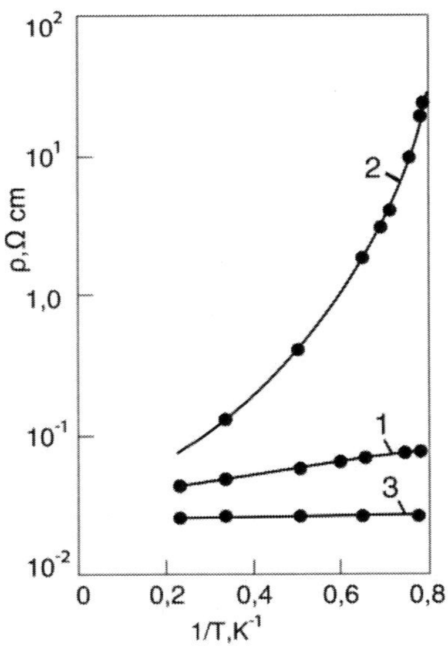

Figure 6. MIT and IMT in uniaxially stressed p-InSb(Mn) crystal at N_{Mn}=1,7×10^{17}cm^{-3}; 1-X=0, B=0; 2-X=2,1kbar, B=0; 3-X=2,1kbar at B=8T; data adapted from [14].

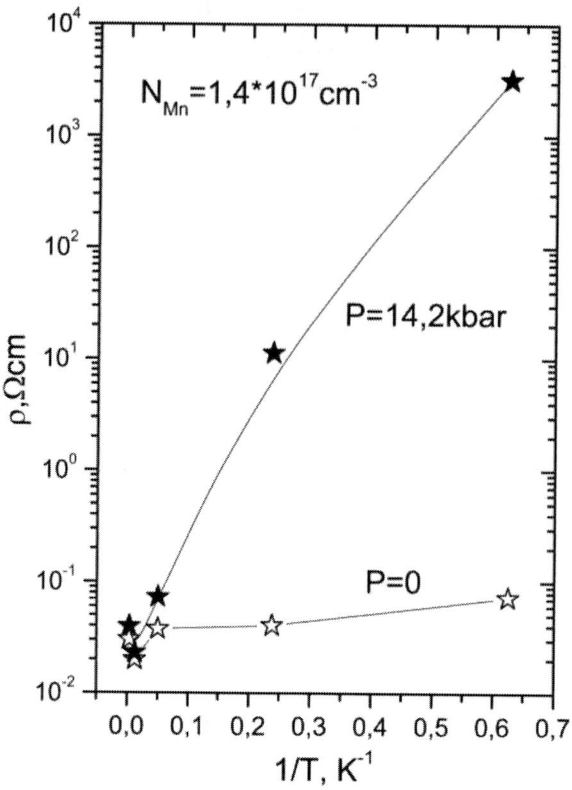

Figure 7. Pressure driven MIT in p-InSb(Mn), data adapted from [15].

Under uniaxial stress p-InSb(Mn) crystals demonstrated not only transition from quasi-metal conductivity to semiconductor state but also reverse transition in magnetic field, i.e. transition from insulator to quasi-metal state [14] as it is shown in Fig.6.

Hydrostatic pressure can also induce MIT in doped semiconductors. For example in p-InSb(Mn) crystals [15] specific resistivity increased by a factor of 10^4 under pressure of P=14,2kbar, at T=1,6K which led to the transition from quasi-metal to semiconductor state as it is shown in Fig.7.

1.2. Theoretical Description of MIT

Description of MIT in doped semiconductors crystals with randomly distributed impurities was done by Mott [16]. One of the major advances in the Mott's theory of disordered systems is the Mott's law which describes the relation between the critical concentration of metal-insulator transition N_{cr} and the radius of Bohr orbit a_B for charge carriers (electrons or holes). This law is presented as equation

$$(a_B)^3 N_{cr}=0,25 \quad (1)$$

This equation has been experimentally proved for many doped semiconductors – Ge, Si, A^3B^5 compounds and etc. For example, from Eq.1 for p-Ge(Ga) and n-Si(P), N_{cr} was determined as $N_{cr}=2\times10^{18}cm^{-3}$ and $N_{cr}=2\times10^{17}cm^{-3}$ respectively (Fig.2). Strong external magnetic field can drive semiconductors to the insulator state which can be explained as the compression of isolated impurity atom Bohr orbits. This effect named magnetic freezeout, led to Giant Positive Magnetoresistance effect (see Fig. 3).

Another important result of the theory of disordered systems was the development of Hopping Conductivity Model which successfully described the low temperature conductivity in DS as well as in other disordered systems. It was shown [17] that T-dependence of specific resisitivity at low temperatures follows the Variable Range Hopping (VRH) equation

$$\rho=\rho_0 \exp (T_0/T)^{0,25} \quad (2)$$

where T_0 and ρ_0 are temperature independent parameters.

Further development of VRH model was presented in the works of Efros and Shklovskii [18] who suggested that coulomb gap formed in impurity band is responsible for semiconductors ρ-T dependence, for example in p-Ge [19], resistivity behaves as

$$\rho=\rho_0 \exp (T_0/T)^{0,5} \quad (3)$$

However, studies of resistivity-temperature dependence in p-InSb(Mn) crystals at manganese concentration $N_{Mn}\leq N_{cr}=2\times10^{17}cm^{-3}$ revealed the conductivity mechanism different from VRH conductivity observed in semiconductors doped with non-magnetic impurities [11,20,21]. Instead of VRH conductivity, described by Eq.3, ρ-T dependence in p-InSb(Mn) between 10K and 1,5K follows the stretched exponential function

$$\rho = \rho_{01} \exp(\Delta_1/kT)^2 \quad (4)$$

where Δ_1 is the activation energy and k the Boltzmann's constant. At temperature below ~1,5K resistivity dependence on temperature follows

$$\rho=\rho_0 \exp(\Delta/kT) \quad (5)$$

where $\Delta\approx3\Delta_1$ [ref. 21].

Magnetic field in p-InSb(Mn) crystals induced Giant Negative Magnetoresistance effect (Fig.4) and reduced both Δ_1 and Δ and thus caused IMT. It was supposed in [20,21] that transport and magnetotransport anomalies observed in p-InSb(Mn) could be explained in the framework of excitonic insulator phase model [22]. Excitonic insulator model describes MIT in narrow gap semiconductors as the result of electron-hole insulating pares formation, i.e. generation of permanent excitons.

Excitons are often related to bosons, subatomic particles having integer spin and possibly forming super fluid liquid, and therefore they can be responsible for the superconductivity [23]. Search for the experimental evidence of the excitonic insulator phase in semiconductors is one of the challenges in the present-day physics of semiconductors. Excitonic insulator realization would open new opportunities for the manipulation with macroscopic quantum coherence and for the understanding of superconductivity. Sharp transition from metal to semiconductor state under pressure-driven energy gap can be the proof of the permanent excitonic-insulator phase existence in semiconductors.

An unusual peak structure in the resistivity-pressure dependence observed in narrow-gap semiconductor $TmSe_{1-x}Te_x$(~60meV) and measurements of the Hall constant showed that resistivity rise under pressure is caused by the condensation of free carriers which strongly supports the concept of excitonic insulator [24].

In the case of p-InSb(Mn) crystals the MIT driven by high pressure [15] could be the evidence of excitonic insulator phase. The important fact is that the ratio of heavy hole and electron masses in InSb is about m_{hh}/m_e=0.5/0.013~ 40, i.e. the condition close to excitonium formation [23].

To summarize, p-InSb crystals provide unique possibility to study practically all types of MIT and IMT in a single semiconductor crystal at impurity concentration around $N_{cr}=2\times10^{17}cm^{-3}$, i.e. magnetic field driven IMT, MIT caused by impurity concentration variations, MIT in uniaxially stressed and hydrostatic pressured crystals. Thus, the CHAPTER aims to discusses both MIT and IMT in InSb crystals doped with manganese, germanium and co-doped with Mn and Te crystals.

2. Scope of the Material

2.1. The History of InSb Discovering and Crystal Grows Methods

Synthesized in 1952, InSb was the first A^3B^5 semiconductors material [25]. Detailed studies of p-InSb transport and magnetotransport properties were carried out in the middle of 50-s of the last century [26]. Due to its small direct band gap E_g=0,17meV, low effective electron mass m_e= 0,013m_0, and high room-temperature electron mobility InSb serves as a material for the variety of device applications, including infrared sources, detectors, and magnetoresistive sensors.

Manganese and Germanium, as impurities in indium antimonide crystals, induce shallow acceptor levels at energy E_a~9meV [ref. 27]. It was shown that InSb crystals can be doped with Mn in a wide range of concentrations $10^{14}<N_{imp}<10^{19}cm^{-3}$ [ref.28]. The critical concentration of MIT both in p-InSb(Mn) and p-InSb(Ge) crystals is $N_{cr}=2\times10^{17}cm^{-3}$. It should be mentioned that in Germanium single crystals impurity of Ga introduces a shallow acceptor level with energy E_{Ga}~7meV and N_{cr} in p-Ge(Ga), p-InSb(Ge) and p-InSb(Mn) with the same $N_{cr}=2\times10^{17}cm^{-3}$. Telluride as impurity forms a shallow donor separated from the conduction band at E_d~1meV [ref.27]. All above mentioned acceptors and donors, i.e. Mn, Ge and Te were used in our comparative study.

The first InSb crystals were obtained by zone purification method at N_d-N_a~ $10^{13}cm^{-3}$ [ref.29]. Since 1950th up to nowadays Chokhralsky method remains the conventional method of InSb single crystals growth for the research and industry. Molecular Beam Epitaxy (MBE) methods also gave good results for InSb crystals growing [28]. Metal-organic vapor-phase epitaxy (MOVPE) technology is usually used for the fabrication of $In_{1-x}Mn_x$ Sb Diluted Magnetic Semiconductor [30].

Co-doping of InSb crystals with Mn and Te gives possibility to grow appropriate crystals for MIT studies as this set of impurities both maintains concentration of holes around p=N_{cr} and significantly increases Manganese concentration. For example, if during InSb crystal growth we add Manganese at the concentration N_{Mn} ~$7\times10^{17}cm^{-3}$ and Telluride at N_{Te}

$\sim 5\times10^{17}cm^{-3}$ we get crystals with the hole concentration around N_{cr} where $p=7\times10^{17}\div5\times10^{17}=2\times10^{17}cm^{-3}\sim N_{cr}$. So by the variation of compensation factor $K=N_{Te}/N_{Mn}$ we can study the influence of Manganese concentration on MIT in p-InSb(Mn,Te) crystals when N_{Mn} significantly exceeds N_{cr}. Co-doping of InSb with Mn and Te can be especially helpful when we compare the properties of InSb crystals doped with Mn and Diluted Magnetic Semiconductor $In_{1-x}Mn_xSb$.

2.2. InSb Band Structure Parameters

Band structure of InSb was calculated by Kane [31]. The results of these calculations are shown in Fig.8. The effective mass of electrons is $m_e=0{,}013m_0$, effective mass of light holes is $m_{lh}=0{,}015m_e$ and heavy holes mass is $m_{hh}=0{,}5m_e$ [ref. 26, 27]. Uniaxial compression splits valence bands as it is shown in Fig.8b. for stress along axis [100]. Band splitting at **k =0** is denoted for [100] compressional stress $E_s=2b(S_{11}-S_{12})X$ and for [111] $E_s=(S_{44}d/\sqrt{3})X$ where S_{11},S_{12}, and S_{44} are elastic constants and b=-2 and d=-4,9 are deformation potentials [32].

Under hydrostatic pressure band gap in InSb increases according to expression $E_g=E_{g0}+13{,}7\times10^{-3}P-3{,}6\times10^{-5}P^2$, where P is pressure in kbar.

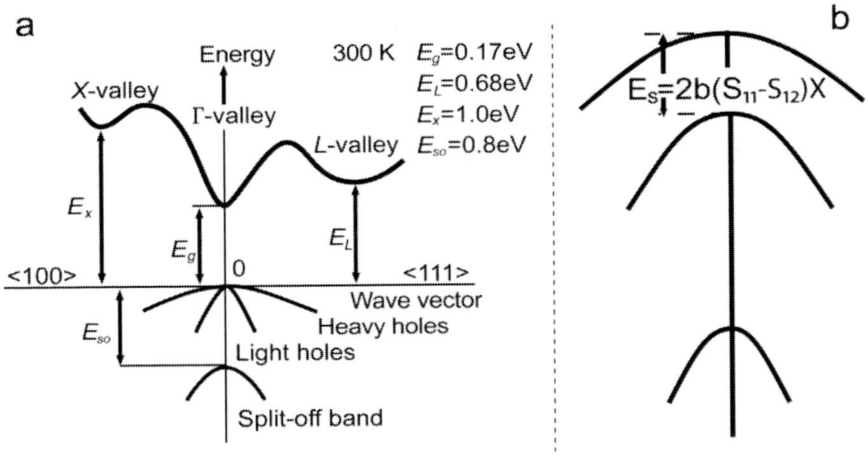

Figure 8. InSb band structure in the absence of stress (a) and valence band in uniaxially compressed crystal (b).

2.3. InSb Transport Characteristics

Crystals of p-InSb were grown by Chokhralsky method. The diameter of single crystals was about 3-4cm, with the length of about 5-10cm. Doping level varied from 10^{16} to $3\times10^{17}cm^{-3}$ both in manganese and germanium doped crystals. After being etched with the reagent CP-4A, the contacts were soldered with indium. Hall mobility measured at T=77K and B=10T varied in single crystals from 3000 to $500cm^2/Vsec$, when holes concentration increased from 10^{16} to $3\times10^{17}cm^{-3}$. Hall mobility in compensated p-InSb(Mn,Te) crystals at $p\sim2\times10^{17}cm^{-3}$ did not exceed $100\div200cm^2/Vsec$.

3. Metal-Insulator Transition in p-InSb(Ge), p-InSb(Mn) and p-InSb(Mn,Te) Single Crystals Induced by Impurity Concentration Variation

3.1. Magnetotransport Effects in p-InSb(Ge), p-InSb(Mn) and p-InSb(Mn,Te) Crystals Around N_{cr} at the Temperature Range T=20÷280K

Specific resistivity and Hall constant in p-InSb(Mn) and p-InSb(Ge) crystals do not significantly differ within the temperature range 280K - 20K as it can be seen from the comparison of the results in Fig.9 and Fig.10. Both p-InSb(Mn) and p-InSb(Ge) demonstrate the Hall constant sign inversion from negative to positive value at ambient temperature in magnetic field B~1T. In high magnetic filed at B> 3T the Hall constant is positive at 280K. Between 80K and 20K Hall constant is positive along the range of magnetic field B=0÷10T.

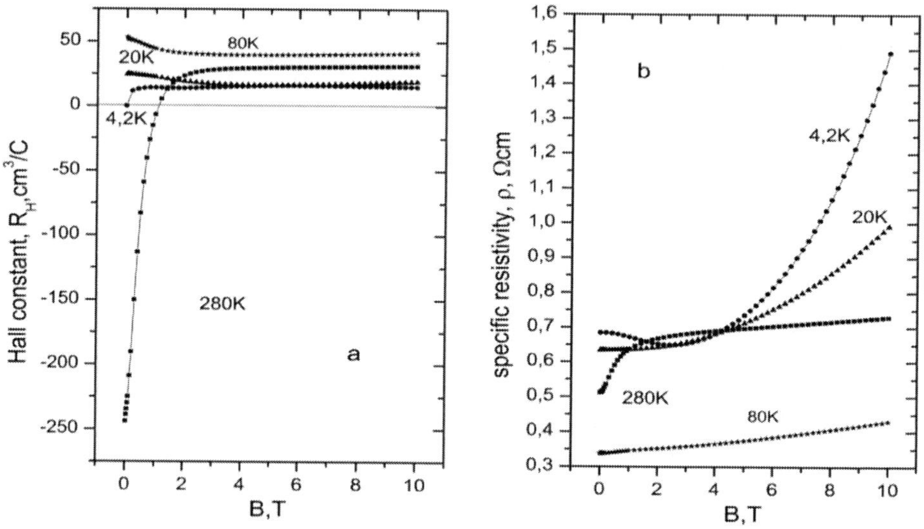

Figure 9. Hall constant (a) and specific resistivity (b) vs magnetic field in p-InSb(Mn) at various temperatures, $N_{Mn}=1,4\times10^{17}cm^{-3}$.

Magnetoresistance is positive between 280 and 20K at B=0÷10T both in p-InSb(Ge) and p-InSb(Mn) crystals as it is shown in Fig. 9b and 10b with the break of ρ-B dependence at B~1T at ambient temperature. Compensated p-InSb(Mn,Te) crystals demonstrate similar to p-InSb(Ge) and p-InSb(Mn) R_H-B and ρ-B dependencies (Fig.11). The Hall constant in p-InSb(Mn,Te) sample demonstrates sign inversion at B~1T and ambient temperature as it is displayed in Fig.11a and magnetoresistance shows a break in the same magnetic field at B~1T (Fig. 11b).

Temperature and magnetic field induced behavior of the Hall constant can be explained if we take into account that InSb is an intrinsic semiconductor at ambient temperatures [26]. High mobility electrons dominate at high temperatures and holes determine conductivity at low temperatures. The change of the slope on ρ vs B curve (Fig. 9b÷11b) and the change of the Hall constant sign at B~1T (Fig. 9÷11a) in these two charge carriers models can be understood as a transition in magnetic field from conductivity determined by the electrons with small effective mass (m_e=0,013m_0) to the conductivity determined by the holes with m_{hh}=0,5m_0. At T=80K crystals conductivity is determined by the holes and therefore the Hall constant and magnetoresistance dependence on magnetic field differ from magnetotransport effects at 280K.

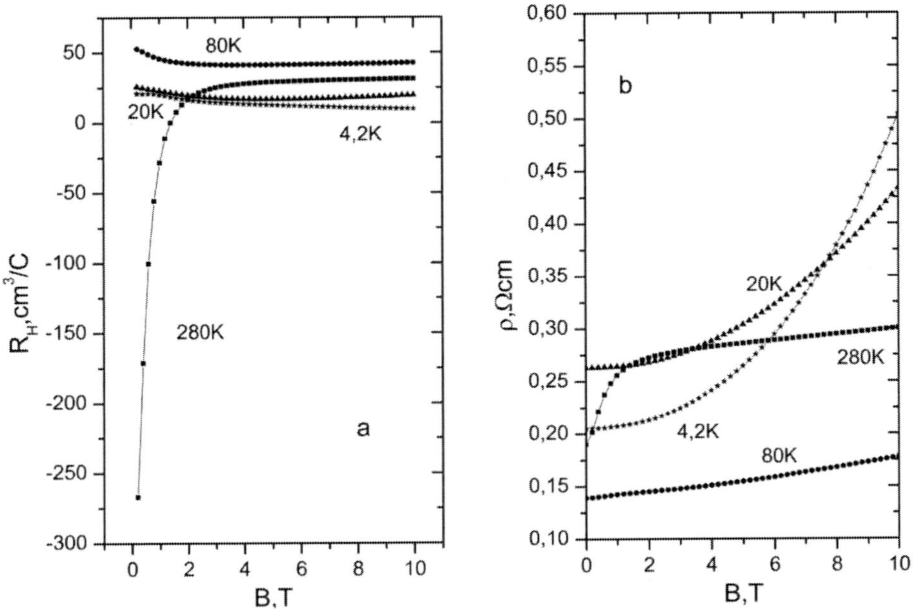

Figure10. Hall constant (a) and specific resistivity (b) vs magnetic field in p-InSb(Ge), N_{Ge}=1,9×10^{17}cm^{-3}.

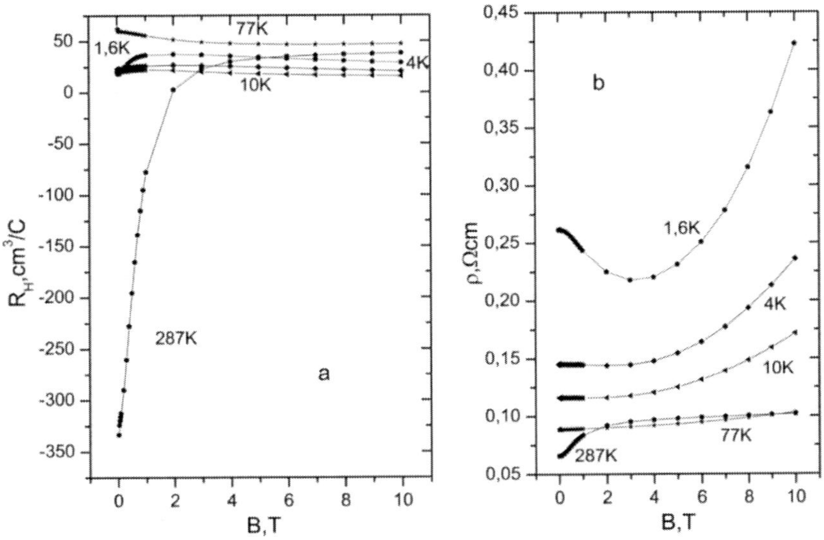

Figure 11. Hall constant (a) and specific resistivity (b) vs magnetic field in the compensated p-InSb(Mn, Te) crystal, p=N_{Mn}-N_{Te}=1,32×10^{17}cm^{-3}, K=0,5.

3.2. Low Temperature (T=4,2K÷0,3K) Conductivity and Magnetoresistance in p-InSb(Ge), p-InSb(Mn) and p-InSb(Mn,Te) Crystals

The dramatic difference in ρ-and R_H functions of temperature and magnetic field was revealed at temperature below 4,2K, when we compared p-InSb(Ge), p-InSb(Mn) and p-InSb(Mn,Te) crystals. Specific resistivity dependence on impurity concentration in p-InSb(Ge) crystals was similar to ρ-N_{imp} dependence observed previously in other semiconductors doped with nonmagnetic impurities, e.g. p-Ge(Ga) and p-Si(B) crystals (Fig.2). Resistivity in p-InSb(Ge) crystals gradually went up as Ge impurity concentration decreased at T=1,1K (Fig.12).

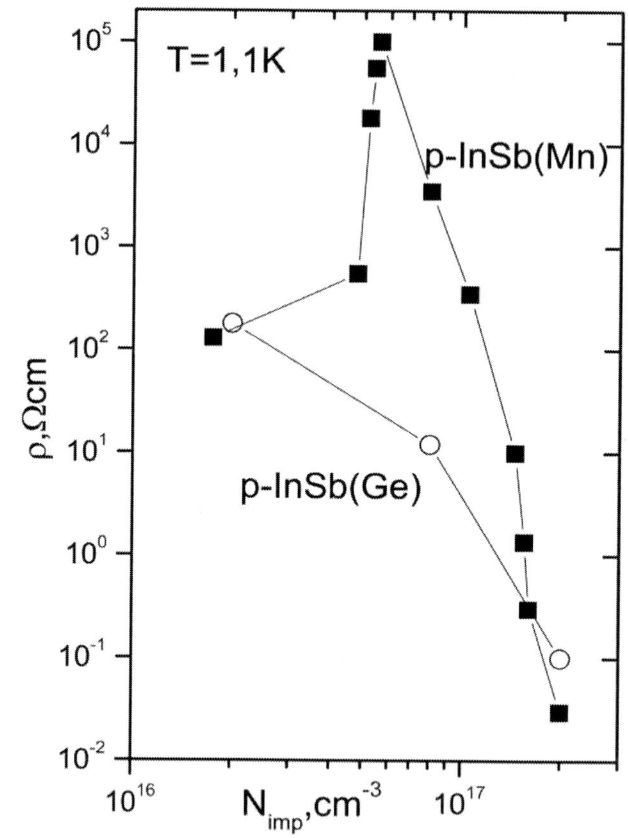

Figure12. MIT in p-InSb(Ge) and p-InSb(Mn) at 1,1K induced by impurity concentration variation, data adapted from [20].

However p-InSb(Mn) crystals around N_{cr} demonstrated unusual for doped semiconductors resistivity dependence on impurity concentration. For impurity concentrations below N_{cr} =2×10^{17}cm^{-3} specific resistivity sharply increased, reached maximum at N_{Mn}~6×10^{16}cm^{-3} and for N_{Mn}<6×10^{16}cm^{-3} dramatically fell from ρ=10^5 to 10^2 Ωcm at N_{Mn}=10^{16}cm^{-3}.

At N_{imp}~10^{16}cm^{-3} specific resistivity in p-InSb(Mn) was close to the resistivity observed in p-InSb(Ge) as it is depicted in Fig.12. So the principal difference in MIT for p-InSb(Ge) and p-InSb(Mn) crystals at low temperatures was revealed within the impurity concentration range of $10^{16} \div 2\times10^{17}$cm^{-3}.

Resistivity-temperature dependence within the observed impurity concentration range in p-InSb(Ge) and p-InSb(Mn) crystals also significantly differed [21]. Crystals of p-InSb(Ge) showed typical for doped semiconductors hopping conductivity type described by, $\rho=\rho_0\exp(T_0/T)^{0,25}$ (Eq. 2, paragraph **1.2**) But we can identify two ranges of temperatures where ρ-T dependence in p-InSb (Mn) crystals showed unusual behavior.

The first range is between temperatures 10÷1,5K (Fig.13a). Here experimental data are in good agreement with given above exponential quadratic function $\rho=\rho_{01}\exp(\Delta_1/kT)^2$ (Eq. 4, paragraph **1.2.**) where Δ_1 goes up from $\Delta_1\approx0$ to $\Delta_1=0{,}3$meV with the decrease in manganese concentration from 2×10^{17} to $1{,}08\times10^{17}$cm^{-3} (Fig.14a). Approximation of Eq. 4 to 1/T =0 gives for resistivity ρ_{01} variation from ~0,04Ω to 0,1Ωcm in the same range of manganese concentration.

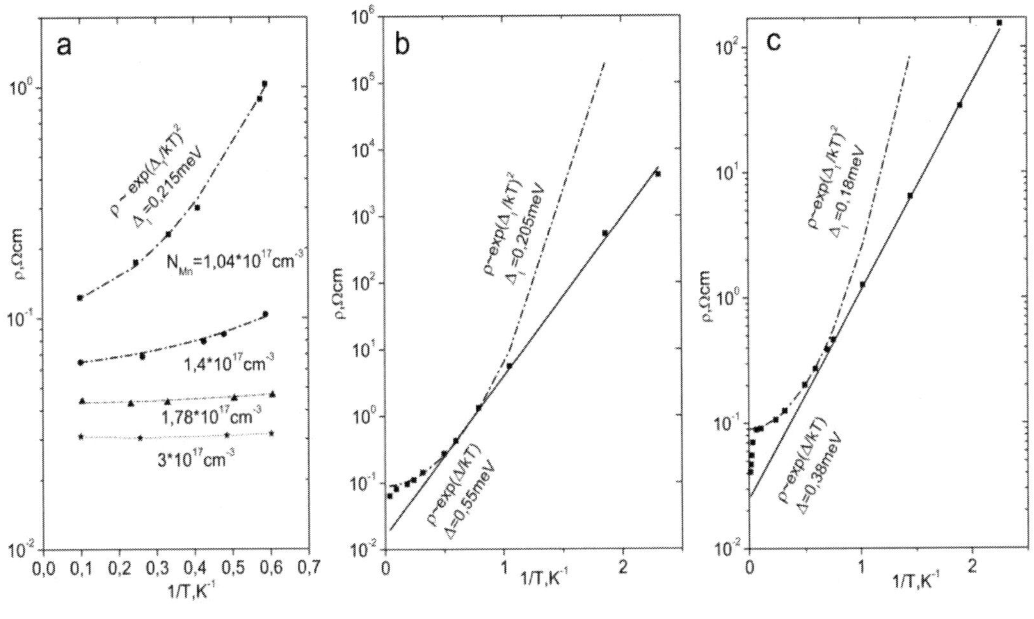

Figure 13. The temperature dependence of the specific resistivity ρ at low temperatures in p-InSb(Mn) crystals around metal-insulator transition ; a- data adapted from [21], b-$N_{Mn}=1.08\times10^{17}$cm^{-3}, c-$N_{Mn}=1{,}2\times10^{17}$cm^{-3}; dash line is a fit to Eq.4 [$\rho=\rho_{01}\exp(\Delta_1/kT)^2$], solid lines is a fit to Eq. 5 [$\rho=\rho_0\exp(\Delta/kT)$]

At temperatures below 1,5K (Fig.13b and Fig.13c) the ρ-T dependence showed transition to another type of conductivity which could be described with the exponential expression $\rho=\rho_0\exp(\Delta_0/kT)$ (Eq.4, from **1.2**) where Δ_0 decreased from 0,7meV to zero with the increase of manganese concentration from $1{,}08\times10^{17}$ to 2×10^{17}cm^{-3} as it is shown in Fig. 14a. Thus, in p-InSb(Mn) crystals around N_{cr} at temperature below 10K we did not reveal Variable Range Hopping conductivity described by Eq. 2, $\rho=\rho_0\exp(T_0/T)^{0,25}$ (paragraph 1.2).

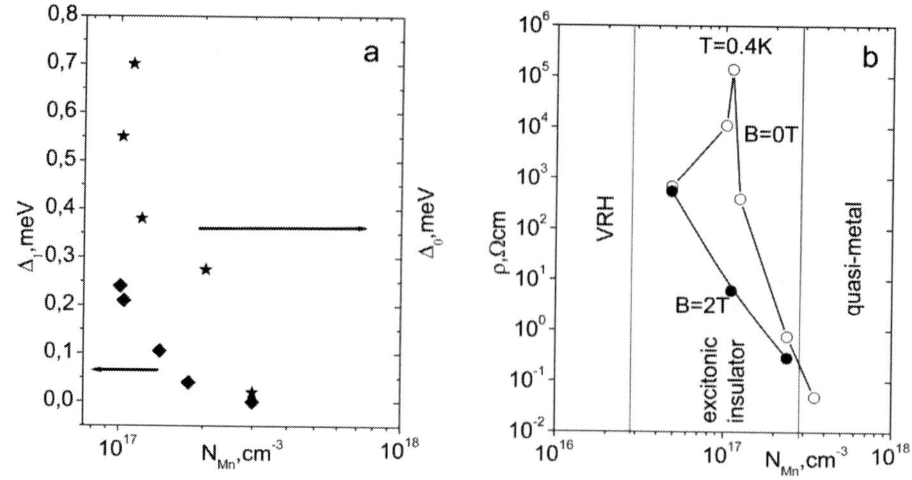

Figure 14. Metal-insulator transition in p-InSb(Mn) induced by manganese concentration variation, data adapted from [21]. Activation conduction energy Δ_1 (**solid rhombus) and** Δ_0 (solid stars) as a function of manganese concentration (a). An anomaly at specific resistivity dependence on manganese concentration around N_{cr} **at T=0,4K (b). Open circles are the ρ at B=0 and solid circles are the ρ at** B=2T.

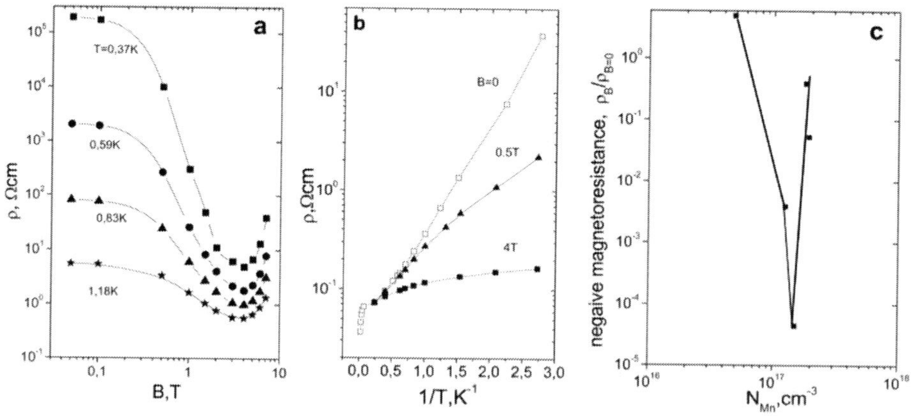

Figure 15. Colossal magnetoresistance in p-InSb(Mn), data adapted from [21]; a- **specific resistivity ρ** *vs* magnetic field B in the temperature range 1,18÷0,37K, N_{Mn}=1,05×10^{17}cm^{-3}; b-specific resistivity as a function of temperature in a magnetic field, N_{Mn}=1,5 ×10^{17}cm^{-3}; c-Negative Magnetoresistance $\rho_{B=2T}/\rho_{B=0}$ as a function of manganese concentration at T=0,4K, B=2T.

In an external magnetic field the specific resistance of p-InSb(Mn) crystals around N_{cr} at T<10K demonstrated different types of behavior depending on the magnetic field. At low magnetic field the resistance remained independent of the magnetic field value within the range B=0÷0,1T (Fig. 15a). At B=0,1÷2T the resistance dramatically decreased (by a factor of 10^4 at T~0,4K and N_{Mn}=1,05×10^{17}cm^{-3}). Within the increasing magnetic field the specific resistivity declined, reached its minimum at B~2T, and then increased in magnetic field B>4T as it is shown in Fig. 15a. A strong negative magnetoresistance at B~2T induced transition to VRH conductivity as it is seen in the sample with N_{Mn}=1,5×10^{17}cm^{-3} (Fig. 15b).

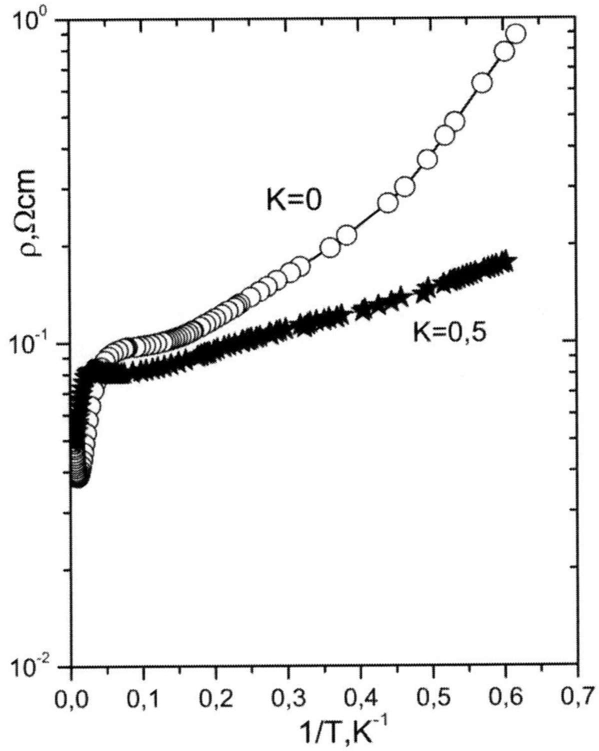

Figure 16. ρ-T dependences in uncompensated p-InSb(Mn) and compensated p-InSb(Mn,Te) crystals; $N_{Mn}=1,27\times10^{17}cm^{-3}$ at K=0, $p=1,5\times10^{17}cm^{-3}$ at K=0,5.

The ratio $\rho_B/\rho_{B=0}$, which determines the effect of negative magnetoresistance, decreased with the decrease of Mn concentration from $2\times10^{17}cm^{-3}$ and reached its minimal value ($\rho_B/\rho_{B=0}\sim10^{-4}$) at $N_{Mn}\sim10^{17}cm^{-3}$ (Fig. 15c). At the lower levels of manganese concentration ($N_{Mn}<6\times10^{16}cm^{-3}$) the magnetoresistance sign demonstrated change from negative to positive ($\rho_B/\rho_{B=0}>1$) when the VRH conductivity appeared (Fig.14b and Fig. 15b).

Thus, Fig. 12 and Fig. 14b, being compared, demonstrate that an external magnetic field can eliminate the difference between p-InSb(Mn) and p-InSb(Ge) specific resistivity dependence on N_{imp}. Although InSb(Mn) crystals belong to the class of nonmagnetic materials, still, the level of negative magnetoresistance revealed here was comparable to the values of CMR observed previously in magnetic materials like $Pr_{0,7}Ca_{0,26}Sr_{0,04}MnO_3$ [Ref. 33] and p-$Hg_{1-x}Mn_xTe$ [Ref. 9], with their Mn concentration 10^5 as large as Mn concentration in p-InSb(Mn) crystals.

In p-InSb(Mn,Te) crystals with compensation ratio K=0,5 around $p_{cr}=N_{Mn}-N_{Te}\approx2\times10^{17}cm^{-3}$ we also have not observed VRH conductivity. Instead, we revealed exponential increase of specific resistivity at temperatures below ~10K, as it is demonstrated for p-InSb(Mn,Te) in Fig. 16. Similar to p-InSb(Mn) uncompensated crystals, compensated p-InSb(Mn,Te) crystals demonstrated the effect of negative magnetoresistance at T<10K, as it is depicted above in Fig. 11.

3.3. Low Temperature (T=4,2K÷0,3K) Hall Effect in p-InSb(Mn) Crystals

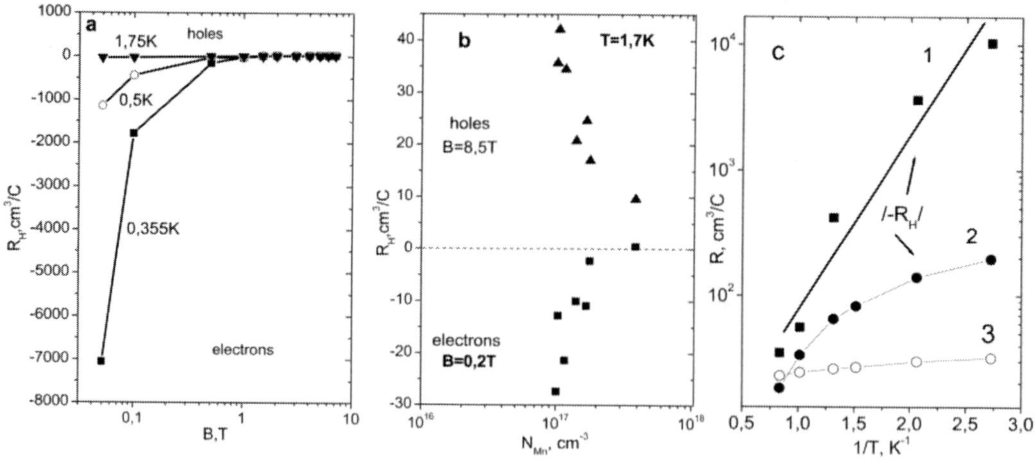

Figure 17. a- Hall constant in p-InSb(Mn) at low temperatures. Hall constant sign inversion in a magnetic field at $N_{Mn}=1,5\times10^{17}cm^{-3}$; b-the Hall constant *vs* manganese concentration in high (B=8,5T) and low (B=0,2T) magnetic fields at T=1,7K ; c-exponential increase of /-R_H/ with the decrease of temperature (1-B=0,05T, 2-B=0,5T, 3-B=7T; Δ=0,27meV)

In p-InSb(Mn) crystals at $10^{17}cm^{-3}\leq N_{Mn}\leq 2\times10^{17}cm^{-3}$ and at temperature T~4,2K the Hall constant was positive but showed a tendency to sign inversion as it is depicted in Fig. 9a at $N_{Mn}=1,4\times10^{17}cm^{-3}$. As temperature was reduced below 4,2K the Hall coefficient at B<1T decreased and changed its sign to negative as it is demonstrated in Fig. 17a. At temperature T=0,35K the value of the Hall constant reached -7000cm³/C which corresponds to the electron concentration of about $n\approx10^{15}cm^{-3}$. In high magnetic field the Hall coefficient remained positive and increased in the range $N_{Mn}=(1\text{-}2)\times10^{17}cm^{-3}$ as it is displayed in Fig. 17b, while the temperature range of 4,2K÷0,3K did not demonstrate any significant Hall coefficient changes (Fig. 17c).

From the Hall effect measurements we can conclude that the low temperature conductivity in p-InSb(Mn) crystals with $N_{Mn}=1\div2\times10^{17}cm^{-3}$ at T≤4K is determined by electrons which are responsible for the R_H negative sign. On cooling at T < 1 K the value of /-R_H/ increased exponentially (Fig. 17c). Thus, the R_H-T dependence can be expressed in the millikelvin range of temperature by:

$$R_H = -R_{H0}\exp(\Delta/kT) \qquad (6)$$

where the value of parameter Δ is dependent on manganese concentration and R_{H0} is the temperature independent parameter. The accuracy of the Hall effect measurements did not exceed ~20% that gave no possibility to find essential difference between Δ values obtained from R_H and those obtained from ρ–T measurements. Hall mobility, Rσ, remained practically independent of temperature in the range 4.2–0.3 K. The high magnetic field Hall constant at B~10T is of positive sign indicating that the low temperature conductivity in the strong magnetic field is determined by holes (Fig. 17a-c). It should also be mentioned that a similar Hall constant sign inversion in magnetic field was observed earlier in a Diluted Magnetic Semiconductor p-$Hg_{1-x}Mn_xTe$ [Ref.34].

3.4. Magnetotransport Effects in p-InSb(Mn) and p-InSb(Mn, Te) Crystals in Super-Low Temperature Range (T=300mK÷10mK)

The results of temperature dependence measurements for specific resistivity in p-InSb(Mn) in the temperature range 300mK÷10mK for three p-InSb(Mn) samples is shown in Fig. 18. The exponential increase of the resistivity in the millikelvin range of temperature is described by Eq.5, $\rho=\rho_0\exp(\Delta/kT)$, where Δ varies in manganese concentration range $N_{Mn}=2{,}03\times10^{17}cm^{-3}\div2{,}14\times10^{17}cm^{-3}$ from Δ= 0,1meV to Δ=0,07 meV. Specific resistance in magnetic field at B~3T demonstrated Colossal Magnetoresistance effect when resistivity decreased by a factor of >10^3 at T=80mK as it is shown in Fig. 19a and induced IMT (Fig. 19b).

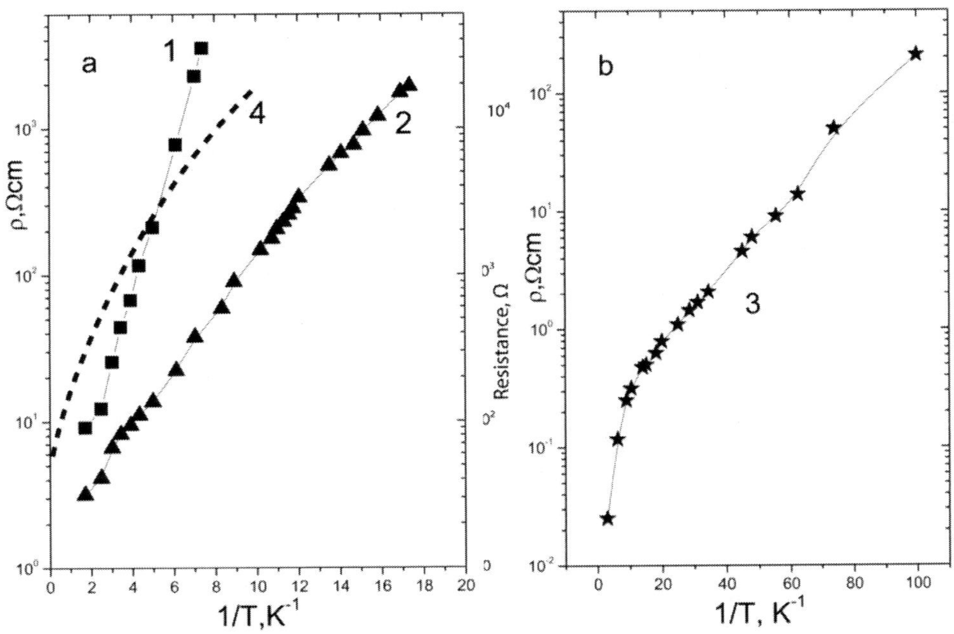

Figure 18. Specific resistivity in p-InSb(Mn) at millikelvin range of temperature.

1 – $N_{Mn}=2{,}03\times10^{17}cm^{-3}$, Δ=0,1meV;

2 – $N_{Mn}=2{,}1\times10^{17}cm^{-3}$, Δ=0,035meV;

3 – $N_{Mn}=2{,}14\times10^{17}cm^{-3}$, Δ=0,007meV;

4 – Lake Shore thermoresistor.

Compensated p-InSb(Mn, Te) crystals at K=0,5 demonstrated jumps both of the specific resistivity and activation conduction energy at low temperatures (Fig. 20). The steps (two-fold jump of resistivity) at T=1K and 0,5K could be explained as a phase transition in the e-h system.

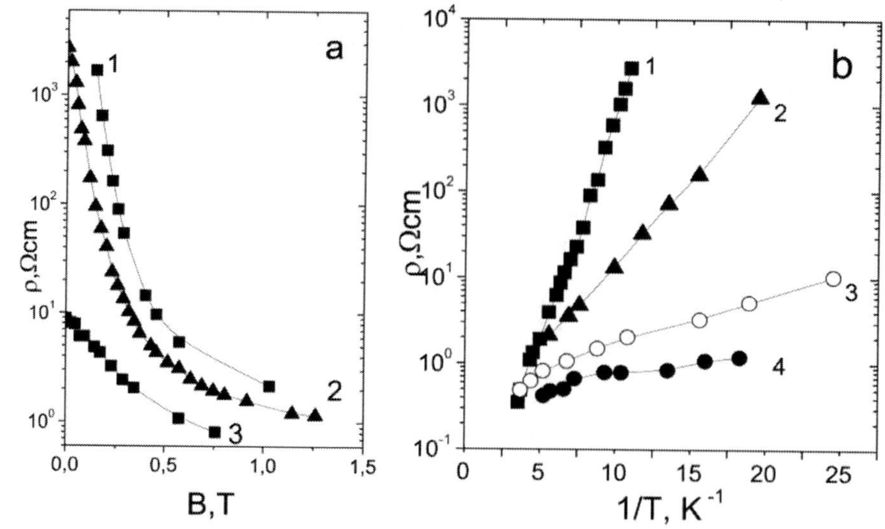

Figure 19. MIT in p-InSb(Mn) at N_{Mn}= 1,13×$10^{17}$$cm^{-3}$ in millikelvin temperature range adapted from [11]; a- CMR, curve 1-T=80mK, curve 2-129mK, curve 3-257mK ; b- IMT in p-InSb(Mn) induced by magnetic field (curve 1-B=0T; curve 2- B=0,27T; curve 3-0,74T; curve 4-2,8T).

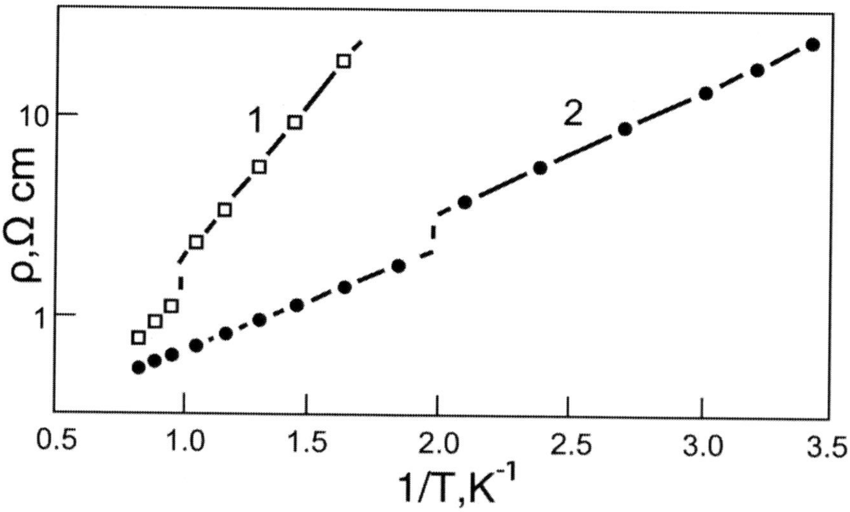

Figure20. Resistivity jumps in p-InSb(Mn,Te) crystals at low temperatures.

3.5. Discussion

There should be made a few important remarks arising from our observation of transport and magnetotransport effects in p-InSb crystals doped with Ge, Mn and compensated p-InSb(Mn,Te) crystals at MIT critical concentration. Firstly, unusual transport and magneto-transport properties of p-InSb(Mn) crystals, i.e. the exponential quadratic ρ-T dependence (Eq. 4, paragraph **1.2**), the maximum in ρ-N_{Mn} dependence graph (Fig. 14b), CMR (Fig. 15b) and the Hall constant sign inversion in magnetic field (Fig. 17a) were observed at about N_{cr} on the insulator side of transition. Secondly, the effects involved charge carriers interplay, i.e. p-InSb(Mn) shows mixed conduction behavior (Fig.17). These new findings suggest that the excitonic insulator phase model [22] could be the most appropriate concept to explain p-InSb(Mn) low temperature conductivity and magnetotransport properties. We propose the scenario when in p-InSb(Mn) crystals on the insulator side of MIT at $N_{Mn}<2\times10^{17}cm^{-3}$ intrinsic electrons and impurity holes can form stable, "permanent" excitons at temperature below 10K. At temperature below 1,6K we observed the Bose condensation of excitons, i.e. the transition to the excitonic insulator phase which causes an increase in specific resistivity around N_{cr} (Fig.14b). In the framework of excitonic insulator model the appearance of conduction activation energy Δ_1 (Eq. 4) can be associated with exciton binding energy and Δ_0 (Eq. 5) is a gap in the energy spectrum of excitonic insulator. In the framework of excitonic insulator model resistance decrease in the magnetic field, i.e. CMR, can be described as a drop of e-h insulating pairs concentration in an external magnetic field caused by magnetic field influence on spin holes and electrons that form excitons.

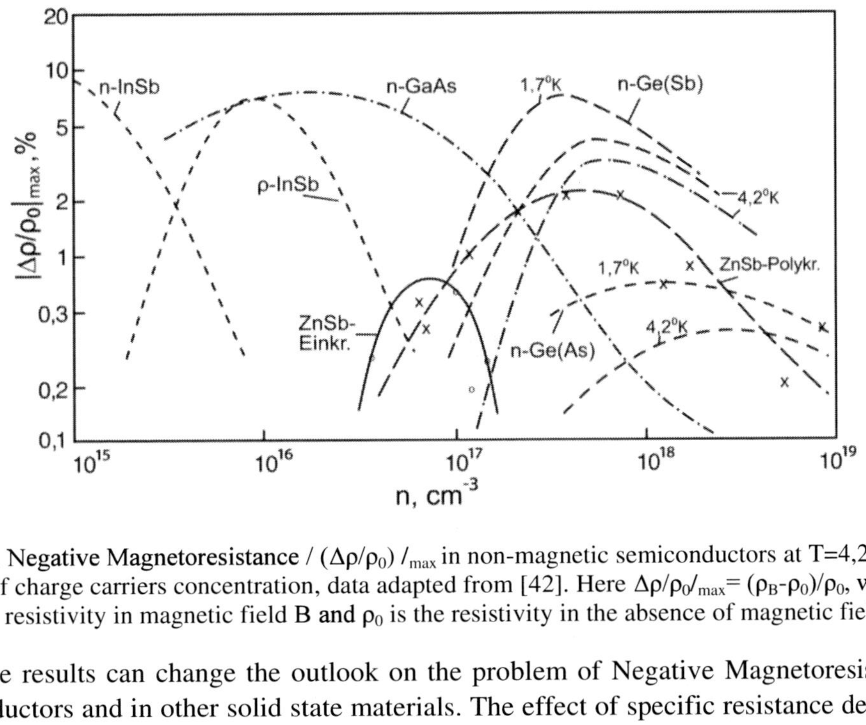

Figure 21. Negative Magnetoresistance / $(\Delta\rho/\rho_0)$ $/_{max}$ in non-magnetic semiconductors at T=4,2K as a function of charge carriers concentration, data adapted from [42]. Here $\Delta\rho/\rho_0/_{max}=(\rho_B-\rho_0)/\rho_0$, where ρ_B is specific resistivity in magnetic field B and ρ_0 is the resistivity in the absence of magnetic field.

These results can change the outlook on the problem of Negative Magnetoresistance in semiconductors and in other solid state materials. The effect of specific resistance decrease in magnetic field, known as Negative Magnetoresistance (NM), was first observed in iron by Lord Kelvin in 1857 [ref. 35]. Since then, NM has been observed and studied for over 150 years in most types of solid state materials - semiconductors, superconductors, perovskites, graphite and etc. But despite the considerable attention to the phenomenon of NM and its potential application in magnetic devices [36] the physical nature of NM effect still remains the matter of discussion.

If we overview numerous NM investigations we can see that different types of solid state materials demonstrate drastic variety in NM values. On the one hand, the maximal resistance decrease of $(\rho_B/\rho_{B=0})_{min}\sim0.5$ in magnetic field $B\sim1\div10T$ was revealed in ferromagnetic metals

[37], high critical temperature superconductor (HTS) cuprates [38], graphite [39] and nonmagnetic semiconductors - InSb [40], n-Ge [41], ZnSb [42], GaAs [43]. Figure 21 represents Negative Magnetoresistance in a few non-magnetic semiconductors. It follows from this figure that the maximum value of NM was revealed around critical concentration of metal-insulator transition in semiconductors in the range of charge carrier concentration between 10^{16} and $10^{18}cm^{-3}$. On the other hand Magnetic Semiconductors, e.g. EuSe [44], $Gd_{1-x}V_xS_4$ [10], EuTe [45] and (DMS)- p-$Hg_{1-x}Mn_xTe$ [9] showed Giant Negative Magnetoresistance (GMR) with $\rho_B/\rho_{B=0}$~10^{-5}÷10^{-7}. Later in perovskite manganites the effect of Colossal Negative Magnetoresistance (CMR) was observed with $\rho_B/\rho_{B=0}$~10^{-15} for $Pr_{0,7}Ca_{0,26}Sr_{0,04}MnO_3$ [ref. 33]. Fig. 22 displays GNM in MS and DMS.

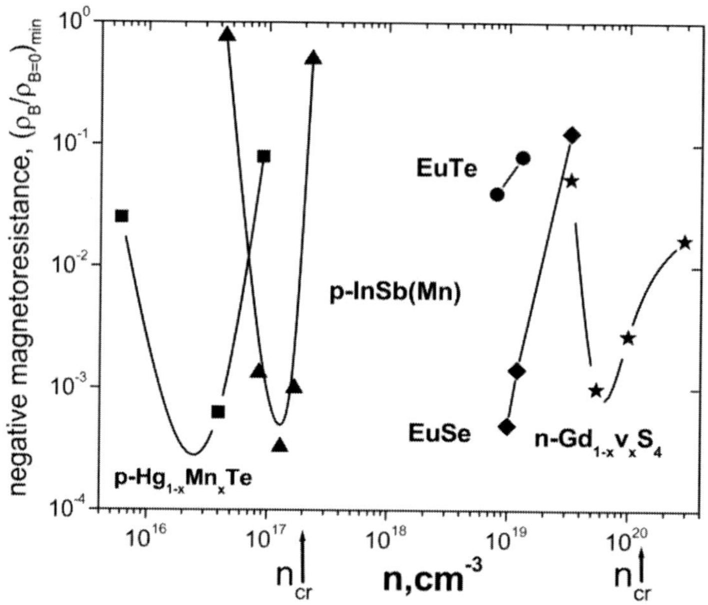

Figure 22. Giant Negative Magnetoresistance in p-InSb(Mn), DMS and MS; p-$Hg_{1-x}Mn_xTe$ at B=3T, T=4,2K [ref. 9]; n-EuTe at 6T, T=4,2K [ref. 45]; n-$Gd_{1-x}v_xS_4$ at 3,2T, T=4K [ref. 10]; n-EuSe at B=5T, T=15K [ref. 44]; p-InSb(Mn) at B=5T, T=0,5K.[ref.20].

This diversity of NM effect values in solid state materials gave the ground for different physical models explaining resistance decrease in magnetic field. If we review the works on nonmagnetic semiconductors describing small values of NM ($\rho_B/\rho_{B=0}$>0.5) we can highlight the most popular physical models:

- the model of localized magnetic moments at impurity centers [46]
- the model of insulator-metal transition induced by magnetic field [47]
- the quantum correction model [48].

GMR in MS and DMS was explained in the framework of magnetic polaron model [49, 50]. This model described resistance decrease in magnetic field as the result of less intensive charge carriers spins scattering on magnetic host ions clusters. CMR in manganites was described as the transition from antiferromagnetically charge-ordered insulating phase to ferromagnetic metal phase in magnetic field [51,52]. The model of magnetic polarons was also used for the explanation of NM in HTS ruthenocuprates $RuSr_2(Nd,Y,Ce)_2Cu_2O_{10-\delta}$ [ref. 53]. Energy gap suppression by magnetic field was also suggested for GMR explanation ($\rho_B/\rho_{B=0}<0.1$) observed in granular Pb films [54]. In the case of diluted magnetic A^3B^5 semiconductors, for example in $Ga_{1-x}Mn_xAs$ [ref. 55], the observed $\rho_B/\rho_{B=0}\sim0.2$ was explained as the reduction of charge carriers spin scattering due to d-electrons spins aligning in magnetic field direction. Similar spin-dependent scattering model was also proposed for the explanation of NM in ferromagnetic metals [56].

Negative Magnetoresistance, revealed in p-InSb(Mn) crystals, is too large to be explained in the frameworks of above models used for NM explanations in nonmagnetic semiconductors. But the models of magnetic polaron or description of CMR in p-InSb(Mn) as the transition from antiferromagnetic insulator to ferromagnetic metal phase also is not applicable, as a InSb crystal doped with Mn is not a magnetic material. In this situation we proposed that the model of exciton insulator could be the universal model for CMR explanation [20, 21]. In the framework of excitonic insulator model the CMR in p-InSb(Mn) can be described as a drop of e-h insulating pairs concentration in an external magnetic field resulting from magnetic field influence on spin holes and electrons forming excitons. This scenario explains the decrease of the Hall constant in magnetic field, i.e. the increase of charge carrier concentration observed in p-InSb(Mn) at low temperatures (Fig. 17).

4. Influence of Uniaxial Stress on MIT in p-InSb(Ge) and p-InSb(Mn) Crystals

4.1. Piezoresistance in Uniaxially Stressed p-InSb, p-Ge(Ga) and p-Si(B) Crystals

Previously we discussed MIT in p-InSb(Mn) induced by impurity concentration variation and magnetic field. Another way to induce MIT in p-InSb crystals is to apply uniaxial compression along one of crystallographic axes. The first observation of uniaxial stress influence on conductivity and magnetoresistance of p-doped semiconductor crystals at low temperatures was carried out by Sigiyama and Kobayashi [57] in 1963 in Ge single crystals. They observed the increase of specific resistance and induction of negative magnetoresistance in p-Ge under uniaxial compression. Later the detailed study of uniaxial stress influence on p-Ge(Ga) conductivity at temperature 300-1,2K and impurity concentration $9\times10^{14}<N_A<9\times10^{16}cm^{-3}$ was done by Pollak [13]. In p-Ge(Ga) at Gallium impurity concentration $N_{Ga}\sim10^{17}cm^{-3}$ he observed the gigantic increase of specific resistivity - $\rho_X/\rho_0\sim10^4$ under uniaxial stress X~3kbar and T~1,7K. Negative Magnetoresistance [$(\rho_B - \rho_0)/\rho_0$] of -50% was observed in uniaxial stressed p-Ge(Ga) [58]. In p-Si(B) [12] sample with the metallic type of conductivity at acceptor concentration $N_A\sim2,8\times10^{18}$ uniaxial compression induced transition from metal to semiconductor state. The influence of uniaxial stress on MIT p-InSb(Ge) and p-InSb(Mn) crystals was studied in [8,14]. The survey of these works demonstrates that specific resistance dependence on temperature and uniaxial compression in p-InSb crystals at the impurity concentration about N_{cr} is qualitatively similar to the ρ-X dependence observed previously in p-Ge(Ga) and p-Si(B) at N_{cr} as it is shown in Fig. 5. The specific resistance in all three semiconductors increased under unaxial stress, reached maximum and then decreased under high stress. Maximal value of $(\rho_X/\rho_0)_{max}$ and stress at which ρ_X/ρ_0 reached maximum was dependent on temperature and impurity concentration around N_{cr} in all observed semiconductors. It was proposed in [13] that observed high level of specific resistivity growth in p-Ge(Ga) could be explained if we assumed that holes Borh orbit shrinked under compression. Obviously, to explain such a complicated behavior of resistivity as a function of stress and its dependence on impurity concentration using only information about conductivity, as it was done in the case of p-Ge and p-Si, is not sufficient. To get information about charge carriers concentration and their mobility we need to have also results of the Hall and magnetoresistance measurements in uniaxially stressed crystals. But in literature we failed to find the information about systematically studied Hall effect and magnetoresistance neither in uniaxially stressed p-Ge(Ga) nor in p-Si(B) around MIT. Therefore we studied magnetoresistance and Hall constant in uniaxially stressed p-InSb and p-Ge (Ga) crystals. Comparison of magnetotransport properties in p-InSb(Ge), p-InSb(Mn) and p-Ge(Ga) crystals is of special interest as MIT in these crystals occurs at the same critical concentration $N_{cr}=N_{Ge}=N_{Ga}=N_{Mn}=2\times10^{17}cm^{-3}$.

4.2. Uniaxial Stress Influence on Conductivity and Magnetoresistance of p-InSb(Ge) and p-Ge(Ga) Crystals

Figures 23-25 represent the specific resistivity as a function of temperature and the magnetonetoresistance versus magnetic field in p-InSb(Ge) crystals at compressional stress in the concentration range $4\times10^{15}\leq N_{Ge}\leq2\times10^{17}cm^{-3}$. Specific resistivity in the low doped crystal at $p=N_{Ge}=4\times10^{15}cm^{-3}$ slightly increases under stress, reaches maximum $\rho_X/\rho_0\sim2$ at X~1kbar and then decreases under the growing uniaxial compression till $\rho_X/\rho_0\sim0,01$ at X~3kbar. Magnetoresistance in unstressed sample is positive (Fig.23c, curve 1). A small negative magnetoresistance ($\rho_B/\rho_0\sim0,5$) appears in magnetic field B =3T at uniaxial stress X>1kbar.

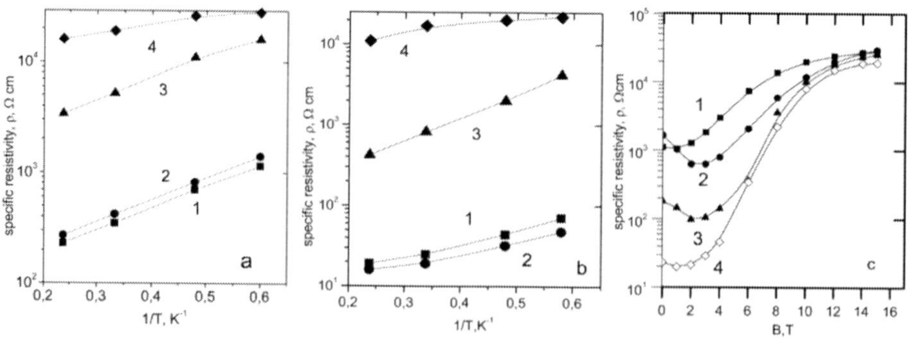

Figure 23. Specific resistivity in p-InSb(Ge) at $N_{Ge}=4\times10^{15}cm^{-3}$ as afunction of temperature, uniaxial compression and magnetic field, X//B//[111].

a and b –curve 1 - B=0, curve 2- B=3,3T, curve 3- B=8.5T, curve 4 - B=15T; a-X=0, b-X=2.5kbar; c-curve 1- X=0, curve 2 - X=0.96kbar, curve 3 -X=1.9kbar, curve 4-X=3kbar, T=1.72K.; Here X is stress direction, B- magnetic field direction and [111] – crystallographic axis.

In unstressed sample in the intermediate range of impurity concentration at $N_{Ge}=8\times10^{16}cm^{-3}$ magnetoresistance is positive (Fig. 24a and Fig. 24c, curve 1). Under uniaxial stress specific resistance increases and strong temperature dependent negative magnetoresistance appears (Fig. 24b,c). With the increase of magnetic field the ratio ρ_B/ρ_0 at T=1,84K reaches its minimal value $\rho_B/\rho_0=0,1$ at B~ 7T (Fig. 24c).

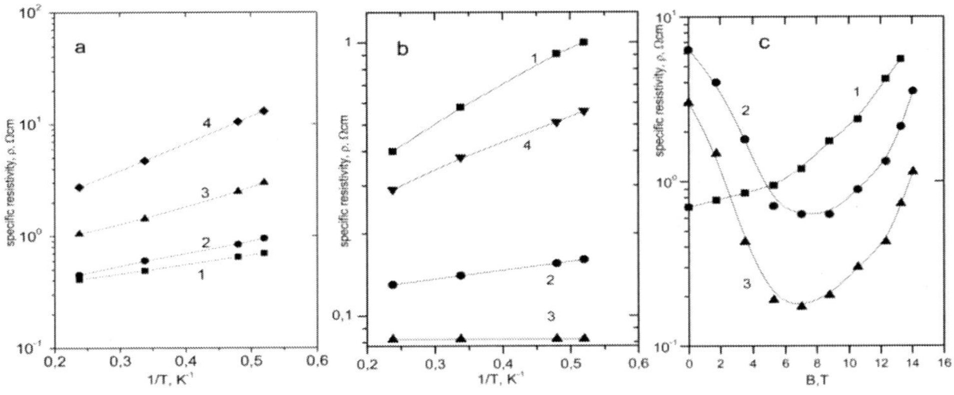

Figure 24. Specific resistivity in p-InSb(Ge) crystal at $N_{Ge}=8\times10^{16}cm^{-3}$ vs temperature and magnetic field at various uniaxial stress values, X//B//[111].
a and b –curve 1, B=0; curve 2- B=3,3T; curve-3- B=8,5T; curve 4- B=15T; a-X=0, b- X= 3.2kbar; c- curve 1, X=0, curve 2-X=0.85kbar, curve 3-X=2.4kbar, T=1,84K.

A sample at $N_{Ge}=N_{cr}\approx2\times10^{17}cm^{-3}$ demonstrates the metallic type of conductivity in the absence of stress but uniaxial stress induces MIT transition, similar to the MIT observed in uniaxially stressed p-Si(B) [12] as it is seen from Fig. 25a. Contrary to MIT induced by uniaxial compression, an external magnetic field induces IMT transition as it is depicted in Fig. 25b. To compare the influence of uniaxial compression on specific resistivity of p-InSb(Ge) and p-Ge(Ga) crystals at about N_{cr} we measured specific resistivity in p-Ge(Ga) crystals as a function of temperature between 4,2 and 1,5K in magnetic field B=0÷15T, under uniaxial compression X=0÷5kbar and impurity concentration range $2\times10^{16}cm^{-3}<N_{Ge}<2\times10^{17}cm^{-3}$. The crystal of p-Ge(Ga) of metal conductivity type demonstrates strong positive magnetoresistance ($\rho_B/\rho_0>500$ at B=15T) in the absence of stress while uniaxial stress induces both MIT (Fig. 26a) and transition from positive to negative magnetoresistance (Fig. 26c). Maximum of p-Ge(Ga) piezoresistance $(\rho_X/\rho_0)_{max}$ and minimum of negative magnetoresistance $(\rho_B/\rho_0)_{min}$ occured at the same concentration $N_{Ge}\sim9\times10^{16}cm^{-3}$ as it is depicted in Fig. 27. It should be stressed that piezoresistance in p-Ge(Ga) crystals exceeded the ρ_X/ρ_0 in p-InSb(Ge) samples with the equal impurity concentration, but the ratio ρ_B/ρ_0 was smaller for p-InSb(Ge) crystals. Temperature-resistivity dependence at temperature between 4,2K and 1,5K in p-InSb(Ge) samples in the range of impurity concentration $8\times10^{16}cm^{-3}<N_{Ge}<2\times10^{17}cm^{-3}$ and p-Ge(Ga) sample at $N_{Ga}=1,5\times10^{17}cm^{-3}$ can be expressed by Eq. 5 (paragraph **1.2**). It can be seen from the Figures 23-27 that piezoresistance and negative magnetoresistance was mainly determined by the variation of the activation conduction energy Δ in magnetic field and under stress. The dependence of Δ on uniaxial compression and on magnetic filed is represented in Fig. 28.

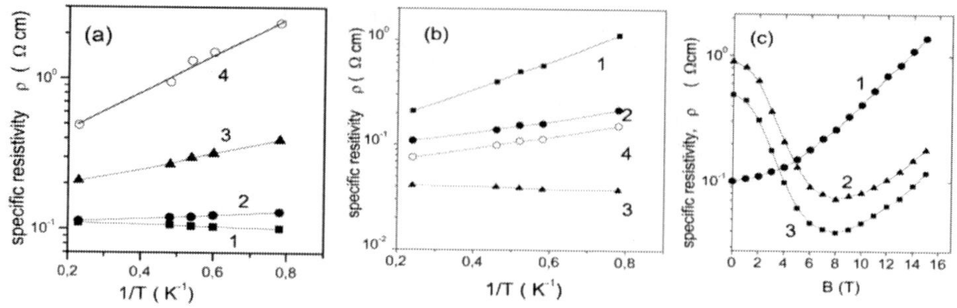

Figure 25. Specific resistance as a function of temperature (a, b) and magnetoresistance (c) in p-InSb(Ge) at N_{Ge} =2×10^{17}cm^{-3}, X//B//[111].
a and b – curve 1, B=0 ; curve 2- B=3.3T ; curve 3- B=8.5T ; curve 4- B=15T ; a-X=0, b -X=2.4kbar; c-curve 1- X=0kbar; curve 2- X=1.58kbar; curve 3- X=2.4kbar, T=1,84K,

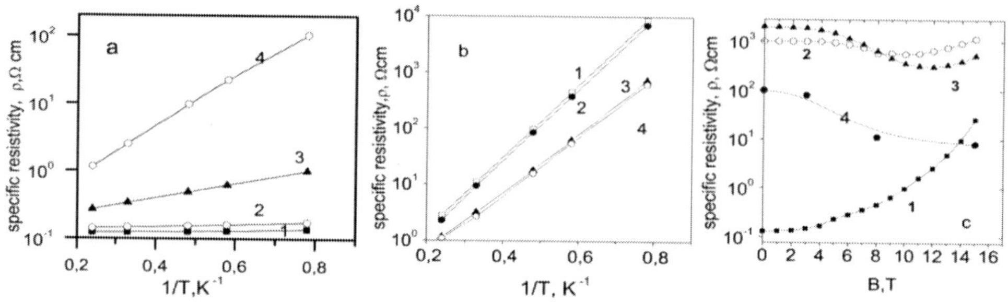

Figure 26. MIT and magnetoresistance in uniaxially compressed p-Ge(Ga) crystal at N_{Ga}≈N_{cr}, N_{Ga}=1.5×10^{17}cm^{-3}, X//B//[111].
a and b –curve 1, B=0; curve 2, B=3.3T; curve 3, B=8.5T; curve 4, B=15T; a- X=0kbar, b-X= 4.8kbar; c-curve 1, X=0kbar; curve 2, X=1.8kbar; curve 3, X=3kbar; curve 4, X=6kbar, T=1,7K.

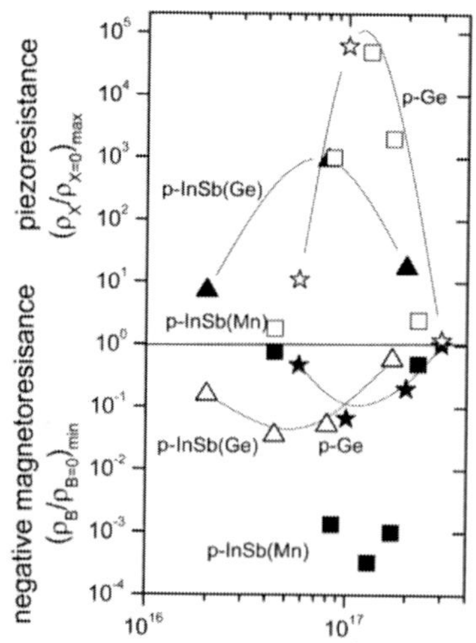

Figure 27. Piezoresistance and negative magnetoresistance in p-InSb(Ge), p-Ge(Ga) and p-InSb(Mn) at T=1,25K around N_{cr}.

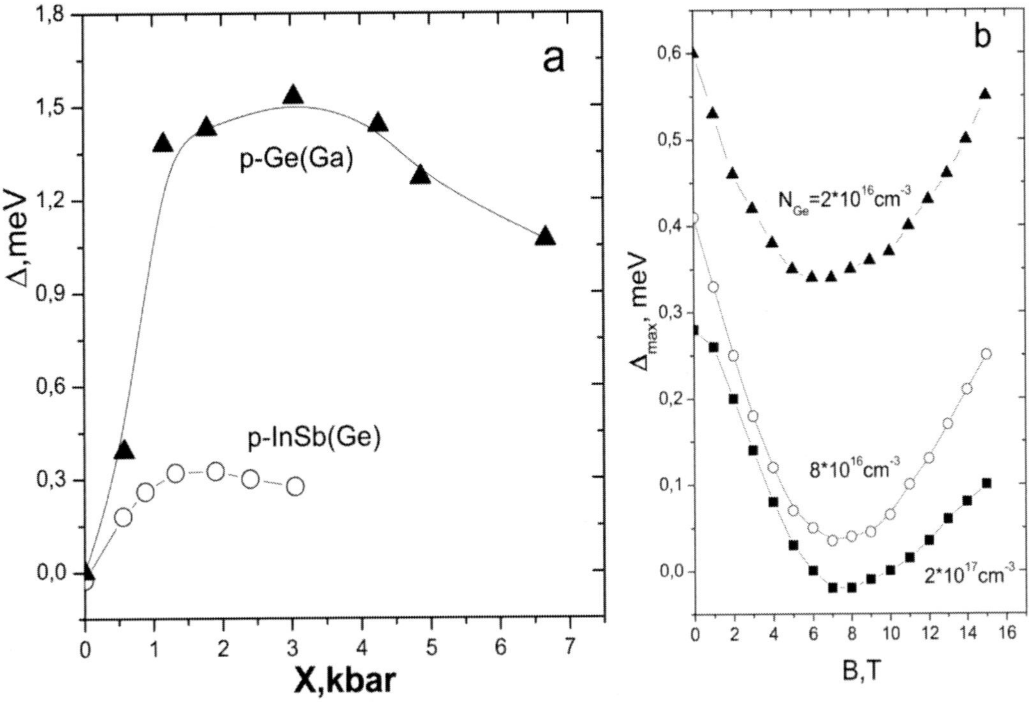

Figure 28. Activation conduction energy variations in p-InSb(Ge) and p-Ge(Ga) under uniaxial compression (a) and magnetic field (b).

4.3. MIT in p-InSb(Mn) Crystals Induced by Uniaxial Compression

The influence of uniaxial compression on resistivity-temperature dependence of p-InSb(Mn) crystal at $N_{Mn}=1{,}8\times10^{17}cm^{-3}$ is displayed in Fig. 29. Without stress p-InSb (Mn) sample demonstrates a quasi-metal conductivity type (Fig. 29a). Under uniaxial compression X=1,7kbar resistivity increased by a factor of 10^3 at T=1,25K.

In compressed crystals ρ-T dependence can be described by the stretched exponent $\rho=\rho_{01}\exp(\Delta_1/kT)^2$ (Eq. 4). The same exponential dependence was revealed for unstressed p-InSb(Mn) crystals (paragraph **3.2**). The increase of specific resisitivity under stress was caused mainly by the increase of Δ_1. External magnetic field induced IMT in uniaxially stressed p-InSb(Mn) crystals as it is shown in Fig. 29b.

Negative Magnetoresistance in uniaxially stressed p-InSb(Mn) crystals increased under the growing stress and the decrease of temperature as it is shown in Fig. 30a and 30c. External magnetic field radically changed the character of piezoresistance, i.e. the maximum in ρ_X/ρ_0 – stress dependence disappeared in strong magnetic field (Fig. 30b). Similar qualitative results were obtained in p-Ge(Ga) [58], but ρ_B/ρ_0 at $N_{Ga}=4{,}6\times10^{16}cm^{-3}$ was only about 0,5.

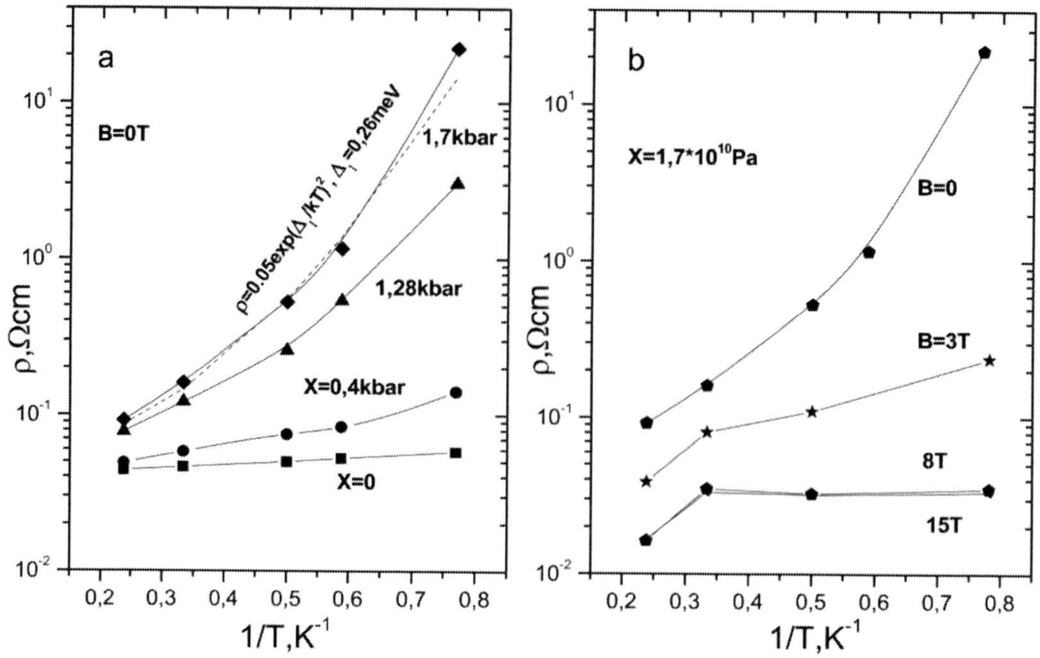

Figure 29. MIT induced by uniaxial compression (a) and IMT in magnetic field (b) in p-InSb(Mn) sample at N_{Mn} =1,5×10^{17}cm^{-3}. Dash line is a fit to $\rho=0.05\exp(\Delta_1/kT)^2$ at Δ_1=0,26meV.

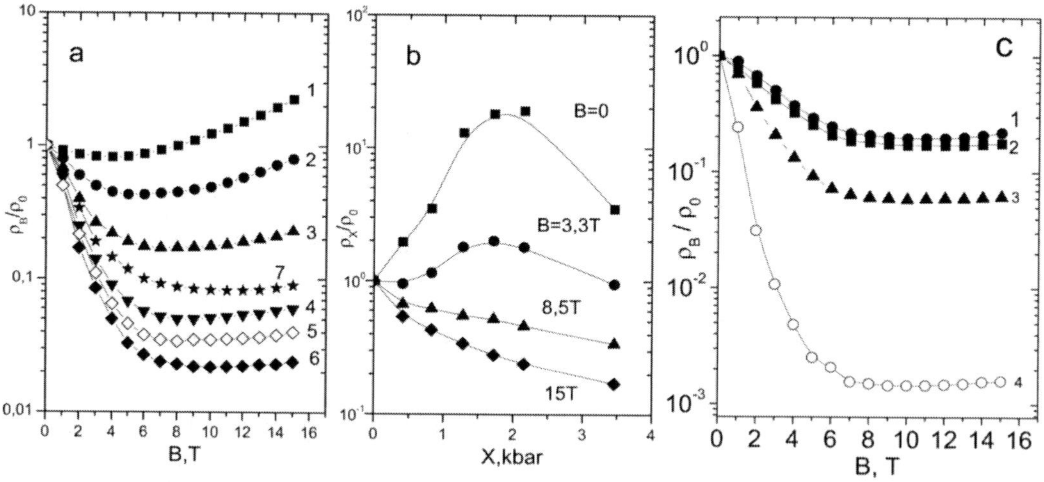

Figure 30. Negative Magnetoresistance and Piezoresistance in p-InSb(Mn) as a function of magnetic field and temperature at N_{Mn}=1,5×10^{17}cm^{-3}.

a,b -T=1,7K, curve 1-X=0kbar, curve 2-X=0,4kbar, curve 3-X=0,8kbar, curve 4- X=1,2kbar, curve 5-X=1,7kbar, curve 6-X=2,2kbar, curve 7-X=2,3kbar.

c- X=1,7kbar; curve 1- T= 4,2K, curve 2-T=3K, curve 3-T=2K, curve 4-T=1,28K, X//[111]//B.

4.4. Hall Effect In Uniaxially Compressed p-InSb(Ge), p-Ge(Ga) and p-InSb(Mn) Crystals

Hall constant dependence on magnetic field in two uniaxially stressed samples of p-InSb(Ge) and p-Ge(Ga) at impurity concentration on insulator side of MIT is displayed in Fig.31.

It can be seen from Fig. 31 that whereas R_H remained practically independent on magnetic field in unstressed p-InSb(Ge) and p-Ge(Ga) samples, R_H in uniaxially stressed crystals increased with the increase of magnetic field. Maximal ratio of $(R_H)_{B=2,5T}/(R_H)_{B=0}$ was about 11 for p-InSb(Ge) and $(R_H)_{B=5T}/(R_H)_{B=0}$ was about 10 for p-Ge(Ga).

In p-InSb(Mn) crystals with the impurity concentration $N_{Mn}=1{,}7\times10^{17}cm^{-3}$ in the absence of uniaxial stress the Hall constant in the range of magnetic field B=0,1-1T was positive as it is depicted in Fig.32. Application of uniaxial compression at X=0,1kbar changed the Hall sign to positive. An absolute value of $/-R_H/$ in uniaxially stressed crystal increased with the decrease of temperature. Similar $/-R_H/$dependence on temperature was observed in a single unstressed p-InSb(Mn) crystal with manganese concentration $N_{Mn}=1{,}5\times10^{17}cm^{-3}$ at T below 1K (Fig. 17c, paragraph **3.3**).

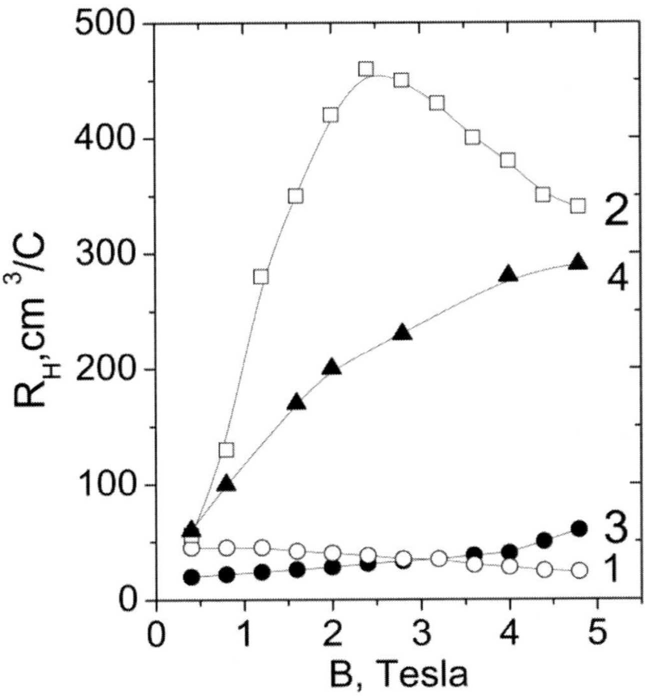

Figure 31. Hall constant in uniaxially stressed p-InSb(Ge) and p-Ge(Ga) crystals X//I//[111], B//[110]

curves 1,2-p-InSb(Ge), $p=8\times10^{16}cm^{-3}$, T=1.7K; curve 1-X=0, curve 2-X=2kbar

curves 3,4-p-Ge, $p=1\times10^{17}cm^{-3}$, T=4.2K; curve 3-X=0, curve 4 - X=1.7kbar.

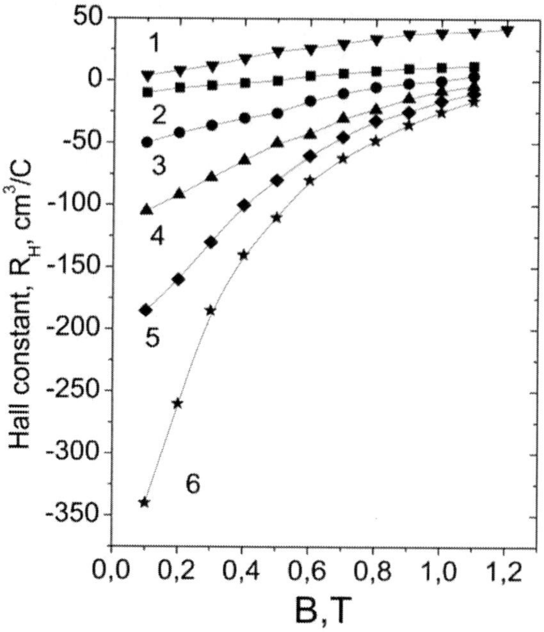

Figure 32. Hall constant sign inversion under uniaxial compression in p-InSb(Mn).
curve 1-X=0, curves 2-6 -X=0,1kbar, curves 1,2-T=1.75K, curve 3-T=1.38K, curve 4-T=0.91K, curve 5-T=0.52K; curve 6-T=0.488K.

4.5. Discussion

It follows from magnetotransport studies that in uniaxialy stressed p-InSb crystals at impurity concentration around N_{cr} at low temperatures:

- iniaxial compression induces MIT in p-InSb(Ge), p-InSb(Mn) and p-Ge(Ga) crystals at temperature below 10K.
- effect of specific resistivity increases under stress caused by the decrease of charge carrier concentration.
- magnetic field induces IMT in uniaxially stressed p-InSb(Mn) and p-InSb(Ge)crystals and significantly decreases resistance in p-Ge(Ga)

The most important result of above experiments is that from the one hand anomalies of resistivity dependence on impurity concentration and negative magnetoresistance can be driven by uniaxial stress in p-InSb(Ge) and p-Ge(Ga) crystals doped with nonmagnetic impurities, but from the other hand similar anomalies in ρ-N_{imp} dependence and Colossal Negative Magnetoresistance can be observed in unstressed InSb crystals doped with Mn as a magnetic impurity. Such a coincidence in the properties of unaxially stressed p-InSb(Ge), p-Ge(Ga) and unstressed p-InSb(Mn) crystals put forward the idea that both unaxial stress and manganese doping induce similar changes in the band structure of InSb. In unstressed InSb and Ge crystals the valence band at k=0 is degenerate. The shear component of uniaxial stress removes this degeneracy and bands are separated by the energy proportional stress (Fig. 8).

Obviously the splitting of a valence band under stress could be the main cause of unusual transport and magnetotransport properties both in InSb and Ge. We suggest that similar influence of Mn ions on InSb valence band can be expected due to the fact that Mn^{++} ions can split valence band by Jahn-Teller distortion [59] and so eliminate degeneracy. This situation erases the difference between valence band structure of unstressed p-InSb(Mn) and uniaxial stressed p-InSb(Ge) crystals. Another suggestion is that in uniaxially stressed p-InSb(Ge) and p-Ge(Ga) crystals the split of the valence band stimulates excitons formation which leads to the resistivity increase under stress. Contrary, an external magnetic field decreases excitons binding energy and destroys excitons, i.e. induces CMR and IMT transition.

Similar approach based on excitonic model we could apply to CMR interpretation in DMS and MS where, as it follows from literature, the decrease of activation conduction energy is responsible for the resistivity fall in magnetic field (Fig. 33) exemplified by p-$Hg_{1-x}Mn_xTe$ [ref. 9] and n-$Gd_{1-x}V_xS_4$ [ref.10].

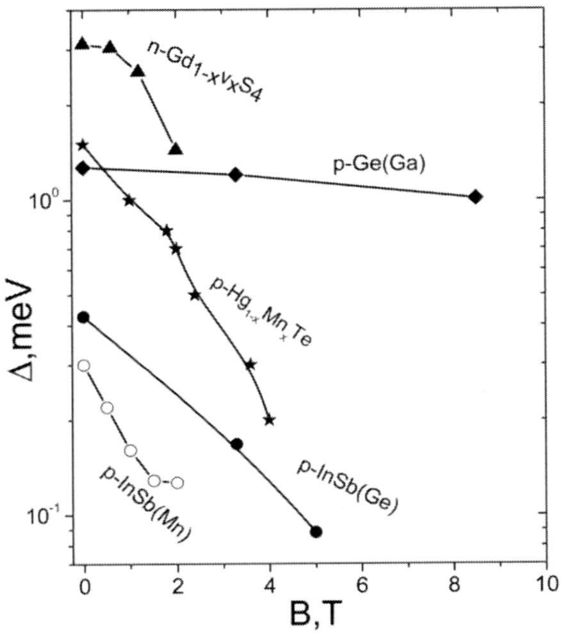

Figure 33. Activation conduction energy decrease in magnetic field in uniaxially stressed p-InSb, p-Ge, DMS and MS. p-InSb(Mn) at X=1,5kbar, p-InSb(Ge) at X=1,58kbar, p-Ge(Ga) at X=6,7kbar, p-$Hg_{1-x}Mn_xTe$ [ref. 9] and n-$Gd_{1-x}V_xS_4$ [ref. 10] at X=0kbar.

5. MIT in p-InSb(Mn) Crystals Induced by Hydrostatic Pressure

In previous paragraphs 3 and 4 we discussed MIT in p-InSb(Ge) and p-InSb(Mn) crystals induced by impurity concentration variations and uniaxial stress. We revealed that MIT caused by the variation of N_{Mn} and uniaxial compression of p-InSb(Mn) crystals is accompanied by anomalous conductivity behavior, negative magnetoresistivity and Hall constant inversion around N_{cr}~$2\times10^{17}cm^{-3}$. These effects radically differ from described in literature transport and magnetotransport effects around N_{cr} in unstressed semiconductor crystals doped with nonmagnetic impurities, e.g. –p-InSb(Ge), p-Ge(Ga) and p-Si(B). It was proposed that the anomalous InSb(Mn) behavior can be explained in the framework of excitonic phase insulator model. Previously, magnetotransport studies of $TmSe_{1-x}Te_x$ under hydrostatic pressure supported the conception of excitonic insulator [24]. Following the ideology of experiments on $TmSe_{1-x}Te_x$ we studied the influence of hydrostatic pressure up to 16kbar on transport and magnetotransport properties of p-InSb(Mn) and p-InSb(Ge) crystals at impurity concentration around N_{cr} and temperature range 280K÷1,6K.

5.1. High Temperature (T=280K) Transport and Magnetotransport Effects in Hydrostatically Pressured p-InSb(Mn) and p-InSb(Ge) Crystals

At ambient temperature (T=280K) in the absence of pressure the Hall constant R_H is negative at B<1,5T but changed its sign to positive at B>1,5T as it is shown in Fig.34a. In a compressed crystal at P>1,5kbar the Hall constant was positive between B=0 and 10T. Specific resistivity under pressure increased about 10% at maximal pressure (P=15kbar) as it is shown in Fig. 34b. Magnetoresistance as a function of magnetic field in uncompressed crystal was described in paragraph 3 where break of ρ-B dependence was related to different contribution of electrons and holes into conductivity in external magnetic field. In low magnetic field positive magnetoresistance sharply increased and reached ρ_B/ρ_0~1,4 at B~1T (Fig. 34c). Between B=2T and 10T the magnetoresistance slowly increased with the increase of magnetic field and reached ρ_B/ρ_0~1,6 at B=10T. Under pressure the curve ρ_B/ρ_0 vs B demonstrated the disappearance of the break on ρ-B dependence (Fig. 34c).

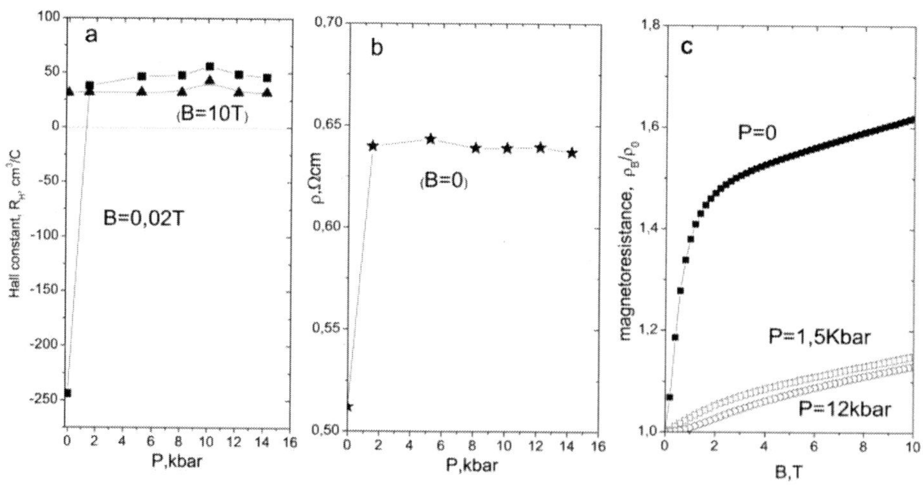

Figure 34. (a) Hall constant, (b) specific reisistivity and (c) magnetoresistnace under high pressure in p-InSb(Mn) crystal at $N_{Mn}=1,4\times10^{17}cm^{-3}$ at T=280K.

Similar magnetotransport effects were demonstrated for p-InSb(Ge) crystals under pressure. Fig. 35 displays p-InSb(Ge) Hall constant sign inversion in low magnetic field under pressure, demonstrates the increase of the specific resisitivity in p-InSb(Ge) sample and changes in magnetoresistance dependence on magnetic field.

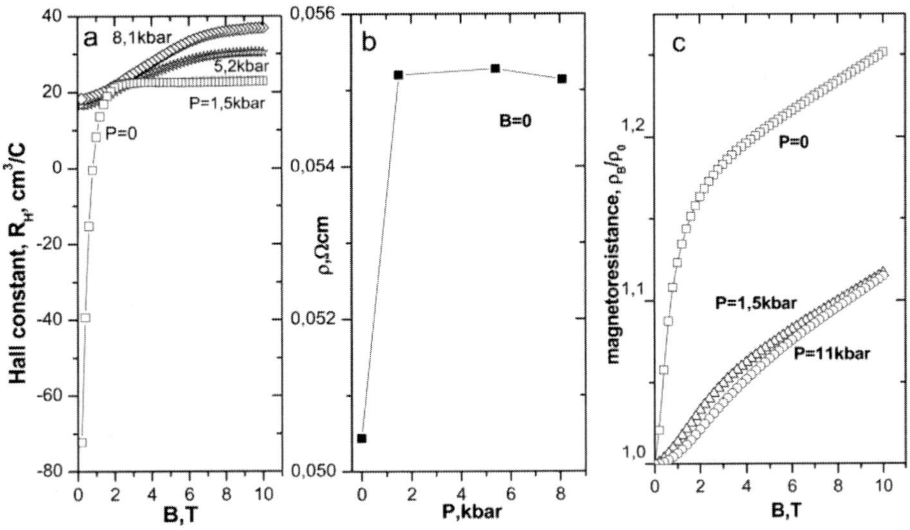

Figure 35. Hall constant (a), specific resistivity (b) and magnetoresistance (c) under high pressure in p-InSb(Ge) at N_{Ge}=1,98×$10^{17}$$cm^{-3}$ at ambient temperature (T=280K).

As it was observed above such a behavior of R_H and ρ_B/ρ_0 in p-InSb(Mn) and p-InSb(Ge) crystals as a function of magnetic field B is the result of the input of two charge carries types into the Hall effect and magnetoresistance, i.e., light electrons and heavy holes. Under hydrostatic compression the contribution of electrons into the Hall effect becomes less important as it is demonstrated in Fig. 34a, c and Fig. 35a,c. These facts can be explained by the increase in elecrton-hole pairs concentration, i.e., permanent excitons. As a result, the increase in nonconducting electron- hole pairs induces the resistance increase (by~10%). (Fig. 34b and 35b).

5.2. Magnetotransport Properties of Hydrostatically Compressed p-InSb(Mn) Crystals at Liquid Nitrogen Temperature (T=80K)

The magnetotransport properties of p-InSb(Mn) crystals under pressure significantly differed at temperatures T=80K and T=280K. It can be seen from Fig.36a that in p-InSb(Mn) the Hall constant in the absence of pressure was positive and remained positive with a slight change of value under pressure (Fig. 36a). Under pressure P=14kbar Hall constant increased by ~10% in the magnetic field B=10T. Magnetoresistance increased by about 10% within B=0 -10T (Fig. 36c.)

The difference between transport and magnetotransport properties at both ambient and liquid nitrogen temperatures can be understood if we remember that conductivity in a manganese-doped p-InSb crystal at T=80K is determined by the holes whereas at T=280K indium antimonide is the intrinsic semiconductor with the conductivity determined by the contribution of electrons and holes [26].

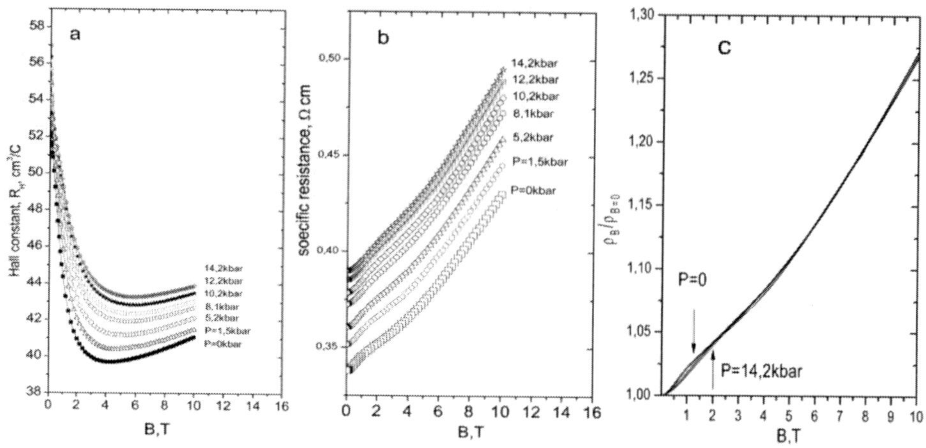

Figure 36. Hall constant (a), specific resistivity (b) and magnetoresistance (c) in p-InSb(Mn) crystal under pressure at 80K.

5.3. Pressure Driven MIT in p-InSb(Mn) Crystals at Low Temperatures (T=20K÷1,6K)

Under hydrostatic pressure and at 1,6K the specific resisitivity of p-InSb(Mn) increased by a factor of 10^4, reached maximum at P~10÷12 kbar and showed the tendency to decline with the further pressure increase. Fig. 37 demonstrates p-InSb(Mn) ρ-P dependence for two samples at T=1,6K. A huge increase of the resistivity by about 10^4 at 1.6 K occured with the increasing pressure in p-InSb(Mn), leading to the growth of the activation energy Δ from about 0.1 meV at ambient pressure to 1.4 meV at 12 kbar (Fig. 38).

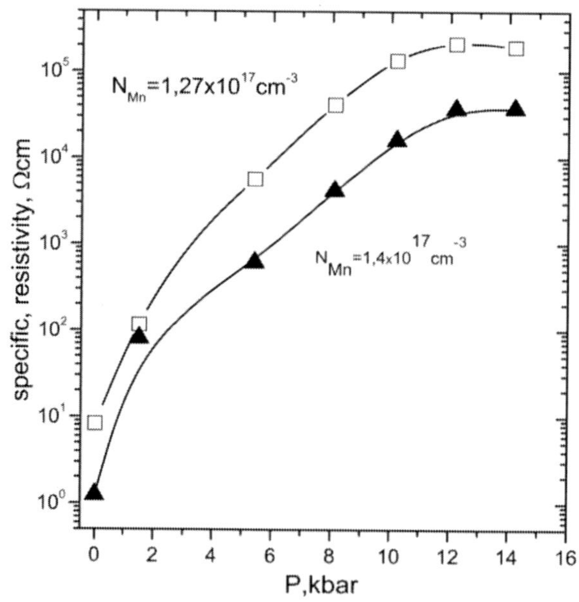

Figure 37. Piezoresistance of p-InSb(Mn) samples at T=1,6K at N_{Mn} around N_{cr}.

Contrary to p-InSb(Mn), in the case of InSb(Ge) only a minor increase of the resistivity not leading to MIT was observed under pressure at T below 10K (Fig. 38c).

If we compare the influence of uniaxial stress and hydrostatic compression on MIT in InSb(Ge) and InSb(Mn) crystals we can conclude that our hypotheses concerning anisotropic structure of InSb(Mn) valence band (4.5) can be supported by the fact that InSb(Ge) crystals practically unchanged their transport properties under pressure. In this situation we can postulate that high pressure and uniaxial stress in InSb(Mn) bring about the same changes as the variation of Mn concentration.

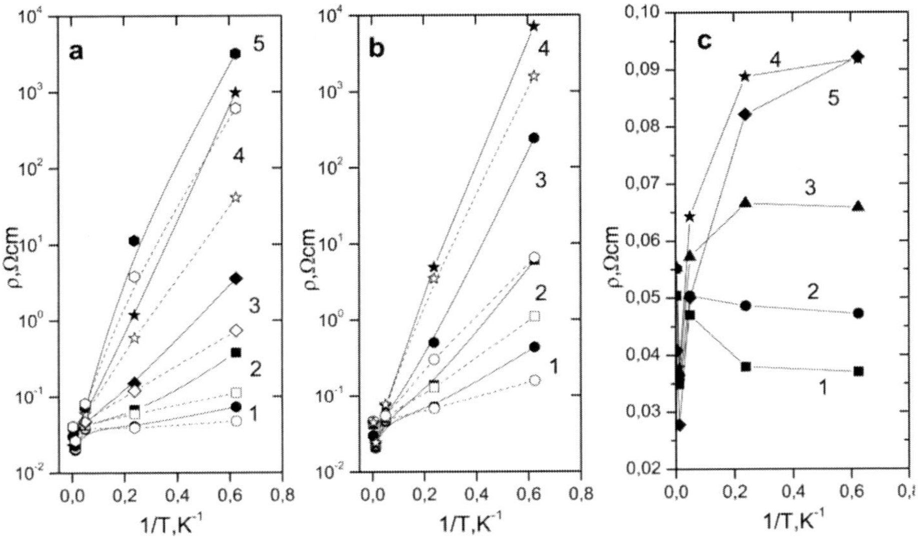

Figure 38. Pressure driven MIT in p-InSb(Mn) crystals (a,b) and ρ-T dependence in p-InSb(Ge) under pressure variations (c).

a,b-solid symbols at B=0, open symbols at B=3T.

a-curve 1- P=0kbar, curve 2-P=1,5kbar, curve 3-P=5,2kbar, curve 4-P=10kbar, curve 5-P=14kbar.

b- curve 1-P=0kbar, curve 2-P=1,5kbar, curve 3-P=8,1kbar, curve 4-p=10,1kbar.

c- curve 1-P=0kbar, curve 2 –P=1,5kbar, curve 3-P=5,2kbar, curve 3-P=8,1kbar, curve 4- P= 11,2kbar.

a- $N_{Mn}=1,4\times10^{17}cm^{-3}$, b-$N_{Mn}=1,27\times10^{17}cm^{-3}$, c-$N_{Ge}=1,9\times10^{17}cm^{-3}$.

5.4. Negative Magnetoresistance in p-InSb(Mn) Crystals Under High Pressure

In p-InSb(Mn) crystals under hydrostatic pressure a negative magnetoresistance effect became more noticeable at $N_{Mn}=1,4\times10^{17}cm^{-3}$(Fig. 39a). The dependence of ρ_B/ρ_0 on pressure is shown in Fig. 39b.

It can be seen that under pressure the ratio ρ_B/ρ_0 reaches its minimal value at P~8kbar at $N_{Mn}=1,4\times10^{17}cm^{-3}$ and at P~5kbar at $N_{Mn}=1,27\times10^{17}cm^{-3}$.

Like in uncompressed crystals the external magnetic field reduces an activation conduction energy Δ, which can be seen from Fig. 38a and 38b which is demonstrated by the decrease of ρ-T curves slope in magnetic field.

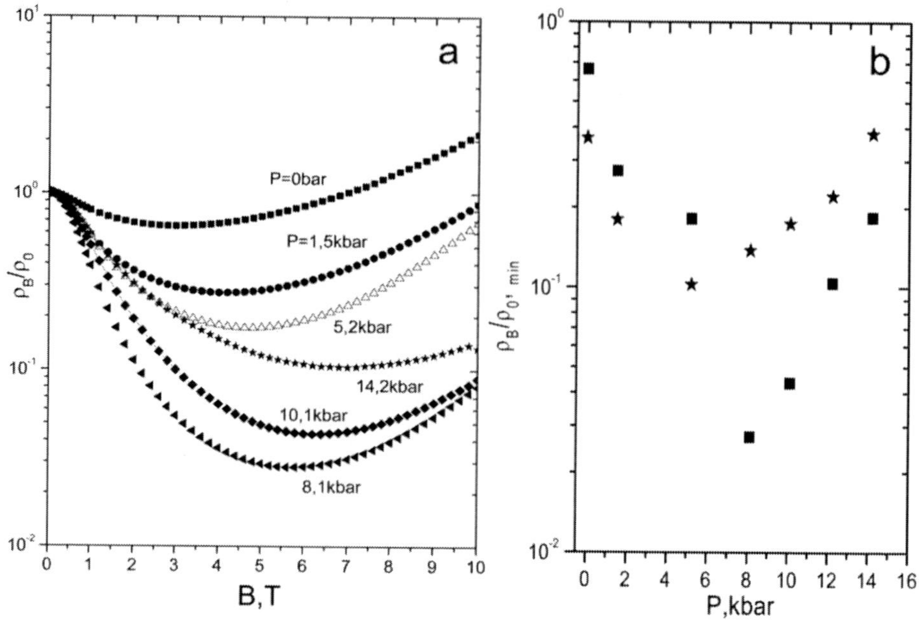

Figure 39. Negative Magnetoresistance in p-InSb(Mn) under pressure at 1,6K a- N_{Mn}=1,4×10^{17}cm^{-3}; b-solid squares a t N_{Mn}=1,4×10^{17}cm^{-3}, solid stars - N_{Mn}=1,27 ×10^{17}cm^{-3}.

5.5. Hall Effect in p-InSb(Mn) Crystals Below 20K Under High Pressure

At temperature T=20K the Hall constant in p-InSb(Mn) increased insignificantly under pressure (Fig. 40a). Also magnetoresistance was slightly dependent on pressure and remained positive (Fig. 40b). So at T=20K the Hall constant and resistivity dependencies in magnetic field under pressure were similar to R_H and ρ at 80K.

At temperature below 20K the unusual behavior of the Hall constant was revealed in p-InSb(Mn) crystals under pressure. Fig. 41 displays R_H vs B under pressure variations in p-InSb(Mn) sample at N_{Mn}=1,4×10^{17}cm^{-3} at T=1,6K. Without pressure R_H-B dependence (Fig.41a) demonstrated typical for the p-InSb(Mn) crystal the Hall constant sign inversion in low magnetic field, the effect discussed above in paragraph **3**. Low magnetic field absolute value /-R_H/ demonstrated negative pressure dependence and decreased with the increase of pressure. R_H changed it sign to positive at P~5-8 kbar and reached R_H=2×10^{5}cm^3/C at P=12kbar (Fig. 41b).

At pressure P>12kbar R_H decreased under the further pressure growth (Fig. 41c, 42a). Similar R_H vs P dependence was obtained for the p-InSb(Mn) crystal at N_{Mn}=1,27×10^{17}cm^{-3} but the Hall constant sign inversion was observed at lower pressure P=1,5kbar. Contrary, the high magnetic field the Hall constant under compression did not change its sign but increased gradually (Fig. 42b).

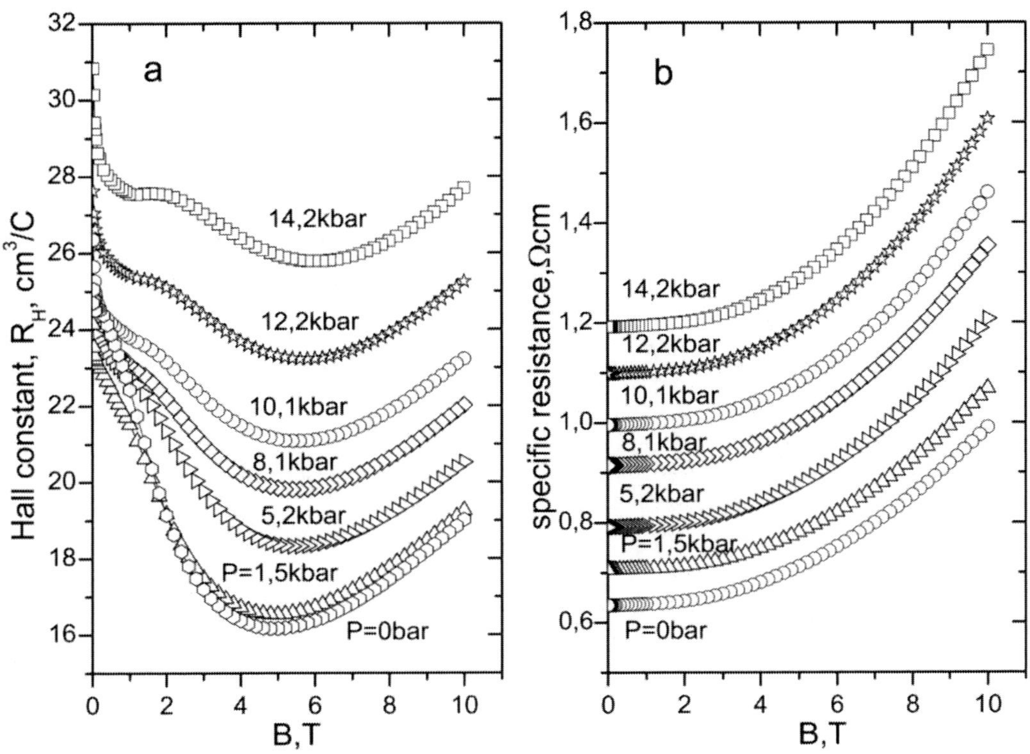

Figure 40. Hall constant (a) and specific resistivity (b) dependence on magnetic field under pressure variations in p-InSb(Mn) at 20K, $N_{Mn}=1,4\times10^{17}cm^{-3}$.

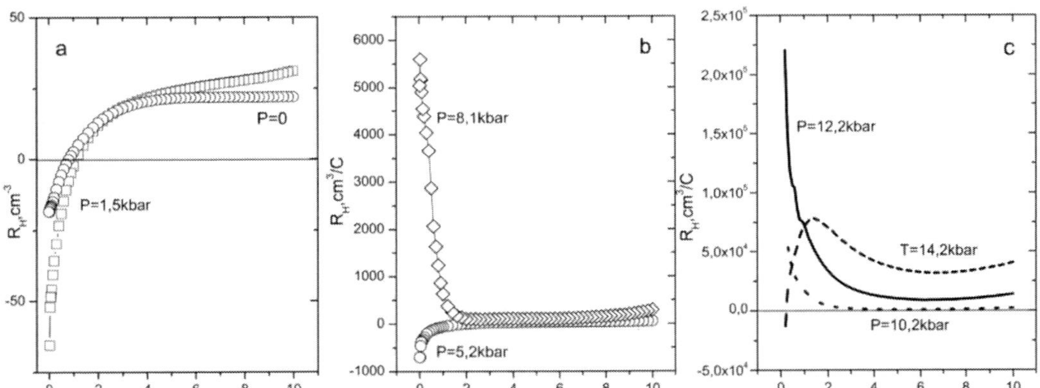

Figure 41. Hall constant sign inversion under pressure in p-InSb(Mn) at T=1,6K in sample at $N_{Mn}=1,4\times10^{17}cm^{-3}$. (a) - P=0÷1,5kbar, (b) - P=5÷8kbar, (c)- P=10÷14kbar.

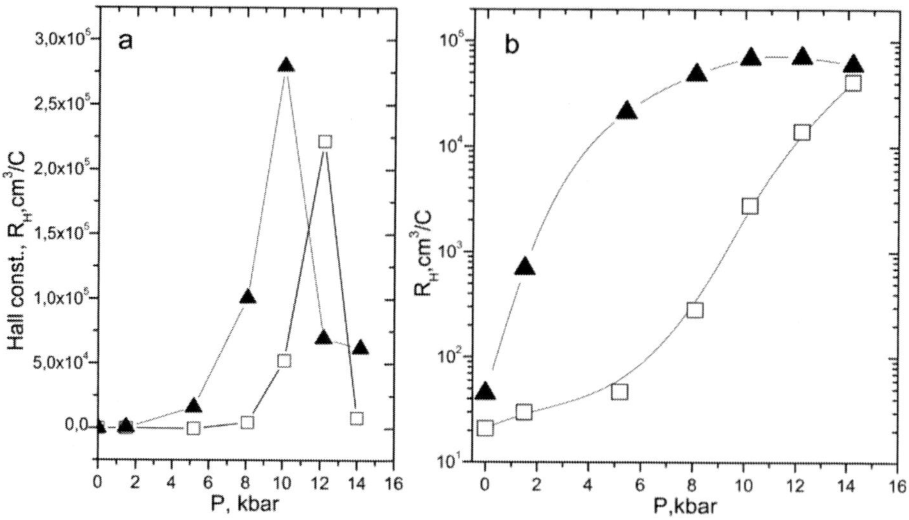

Figure 42. R_H vs P in p-InSb(Mn) samples in low (B=0,2T) (a) and high (B=10T) (b) magnetic fields at T=1,6K. Solid triangles- $N_{Mn}=1{,}27\times10^{17}cm^{-3}$, solid squares -$N_{Mn}=1{,}4\times10^{17}cm^{-3}$.

5.6. Discussion

The results of the Hall constant study in p-InSb(Mn) crystals under hydrostatic pressure give the ground to conclude that pressure driven MIT can be explained by the decrease of the charge carrier concentration but not by the mobility changes under pressure. It can be proved by the following- in R_H-P dependence studies at $N_{Mn}\leq N_{cr}$ the hole concentration determined as $p=1/R_H$ decreased by a factor of ~10^4 at pressure P~12kbar, T=1,6K (Fig. 42). Alternatively, at the same temperature T=1,6K Hall mobility determined from $\mu=R_H/\rho$ decreased only by a factor of 4 under the same pressure. At ambient pressure the Hall mobility was about 2cm^2/Vs whereas at P=12kbar it did not exceed 0,5cm^2/Vs. Strong decrease of charge carrier concentration under pressure can be explained in the framework of excitonic insulator model described in [22]. It was suggested above in paragraph **3.5** that electrons and holes can form permanent excitons without pressure. Electrons which are not bound to holes in the absence of pressure are responsible for the quasi-metal conductivity as it follows from the negative sign of the Hall constant at low magnetic field in the absence of pressure. But, at pressure about 5-8 kbar we observe the Hall constant sign inversion, when all electrons are coupled with holes and, therefore, conductivity is determined by holes. The maximum of binding energy Δ~1,5meV is at high pressure of about 10kbar. Further increase of pressure at P> 12kbar evidently will destroy excitons which leads to VRH conductivity as it is demonstrated in Fig. 38a, b. Similar conductivity transition from quasi- metal via activated conductivity to VRH type is induced in p-InSb(Mn) crystals by the variation of manganese concentration and uniaxial stress as it was discussed in paragraphs 3 and 4.

6. Applications

Magnetotransport studies of InSb crystals doped with manganese at impurity concentration around N_{cr} revealed exceptional physical properties that can be used for advanced cryogenic devices fabrication. A few implementation possibilities could have advantage over the conventional doped semiconductors. To the first group of devices based on InSb properties can be related thermometers and bolometers operated in the millikelvin range of temperatures and to the second group -selective millimeter and radio-frequency detectors.

Doped semiconductors at impurity concentration $N_{imp} \leq N_{cr}$ are widely used in the fabrication of thermoresistors and bolometers operated in Low and Ultra-Low temperature range. Germanium doped with Gallium (Fig.18) and Neutron Transmutation Doped (NTD) Germanium [ref. 19], usually used for thermoresistors and cryobolometers, demonstrated ρ-T dependence described by Eq. 3 for VRH conductivity in the presence of Coulomb gap. The Coulomb interaction between localized electrons has been demonstrated to create a 'soft' gap in the density of states near the Fermi level. It is well known that at super-low temperatures in the materials with hopping conductivity thermosensitivity is tempetature dependent and declines with the decrease of temperature. To reach high sensitivity it is reasonable to implement enhanced resistivity, but such an approach can be limited by the critical power P_c.

The advantage of p-InSb(Mn) as thermosensitive material is that it is not restricted by the limitations of hopping processes at low and super-low temperatures, the quality which noticeably differ from the conventional Ge and Si semiconductors. What is more, in contrast to the conventional thermoresistors, p-InSb(Mn) resistance thermometers show exponential resistivity-temperature dependence described by Eq.5 ($\rho=\rho_0 \exp(\Delta/kT)$, where Δ varies between 0,007 and 1meV. This unique property of p-InSb(Mn) crystals allows to get high thermosensitive materials with low specific resistivity in the millikelvin range of temperature thus to reduce P_c. It was shown in [60, 61] that sensitivity (dR/RdT) of InSb(Mn) resistance at 10mK reached value ~1000K^{-1}. Note that at T below 20mK NTD Germanium thermoresistors practically lose their sensitivity as a result of VRH conductivity type [19].

Another possible practical application arises from magnetotransport measurements where the activation conduction energy Δ is associated with the gap in the excitonic insulator energy spectrum. The existence of this super-narrow gap in the range 0,007-1meV gives an opportunity to develop selective detectors which could operate at temperature below 1K in the spectral window ranging from millimeter to meter wavelength (λ=1mm-1m).

The further exploration of p-InSb(Mn) as a thermosensitive and super-narrow gap material can transcend the limits of semiconductors application in a lot of practical and scientific areas.

Conclusion

Comparative study of MIT in InSb single crystals doped with magnetic (Mn) and nonmagnetic (Ge) impurities has revealed several phenomena unusual for doped semiconductors around N_{cr}. They are:

1. Colossal Magnetoresistance both in uniaxially stressed p-InSb(Ge) and unstressed p-InSb(Mn) crystals.
2. Hall constant sign inversion in magnetic field and under hydrostatic pressure in p-InSb(Mn).
3. ρ-T dependence described in p-InSb(Mn) by exponential quadratic function $\rho=\rho_{01}\exp(\Delta_1/kT)^2$.
4. absence of VRH conductivity type in p-InSb(Mn).
5. MIT driven by hydrostatic pressure in p-InSb(Mn).

If we compare magnetotransport effects in p-InSb(Mn) and uniaxially stressed p-InSb(Ge) crystals we can notice that specific p-InSb(Mn) properties rather relate to the influence of Jahn-Teller crystal lattice distortion caused by Mn^{++} ions than to the interaction between manganese ions. Coexistence of electrons and holes at temperatures around the metal-insulator transition in p-InSb(Mn) crystals gives the ground to explain the above mentioned unusual effects in the framework of excitonic insulator phase model [22]. It was revealed that a gap in excitonic insulator electronic spectra can vary between $\Delta\sim10^{-2}$meV and $\Delta=0{,}7$meV with the decrease of manganese concentration from $N_{Mn}=2\times10^{17}cm^{-3}$ to $10^{17}cm^{-3}$ and under hydrostatic pressure it can reach $\Delta\sim1{,}5$meV. An external magnetic field destroys permanent excitons in InSb which brings about the increase in charge carrier concentration, i.e. magnetic field induces CMR and drives IMT transition. The existence of CMR in nonmagnetic semiconductor supports the hypothesis that the origin of CMR in DS, DMS, MS and Manganite Perovskites has the same physical nature, i.e. can be described by excitons behavior in magnetic field. An existence of the gap energy Δ in InSb(Mn) in temperature millikelvin range could be used as a bases for advanced cryogenic devices creation - thermometers, bolometers and sensors for millimeter and radio-frequency wavelengths.

ACKNOWLEDGMENTS

The author gratefully acknowledges the financial support from Alexander von Humboldt Foundation (Germany) and FAPESP (Brazil). The author expresses his gratitude to Lake Shore Cryotronics (Westerville, Ohio) for measurements in a dilution refrigerator. The author greatly appreciates valuable assistance of Prof. N. F. Oliveira Jr., Prof. W. Heimbrodt, Prof. P. J. Klar, Dr. J. Teubert.

REFERENCES

[1] Haller, E.E., Palaio, N.P., Rodder, M., Hansen, W.L., & Kreysa, E. Ntd Germanium: A Novel Material for Low-Temperature Bolometers. *Proceedings of the Fourth International Neutron Transmutation Doping Conference, National Bureau of Standards, Gaithersburg, MD*, June 1-3 (1982), 1-16.

[2] Le Hir, M.J.F. Coefficient de Hall et resistivite du germanium dope a l'arsenic en fonction de la temperature. *Journal de Physique, Juillet* (1967), 28, 563-572 (in French).

[3] Obukhov, S.A. Metal–insulator transition in manganese-doped InSb crystals. *Phys. Stat. Solidi.* (b), №6, (2005), 242, 1298-1306.

[4] Ftitzsche, H. & Cuevas, M. Impurity Conduction in Transmutation-Doped p- Type Germanium. *Phys. Rev.,* (1960), 119, 1238-1245.

[5] Alexander, M.N. & Holcomb, D.F. Semiconductor-to-Metal Transition in *n*-Type Group IV Semiconductors. *Rev. Mod. Phys.,* (1968), 40, 815-829.

[6] Westervelt, R.M., Burns, M.J., Hopkins, P.F., Rimberg, A.J., & G.A. Thomas. Magnetic-filed- induced metal-insulator transition in degenerately doped n-type Ge. *Proceedings of the University of Tokyo Symposium on Anderson Localization, Tokyo,* August 16-18 (1987), 1-7.

[7] Ferre, D., Dubois, H. & Biskupski, G. Structure of the impurity band and magnetic-filed-induced metal-nonmetal transition in n-type InSb. *Phys. Stat. Solidi,* (1975), *B* 70, 81- 90.

[8] Obukhov, S.A. Giant negative magnetoresistance in uniaxially stressed p-Ge and p-InSb single crystals. *Phys. Stat. Sol.*(b), (2001), 223, 535-538.

[9] Wojtowicz, T. & Mycielski, A. An unusual magnetic field dependence of the acceptor ionization energy p-$Hg_{1-x}Mn_xTe$. *Acta Physica Polonica,* (1985), A67, 363-367.

[10] von Molnar, S. & Holtzberg, F. The effect of coulombic and magnetic disorder on transport in magnetic semiconductors. *Magnetism and Magnetic Materials – 1972,* edited by C.D. Graham, Jr., & J.J. Rhyne, AIP Conference Proceedings №10 (American Institute of Physics, New York), (1973), 1259-1273.

[11] Obukhov, S.A., Neganov, B.S., Kiselev, Yu.F., Chernikov, A.N., Vekshina,V.S., Pepik, N.I. & Popkov, A.N. Low temperature resistance of p-InSb(Mn). *Cryogenics,* (1991), 31, №10, 874-876.

[12] Staunton, H.F. & F.H. Pollak. Effect of uniaxial compression on impurity conduction in p-silicon. *Physics Letters A,* (1967), 25, 751÷752.

[13] Pollak, F.H. Effect of uniaxial compression on impurity conduction in p-germanium. *Phys. Rev*., (1965), 138, №2A, 618-631.

[14] Averkiev, N.S., Gey, W., Obukhov, S.A. & Rogachev, A.A. Giant negative magnetoresistance in uniaxially deformed, manganese –doped indium fntimonide. *Soviet Phys.-JETP Lett*., (1984), 40, 773-776.

[15] Teubert, J., Obukhov, S. A., Klar, P. J. & Heimbrodt, W. Influence of Magnetic Dopants on the Metal-Insulator Transition in Semiconductors. *Phys. Rev. Letters*, 102, 046404 (2009).

[16] Mott, N.F. *Phil. Mag*., (1969),19, 835-852.

[17] Mott, N.F. & Davis, E.A. *Electronic Processes in Non-Crystalline Materials.* Clarendon Press, Oxford (1979).

[18] Shklovskii, B.I. & Efros, A.L. *Electronic properties of doped semiconductors.* Springer Verlag, Berlin Heidelberg New York Tokyo (1984).

[19] Woodcraft, A. *Low Temperature Thermometry meeting, St. Hugh's College, Oxford*, 24 November (2005).

[20] Obukhov, S.A. Transport and magnetotransport effects in indium antimonide doped with manganese within quasimetal-insulator transition. *Phys. Stat. Sol*., (2012), C 9, No. 2, 247-250.

[21] Obukhov, S.A. A new type of low temperature conductivity in InSb doped with Mn. *AIP ADVANCES*, (2012), 2, 022116(1-5).

[22] Keldysh, L. V. & Kopaev, Yu. V. Possible instability of the semimetallic state toward Coulomb interaction. *Sov. Phys. Solid State*, (1965), 6, 2219-2224.

[23] Abricosov, A.A. *Possible mechanism of high-temperature superconductivity.* Pis'ma Zh. Eksp. Teor. Fiz., (1978), 27, 235-238.

[24] Bucher, B., Steiner, P. & Wachter, P. Excitonic insulator phase in $TmSe_{0.45}Te_{0.55}$. *Phys. Rev. Letters*, (1991), 67, №19, 2717-2720.

[25] Goryunova, N. A. & Obukhov, A. P. *Journ. of Techn. Phys.* (1952), 21, 237.

[26] Madelung, O. *Physics of III-V Compounds* (John Willey and Sons, Inc., New York-London-Sydney, (1964).

[27] Bimberg, D. in *Semiconductors - Physics of Group IV Elements and III-V Compounds*, edited by O. Madelung, Landolt-Bornstein, New Series, Vol. 17a (Springer, Berlin, (1982).

[28] Partin, D.L., Heremans, J. & Thrush, C. M. Indium antimonide doped with manganese grown by molecular beam epitaxy. *J. Cryst. Growth*, (1997), 175–176, 860-867.

[29] Volokobinskaya, N.I., Galavanov, V.V. & Nasledov, D.N. Fiz. Tverd. Tela,(**1959**) 1, 755, *Solid State Physics*, (1959), 1, 755-760.

[30] Peters, J.A., Parashar, N.D.,Rangaraju, N. & Wessels, B.W. Spin-dependent magnetotransport in a p-InMnSb/n-InSb magnetic semiconductor heterojuction. *Appl. Phys. Lett.*, (2011), 98 193506(1-3).

[31] Kane, E.O. Band structure of indium antimonide. *Journ. Phys. Chem. Solids*, (1957)1, 249-261.

[32] Pollak F.H. et al.- Bull. *Amer. Phys. Soc.*, (1968),14, 433.

[33] Raveau, B. Les oxydes des metaux de transition des materiaux fonctionnels tres prometteurs. *La lettre de l'Academie des sciences.* (2003), №10/hiver, 8-10. (in French).

[34] Germanenko, A.V., Zverev, L.P., Kruzhaev, V.V., Min'kov, G.M. & Rut', O.E. Pecularities of galvanomagnetic effects and frees-out on acceptor in magnetic field in zero-gap p-$Hg_{1-x}Mn_xTe$ at low temperatures. *Soviet. Phys. Techn. Semicond.,* (1984), 20, №1, 80-89.

[35] Lord Kelvin. *Proceedings of the Royal Society of London.* (1857), 8, 546.

[36] Tokura, Y. Correlated-electron physics in transition-metal oxides. *Physics Today*, (2003), July, 50-55.

[37] Englert, E. Ferromagnetismus und electrische Eigenschaften VI. Mitteilung: Longitudinale und transversale ferromagnetische Widerstandsanderung. *Annalen der Physik*, (1932), 406, 5, 589-612. (in German).

[38] Mclaughlin, A. C., Sher F. & Attfield, J. P. Negative lattice expansion from the superconductivity–antiferromagnetism crossover in ruthenium copper oxides. *Nature*, (2005), 436, 829-832.

[39] Zhang, X., Xue, Q.Z. and Zhu D.D. Positive and negative linear magnetoresistance of graphite. *Physics Letters*, (2004), *A* 320, 471-477.

[40] Frederikse, H.P.R. and Hosler, W.R. Galvanomagnetic effects in p-type indium antimonide. *Phys. Rev.*, (1957), 108, №5, 1136-1145.

[41] Sasaki W. Negative magnetoresistance in the metallic impurity conduction of n-type germanium. *J. Phys. Soc. Japan*, (1965), 20, 825-833.

[42] Peters, F. & Schneider G. Negative magnetische Widersandsanderung von ZnSb. *Zeitschrrift fur Naturforsch.*, (1969), 24A, 620-628 (in German).

[43] Zavaritskaya, E.I., Voronova, I.D. and Rozhdestvenskaya, N.V. Negative magnetoresistance in compensated gallium arsenide at low temperatures. *Soviet. Phys. Techn. Semicond.*, (1972), 6, №10, 1945-1953.

[44] Shapira, Y., Foner, S., Oliveira, N.F. Jr. and Reed T.B. Resistivity and Hall effect of EuSe in fields up to 150 kOe. *Phys. Rev. B,* (1974), 10, 4765-4780.

[45] Shapira, Y., Foner, S., Oliveira, N.F. Jr. & Reed, T.B. EuTe.II. Resistivity and Hall effect. *Phys. Rev. B,* (1972), 8, 2647-2657.

[46] Toyazawa, Y. Theory of localized spins and negative magnetoresistance in the metallic impurity conduction. *J. Phys. Soc. Japan*, (1962), 17, 986-1004.

[47] Mott, N.F. & Zinamon, Z. The metal-nonmetal transition. *Rep. Prog. Phys.,* (1970), 33, 881-940.

[48] Altshuler, B.L. & Aronov A.G. in *Electron-Electron Interaction in Disordered Systems*, edited by A.L.Efros and M.Pollak. North-Holland, Amsterdam (1985).

[49] Nagaev, E.L. *Colossal magnetoresistance and phase separation in magnetic Semiconductors*. Imperial College Press, London (2002).

[50] Furdyna, J.K. and Kossut, J. *Diluted Magnetic Semiconductors, 25, Semiconductors and Semimetals,* Academic Press, Inc. (1988).

[51] Cengiz, Sen C., Alvarez, G. and Dagotto, E. Competing ferromagnetic and charge-ordered states in models for manganites: The origin of the colossal magnetoresistance effect. *Phys.Rev. Letters,* (2007), 98, 127202(1-4).

[52] Coey, J. M. D., Viret M. & von Molnar S. Mixed-valence manganites. *Advances in Physics*, (1999), 48, 167-293.

[53] Yan, Y.F., Matl, P., Harris J.M. & Ong N.P. Negative magnetoresistance in the *c* –axis resistivity of $Bi_2Ca_2Cu_2O_{8+\delta}$ and $YBa_2Cu_3O_{6+\chi}$. *Phys.Rev.B*, (1995), 52, №2, R751-R754.

[54] Barber, R.P.Jr., Hsu, Shih-Ying, Valles, J.M., Dynes, R.C. & Glover, III R.E. Negative magnetoresistance, negative electroresistance, and metallic behavior on the insulating side of the two-dimensional superconductor-insulator transition in granular Pb films. *Phys. Rev. B,* (2006), 73, 134516 (1-5).

[55] Matsukura, F., Sawicki, M., Dietl, T., Chiba, D. and Ohno, H. Magnetotransport properties of metallic (Ga,Mn)As films with compressive and tensile strain. *Physica E,* (2004), 21, 1032–1036.

[56] Xiao, John Q., Jiang, J. Samuel & Chien, C. L.. Giant magnetoresistance in nonmultilayer magnetic systems. *Phys. Rev. Lett.,* (1992), 68, 3749-3752.

[57] Sigiyama, K._& Kobayashi, A. Piezoresistance and magnetoresistance in impurity conduction of germanium. *J. Phys. Soc. Japan,* (1963), 18, 163-174.

[58] Chroboczek, J., Fritzsche, H., Jiang, C.-L. & Pollak, M. Piezomagnetoresistance studies of p-type germanium in the hopping conduction region. *Proceeding of the 13th International Conference on Physics of Semiconductors, Rome, North-Holland, Amsterdam.* (1976), 1113-1116.

[59] H. Jahn and E. Teller. Stability of Polyatomic Molecules in Degenerate Electronic States. I. Orbital Degeneracy. Proceedings of the Royal Society of London. Series A, *Mathematical and Physical Sciences* (1934-1990) 161 (905): 220–235.

[60] Obukhov, S.A. New type of cryogenic semiconductor resistance thermometer. *Cryogenics*, (1994), 34, №3, 237-240.

[61] Obukhov, S.A. & Trofimov, V.N. New Low-Temperature Thermistors InSb:Mn for Nuclear Cryogenic Detectors. *Journal de Physique IY, Colloque 3, Supplement au Journal de Physique III,* (1996), 6, avril, C3, 169-174.

In: Indium
Editors: Hsaio G. Woo and Huang Tsai Choi
ISBN: 978-1-62257-696-8

Chapter 4

ALKYNYL, ALKENYL AND ARYL INDIUM REAGENTS

Jacques Augé, Nadège Lubin-Germain and Jacques Uziel
Université de Cergy-Pontoise, 5 mail Gay-Lussac,
Neuville-sur-Oise, Cergy-Pontoise, France

ABSTRACT

Indium-mediated organometallic reactions have elicited considerable attention from the discovery of the remarkable reactivity of the metal in organic or aqueous media. Organometallic species may be postulated when In(0) or In (I) is inserted into a carbon-halogen bond. Organoindium reagents may also be prepared by transmetalation. This chapter is specifically devoted to alkynyl, alkenyl and aryl indium reagents and their use in organic synthesis.

1. INTRODUCTION

Indium-mediated organometallic reactions have elicited considerable attention from the discovery of the remarkable reactivity of the metal in organic [1] or aqueous media [2]. Several authors have highlighted the special interest of using indium metal in aqueous media since it is not affected by water or by air [3-7]. Several reviews describe different aspects of indium (0) and indium(III) chemistry [8-15]. Main attention was focused on allylation reactions, since In(0) or In(I) is easily inserted into the carbon-halogen bond of allyl halides [16]. Apart the Barbier methodology, organoindium reagents can be generated through transmetalation with $InCl_3$ as the source of indium [17]. Another mechanism, which prevails with alkyl halides, goes through a radical intermediate directly from indium (0) as the reducer agent [18] or In(III) in the presence of hydrides [19]. This chapter is specifically devoted to reactions in which alkynyl, alkenyl or arylindium species may be postulated as transient species.

2. Alkynyl Indium Reagents

2.1. Indium (0) Mediated Carbonyl Alkynylation and Related Alkynylation

Barbier-type alkynylation, *i. e.* the addition of alkynyl iodide upon carbonyl compounds in the presence of a metal was discovered by Augé *et al.* in 2002 with the use of indium [20]. To carry out such indium(0) mediated carbonyl alkynylation, alkynyl iodides are previously prepared from alkynes and iodine in the presence of morpholine. With an excess of alkynyl iodide and indium metal, aldehydes and ketones afford the corresponding propargylic alcohols in refluxing dichloromethane (Scheme 1). The alkynylation reaction of ketones is generally slower than the same reaction with aldehydes. Owing to a competitive carboindation leading to an enyne, an excess of alkynyliodide is required to get a good yield; 2 equiv of alkynyl iodide are therefore engaged in the reaction which proceeds nicely except with aromatic ketones for the previously invoked reason. When the aldehyde is in excess, a propargylic ketone is formed, as a by-product arising from an Oppenauer-type oxidation of the indium(III) alkoxide intermediate (Scheme 1). To prepare in good yields such propargylic ketones from non enolisable aldehydes, 2 equiv of aldehyde are used in refluxing dichloroethane rather than dichloromethane. Propargylic ketones can also be prepared from indium-mediated alkynylation of acyl chlorides [21].

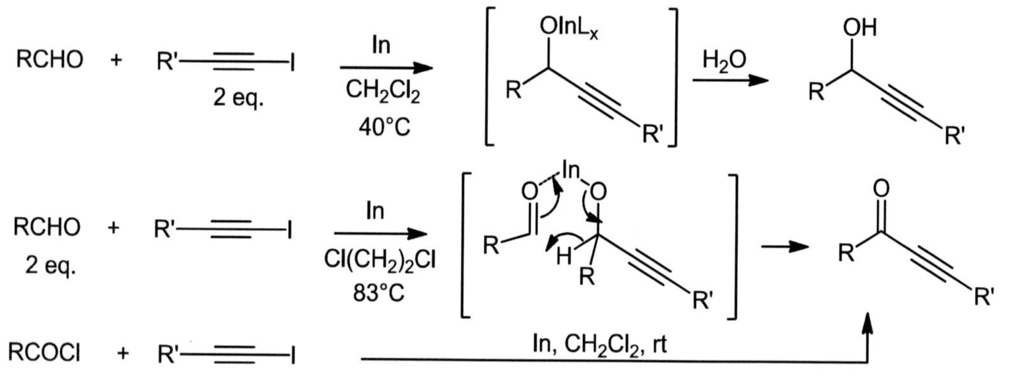

Scheme 1 Indium-mediated alkynylation of carbonyl compounds

This Barbier-type indium mediated alkynylation of aldehydes was applied to the alkynylation of perbenzylated and peracetylated formylglucose affording the corresponding propargylic alcohols in 87% and 71% respective yields, both of them obtained as a mixture of diastereoisomers in a ratio 65/35 (Scheme 2). In these cases no further Oppenauer oxidation of the propargylic alcohols to the corresponding propargylic ketones was observed [22]. The indium-mediated alkynylation of perbenzylated formylglucose was also carried out with functionalized amino acids precursors [23].

In the indium(0) mediated formation of quinolines (Scheme 3), nitroarenes are first reduced to anilines by indium/hydrochloric acid. Whereas indium is oxidised to $InCl_3$, the transient anilines react immediately with aldehydes, producing the corresponding imines. These imines may undergo the addition of indium phenylacetylide, as a transient species arising from phenylacetylene and $InCl_3$ [24]. The further cyclization may be envisioned as an arylindation of alkynes (see paragraph 4.6.).

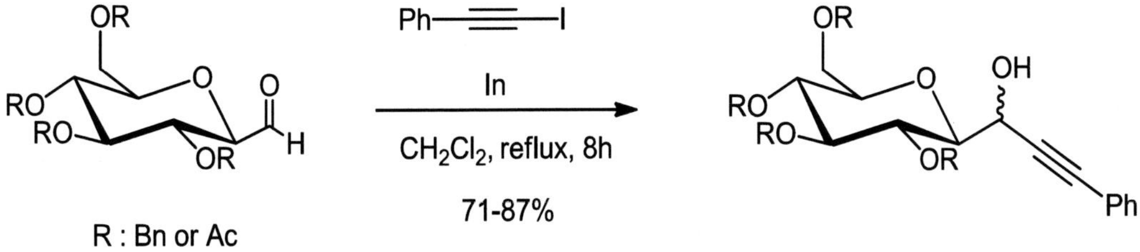

Scheme 2 Indium-mediated alkynylation of formylglucose derivatives

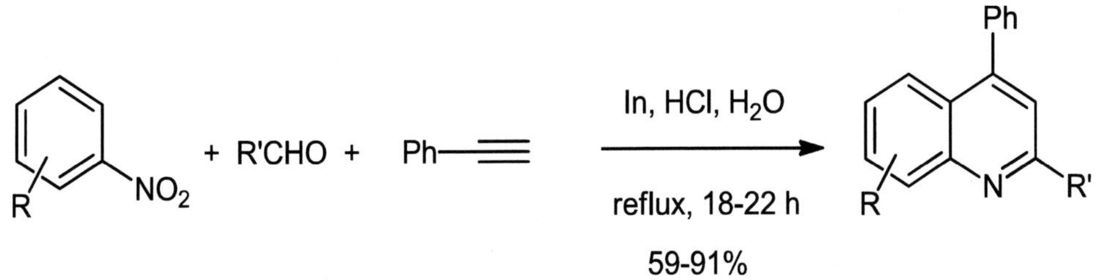

Scheme 3 Synthesis of quinoline derivatives

2.2. Indium(III) Catalyzed Carbonyl Alkynylation and Related Alkynylation

The indium(0) mediated carbonyl alkynylation requires the previous formation of alkyne iodide but the advantage of this reaction lies on the fact that the alkynylation proceeds in neutral conditions. Another strategy is based on the transmetalation of alkynylstannanes giving rise to alkynyldichloroindium which smoothly adds to aldehydes in acetonitrile (Scheme 4); a catalytic amount of $InCl_3$ (0.1 equiv) is sufficient owing to the presence of chlorotrimethylsilane as an oxophile regenerating the indium salt [25].

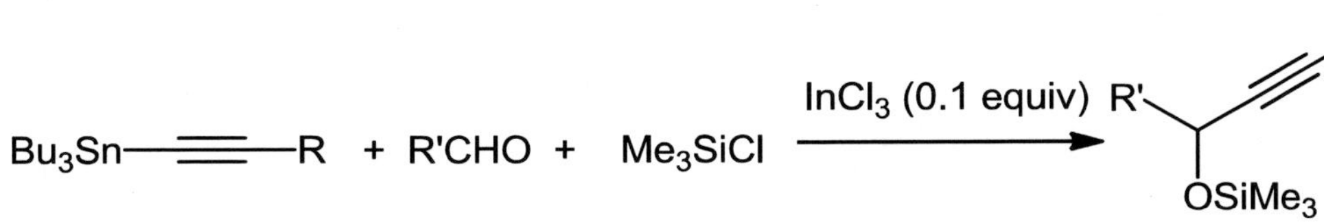

Scheme 4 Transmetalation of alkynylstannanes

In the silane series, chlorotrimethylsilane can be omitted, so that alkynylsilanes undergo coupling with acid chlorides in the presence of 0.05 equiv of $InBr_3$ at 0°C to afford the corresponding α,β-acetylenic ketones (Scheme 5) [26].

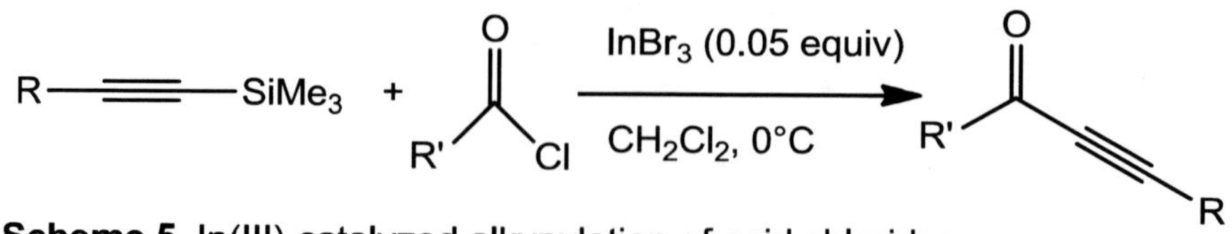

Scheme 5 In(III)-catalyzed alkynylation of acid chlorides

A new challenge appeared with the direct use of terminal alkynes which could be metalated by indium (III) salts under basic conditions, in order to be added to carbonyl compounds according to a mechanism previously described by Carreira in zinc chemistry [27]. Since the acetylenic hydrogen of phenylacetylene is more acidic than that of aliphatic acetylenes, C_{sp}-H bond of phenylacetylene can be activated using Ru(III)-In(III) catalytic system, giving a ruthenium intermediate which undergoes Grignard-type addition to aldehydes under aqueous conditions [28]. The reaction is however limited to aromatic aldehydes and aliphatic aldehydes without α-hydrogen, probably because of the use of amine as additive (Scheme 6).

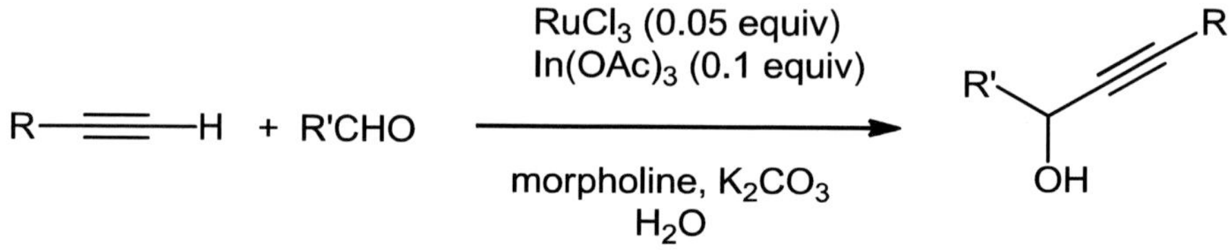

Scheme 6 Ru-In Grignard type addition

Indium acetylide is formed when stoichiometric amounts of $InBr_3$ and Et_3N are mixed to an alkyne in ether giving access to propargylic alcohols from aromatic aldehydes and aliphatic non-enolisable aldehydes [29, 30]. Interestingly, propargylic amines can be prepared from *N,O*-acetals in the same conditions, although the alkynylation proceeds faster (Scheme 7).

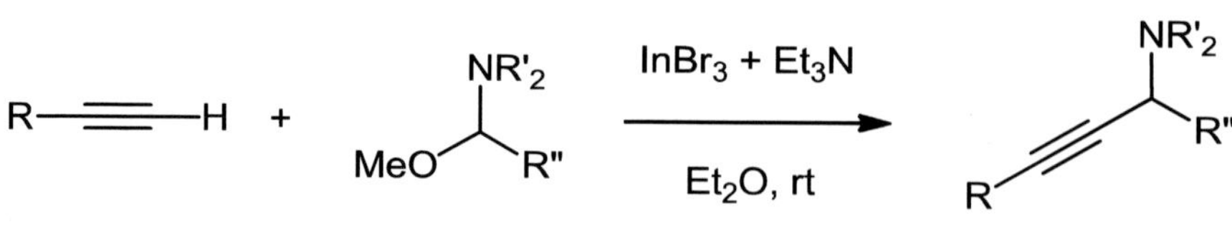

Scheme 7 $InBr_3$ - Et_3N promoted alkynylation

A catalytic version of the In(III)/tertiary amine reaction of alkynes with aldehydes was recently carefully investigated [31]. Under the following conditions (0.1 equiv of $InBr_3$, 0.2 equiv of *i*-Pr_2NEt, 2 equiv of terminal alkynes) a variety of aromatic aldehydes were smoothly converted into the corresponding propargylic alcohols (Scheme 8). In contrast, aliphatic aldehydes, except cyclohexane carboxaldehyde, gave unsatisfactory results due either to their self-condensation (for aldehydes with a primary alkyl group) or to their low reactivity (for aldehydes with a tertiary alkyl group). Changing the indium source from $InBr_3$

to $In(OTf)_3$ greatly improved the reactivity of cyclohexanone as model substrate. Under the optimized reaction conditions (0.2 equiv of $In(OTf)_3$, 0.5 equiv of *i*-Pr_2NEt at 40°C in DME) a large range of ketones were successfully converted to the tertiary corresponding propargylic alcohols. *In situ* IR and NMR spectroscopic studies showed that dual activation of both terminal alkynes and carbonyl group by In(III) salts occurred.

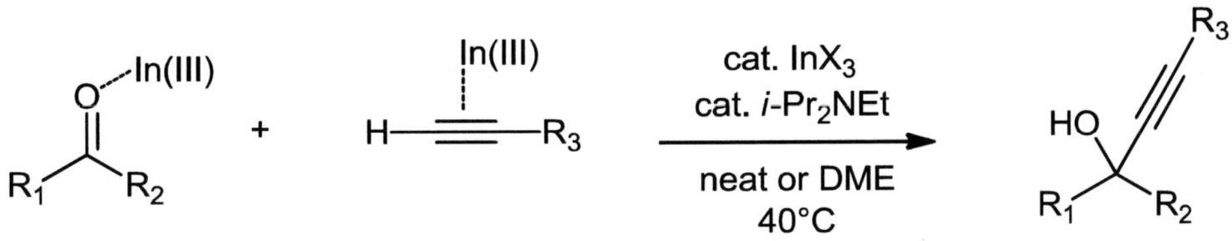

Scheme 8 Dual activation in the catalyzed-alkynylation of carbonyl compounds

The $In(OTf)_3$-catalyzed alkynylation of ketones was applied to 2-aminoaryl ketones in solvent-free conditions, using microwave irradiation at 110°C for a very short time (4-5 min). The resulting arylpropargylic alcohols underwgo a Meyer-Schuster rearrangement affording 2,4-disubstituted quinolines [32].

A slight modification of the catalytic system with the addition of a chiral ligand made enantioselective the alkynylation of aldehydes via an In(III)/BINOL complexe [33, 34]. Excellent enantioselectivities of propargylic alcohols (up to 99%) were obtained with various aldehydes and alkynes. The reaction was performed in CH_2Cl_2 at 40°C using Cy_2NMe (0.5 equiv) as a base, and $InBr_3$ (0.1 equiv) as a catalyst. For the easily enolisable aldehydes, slow addition of the aldehyde prevented self-condensation. (*R*)-BINOL (0.1 equiv) induced the nucleophilic attack of the indium acetylide on the *Re* face of the aldehyde; strong positive non linear effects were observed suggesting a bimetallic mechanism in the catalytic cycle. The accelerating effect of biaryldiol ligands, such as 2,2'-biphenol and BINOL was particularly underlined in the indium(III)-catalyzed alkynylation of aldehydes with 2-methyl-3-butyn-2-ol as an ethyne equivalent donor [34, 35]. By changing the amine from *i*-Pr_2NEt to Cy_2NMe the reactivity was further improved. With (*S*)-BINOL as a chiral ligand, the reaction (Scheme 9) proceeded smoothly in moderate to good yields (up to 97%) and high enantioselectivities (up to 99% ee).

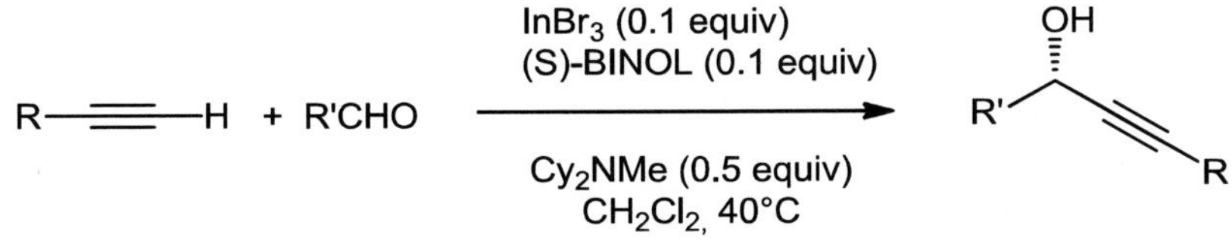

Scheme 9 $InBr_3$/BINOL complex-catalyzed asymmetric alkynylation

A related indium-catalyzed alkynylation is the three-component coupling of an aldehyde, an alkyne and an amine via C-H bond activation by $InCl_3$ (Scheme 10) [36] This A^3-coupling currently performed under copper, silver or gold catalysis [37] is a versatile method to prepare propargyl amines with a high atom economy (H_2O is the only by-product).

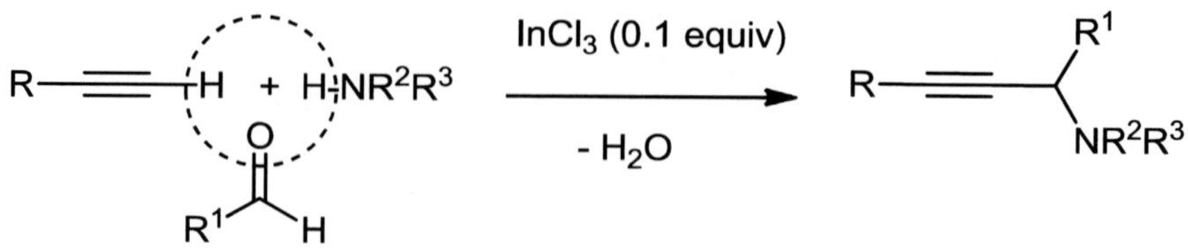

Scheme 10 A^3-Coupling Reaction

In a similar way, *N*-hydroxy-propargylamines could be directly prepared in high yields from nitrones and terminal alkynes in the presence of 0.15 equiv of $InBr_3$ and 0.20 equiv of *i*-Pr_2NEt (Scheme 11) [38].

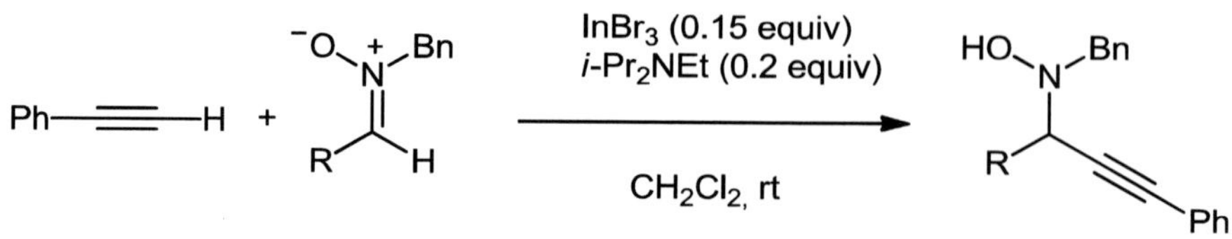

Scheme 11 $InBr_3$ catalyzed alkynylation of aliphatic nitrones

In the absence of base, but in the presence of CuCl (0.12 equiv), phenylacetylene adds to aldimines under aqueous conditions with only 0.12 equiv of $InCl_3$ affording the corresponding propargyl amines [39].

2.3. Indium Mediated Alkynylation by Substitution Reaction

By analogy with the A^3-coupling, the three-component coupling of alkynes, dihaloalkanes and amines (AHA coupling) has been developed to give propargyl amines with the use of nano In_2O_3 as a recyclable catalyst (Scheme 12) [40]. Accordingly, the plausible reaction path goes through the transient formation of an alkynyl indium species which reacts with dichloromethane followed by coupling to amines to afford propargyl amines.

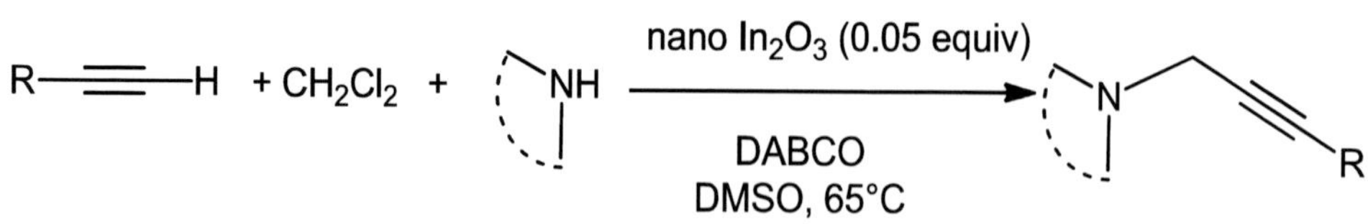

Scheme 12 AHA coupling

An alkynyl indium species is also supposed to be a plausible intermediate in the preparation of alkynyl selenides from terminal alkynes (Scheme 13) [41].

Phenylethynyldichloride preformed from lithium phenylacetylide and $InCl_3$ could add to iodoacetate to give the substitution product in the presence of a radical initiator (Scheme 14)

[42]. The reaction was also carried out with alkenylindium and arylindium reagents. In these reactions, dichloroindium (II) radical is an efficient radical mediator.

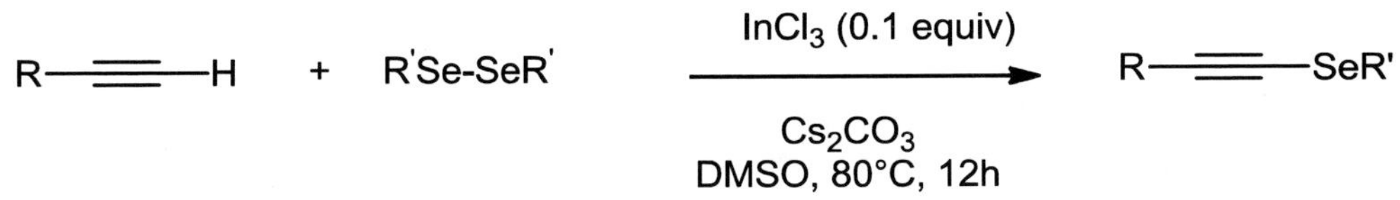

Scheme 13 $InCl_3$ catalyzed alkynylselenides formation

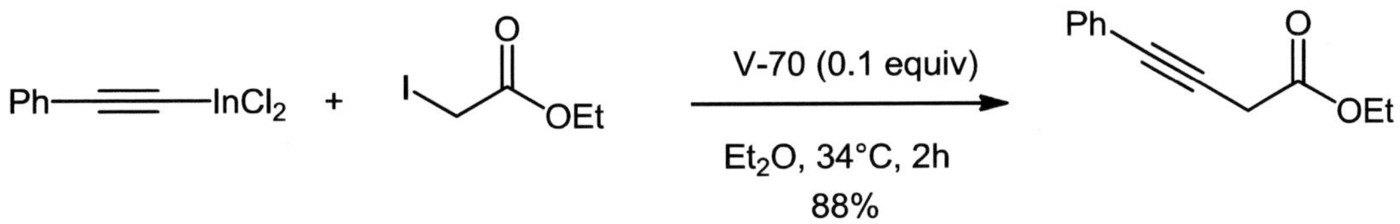

Scheme 14 Radical alkynylationof α-iodo ester

The transmetalation reaction was also used to preform triorganoindium reagents prone to react in stoichiometric amount with tri-O-acetyl glucal affording *C*-aryl, *C*-alkenyl and *C*-alkynyl glycosides [43]. The use of a catalytic amount of indium salts is possible with alkynyl silanes, which react with protected glycals to afford *C*-alkynyl glycosides in excellent yields with high α-selectivity (Scheme 15) [44].

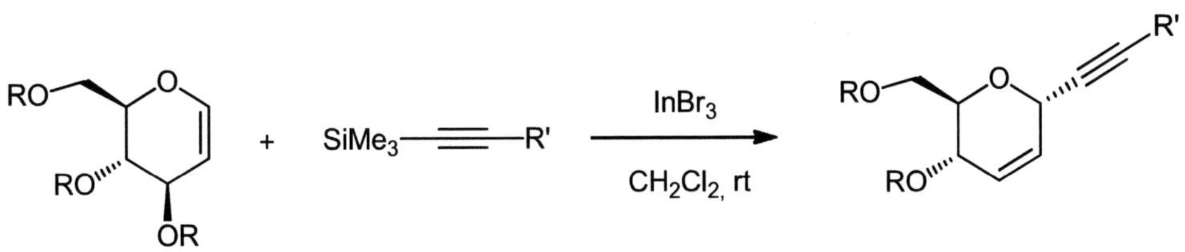

Scheme 15 $InBr_3$ - catalyzed synthesis of alkynylsugars

This indium-mediated Ferrier rearrangement was investigated under Barbier conditions. In (0), In (I) and In(II) could promote this S_N2'-type reaction leading to α-2,3-unsaturated-*C*-glycosides from glycals and iodoalkynes (Scheme 16) [45]. When glycosyliodoalkynes were used, trehalose-derived compounds and α-(1→6)-*C*-disaccharides were obtained. Different precursors of *C*-glycosyl amino acids with α configuration were prepared according to this methodology [46].

Similarly, Baylis-Hillman acetates undergo smooth alkynylation with aryl-substituted iodoalkynes to afford 1,4-enynes (Scheme 17). With indium (0) alone, a 1:1 mixture of S_N2 and S_N2' products are obtained; upon addition of $InBr_3$ (0.1 equiv) the reaction proceeds preferably in the S_N2 manner [47].

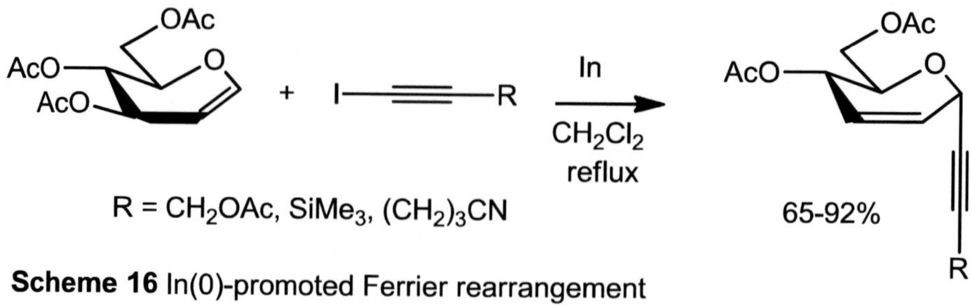

Scheme 16 In(0)-promoted Ferrier rearrangement

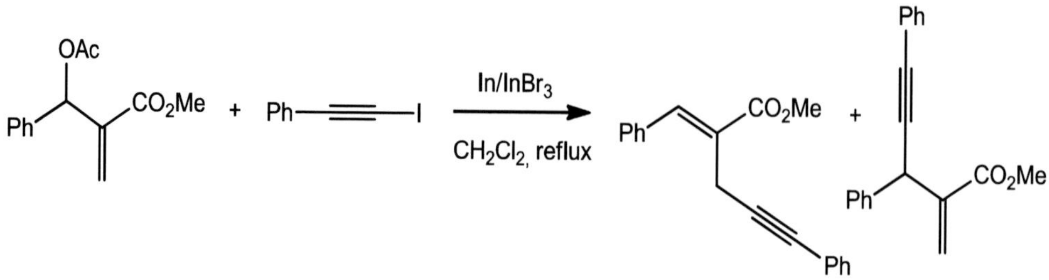

Scheme 17 Indium-mediated alkynylation of Baylis-Hillman acetates

The Barbier-type methodology was also applied for the direct *C*-glycosylation of carbohydrates bearing an acetyl group at the anomeric position. In the pyranose series, good α selectivities were obtained (Scheme 18). The reaction is compatible with various protecting groups except the participating group in the 2-position. This alkynylation reaction was applied to the synthesis of a α-(1→6)-*C*-disaccharide analogue of isomaltoside [48].

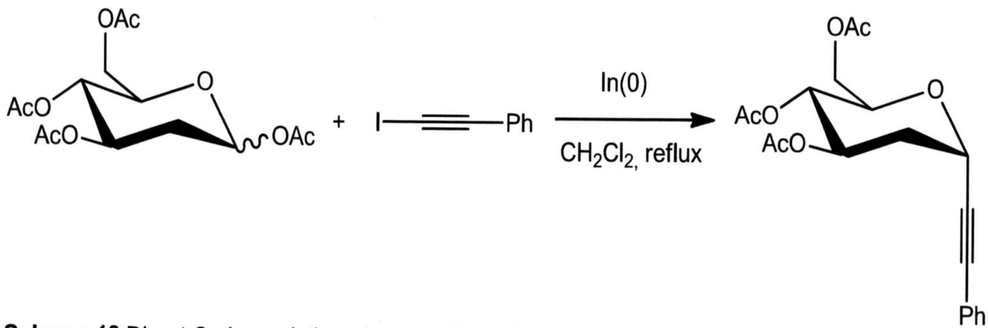

Scheme 18 Direct C-glycosylation at anomeric position

In the furanose series, the selectivity is in favour of the β product, which was exploited in the synthesis of ethyl *C*-ribosylpropiolate as a precursor of ribavirin analogues via a Huisgen cycloaddition reaction (Scheme 19) [49].

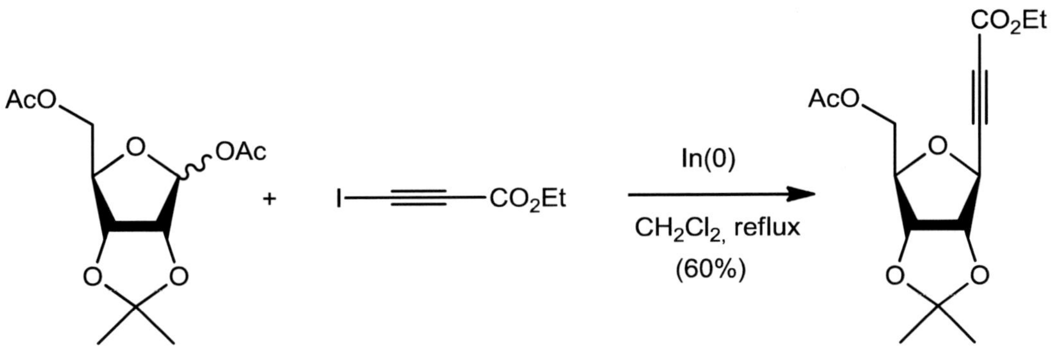

Scheme 19 Synthesis of ethyl *C*-ribosylpropiolate

2.4. Indium (III)-catalyzed Sonogashira-type Alkynylation

Indium acetylides formed in situ from terminal alkynes and a catalytic amount of $InBr_3$ (0.05 equiv) in the presence of piperidine can react with aryl iodides in the presence of Pd $Cl_2(PPh_3)_2$ (0.05 equiv) to afford the corresponding cross coupling products (Scheme 20) [50].

With R= Ph, such Sonogashira coupling was carried out in refluxing benzene in the absence of palladium, using only $InCl_3$ (0.10 equiv) as catalyst [51].

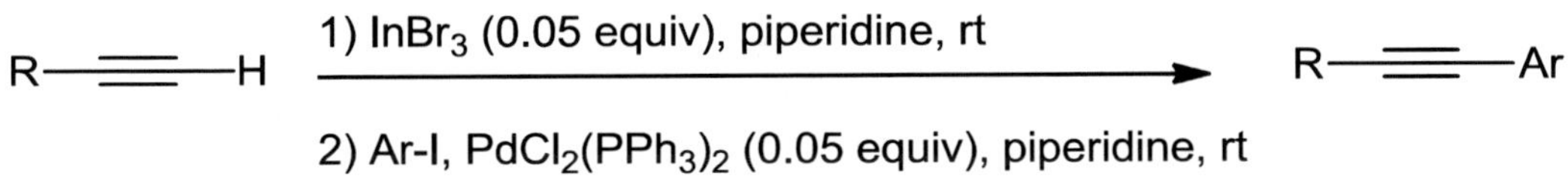

Scheme 20 Palladium-catalyzed coupling reaction in presence of $InBr_3$

3. Aryl Indium Reagents

The first insertion of indium (0) into the sp^2 carbon-iodine bond was mentioned when three equiv of pentafluorophenyl iodide reacted with 4 equiv of indium to afford $(C_6F_5)_3In$ [52]. Unlike In(0), InBr (or In + 0.5 Br_2) could insert into the carbon-bromine bond of C_6F_5Br in THF to afford $C_6F_5InBr_2$.2 THF [53]. For non-fluorinated aryl iodides, the insertion is more difficult and requires the presence of LiCl as a reaction additive [54, 55].

According to the procedure of Knochel *et al.* [54], indium (2-4 equiv), previously activated with chlorotrimethylsilane and 1,2-dibromoethane, is added to LiCl (2-4 equiv) in THF. In these conditions, functionalized aryl and heteroaryl iodides give after 20 min-24 h at 30-50°C the corresponding aryl and heteroaryl indium (III) species in 70-97% yield (Scheme 21). The organoindium reagents which may contain ester, ketone or aldehyde functionalities, undergo cross-coupling reactions under palladium catalysis with a variety of aryl iodides.

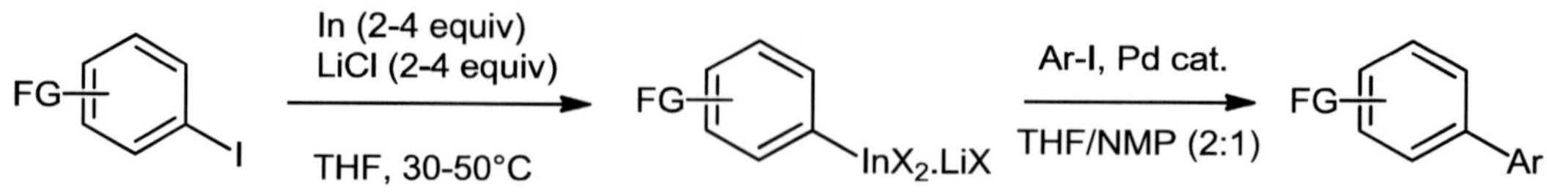

Scheme 21 Aryl indium (III) reagent by direct insertion of indium in presence of LiCl

According to the procedure of Minehan *et al.* [55], a mixture of indium containing 0.1% Mg (1.5 equiv) and LiCl (3 equiv) is activated with stirring in vacuum at 110°C. Aryl iodides (or eventually aryl bromides) with electron withdrawing groups (ester, CN, Cl) undergo the insertion of indium at 120°C for 12 h in DMF (Scheme 22). The resulting aryl indium(III) species can be used in various palladium-catalyzed cross-coupling reactions, such as coupling with aryl iodides, acyl chlorides and allylic acetates.

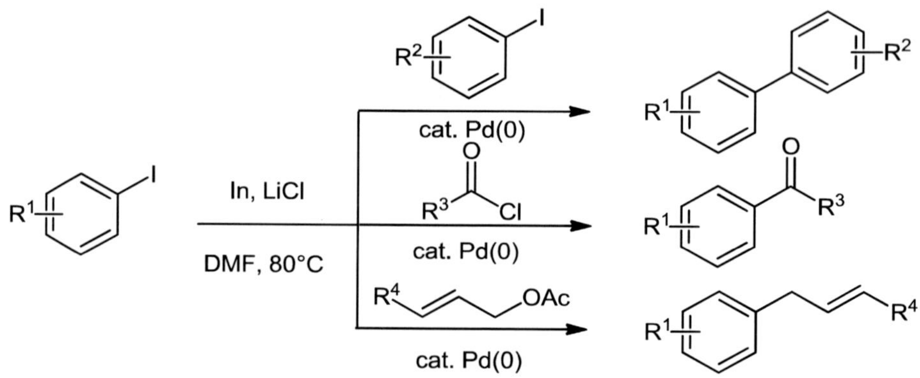

Scheme 22 Palladium catalyzed reactions of arylindium reagents

These two methods to prepare arylindiums are mostly limited to aryl iodides. With aryl and heteroaryl bromides, a cobalt-bathophenanthroline catalyst promotes the insertion of indium into the carbon-bromine bond in the presence of a stoichiometric amount of LiCl (Scheme 23) [56]. ^{1}H and ^{13}C NMR analysis of phenylindium indicates that it mainly consists of two species in an approximative ratio 2:1. The authors have concluded that phenylindium is a mixture of phenylindium(III) dihalide ($PhInX_2$) and diphenylindium (III) halide (Ph_2InX) and/or their aggregate species including phenylindium (III) sesquihalide ($Ph_3In_2X_3$). The thus-formed arylindium reagents also undergo cross-coupling reactions.

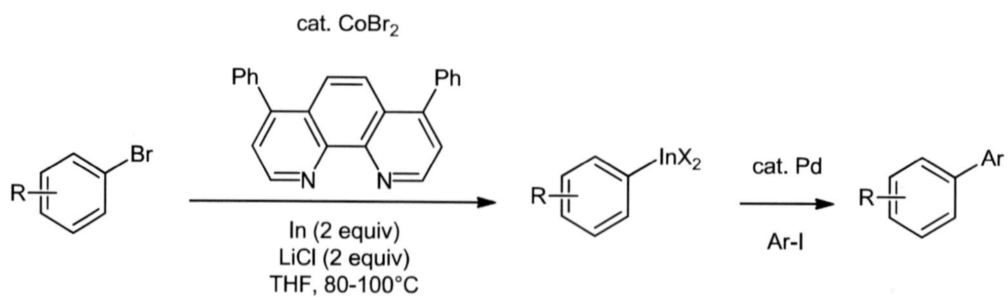

Scheme 23 Cobalt-catalyzed preparation of arylindium reagents

4. ALKENYL INDIUM REAGENTS

To our knowledge, alkenyl indium reagents have never been prepared by the direct insertion of indium metal into the carbon-halogen bond of an alkenyl halide. Alkenyl indium reagents have been evidenced in the indium (0) mediated allylation of C=O and C=N bonds with 3-bromo-1-iodopropene, which are immediately coupled with aryl, alkenyl or allyl halides (Scheme 24) [57].

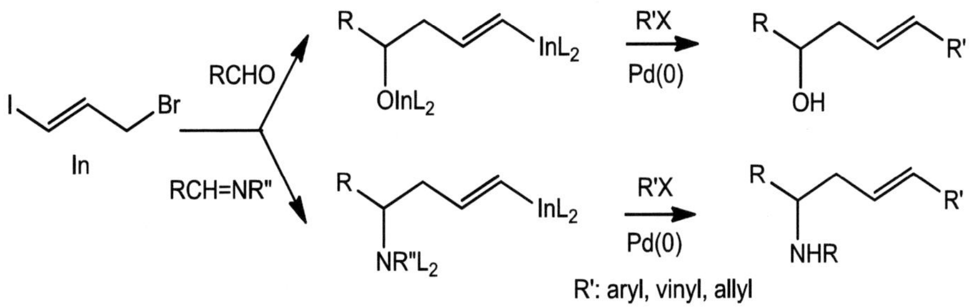

Scheme 24 Cascade coupling reactions

Although transmetalation of alkenyl lithium or magnesium reagents remains the classical method to prepare alkenyl indium reagents (see paragraph 5), a new methodology based on the indation of alkynes leads to the formation of transient alkenyl indium species.

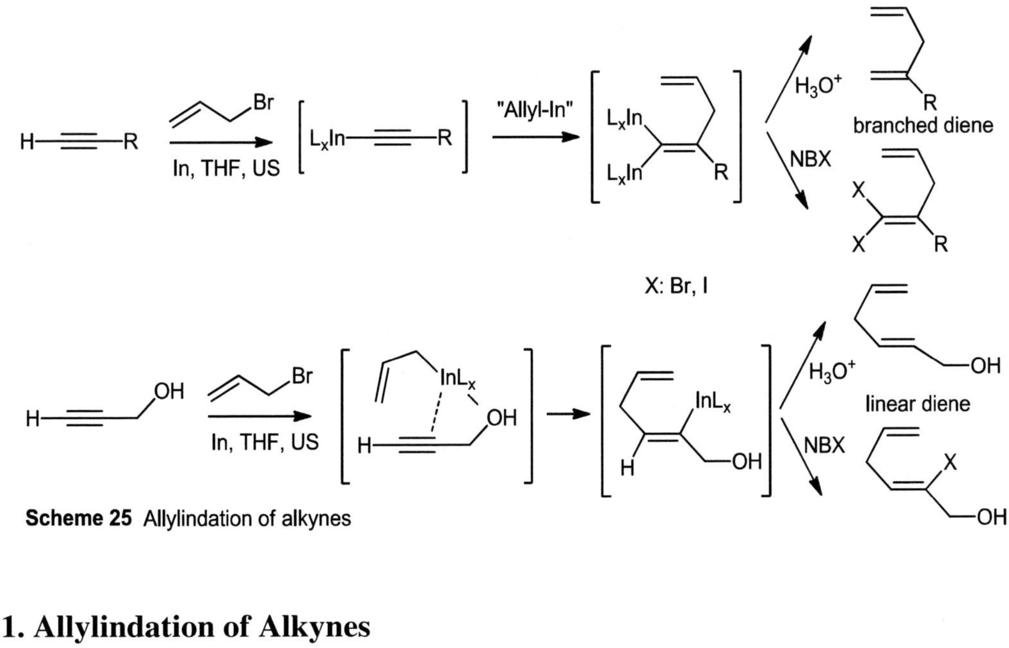

Scheme 25 Allylindation of alkynes

4.1. Allylindation of Alkynes

The first example of allylindation of alkynes was described by Butsugan *et al.* with the addition of allylindium reagent, prepared from allylic bromide and indium metal, onto alkynols in DMF at 100-140°C for 3-6 h [58]. The regioselectivity is dictated by the chelation between hydroxy group and indium leading to the anti-Markovnikov product with *E*-

configuration [59]. Such a chelation was also observed in the selective allylation of 2,3-butadien-1-ol yielding 2,6-heptadien-1-ol in DMF at 140°C; the *E*-configuration of the double bond comes from the attack of the allylindium reagent *syn* to the hydroxymethyl group [60]. Changing the solvent, in THF, even the inactivated terminal alkynes undergo indium (0) mediated allylation with allyl iodide [61] or bromide [62] in smooth conditions. Non-functionalized alkynes, such as hex-1-yne, oct-1-yne and phenylacetylene afford the Markovnikov adducts. As expected, the alkynols lead to the anti-Markovnikov adducts, but interestingly the corresponding ethers or the relevant tertiary amines undergo Markovnikov coupling exclusively [63]. As a possible mechanism of the allylindation of terminal alkynes, it was suggested by Schmid that an indium acetylide is first formed, followed by the addition of 1 equiv of allylindium giving rise to a vinylic α,α-bis-indium intermediate and then to branched 1,4-dienes; the intermediate could be quenched with *N*-bromosuccinimide and *N*-iodosuccinimide to form *gem*-dihalo-1,4-dienes [64]. This mechanism could be applied even to alkynols in which the hydroxyl group is remote from the alkyne function. In contrast, propargylic alcohol gave linear 1,4-dienes suggesting the involvement of bicyclic chelation-controlled transition state; quenching of the intermediate by *N*-halosuccinimide yielded monohalogenated dienes (Scheme 25).

Carbometalation of allenyl alcohols is also possible with propargyl/allenyl indium compounds under sonication, but $InBr_3$ as a catalyst is also required for this propargylation/allenylation reaction [65]. Thus the reaction between prop-2-ynyl bromide and 1,2-disubstituted allenols in the presence of a stoichiometric amount of indium, carried out in THF gives *E*-1-substituted-2-substituted-hepta-2-en-6-yn-1-ol. With but-2-ynyl bromide the reaction leads to *E*-1-substituted-2-substituted-5-methyl-heptatri-2,5,6-en-1-ol. This product can be directly obtained from 2 equiv of but-2-ynyl bromide and the corresponding aldehyde.

Allylindation can proceed via intramolecular addition; thus the cyclization of 1-bromo-2,7- and 1-bromo-2,8-enynes mediated by indium in DMF at 100°C produces five- and six-membered cyclic compounds [66]. The intramolecular cyclization can be performed at room temperature in THF/H_2O or in dry THF in the presence of organic acid; it was also mediated by stoichiometric In(I) or sub stoichiometric In(III) in the presence of Zn as a coreductant [67].

As an alternative, the Barbier-type pathway can be replaced by a transmetalation method. Starting from 8-tributylstannyl-6-octen-1-ynes (allylstannanes bearing an alkynyl group), the addition of 1 equiv of $InCl_3$ gives a transient allyl indium reagent which undergoes cyclization to afford an alkenyl indium species, prone to hydrolysis into 2-allyl-1-methylenecyclopentanes (Scheme 26). With only 0.2 equiv of $InCl_3$ the cyclization leads to a vinylstannane, which can be isolated and hydrolyzed to afford 2-allyl-1-methylenecyclopentanes [68]. With silanes instead of stannanes, the reaction proceeds via electrophilic activation of the triple bond by indium salt [69].

4.2. Alkylindation of Alkynes

Low-valent indium species formed from bromine or iodine with an excess of indium can initiate radical processes [53] This method is well adapted to alkylindation of alkynes to provide alkenyl indium species.

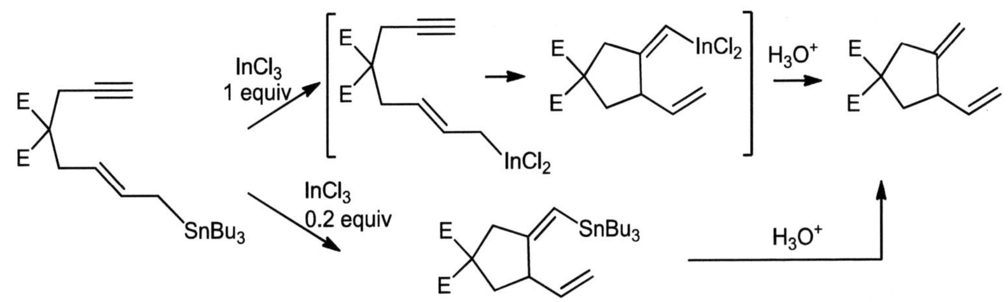

Scheme 26 In(III)-promoted intramolecular addition of allylstannanes to alkynes

Radical cyclization reactions of aliphatic iodoalkynes give alkenyl indium species in the presence of a catalytic amount of indium (0.1 equiv) and iodine (0.05 equiv) (Scheme 27). To demonstrate versatility of these reactions, optically active HIV protease inhibitors were synthetized by this reductive cyclization method [70].

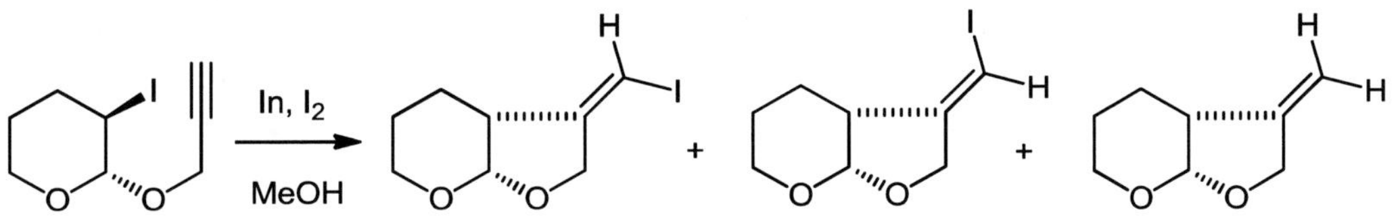

Scheme 27 Radical cyclization of iodoalkynes

One of the key steps in the enantioselective synthesis of (-) aphamorphine is this indium-initiated atom transfer radical cyclization using indium (2 equiv) and I_2 (1 equiv) in DMF (Scheme 28) [71].

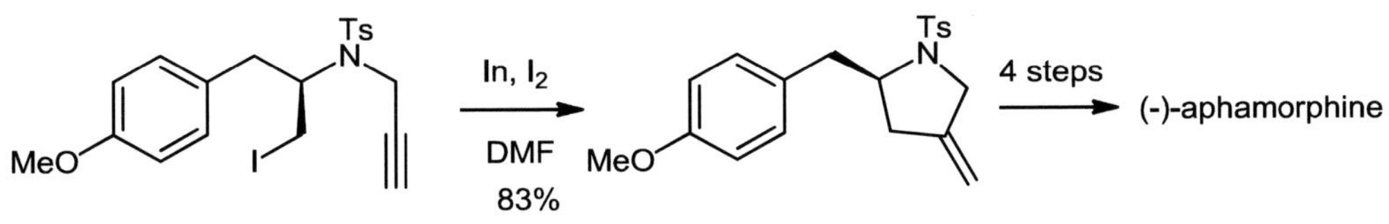

Scheme 28 Synthesis of (-)-aphamorphine

The couple In/I_2 can be replaced by the couple In/Br_2 or even $In/Py.HBr_3$. Such conditions were used in the 5-*exo*-radical cyclization of iodo-ynamides, leading to alkenyl lindium intermediates, which can be quenched by a proton or which can undergo palladium-catalyzed cross-coupling reactions (Scheme 30) [72].

According to a similar mechanism, InI in a stoichiometric amount can promote the cyclization of δ-bromoalkynes giving rise to alkenyl indium intermediate, which is hydrolyzed to 4-methylene tetrahydrofurans [73].

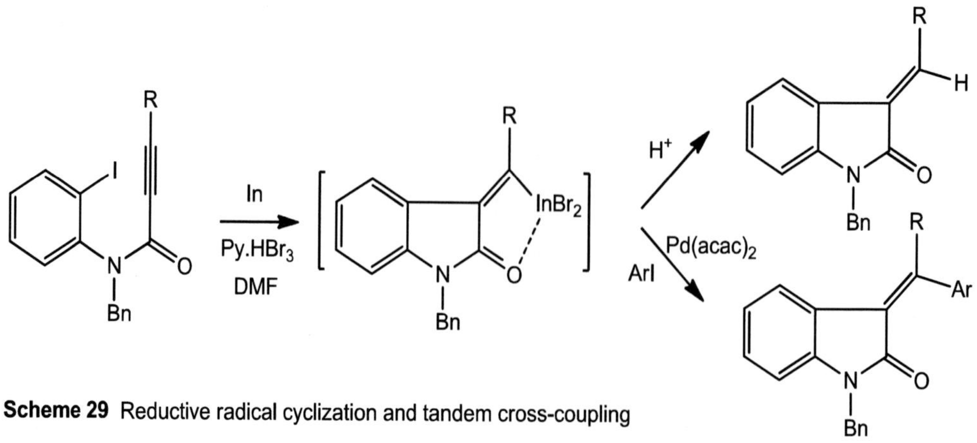

Scheme 29 Reductive radical cyclization and tandem cross-coupling

4.3. Hydroindation of Alkynes

Transmetalation between $InCl_3$ and hydride, such as $NaBH_4$, DIBAH or silanes can generate $HInCl_2$. The transient formation of $InCl_2$ as a radical intermediate is the driving force of the subsequent hydroindation of alkynes to produce vinyl species.

Terminal aryl alkynes can be reduced by the $NaBH_4$/stoich $InCl_3$/CH_3CN system to give transient alkenyl indium species which can couple with terminal akynes [74], aryl iodides (or bromides) [75] and diaryliodonium salts [76] to afford preferentially *E*-enynes and *E*-alkenes, respectively, without the need of a transition metal as catalyst (Scheme 30). No coupling product was obtained with aliphatic terminal alkynes Hydroindation of arylalkynylsilanes requires THF as solvent to afford *E*-alkenyl silanes [73]. Direct hydroindation of 2-arylalkynylphosphonates with the $NaBH_4$/ stoich $InCl_3$/ CH_3CN system gives after hydrolysis *E*-2-arylvinylphosphonates [77].

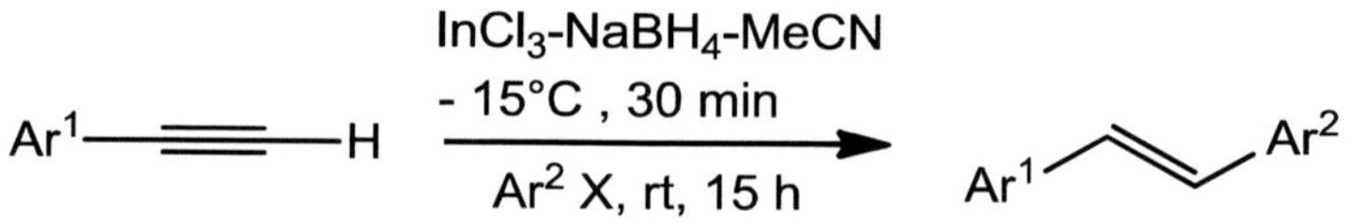

Scheme 30 Hydroindation with $InCl_3$-$NaBH_4$-MeCN system

Treatment of $InCl_3$ in THF with DIBAH at 0°C yields dichloroindium hydride, which can further react with terminal alkynes in the presence of a catalytic amount of Et_3B as a radical initiator [78]. Hydroindation proceeds in an *anti*-fashion, giving *Z*-alkenylindium dichloride and then the corresponding *Z*-product by deuteration, iodolysis or palladium-catalyzed cross-coupling reaction with aryl iodide (Scheme 31). *E*- or *Z*-alkenylindium dichloride prepared by this way can also react with α-halo carbonyl compounds with retention of configuration, which constitutes an alternative to the transmetalation of vinyllithium [79].

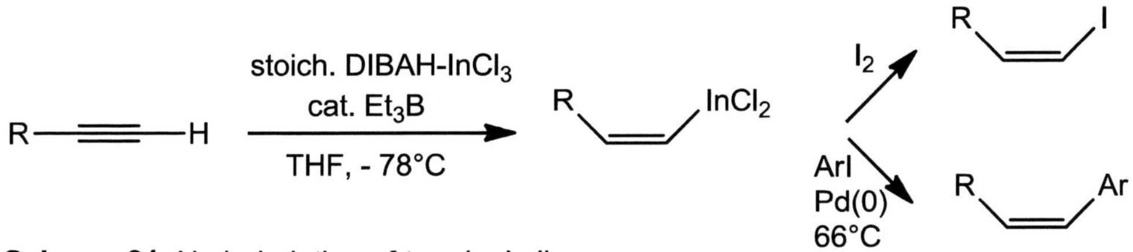

Scheme 31 Hydroindation of terminal alkynes

Another way to produce dichloroindium hydride is the use of Et_3SiH / $InCl_3$. In the presence of a catalytic amount of Et_3B, this reagent can reduce alkynes and induce intramolecular cyclization of 1,6-enynes (Scheme 32) [80]. Such a radical cyclization leading to an alkenyl indium species can occur without the side-reactions, such as hydroboration, observed with the $NaBH_4/InCl_3$ system.

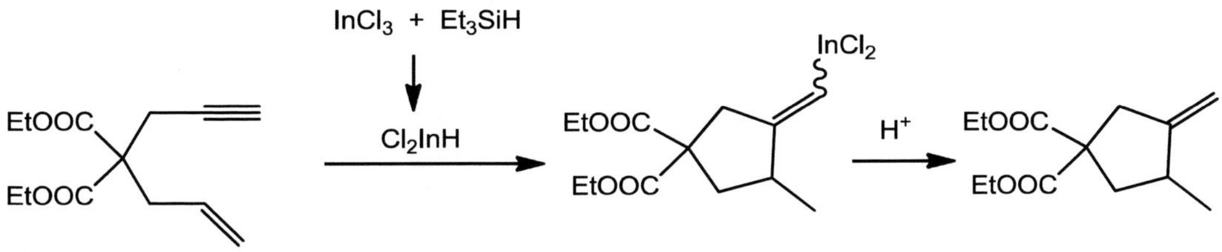

Scheme 32 Triethylsilane-Indium (III) chloride system as a radical reagent

Besides dichloroindium hydride, the reaction between silanes and $InCl_3$ affords chlorosilanes, the Lewis acidity of which is enhanced by combination with $InCl_3$. To overcome this putative problem, $InCl_3$ was replaced by $InCl_2OMe$ in the transmetalation with Ph_2SiH_2 (Scheme 33). In the presence of Et_3B, this system can promote radical cyclizations of halo-alkenes and enynes, along with intermolecular coupling reaction between organic halides and alkynes with activated alkenes, with a special interest with acid-sensitive substrates [81].

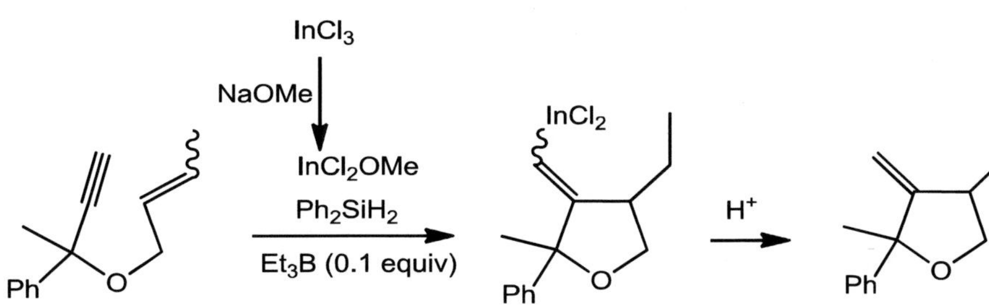

Scheme 33 Intramolecular radical coupling of enynes

4.4. Indium-catalyzed 2-alkenylation of 1,3-dicarbonyl Compounds

The In (III)-catalyzed addition of active methylene to terminal alkynes is a fascinating solventless reaction with a total atom economy (Scheme 34).

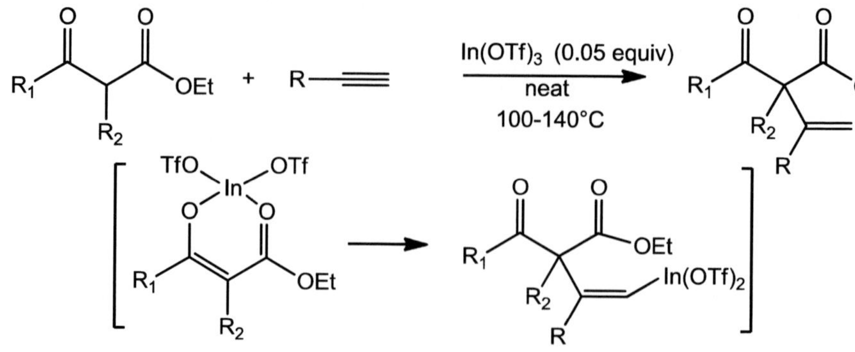

Scheme 34 Indium (III)-catalyzed α-alkenylation of carbonyl compounds

Indeed, this reaction, discovered by Nakamura, is an intermolecular version of the Conia-ene reaction. It is supposed that an indium enolate, generated from β-carbonyl compound and $In(OTf)_3$ adds across the triple bond of the terminal alkyne with a total regioselectivity. This carboindation process was evidenced by the higher reactivity of electron-rich alkynes suggesting a coordination of the indium atom to the alkyne [82]. It is possible to apply this methodology in the condensation of β-ketoester with α,ω-diyne leading to [1+n] annulation products, 1,3-dimethylenecycloalkane derivatives [83]. In that case, 0.01 eq of $In(NTf_2)_3$ is used as the catalyst in toluene at 150°C.

Silyl acetylene undergoes the coupling reaction with completely reversed regioselectivity (Scheme 35) giving an *E*-alkene, indicating that a *syn* addition takes place [82].

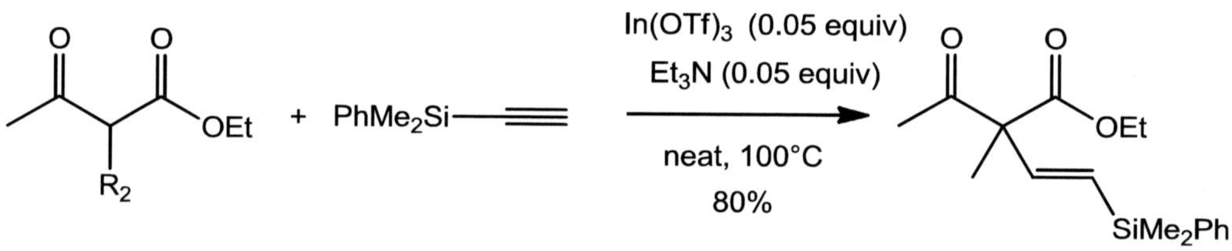

Scheme 35 Coupling of a silyl acetylene

The reaction tolerates a wide variety of functional groups including ester, ether, allylic halide, furan, thiophene and protected amine [84]. With acetylene gas, which is sensitive to acidity, this procedure does not operate, but is successful after addition of molecular sieve, allowing an easy access to α-vinyl keto esters [85]. With 1-iodo-1-alkynes, the reaction takes place in a *syn* fashion to afford *E*-alkenyl iodides, which can be coupled under palladium catalysis with phenylboronic acid with retention of configuration (Scheme 36) [86].

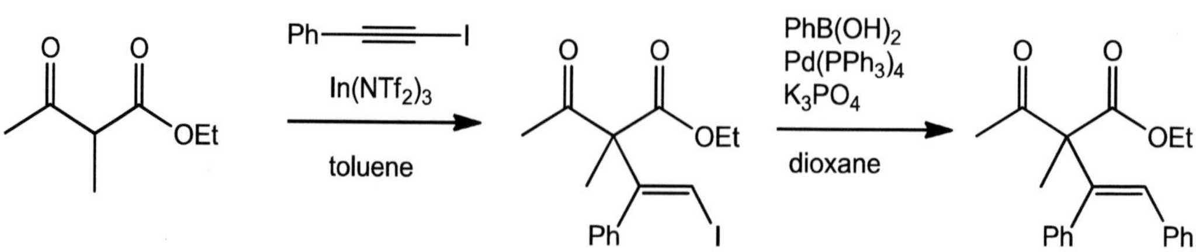

Scheme 36 Stereoselective synthesis of trisubstituted *E*-iodoalkenes

The In^{3+}-exchanged ultrastable Y zeolithe and montmorillonite were also demonstrated to be efficient catalysts in the α-alkenylation of 1,3-dicarbonyl compounds; remarquably is the recyclability of these heterogeneous catalysts [87].

When an unsubstituted malonate is used with aliphatic terminal alkynes in the presence of a catalytic amount of $InCl_3$ (1-3 mol%), only the Knoevenagel product is isolated due to a possible isomerisation of the classical Markovnikov α-adduct; with phenylalkyne the Knoevenagel product is accompanied by a small amount (8-22%) of the α-adduct [88].

With terminal conjugated diynes and β-keto esters, the reaction proceeds at the terminal triple bond in a perfect regioselectivity to give conjugated enynes in high yield if $In(NTf_2)_3$ is used as catalyst [89].

An intramolecular version of the Nakamura reaction, which is a $In(NTf_2)_3$-catalyzed Conia ene reaction was developed to cyclize ω-alkynyl-β-ketoesters into six to fifteen-membered all-carbon or nitrogen-containing rings; note that *in situ* isomerisation of the *exo* olefinic product (the α-adduct) into the *endo* olefinic enone (the Knoevenagel product) generally occurs in the reaction conditions (Scheme 37) [90]. The cyclization of α-(ω'-alkynyl)-β-ketoesters was used with great efficiently in the synthesis of five- and six-membered rings, the formation of seven-membered ring proceeding slowly [91].

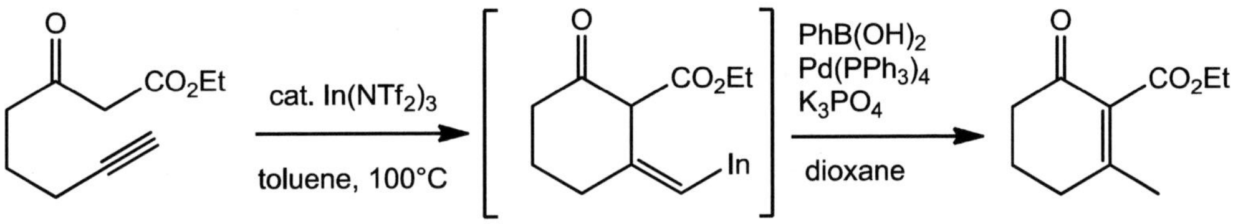

Scheme 37 Cycloisomerization of alkynyl-β-ketoesters

When propargyl amines or alcohols add to ethenetricarboxylates according to a Michael process, the indium enolate intermediate undergoes a cyclization via the activation of the triple bond by $InBr_3$, providing methylenepyrrolidine or methylene tetrahydrofuran derivatives (Scheme 38) [92].

Diethyl acetamidomalonates add to terminal alkynes in the presence of $In(OTf)_3$ (0.1 equiv) and N-methylmorpholine (0.05-0.2 equiv) giving the Markovnikov α-adducts in yields ranging from 54 to 88% [93]. Hydrogenation and decarboxylation of these adducts lead to β-branched α-amino acids. An intramolecular version of the addition of amidomalonates to alkynes was performed to prepare pyrrolidinones by 5-*exo*-dig cyclization as a key step in the total synthesis of (-)- salinosporamide A (Scheme 39) [94].

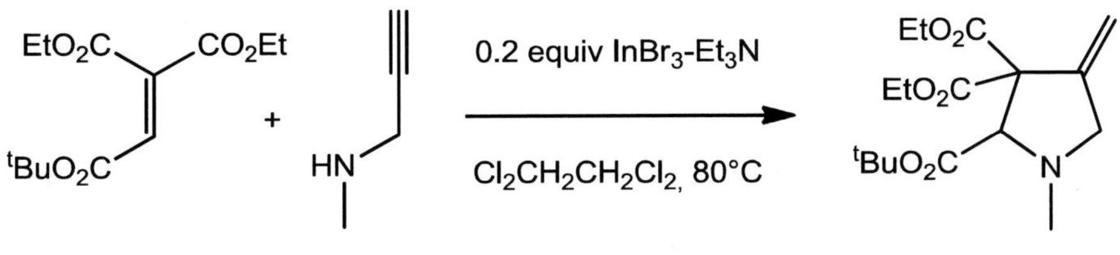

Scheme 38 Indium-promoted conjugate addition-cyclization reactions

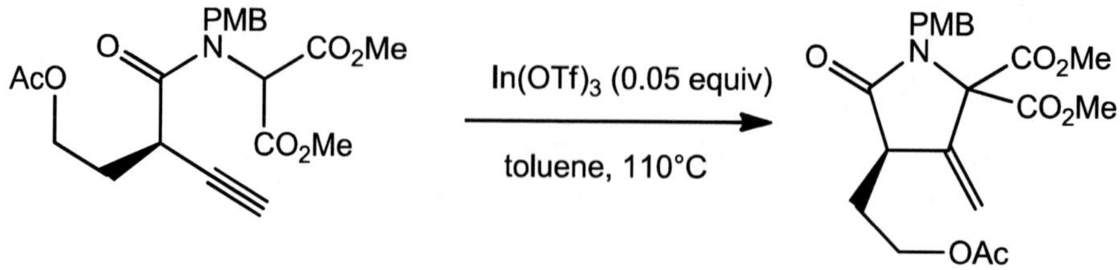

Scheme 39 Synthesis of (-)-salinosporamide A

In the synthesis of 1-oxadecalins, a 6-*exo*-dig cyclization occurs after a Knoevenagel condensation between a β-keto ester and hex-5-ynal, followed by an intramolecular oxa-Michael addition (Scheme 40) [95].

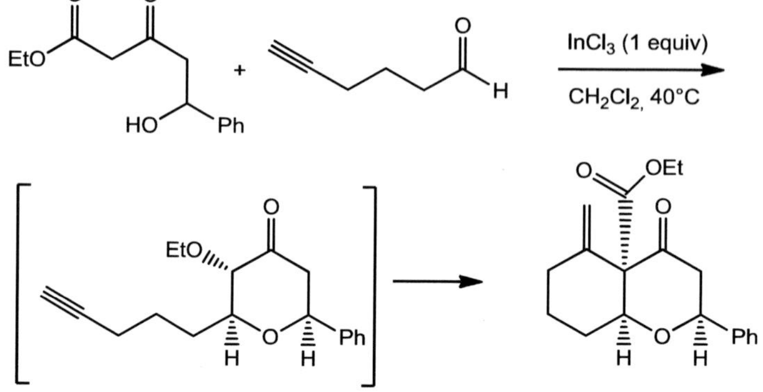

Scheme 40 $InCl_3$ mediated cascade reactions

Finally, an indium-catalyzed asymmetric α-alkenylation of β-keto esters was successfully performed by converting the 1,3-dicarbonyl compound into a chiral enamine (Scheme 41). One of the two diastereotopic faces of the enamine is then alkenylated with a terminal alkyne according the process induced by $In(OTf)_3$ in the presence of a catalytic amount of base (0.1 equiv of BuLi) [96].

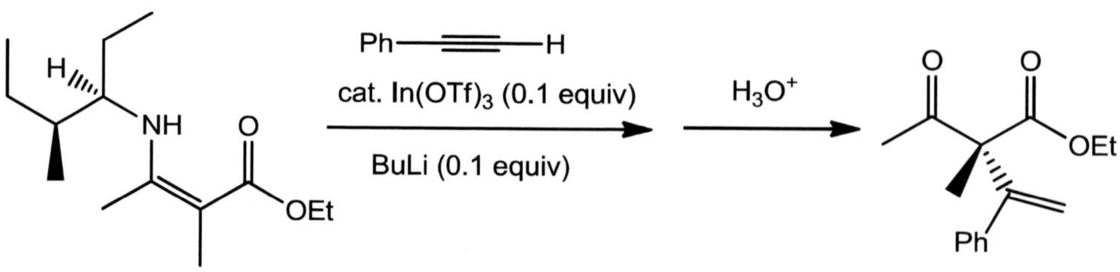

Scheme 41 Indium-catalyzed asymmetric α-alkenylation of β-ketoesters

4.5. Indium-catalyzed α-alkenylation of Aldehydes and Esters

In the Nakamura reaction, 1,3-dicarbonyl compounds are involved. What occurs if the 1,3-dicarbonyl compounds is replaced by α-disubstituted aldehydes? Intramolecular addition onto alkynes may occur with a combination of $InCl_3$ and amine as co-catalysts [95, 96]. Optimized conditions to produce *exo*-methylene cyclopentanes use $InCl_3$ (0.10 equiv) with cyclohexylisopropylamine (0.2 equiv) (Scheme 42). The reaction goes through the formation of an enamine which adds onto alkynes via an *anti* carboindation mechanism [96]. With a previous system of co-catalysts ($InCl_3$/iPrCyNH) [97], another mechanism may, in addition, take place through an enolate/*syn* carboindation pathway [98].

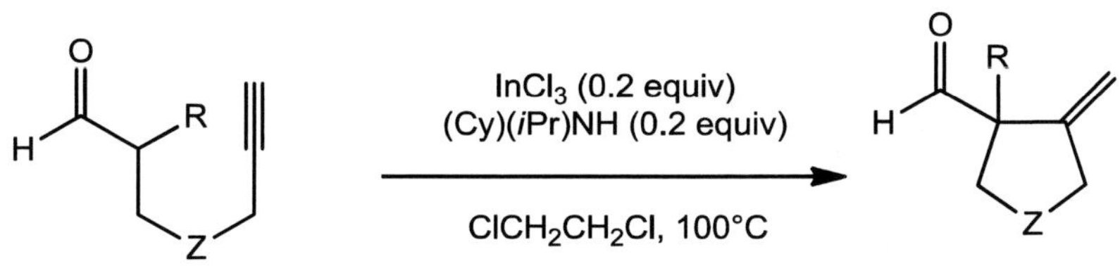

Scheme 42 Combined $InCl_3$ and amine catalyzed intramolecular addition reaction

More generally, with 1 equiv of $InBr_3$, without any base, ketene silyl acetals may add to alkynes with high regio- and stereoselectivity to give α-alkenylated esters (Scheme 43). This intermolecular reaction proceeds in an *anti* fashion as opposed to the *syn* addition observed in the Nakamura reaction. Alkenylindium species as a carboindation product was fully characterized by X-ray and ^1H NMR spectroscopies [99].

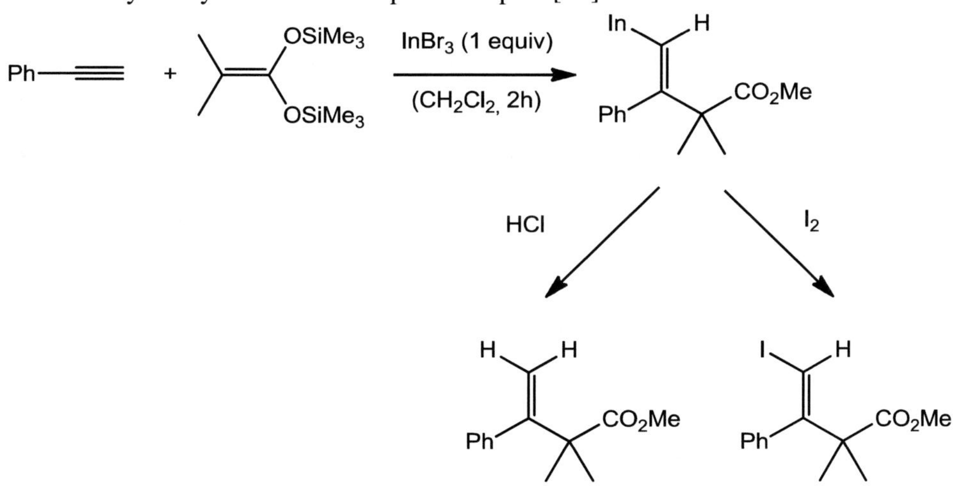

Scheme 43 Reaction with $InBr_3$, alkyne and ketene silyl acetal

4.6. Arylindation of Alkynes

The Friedel-Crafts alkenylation of arenes using alkynes, including internal alkynes, is catalyzed by $In(OTf)_3$, affording 1,1-diarylalkenes in high yields (Scheme 44) [100]. In this first communication, arenes were both used as the reagent and the solvent. The ionic liquid [BMIM] [SbF_6] was proved to be an effective solvent for that type of reaction [101].

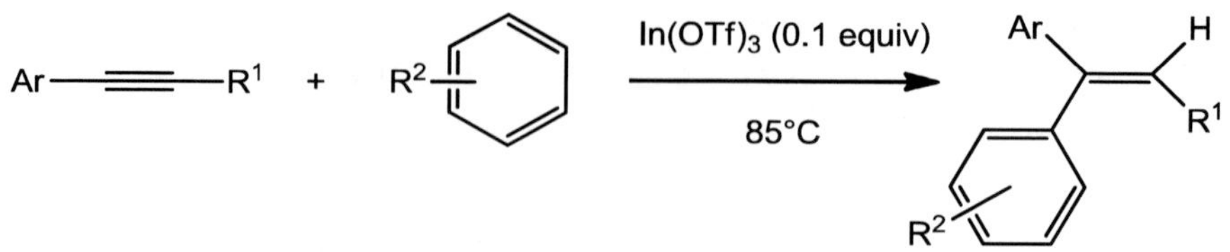

Scheme 44 Friedel-Crafts alkenylation of arenes

An intramolecular version of the Friedel-Crafts alkenylation of arenes was described in the $InCl_3$-catalyzed synthesis of phenanthrenes from biphenyl derivatives containing an alkyne unit at one of their *ortho*-positions This 6-*endo*-dig cyclization was supposed to proceed through initial π-complexation of the triple bond followed by interception of the resulting η^2-indium species by the adjacent arene ring. [102]. Biaryls bearing halo-alkyne units afford to 10-halophenanthrenes. DFT/M06 calculations have confirmed that π-complexation of the triple bond is exergonic by 8.6 kcal.mol-1 and is followed by a 6-*endo*-dig cyclization. The retention of the halogen on C10 of the product (which does not occur with AuCl instead of $InCl_3$)is then attributed to the intermolecular hydrogen shifts of two Wheland-type intermediates with the assistance of the chloride ligand [103].

The hydroarylation of alkynes catalyzed by $InBr_3$ can be carried out with heterocyclic arenes such as indoles at refluxing toluene, leading exclusively to 3-vinylindoles [104].

Double addition may occur with furans, thiophenes and pyrroles to afford 2:1 adducts where two heterocyclic arenes regioselectively attack the same carbon atom of alkynes Preferential formation of *gem*-α,α'-adducts were observed with 2-methylfuran and pyrrole. In contrast, *gem*-β,β'-dipyrrolylalkanes were formed predominantly in the reaction of *N*-substituted pyrroles with bulky substituents (Scheme 45) [105]. With propargyl ethers 2-aryl- and 2-heteroarylindoles give aryl- and heteroaryl-annulated[a]carbazoles with the use of a stronger Lewis acid than $In(OTf)_3$, such as indium (III) nonaflate [106].

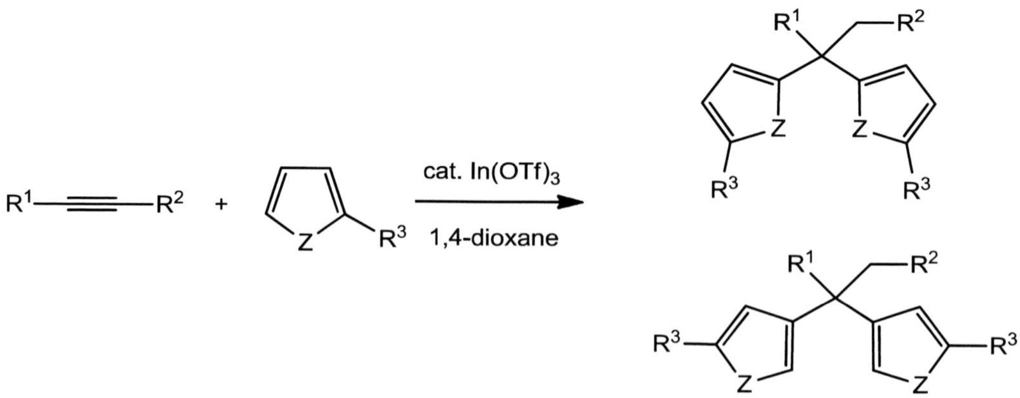

Scheme 45 Indium triflate-catalyzed double addition of heterocyclic arenes

Under the conditions of the reaction (Scheme 46), the C(sp^3)-C(pyrrolyl) bond can be cleaved, giving a transient cationic species, which is trapped by carbon nucleophiles (such as furans, thiophenes, styrenes, allylstannannes, cyanides) [107] or hydrides ($HSiEt_3$) [108].

A powerful construction of complex chiral polycycles can be realized by InI_3 or $InBr_3$-catalyzed cationic cascade where the activation of the triple bond by indium is the driving force (Scheme 47) [109].

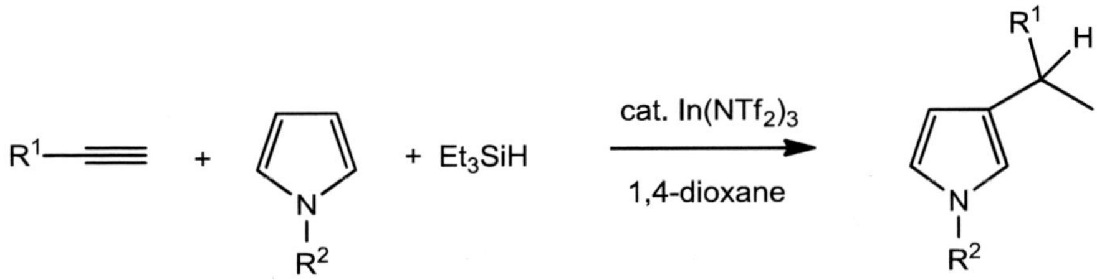

Scheme 46 Indium catalyzed reductive alkylation of pyrroles

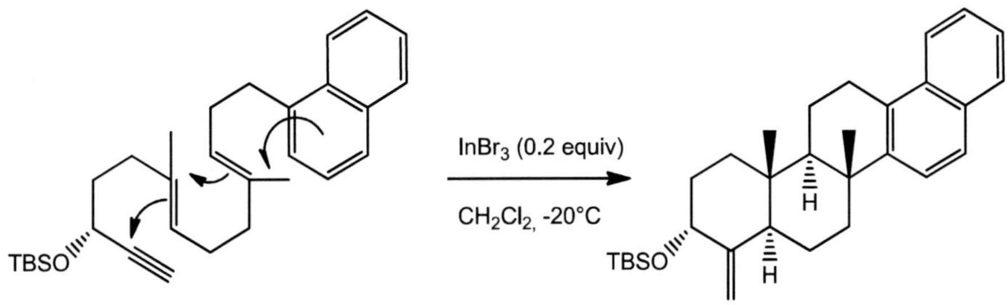

Scheme 47 Indium catalyzed cationic cascade reaction

4.7. Alkynylindation of Alkynes

The indium (III)-catalyzed hydrative cyclization of 1,7-diynyl ethers may be regarded as an alkyne indium activation prior to the formation of an alkenyl indium species via a 7-*endo*-dig cyclization (Scheme 48). Using InI_3 and *p*-toluenesulfonic acid (p-TSA) as cocatalysts, several 2,2-disubstituted tetrahydrofurans with exocyclic enone appendages were prepared in moderate yields [110].

4.8. Heteroindation of Alkynes

The triple bond of 2-alkynylanilines can be activated by $InBr_3$ or $InCl_3$ followed by to 5-*endo*-dig cyclization leading to polysubstituted indoles [111]. When this azaindation of alkynes is carried out using a substrate with a trimethylsilyl group or with no substituent on the terminal carbon, dimerization occurs leading to quinoline derivatives (Scheme 49) [112].

In the presence of β-keto esters, 2-alkynylanilines give the corresponding β-enamino esters under $InBr_3$ catalysis. Such intermediates were not isolated, since they directly lead to β-(*N*-indolyl)-α,β-unsaturated esters owing to the indium(III)-catalyzed 5-*endo*-dig cyclization. *Z*-alkenes are selectively formed due to the stability of β-enamino esters as intermediates of the tandem reaction (Scheme 50) [113].

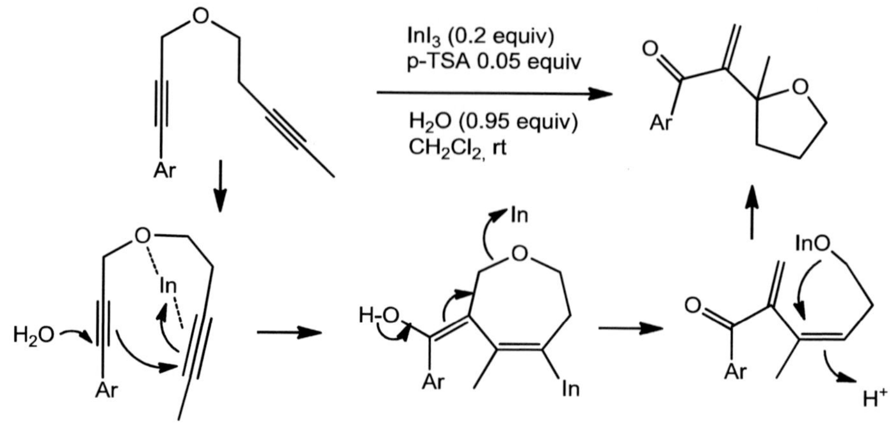

Scheme 48 Indium(III)- catalyzed hydrative cyclization of 1,7-diynyl ethers

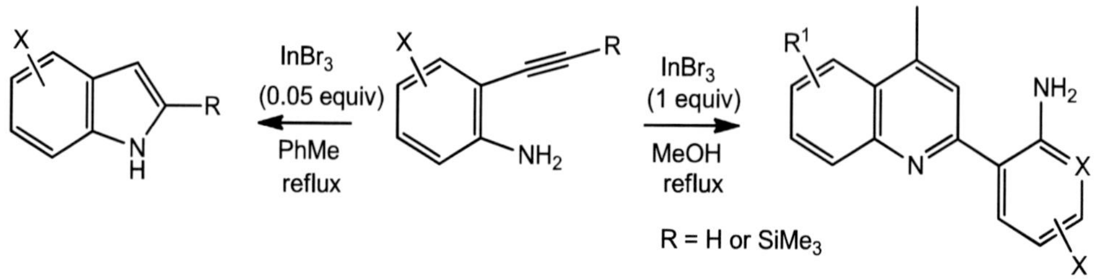

Scheme 49 Indium promoted synthesis of polysubstituted indoles

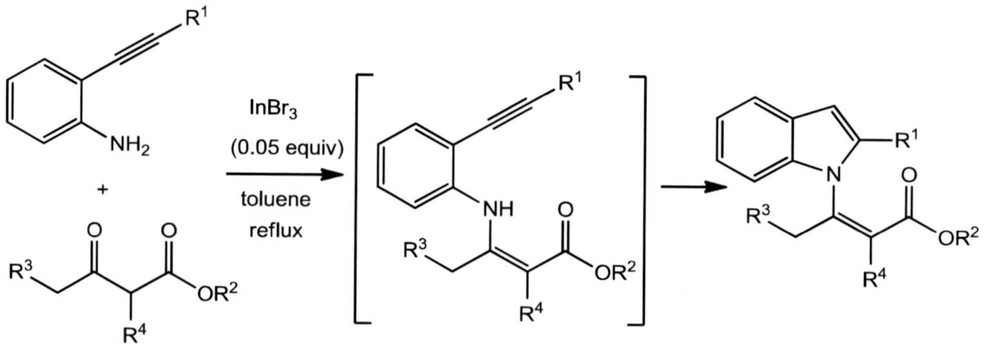

Scheme 50 Tandem β-enamino ester formation and cyclization catalyzed by $InBr_3$

The indium catalytic system successfully accommodates a 6-*endo*-dig cyclization of other arylalkyne skeletons involving a carbocyclic acid and an amide group [114]. In the $InBr_3$–catalyzed cyclization of 2-alkynyl-6-methoxy-*N*-sulfonylanilines, the major product is issued from 1,7-migration of the sulfonyl to produce 6-sulfonylindoles [115].

Under $In(OTf)_3$ catalysis, nucleophilic addition of allylstannanes onto *ortho*-alkynylarylaldimines affords amines, which can then undergo 6-*endo*-dig cyclizations leading to 1,2-dihydroisoquinolines (Scheme 51); ^{13}C NMR investigations support that dual activation of both imine nitrogen atom and alkyne π interactions by $In(OTf)_3$ occurs [116]. The same authors have described an indium (III)-catalyzed tandem reaction with *ortho*-alkynylbenzaldehydes and *ortho* alkynylanilines to heteroaromatic compounds [117]. After the coordination of the triple bond of *ortho*-alkynylbenzaldehydes to $In(OTf)_3$, the nucleophilic attack of the carbonyl oxygen to the deficient alkyne takes place affording a 6-endo-dig cyclized pyrylium cation intermediate, which undergoes cycloaddition with *ortho* alkynylanilines and then dehydration reaction.

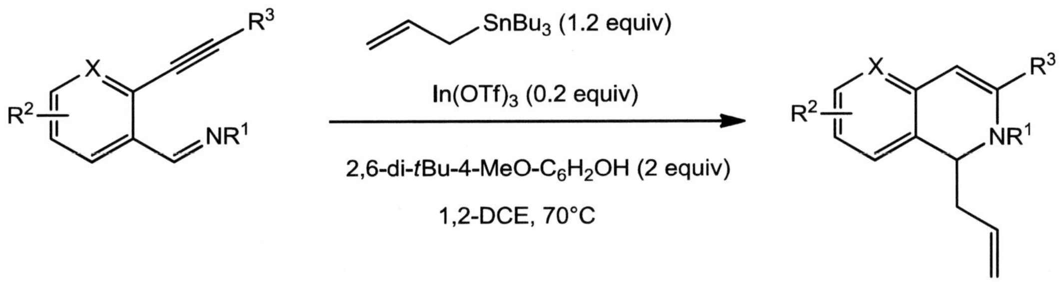

Scheme 51 Indium triflate-catalyzed tandem nucleophilic addition-cyclization

Intermolecular azaindation occurs when adding various anilines to phenylacetylene in the presence of $In(OTf)_3$ (0.10 equiv) in refluxing toluene [118]. Formation of the corresponding imines follows the Markonikov rule. If phenylacetylene is used in excess, a second molecule of phenylacetylene can add to the imine under identical conditions. This second reaction is referred as an indium-catalyzed hydroalkylation (Scheme 52).

The same authors have recently observed that, in the same conditions, the Markovnikov addition product is obtained in good yields when heteroaromatic thiols are reacting with arylacetylenes. Quite surprising is the switch in selectivity with aliphatic or aromatic thiols. Exclusive anti-Markovnikov addition products were then obtained giving rise to *E*- and *Z*-vinyl sulfides [119].

An innovative one-pot synthesis of tetrasubstituted furans is based on a tandem reaction starting with an indium chloride-catalyzed condensation of propargylic alcohol with 1,3-diketones (Scheme 53). The intermediate then undergoes an oxaindation through a 5-*exo*-dig cyclization induced by activation of the triple bond by $InCl_3$ [120].

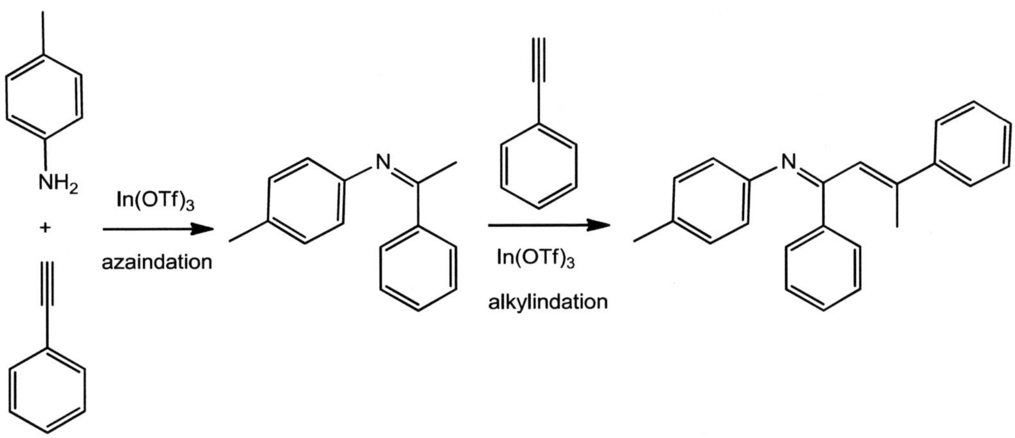

Scheme 52 Indium catalyzed tandem azaindation/alkylindation of terminal alkynes

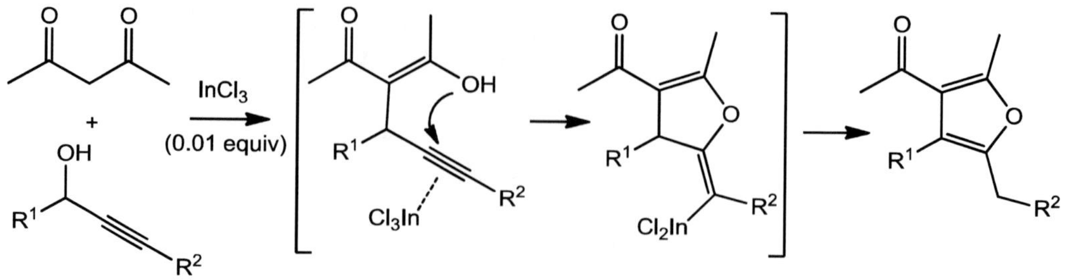

Scheme 53 Synthesis of tetrasubstituted furans via In(III)-catalyzed tandem process

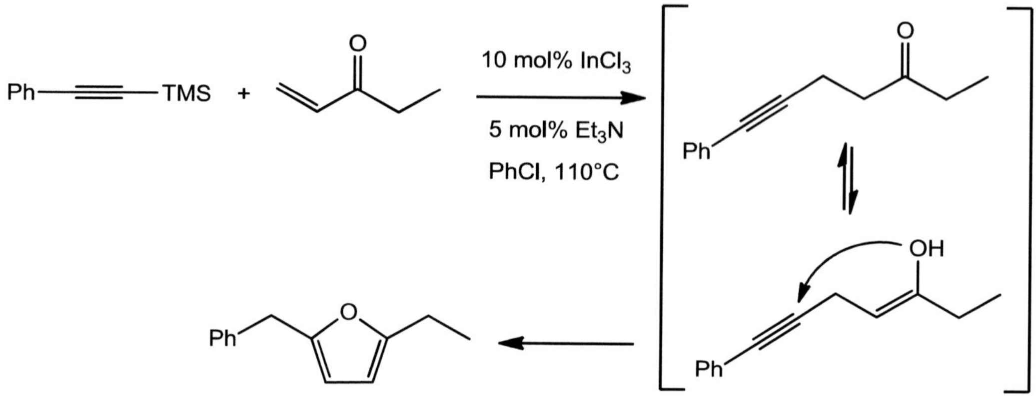

Scheme 54 $InCl_3$ catalyzed conjugate addition of alkynylsilane to enones

In a similar way, $In(OTf)_3$ and $In(NTf_2)_3$ as catalysts were proved to be very effective for cycloisomerization of α-propargyl-β-keto esters and their imine analogues to afford in mild conditions trisubstituted furans and pyrroles via oxaindation and azaindation respectively [121].

Indium (III) chloride can catalyze conjugate addition reaction of alkynylsilanes to acrylate esters and ketones [122]. With a ketone, the reaction leads to 2-5-disubstituted furans resulting also from a further 5-*exo*-dig cyclization induced by activation of the triple bond by $InCl_3$ (Scheme 54). Intramolecular oxaindation of alkynes was also observed in alkynylcarbamates [46].

5. Coupling Reactions with Transmetalated Alkynyl, Alkenyl and Aryl Indium Reagents

5.1. Cross-coupling Reactions

Transmetalation of $InCl_3$ with organolithium and organomagnesium reagents to afford triorganoindium reagents for coupling reactions was pioneered by Nomura *et al.* [17]. The palladium-catalyzed cross-coupling of triorganoindium reagents, including aryl, alkenyl and alkynyl groups, with aryl iodides and aryl/vinyl triflates were first described by Sarandeses [123].

The palladium-catalyzed reaction was applied to various electrophiles including benzyl bromides or acyl chlorides; in contrast less reactive aryl chlorides require Ni(0) catalysis [124]. All three of the organic groups attached to the metal are transferred, as shown by the necessity of using only 0.34 equiv of indium. Multifold and sequential palladium-catalyzed cross-coupling reactions can be performed with oligoarenes using only a small excess of triorganoindium reagents [125]. It is possible to take advantage of the higher reactivity of the aryl-iodine bond over the aryl-bromine bond in sequential cross-coupling reactions to afford unsymmetrical alkyl-, aryl-, and alkynyl-substituted aromatic compounds in one pot (Scheme 55) [126].

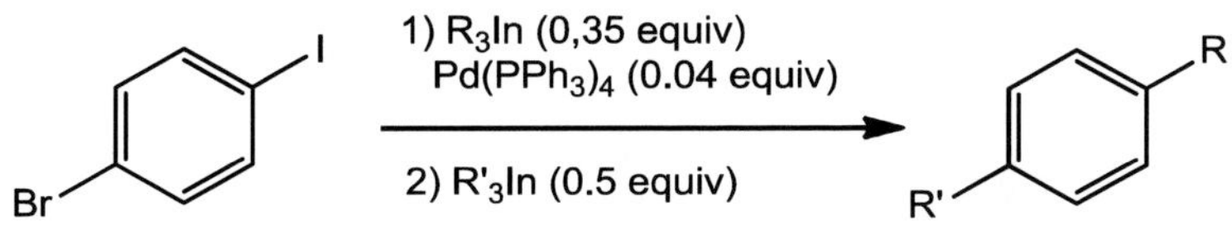

Scheme 55 One-pot sequential palladium-catalyzed cross-coupling reaction

The palladium-catalyzed cross-coupling reactions between vinyl indium $(CH_2{=}CH)_3In$ and two substituted benzyl bromides was successfully applied in the enantioselective synthesis of two proansamitocin derivatives as new macrolactam antibiotics [127]. Similarly the reaction between alkynyl indium reagent and substituted benzyl bromide constitutes one of the key steps in the enantioselective synthesis of the *N*-acetylcysteamine thioester of seco-proansamitocin [128].

Arenediazonium *o*-benzenedisulfonimides as innovative electrophiles in palladium-catalyzed cross coupling reactions were tested in the aryl-aryl coupling with triaryl indium compounds; biaryls (16 examples) are obtained in good yields (Scheme 56) [129]. Noteworthy is the presence of diaryldiazenes as by-products; these compounds can be prepared without catalyst by direct addition of triaryl indium derivatives as nucleophiles to arenediazonium *o*-benzenedisulfonimides as electrophiles.

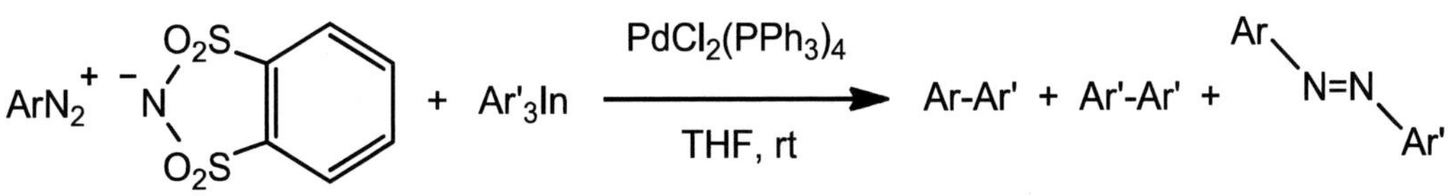

Scheme 56 One-pot sequential palladium-catalyzed cross-coupling reaction

The regio- and stereoselectivity of the palladium-catalyzed cross-coupling of triorganoindium reagents (R_3In; R= alkynyl, alkenyl, aryl and alkyl) with 1-haloalkenes and 1,1-dibromoalkenes has been carefully investigated [130]. With 1-iodoalkenes or 1-bromo alkenes as electrophiles, the reaction is stereoselective affording *E*-alkenes regardless the (*Z*) or (*E*) configuration of the starting haloalkenes. With 1,1'-dibromoalkenes a double coupling occurs when 0.9 equiv of triorganoindium derivatives is used. When only 0.40 equiv of aryl, vinyl- and alkynyl indium derivatives is used, a *trans*-selective monosubstitution occurs leading to (*Z*)-monobromoalkenes, which can be further functionalized by palladium-catalyzed cross-coupling with R_3In (R = alkyl, aryl, alkynyl) to provide trisubstituted olefins (Scheme 57).

The Sarandeses protocol was applied to the synthesis of *C*-aryl glycals through the formation of dihydropyranylindium reagents [131].

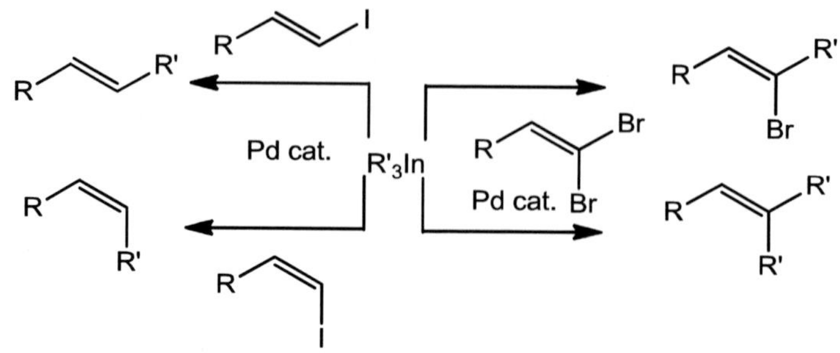

Scheme 57 Palladium-catalyzed cross-coupling reactions

An interesting observation was mentioned concerning the reaction in aqueous media: two of the three aryl, vinyl or alkyl groups in triorganoindium reagents can survive in aqueous THF at reflux, which prevailed a new protocol using the diaryl-, divinyl- or dialkylindium reagents with aryl halides [132]. A wide range of functional groups, including a hydroxyl group or a formyl group are compatible with this reaction.

Noteworthy is the beneficial influence of $InCl_3$ as a cocatalyst in the cross-coupling of alkenyl aluminium and zirconium, prepared by hydrometalation of alkynes, with vinyl halides [133].

The cross coupling reaction with alkenyl bromides [134] and aryl halides or triflates [134, 135] was also carried out under a carbon monoxide atmosphere leading to a carbonylative coupling product. It is also possible to take advantage of the higher reactivity of the aryl-iodine bond over the aryl-bromine bond in sequential cross-coupling reactions to afford unsymmetrical alkyl-, aryl-, and alkynyl-substituted aromatic compounds in one pot [126, 135].

Palladium-catalyzed oxidative carbonylation of alkyl and aryl indium reagents (R_3In) with CO in methanol, ethanol or butanol gave the corresponding esters RCOOR' (R'= Me, Et or Bu) [136]. This novel reaction requires a stoichiometric amount of 2-chloro-1,2-diphenylethanone as an oxidant (Scheme 58). If an amine is used as nucleophile instead of an alcool, an amide is obtained instead of an ester.

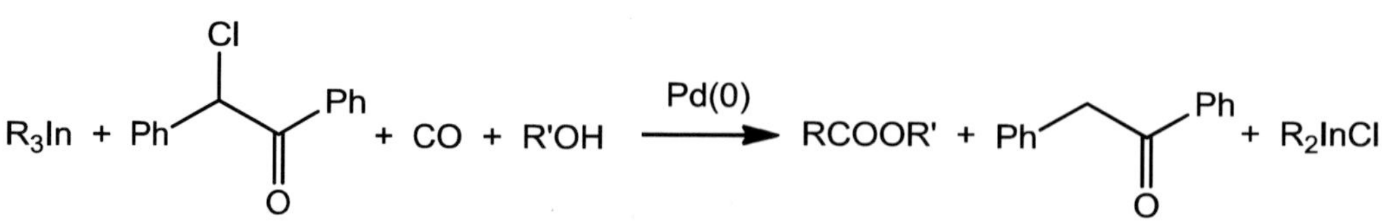

Scheme 58 Palladium-catalyzed oxidative carbonylation of indium reagents

The palladium-catalyzed aryl/aryl cross coupling has been particularly investigated. Thus a variety of polyaromatic compounds bearing 1- and 2-naphthyl groups were prepared from the reactions of the corresponding halides with tri(1- and 2-naphthyl) indium reagents for their optical and electronic properties [137].

Various heterobiaryls were prepared from electron-rich and electron-poor heterocycles. Heteroarylindium derivatives, including tristhienylindium, trispyridylindium, tris (THP-protected pyrazol-4-yl)indium, tris(1-methylpyrazol-5-yl)indium were coupled with (hetero) aromatic electrophiles, such as iodo(bromo)pyridines, iodothiophenes, iodopyrazoles [138]. Sarandeses took advantage of directed metalation groups (OMe, $OCONEt_2$, Boc, F) to initiate *ortho*-lithiation followed by transmetalation with $InCl_3$ and cross-coupling with (hetero)aryl halides (Scheme 59) [139]. Thus various 2-arylindoles were easily prepared from tri-(N-Boc-2-indoyl)indium and aromatic or heteroaromatic halides. The excellent results obtained in these *ortho*-directed cross coupling reactions have prompted the authors to extend the scope of these reactions to non aromatic electrophiles, such as vinyl triflates. Thus *ortho*-substituted arylindium organometallics were successfully coupled with alkenyl triflates.

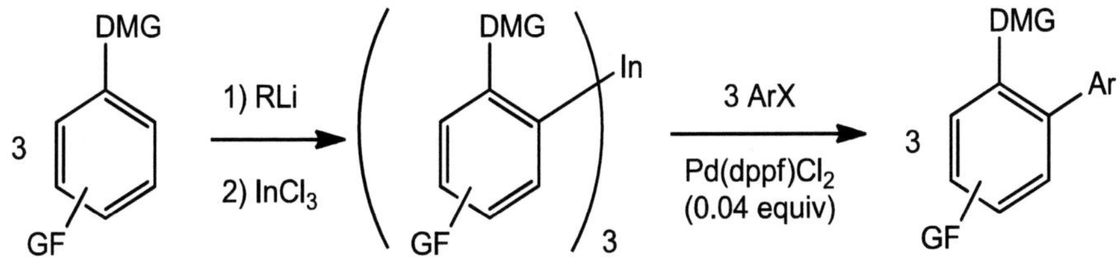

Scheme 59 Palladium-catalyzed aryl-aryl cross-coupling reaction

To prepare hyrtinadine A, a novel bis-indole alkaloid with biological properties, Sarandeses uses the two-fold reactivity of tri(3-indolyl)indium reagent toward 5-bromo-2-chloropyrimidine (Scheme 60) [140]. It is worth to note that 5-bromo-2-chloropyrimidine can undergo a sequential palladium-catalyzed cross-coupling as previously mentioned for other dihalo aromatic substrates.

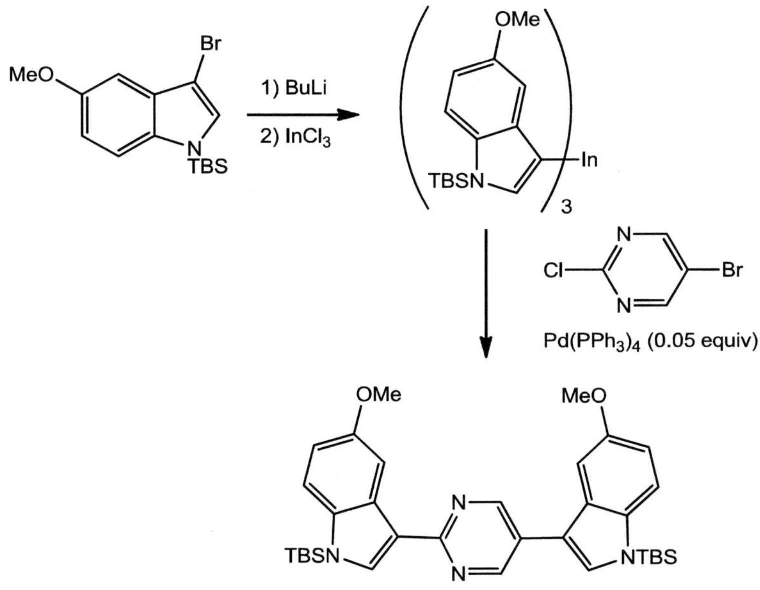

Scheme 60 Synthesis of Hyrtinadine A

Sequential one-pot or stepwise palladium-catalyzed cross coupling reactions were also used to prepare 3,4-disubstituted maleimides from 3,4-dihalomaleimides (Scheme 61) [141].

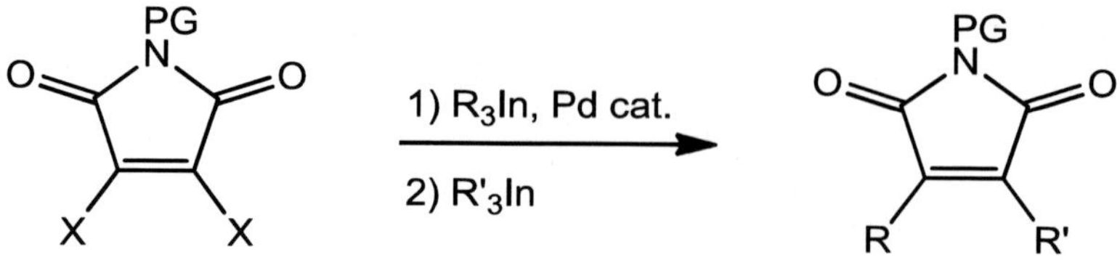

Scheme 61 Synthesis of 3,4-disubstituted maleimides

5.2. Addition and Substitution Reactions

The conjugate addition of triorganoindium reagents to α,β-unsaturated ketones, esters and nitriles was discovered by Sarandeses; in this nickel-catalyzed addition, only one of the three groups attached to indium is transferred [142]. In an effort to improve the atom economy of the reaction, arylindium dichlorides were prepared by transmetalation with only one equimolecular amount of organometallic reagent and $InCl_3$. This new reagent was proved to react in good yields with α,β-unsaturated ketones and lactones under rhodium catalysis (Scheme 62) [143].

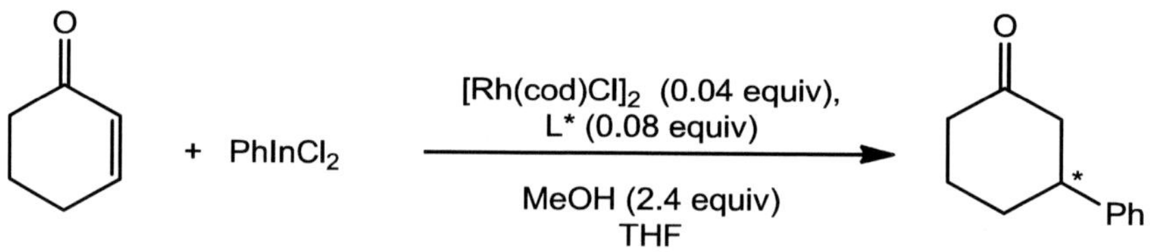

Scheme 62 Rhodium-catalyzed conjugate addition of arylindium reagents

The addition of aryl and alkyl organoindium reagents to thiol esters under palladium catalysis gives ketones in moderate to excellent yields (Scheme 63) [144].

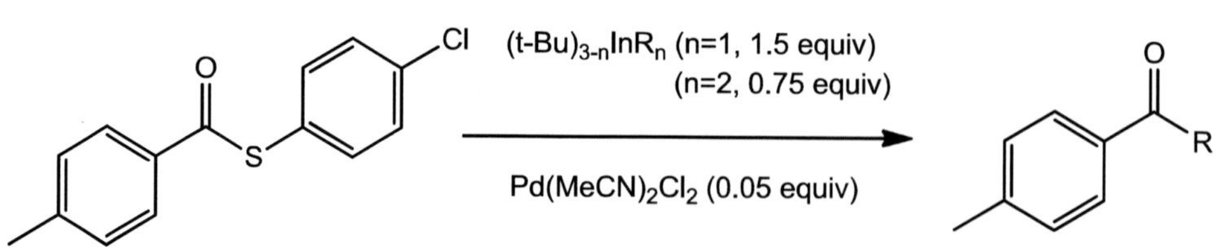

Scheme 63 Palladium-catalyzed coupling of thiol esters

The palladium-catalyzed nucleophilic substitution of allylic substrates, known as the Tsuji-Trost reaction, has been recently extended to organoindium reagents as new nucleophiles. Triorganoindium and tetraorganoindates with aryl, alkenyl and methyl groups react with cinnamyl and geranyl chlorides, bromides and acetates in the presence of 0.5 equiv of $Pd_2(dba)_3$ to afford exclusively the S_N2 products (Scheme 64) [145]. For optimal conditions, a stoichiometric amount of indium is required. The reaction proceeds with net inversion of configuration, which is consistent with the mechanistic pathway of oxidative addition, transmetalation and reductive elimination. The β-hydride elimination was observed with tributylindium compounds.

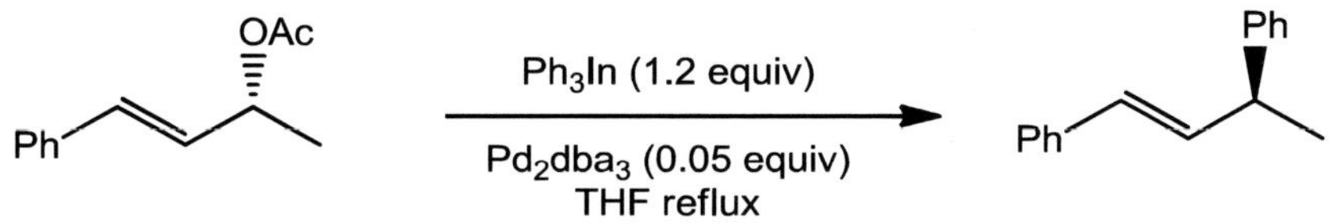

Scheme 64 Palladium-catalyzed cross-coupling of allylic acetates

In an optimized procedure, 0.5 equiv of triarylindium reagent was used in the nucleophilic substitution of allylic acetates into carbocyclic derivatives [146]. The reaction proceeds with inversion of configuration in the presence of 0.01-0.03 equiv of $Pd_2(dba)_3$ and 0.04 equiv of PPh_3. When allylic acetates are part of a molecule bearing a triple bond, cyclization may occur [147]

A S_N2' rearrangement occurs when trialkyl or triarylindium derivatives react at -30°C with cinnamyl and geranyl halides and phosphates in the presence of $Cu(OTf)_2$ as a catalyst and triethylphosphite as a copper ligand [148]. Such a S_N2' rearrangement was also observed with acetates of Baylis-Hillman adducts. The addition of triaryl indium and trialkenyl indium reagents were catalyzed with palladium (0), whereas the addition of trialkyl indium reagents were preferentially catalyzed by CuI [149].

Triphenyl indium adds to glycal acetates without the need of a catalyst according to a S_N2' Ferrier mechanism to produce a *C*-aryl glycoside [43].

Another strategy to prepare *C*-aryl glycosides lies on a sequential indium-mediated aldehyde allylation/palladium-catalyzed cross coupling reaction with triarylindium reagents or arylboronic acids (Scheme 65) [150]. The resulting homoallylic alcohols must then undergo a standard olefin oxidative cleavage in order to get either the pyranose or the furanose ring. Finally 2-deoxy-β-*C*-aryl ribofuranosides and 2-deoxy-β-*C*-aryl allopyranosides are obtained in good yields.

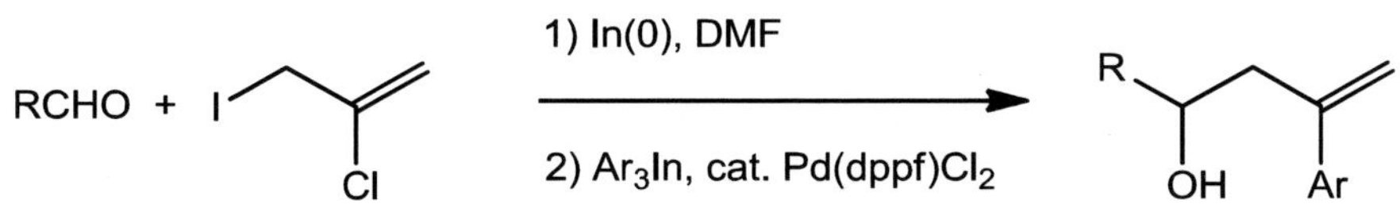

Scheme 65 sequential In mediated aldehyde allylation / Pd catalyzed cross-coupling

In order to prepare allenes, allylic substrates were replaced by propargyl ones. Thus aryl, alkenyl, alkynyl and methyl triorganoindium reagents in stoichiometric amount react with propargylic acetates, benzoates and carbonates under palladium catalysis via a S_N2' rearrangement to give allenes in good yields and with high regioselectivity [151, 152]. The reaction of Ph_3In with enantiopure propargylic esters gave enantiopure allenes, consistent with an *anti*-stereoselectivity (Scheme 66).

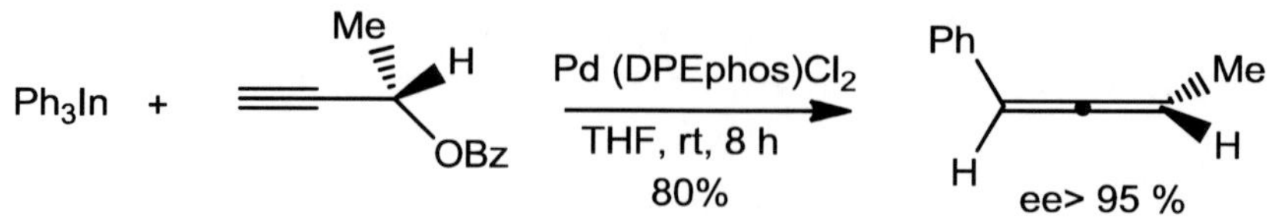

Scheme 66 Palladium-catalyzed cross-coupling reaction of Ph_3In with propargylic esters

5.3. Multicomponent Reactions

A three-component coupling of triorganoindium reagents with imines and acid chlorides, was recently reported under catalysis of CuCl (Scheme 67). No coupling reaction is observed when the triorganoindium reagent is used without any catalyst nor does a palladium catalyst lead to a reaction. All of the three organic groups, such as alkyl, alkenyl or aryl groups, attached to indium atom are transferred to the transient *N*-acyl iminium salt affording α-substituted amines in a mixture of acetonitrile and THF at 45-70°C [153].

This copper-catalyzed reaction was also applied to the multicomponent coupling of triorganoindium reagents with nitrogen-containing aromatic heterocycles and chloroformates. [154]. Scheme 68 shows the plausible mechanism with pyridine and ethylchloroformate, which leads exclusively to 1,4-adducts. Other heterocycles such as isoquinoline derivatives, benzooxazole, benzothiazone or phthalazine lead to the formation of 2-substitutes heterocycles.

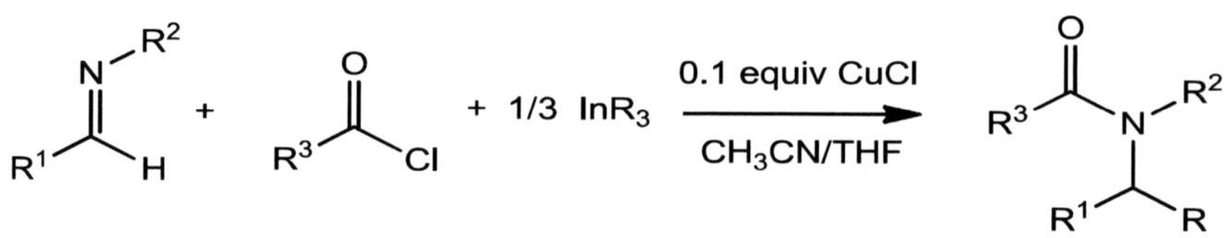

Scheme 67 Three component coupling of organoindium reagents

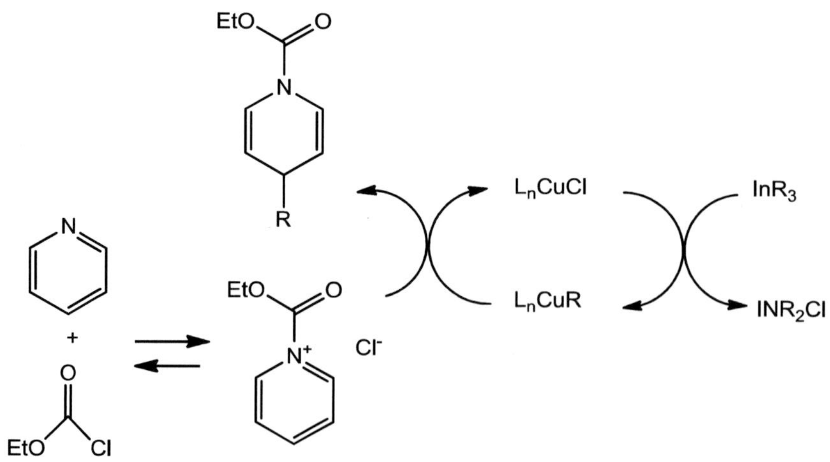

Scheme 68 Proposed mechanism of the Cu-catalyzed coupling of pyridine, chloroformate and organoindium reagents

REFERENCES

[1] Araki, S.; Ito, H.; Butsugan, Y. *J. Org. Chem.* 1988, *53*, 1831-1833.
[2] Li, C.-J., Chan T. H. *Tetrahedron Lett*, 1991, *32*, 7017-7020.
[3] Li, C.-J. *Chem. Rev.* 1993, *93*, 2023-2035.
[4] Li, C.-J. *Tetrahedron* 1996, *52*, 5643-5668.
[5] Lubineau, A.; Augé, J. ; Queneau, Y. *Synthesis* 1994, 741-760.
[6] Paquette, L. A. *Green Chemistry. Frontiers in Benign Chemical Syntheses and Processes*; Anastas, P. T.; Williamson, T. C. Eds.; Oxford University Press: Oxford, 1998, 250-268.
[7] Lubineau, A.; Augé, J. *Topics in Current Chemistry* 1999, *206*, 1-39.
[8] Ranu, B. C. *Eur. J. Org. Chem.* 2000, 2347-2356.
[9] Podlech, J.; Maier, T. C. *Synthesis* 2003, 633-655.
[10] Nair, V.; Ros, S.; Jayan, C. N.; Pillai, B. S. *Tetrahedron* 2004, *60*, 1959-1982.
[11] Augé, J.; Lubin-Germain, N.; Uziel J. *Synthesis* 2007, 1739-1764.
[12] Pardoe, J. A. J.; Downs, A. J. *Chem. Rev.* 2007, *107*, 2-45.
[13] Roy, U. K.; Roy, S. *Chem. Rev.* 2010, *110*, 2472-2535.
[14] Hwan Kim, S.; Seung Lee, H.; Hoon Kim, K.; Hee Kim, S.; Nyoung Kim, J. *Tetrahedron* 2010, *66*, 7065-7076.
[15] Yadav, J. S.; Antony, A.; George, J.; Subba Reddy B. S. *Eur. J. Org. Chem.* 2010, 591-605.
[16] Chan, T. H. ; Yang, Y. *J. Am. Chem. Soc.* 1999, *121*, 3228-3229.
[17] Nomura, R.; Miyazaki, S.-I.; Matsuda, H. *J. Am. Chem. Soc.* 1992, *114*, 2738-2740.
[18] Miyabe,H.; Naito, T. *Org. Biomol. Chem.* 2004, *2*, 1267-1270.
[19] Shibata, I.; Kato, H.; Ishida, T.; Yasuda, M.; Baba, A. *Angew. Chem. Int. Ed.* 2004, *43*, 711-714.
[20] Augé, J.; Lubin-Germain, N.; Seghrouchni, L. *Tetrahedron Lett.* 2002, *43*, 5255-5256.
[21] Augé, J.; Lubin-Germain, N.; Seghrouchni, L. *Tetrahedron Lett.* 2003, *44*, 819-821.
[22] Picard, J.; Lubin-Germain, N.; Uziel, J.; Augé, J. *Synthesis* 2006, 979-982.
[23] Ayed, C.; Picard, J.; Lubin-Germain, N.; Uziel, J.; Augé, J. *Sci. China Chem.* 2010, *53*, 1921-1926.
[24] Das, B.; Jangili, P.; Kashanna, J.; Arawind Kumar, R. A. *Synthesis* 2011, 3267-3270.
[25] Yasuda, M.; Miyai, T.; Shibata, I.; Baba, A.; Nomura, R.; Matsuda, H. *Tetrahedron Lett.* 1995, *36*, 9497-9500.
[26] Yadav, J. S.; Reddy, B. V. S.; Scridhar Reddy M.; Parimala, G. *Synthesis* 2003, 2390-2394.
[27] Anand, N. K.; Carreira, E. M. *J. Am. Chem. Soc.* 2001, *123*, 9687-9688.
[28] Wei, C.; Li, C.-J. *Green Chem.* 2002, *4*, 39-41.
[29] Sakai, N.; Hirasawa, M.; Konakahara, T. *Tetrahedron Lett.* 2003, *44*, 4171-4174..
[30] Sakai, N.; Kanada, R.; Hirasawa, M.; Konakahara, T. *Tetrahedron* 2005, *61*, 9298-9304.
[31] Takita, R.; Fukuta, Y.; Ohshima, T.; Shibasaki, M. *Org. Lett.* 2005, *7*, 1363-1366.
[32] Lekhok, K. C.; Prajapati, D.; Boruah, R. C. *Synlett* 2008, 655-658.
[33] Takita, R.; Yakura, K.; Ohshima, T.; Shibasaki, M. *J. Am. Chem. Soc.* 2005, *127*, 13760-13761.

[34] Takita, R.; Harada, S.; Ohshima, T.; Matsunaga, S.; Shibasaki, M. *Org. Synth.* 2008, *85*, 118-130.
[35] Harada, S.; Takita, R.; Ohshima, T.; Matsunaga, S., Shibasaki, M. *Chem. Commun.* 2007, 948-950.
[36] Zhang, Y.; Li, P.; Wang, M.; Wang, L. *J. Org. Chem.* 2009, *74*, 4364-4367.
[37] Yoo, W.-J.; Zhao, L.; Li, C.-J. *Aldrichim. Acta* 2011, *44*, 43-51.
[38] Ji, D.-M.; Xu, M.-H. *Tetrahedron Lett.* 2009, *50*, 2952-2955.
[39] Prajapati, D.; Sarma, R.; Bhuyan, D.; Hu, W. *Synlett* 2011, 627-630.
[40] Rahman, M.; Bagdi, A. K.; Majee, A.; Hajra, A. *Tetrahedron Lett.* 2011, *52*, 4437-4439.
[41] Rampon, D.; Giovenardi, R.; Silva, T. L.; Rambo, R. S.; Merlo, A. A.; Schneider, P. H. *Eur J. Org. Chem.* 2011, 7066-7070.
[42] Takami, K.; Usugi, S.; Yorimitsu, H.; Oshima, K. *Synthesis* 2005, 824-839.
[43] Price S.; Edwards, S.; Wu, T.; Minehan, T. *Tetrahedron Lett.* 2004, *45*, 5197-5201.
[44] Yadav, J. S.; Reddy, B. V. S. ; Krishnam Raju, A.; Venkateswara Rao, C. *Tetrahedron Lett.* 2002, *43*, 5437-5440.
[45] Lubin-Germain, N.; Hallonet, A.; Huguenot, F.; Palmier, S.; Uziel, J.; Augé, J. *Org. Lett.* 2007, *9*, 3679-3682.
[46] Ayed, C.; Palmier S.; Lubin-Germain, N.; Uziel, J.; Augé, J. *Carbohydr. Res.* 2010, *345*, 2566-2570.
[47] Yadav, J. S.; Subba Reddy, B. V.; Nath Yadav, N.; Pratap Singh, A.; Choudhary, M; Kunwar, A. C. *Tetrahedron Lett.* 2008, *49*, 6090-6094.
[48] Lubin-Germain, N.; Baltaze, J.-P.; Coste, A.; Hallonet, A.; Lauréano, H.; Legrave, G.; Uziel, J.; Augé, J. *Org. Lett.* 2008, *10*, 725-728.
[49] Ramzi, A. Y.; Dos Santos, M.; Roussel, S.; Baltaze, J.-P.; Lubin-Germain, N, Uziel, J. *J. Org. Chem.* 2009, *74*, 4318-4323.
[50] Sakai, N.; Annaka, K.; Konakahara, T. *Org. Lett.* 2004, *6*, 1527-1530.
[51] Borah, H. N.; Prajapati, D.; Boruah, R. C. *Synlett* 2005, 2823-2825.
[52] Deacon, G. B.; Parrott, J. C. *Aust. J. Chem.* 1971, *24*, 1771.
[53] Tyrra, W.; Wickleder, M. S. *J. Organometal. Chem.* 2003, *677*, 28-34.
[54] Chen, Y.-H.; Knochel, P. *Angew. Chem. Int. Ed.* 2008, *47*, 7648-7651.
[55] Papoian, V.; Minehan, T. *J. Org. Chem.* 2008, *73*, 7376-7379.
[56] Adak, L.; Yoshikai, N. *J. Org. Chem.* 2011, *76*, 7563-7568.
[57] Hiroshita, T.; Hayashi, Y.; Mitsui, K.; Araki, S. *J. Org. Chem.* 2003, *68*, 1309-1313..
[58] Araki, S.; Imai, A.; Shimizu, K.; Butsugan, Y. *Tetrahedron Lett.* 1992, *33*, 2581-2582..
[59] Araki, S. ; Imai, A. ; Shimizu, K. ; Yamada, M. ; Mori, A. ; Butsugan, Y. *J. Org. Chem.* 1995, *60*, 1841-1847.
[60] Araki, S. ; Usui, H. ; Kato, M.; Butsugan, Y. *J. Am. Chem. Soc.* 1996, *118*, 4699-4700.
[61] Fujiwara, N.; Yamamoto, Y. *J. Org. Chem.* 1997, *62*, 2318-2319..
[62] Ranu, B. C.; Majee, A. *Chem. Commun.* 1997, 1225-1226.
[63] Fujiwara, N. ; Yamamoto, Y. *J. Org. Chem.* 1999, *64*, 4095.
[64] Klaps, E.; Schmid, W. *J. Org. Chem.* 1999, *64*, 7537-7546.
[65] Miao, W.; Chan, T. H. *Synthesis* 2003, 785-789.
[66] Lee, P. H. ; Kim, S.; Lee, K.; Seomoon, D.; Kim, H.; Lee, S.; Kim, M.; Han, M.; Noh, K.; Livinghouse, T. *Org. Lett.* 2004, *6*, 4825-4828.
[67] Goeta, A.; Salter, M. M.; Shah, H. *Tetrahedron* 2006, *62*, 3582-3599.

[68] Miura, K.; Fujisawa, N.; Hosomi, A. *J. Org Chem.* 2004, *69*, 2427-2430
[69] Miura, K.; Fujisawa, N.; Toyohara, S.; Hosomi, A. *Synlett* 2006, 1883-1886.
[70] Yanada, R.; Koh, Y.; Nishimori, N.; Matsumura, A.; Obika, S.; Mitsuya, H.; Fujii, N.; Takemoto, Y. *J. Org. Chem.* 2004, *69*, 2417-2422.
[71] Yang, X.; Zhai, H.; Li, Z. *Org. Lett.* 2008, *10*, 2457-2460.
[72] Yanada, R.; Obika, S.; Oyama, M.; Takemoto, Y. *Org. Lett.* 2004, *6*, 2825-2828.
[73] Ranu, B. C.; Mandal, T. *Tetrahedron Lett.* 2006, *47*, 2859-2861.
[74] Wang, C. Y; Su, H.; Yang, D.-Y. *Synlett* 2004, 561-563.
[75] Wang, C. Y; Yan, L; Zheng, Z.; Yang, D.; Pan, Y. *Tetrahedron* 2006, *62*, 7712-7717.
[76] Xue, Z. ; Yang, D.; Wang, C. *J. Organomet. Chem.* 2006, *691*, 247-250.
[77] Wang, C.; Pan, Y.; Yang, D. *J. Organomet. Chem.* 2005, *690*, 1705-1709.
[78] Takami, K.; Mikami, S.; Yorimitsu, H.; Shinokubo, H.; Oshima, K. *J. Org. Chem.* 2003, *68*, 6627-6631.
[79] Takami, K.; Yorimitsu, H.; Oshima, K. *Org. Lett.* 2004, *6*, 4555-4558.
[80] Hayashi, N.; Shibata, I.; Baba, A. *Org. Lett.* 2004, *6*, 4981-4983.
[81] Hayashi, N.; Shibata, I.; Baba, A. *Org. Lett.* 2005, *7*, 3093-3096.
[82] Nakamura, M.; Endo, K.; Nakamura, E. *J. Am. Chem. Soc.* 2003, *125*, 13002-13003.
[83] Tsuji, H.; Tanaka, I.; Endo, K.; Yamagata, K.; Nakamura, M.; Nakamura, E. *Org. Lett.* 2009, *11*, 1845-1847.
[84] Endo, K.; Hatakeyama, T.; Nakamura, M.; Nakamura, E. *J. Am. Chem. Soc.* 2007, *129*, 5264-5271.
[85] Nakamura, M.; Endo, K.; Nakamura, E. *Org. Lett.* 2005, *7*, 3279-3281.
[86] Tsuji, H.; Fujimoto, T.; Endo, K.; Nakamura, M.; Nakamura, E. *Org. Lett.* 2008, *10*, 1219-1221.
[87] Kaneda, K.; Motokura, K.; Nakagiri, N.; Mizugaki, T.; Jitsukawa, K. *Green Chem.* 2008, *10*, 1231-1234.
[88] Zhang, J.; Blazecka, P. G.; Angell, P.; Lovdahl, M.; Curran, T. T. *Tetrahedron* 2005, *61*, 7807-7813.
[89] Nakamura, M.; Endo, K.; Nakamura, E. *Adv. Synth. Catal.* 2005, *347*, 1681-1686.
[90] Tsuji, H.; Yamagata, K.; Itoh, Y.; Endo, K.; Nakamura, M.; Nakamura, E. *Angew. Chem. Int. Ed.* 2007, *46*, 8060-8062.
[91] Itoh, Y.; Tsuji, H.; Yamagata, K.; Endo, K.; Tanaka, I.; Nakamura, M.; Nakamura, E. *J. Am. Chem. Soc.* 2008, *130*, 17161-17167.
[92] Morikawa, S.; Yamazaki, S.; Furusaki, Y.; Amano, N.; Zenke, K.; Kakiuchi, K. *J. Org. Chem.* 2006, *71*, 3540-3544.
[93] Angell, P.; Blazecka, P. G.; Lovdahl, M.; Zhang, J. *J. Org. Chem.* 2007, *72*, 6606-6609.
[94] Takahashi, K.; Midori, M.; Kawano, K.; Ishihara, J.; Hatakeyama, S. *Angew. Chem. Int. Ed.* 2008, *47*, 6244-6246.
[95] Peng, W.; Lee, C.-S. *Synlett* 2008, 142-146.
[96] Fujimoto, T.; Endo, K.; Tsuji, H.; Nakamura, M.; Nakamura, E. *J. Am. Chem. Soc.* 2008, *130*, 4492-4496.
[97] Montaignac, B.; Vitale, M. R.; Michelet, V.; Ratovelomanana-Vidal, V. *Org. Lett.* 2010, *12*, 2582-2585.
[98] Montaignac, B.; Vitale, M. R.; Ratovelomanana-Vidal, V.; Michelet, V. *J. Org. Chem.* 2010, *75*, 8322-8325.

[99] Nishimoto, Y.; Moritoh, R.; Yasuda, M.; Baba, A. *Angew. Chem. Int. Ed.* 2009, *48*, 4577-4580.
[100] Tsuchimoto, T.; Maeda, T.; Shirakawa, E.; Kawakami, Y. *Chem. Commun.* 2000, 1573-1574.
[101] Yoon, M. Y.; Kim, J. H.; Choi, D. S.; Lee, J. Y.; Song, C. E. *Adv. Synth. Catal.* 2007, *349*, 1725-1737.
[102] Mamane, V.; Hannen, P.; Fürstner, A. *Chem. Eur J.* 2004, *10*, 4556-4575.
[103] Huang, G.; Cheng, B.; Xu, L. ; Li, Y. ; Xia, Y. *Chem. Eur. J.* 2012, *18*, 5401-5415.
[104] Bhaskar, G.; Saikumar, C.; Perumal, P. T. *Tetrahedron Lett.* 2010, *51*, 3141-3145.
[105] Tsuchimoto, T.; Hatanaka, K.; Shirakawa, E.; Kawakami, Y. *Chem. Commun.* 2003, 2454-2455.
[106] Tsuchimoto, T.; Matsubayashi, H.; Kaneko, M.; Nagase, Y.; Miyamura, T.; Shirakawa, E. *J. Am. Chem. Soc.* 2008, *130*, 15823-15835.
[107] Tsuchimoto, T.; Ainoya, T.; Aoki, K.; Wagatsuma, T.; Shirakawa, E. *Eur. J. Org. Chem.* 2009, 2437-2440.
[108] Tsuchimoto, T.; Wagatsuma, T.; Aoki, K.; Shimotori, J. *Org. Lett.* 2009, *11*, 2129-2132.
[109] Surendra, K.; Qiu, W.; Corey, E. J. *J. Am. Chem. Soc.* 2011, *133*, 9724-9726.
[110] Gibeau, A. L.; Snyder, J. K. *Org. Lett.* 2011, *13*, 4280-4283.
[111] Sakai, N.; Annaka, K.; Konakahara, T. *Tetrahedron Lett.* 2006, *47*, 631-634.
[112] Sakai, N.; Annaka, K.; Konakahara, T. *J. Org. Chem.* 2006, *71*, 3653-3655.
[113] Murai, K.; Hayashi, S.; Takaichi, N.; Kita, Y.; Fujioka, H. *J. Org. Chem.* 2009, *74*, 1418-1421.
[114] Sakai, N.; Annaka, K.; Fujita, A.; Sato, A.; Konakahara, T. *J. Org. Chem.* 2008, *73*,4160-4165.
[115] Nakamura, I.; Yamagishi, U.; Song, D.; Konta, S.; Yamamoto, Y. *Angew. Chem. Int. Ed.* 2007, *46*, 2284-2287.
[116] Yanada, R.; Obika, S.; Kono, H.; Takemoto, Y. *Angew. Chem. Int. Ed.* 2006, *45*, 3822-3825.
[117] Yanada, R.; Hashimoto, K.; Tokizane, R.; Miwa, Y.; Minami, H.; Yanada, K.; Ishikura, M.; Takemoto, Y. *J. Org. Chem.* 2008, *73*, 5135-5138.
[118] Sarma, R.; Prajapati, D. *Chem. Commun.* 2011, *47*, 9525-9527.
[119] Sarma, R.; Rajesh, N.; Prajapati, D. *Chem. Commun.* 2012, *48*, 4014-4016.
[120] Feng, X.; Tan, Z.; Chen, D.; Shen, Y.; Guo, C.-C.; Xiang, J.; Zhu, C. *Tetrahedron Lett.* 2008, *49*, 4110-4112.
[121] Tsuji, H.; Yamagata, K.; Ueda, Y.; Nakamura, E. *Synlett* 2011, 1015-1017.
[122] Xu, Y.; Pan, Y.; Liu, P; Wang, H.; Tian, Su, G. *J. Org. Chem.* 2012, *77*, 3557-3562.
[123] Pérez, I.; Pérez Sestelo, J.; Sarandeses, L. A. *Org. Lett.* 1999, *1*, 1267-1269.
[124] Pérez, I.; Pérez Sestelo, J.; Sarandeses, L. A. *J. Am. Chem. Soc.* 2001, *123*, 4155-4160.
[125] Pena, M. A.; Pérez, I.; Pérez Sestelo, J.; Sarandeses, L. A. *Chem. Commun.* 2002, 2246-2247.
[126] Pena, M. A.; Pérez Sestelo, J.; Sarandeses, L. A. *Synthesis* 2005, 485-492.
[127] Meyer, A.; Kirschning, A. *Synlett* 2007, 1264-1268.
[128] Frenzel, T.; BrUnjes, M.; Quitschalle, M.; Kirschning, A. *Org Lett.* 2006, *8*, 135-138.
[129] Barbero, M.; Cadamuro, S.; Dughera, S.; Giaveno, C. *Eur. J. Org. Chem.* 2006, 4884-4890.

[130] Riveiros, R.; Saya, L.; Pérez Sestelo, J.; Sarandeses, L. A. *Eur. J. Org. Chem.* 2008, 1959-1966.
[131] Lehmann, U.; Awasthi, S.; Minehan, T. *Org. Lett.* 2003, *5*, 2405-2408.
[132] Takami, K.; Yorimitsu, H.; Shinokubo, H.; Matsubara, S.; Oshima, K. *Org. Lett.* 2001, *3*, 1997-1999.
[133] Qian, M.; Huang, Z.; Negishi, E. *Org. Lett.* 2004, *6*, 1531-1534.
[134] Pena, M. A.; Pérez Sestelo, J.; Sarandeses, L. A. *Synthesis* 2003, 780-784.
[135] Lee, P. H.; Lee, S. W.; Lee, K. *Org. Lett.* 2003, *5*, 1103-1106.
[136] Zhao, Y.; Jin, L.; Li, P.; Lei, A. *J. Am. Chem. Soc.* 2008, *130*, 9429-9433.
[137] Lee, W.; Kang, Y.; Lee, P. H. *J. Org. Chem.* 2008, *73*,4326-4329.
[138] Font-Sanchis, E.; Céspedes-Guirao, F. J.; Sastre-Santos, A.; Fernandez-Lazaro, F. *J. Org. Chem.* 2007, *72*, 3589-3591.
[139] Pena, M. A.; Pérez Sestelo, J.; Sarandeses, L. A. *J. Org. Chem.* 2007, *72*, 1271-1275.
[140] Mosquera, A.; Riveiros, R.; Saya, L.; Pérez Sestelo, J.; Sarandeses, L. A. *Org. Lett.* 2008, *10*, 3745-3748.
[141] Bouissane, L.; Pérez Sestelo, J.; Sarandeses, L. A. *Org. Lett.* 2009, *11*, 1285-1288.
[142] Pérez, I.; Pérez Sestelo, J.; Maestro, M.; Sarandeses, L. A. *J. Org. Chem.* 1998, *63*, 10074-10076.
[143] Tato, R.; Riveiros, R.; Pérez Sestelo, J. ; Sarandeses, L. A. *Tetrahedron* 2012, *68*, 1606-1611.
[144] Fausett, B. W.; Liebeskind, L. S. *J. Org. Chem.* 2005, *70*, 4851-4853.
[145] Rodriguez, D.; Pérez Sestelo, J.; Sarandeses, L. A. *J. Org. Chem.* 2004, *69*, 8136-8139.
[146] Baker, L.; Minehan, T. *J. Org. Chem.* 2004, *69*, 3957-3960.
[147] Metza, J. T. ; Terzian, R. A.; Minehan, T. *Tetrahedron Lett.* 2006, *47*, 8905-8910.
[148] Rodriguez, D.; Pérez Sestelo, J.; Sarandeses, L. A. *J. Org. Chem.* 2003, *68*, 2518-2520.
[149] Ranu, B. C.; Chattopadadhyay, K.; Jana, R. *Tetrahedron Lett.* 2007, *48*, 3847-3850.
[150] Moral, J. A.; Moon, S.-J.; Rodriguez-Torres S.; Minehan, T. G. *Org. Lett.* 2009, 11, 3734-3737.
[151] Riveiros, R.; Rodiguez, D.; Pérez Sestelo, J.; Sarandeses, L. A. *Org. Lett.* 2006, *8*, 1403-1406.
[152] Riveiros, R.; Pérez Sestelo, J.; Sarandeses, L. A. *Synthesis* 2007, 3595-3598.
[153] Black, D. A.; Arndtsen, B. A. *Org. Lett.* 2006, *8*, 1991-1993.
[154] Beveridge, R. E.; Black, D. A.; Arndtsen, B. A. *Eur. J. Org Chem.* 2010, 3650-3656.

In: Indium
Editors: Hsaio G. Woo and Huang Tsai Choi
ISBN: 978-1-62257-696-8

Chapter 5

RELIABILITY OF INDIUM TIN OXIDE THIN FILMS: PROPERTIES UNDER MECHANICAL AND THERMAL LOADS

Mohammad M. Hamasha, Tara P. Dhakal, Khalid Alzoubi and Charles R. Westgate*
Center for Autonomous Solar Power (CASP),
State University of New York at Binghamton, Binghamton, NY, US

ABSTRACT

One of the primary applications of indium has been in the form of tin doped indium oxide (ITO). ITO is the preferred transparent conducting layer in many applications, such as liquid crystal displays, plasma displays, touchscreens, and photovoltaics. The high electrical conductivity, optical transmittance and moisture resistance make ITO a dominant transparent conductor. Recently, there has been a tremendous rise in production of portable electronic devices and a necessity of flexible devices is increasing more than ever. To produce flexible devices, flexible substrates should replace conventional rigid substrates. Although ITO has a low sheet resistance and very good transparency, ITO thin film deposited on a flexible substrate produces cracks at low tensile strains. This failure is due to the brittleness of ITO. Therefore, the mechanical stability of ITO on flexible substrates should be thoroughly considered in the design and manufacturing of flexible devices. Furthermore, the thermal expansion coefficient mismatch between the ITO film and the substrate creates cyclic internal stress on the film when the device is subjected to fluctuating temperatures. This could create micro-cracks and consequently damage the film as the micro-cracks propagate throughout the film over a prolonged exposure to the fluctuating temperatures. In this chapter, the effects of mechanical loads such as bending and tensile loads are investigated and discussed. In addition, the effect of mechanical and electrical properties of the ITO films exposed to thermal and isothermal conditions are discussed. The film-response to the mechanical and/or thermal loads is detailed in terms of different measures, namely electrical resistance, surface morphology, film

* Center for Autonomous Solar Power (CASP), State University of New York at Binghamton, Binghamton, NY 13902, US.

composition, film crystallinity, transmittance and failure modes, such as cracks and thermal damage.

2. Introduction

Electronic devices have been used in photovoltaic industry, sensors, medical electronics, and many other applications [1, 2]. For decades, many researchers have been working on developing new conductive thin films substrates and encapsulation materials for electronics applications [3]. However, it is still challenging to develop new thin films that perform well under mechanical and/or environmental loadings over a long time [4, 5]. Two parameters need to be carefully considered in the design stage for the performance of thin films under mechanical loading, namely, the maximum stress on the thin film before failure under one-time loading and the maximum stress on thin films before failure under repeated loading for long periods of time. Similarly, two parameters need to be carefully considered in the design stage for the performance of thin films under environmental loadings, namely, the maximum temperature and/or relative humidity that are applied on the thin film continuously without failing, and the maximum amplitude of fluctuating temperature and/or relative humidity that can be applied on the thin film without being failed.

Tensile and compressive stresses can lead to a distortion of the thin films and degradation of the device [6]. However, the problem becomes severe at elevated temperatures. Many tests have been used to study the behavior of the thin film when subjected to mechanical loadings. However, the most popular are tensile and cyclic bending tests. In the tensile test, the structure of thin film and substrate is stretched while the electrical resistance is being measured. In the cyclic bending test, the structure of thin film and substrate is bent cyclically between two parallel cylindrical bending dies while the electrical resistance is being measured. Images are always taken after both tests to study developed cracks [6].

The failure in the thin film takes place with initiation of many cracks and their slow propagation leading to an increase of the electrical resistance of the thin film. However, in the bulk materials, failure takes place mainly due to development of a single crack that leads to a sudden fracture. [7]. Other failure modes can take place due to the mechanical loads on the thin film, such as intrusions and extrusions [8].

While the thin film device is being used, thin films can be subjected to isothermal aging stress due to continuous operation of electronic devices. Constant high temperature can change the physical, mechanical, and/or electrical properties of the thin films over time. Moreover, temperature fluctuation between the operating and turning off times of an electronic device can cause fatigue and crack the thin film, as in the following scenario. Due to the difference in the thermal expansion mismatch between the thin film and the substrate, both of them do not expand in the same amount under high temperature, and the thin film is either stretched out or compressed. However, when the machine is turned off, the temperature of the thin film and substrate decreases gradually and then removes the stress that was created by the heat. The repetition of straining and unstraining during cyclic heating and cooling fatigues the film gradually and creates micro-cracks.

The effect of constant high temperatures can be studied through the isothermal aging test, where the samples are exposed to high temperature for a certain period of time, and the response of the thin film is monitored. The response measurements vary depending on the researcher's interests, such as surface morphology, electrical conductivity, material composition and material structure. Similarly, the effect of fluctuating temperatures can be studied though a thermal cycling test, where the samples are exposed to thermal cycling with a certain high temperature, low temperature and dwell time.

Indium tin oxide (ITO) is a transparent conductive oxide used widely in applications such as heat reflecting coatings, plasma display, solar cell, energy efficient windows, gas sensors, field emission displays and touch panels [9]. Other applications include the deposition of ITO on flexible organic substrate for organic light-emitting diodes and liquid crystal display applications [10].

ITO films can be produced using many methods, such as radio frequency sputtering, direct current (DC) sputtering, ion Beam sputtering, reactive evaporation, spray pyrolysis, thermal evaporation, screen printing techniques and pulsed laser deposition [11, 12]. Furthermore, ITO can be produced using high speed roll-to-roll manufacturing [13].

In the following Section, research on the reliability of general thin films under mechanical and/or environmental loads will be reviewed briefly.

2.1. A Brief Review on the Effects of Mechanical and Environmental Loads on Thin Films

The following researchers have studied the behavior of different thin films under mechanical stresses. Sierros et al. [14] studied the behavior of carbon nanotube (CNT) film under tensile stress. In their research, two experiments were conducted, tensile and tensile fatigue experiments. In the tensile experiment, the samples were stretched up to 25% of the original length, and in the tensile fatigue experiment, a fluctuating strain of 0-5 % of the original length was applied on the samples. They found that the CNT was relatively unstable under the tensile fatigue condition; however, it was stable under tensile condition. Renault et al. [15] performed X-ray diffraction analysis on 320 nm thick gold films on a polymer substrate under tensile and compression loads. They concluded from the experimental results that crack initiation takes place at stresses higher than 670 MPa. Furthermore, they noticed two types of straight-sided buckling patterns laying perpendicular to the tensile axis. Tsuchiya et al. [16] studied the fracture strain and tensile strength of sputtered titania thin film on flexible substrate. They measured the modulus of elasticity and the tensile strength at different lengths and widths, and reported the results.

The stability of most thin films on flexible substrates under different environmental loads is still a challenge [17-18]. Many authors studied the effect of constant high temperatures (isothermal condition) on the thin film for different materials. Chou, et al. [19] studied the stability of hybrid phlogopite mica thin films under isothermal conditions for 500 and 1000 hours, where they found that the films are unstable at elevated temperatures. Rizvi et al. [20] performed isothermal aging experiments on anisotropic conductive film (ACF) joints at temperatures of 60, 100, 160, 200 and 240°C over time. In their research, they studied the failure rate and the failure mechanism. Sebastain [21] studied the effect of the isothermal aging of CdSe thin films. He discovered that oxygen atoms were attached in the materials after aging. Tosser et al. [22] claimed that the isothermal aging enhance the grain characteristics of Zn and Al depending on the film thickness. Moreover, much research has been conducted to characterize the effect of the thermal cycling on different thin films.

Kaouache et al. [23] investigated the effect of thermal cycling on the microstructure of Al, 0.5% Cu, and 1% Si thin films that deposited onto oxidized silicon substrates. They discovered that dislocation activity and grain growth start almost simultaneously and cooperate throughout the plastic regime of the stress–temperature curve to generate bamboo-type grains with low dislocation density. Schmidt et al. [24] conducted thermal cycling testing on Cu thin films with Ta or Ag interlayer. However, they found that the Ta interlayer, which was expected to enhance Cu adhesion, led to higher stress than Ag.

3. Effect of Mechanical and Thermal Loads on the ITO Films

In this section, the effect of mechanical and thermal stresses on thin films of ITO grown on flexible substrate is presented. In Section 3.1, the effect of cyclic bending fatigue is reviewed using the experiments conducted by Alzoubi, et al. [25] and Hamasha, et al. [26] papers. In subsection 3.2, the effect of stretching is reviewed using the experiments conducted by Hamasha, et al. [27]. In Section 3.3, the effect of both isothermal aging and thermal cycling is reviewed using the experiments conducted by Hamasha, et al. [28].

3.1. Bending Fatigue of Sputtered ITO on PET Substrate

Although ITO film has excellent electrical resistance and optical properties, it is brittle and can fail at very low tensile stress. Therefore, the mechanics of ITO on flexible substrates are carefully considered in the design of flexible devices. For example, in the roll-to-roll manufacturing, stresses that formed because of bending of the web over the rollers can affect the film quality even if the stress is below the yield strength. In this Section, the effect of cyclic bending fatigue on 60 Ω per square sputter-deposited ITO films on 127 -μm Poly Ethylene Terephthalate (PET) substrate is presented. The effects of bending diameter, bending frequency, sample width, temperature and relative humidity on the percentage change in electrical resistance are considered. Optical microscope images are obtained to study crack initiation and propagation in the thin film.

3.1.1. Effect of Bending Diameter, Frequency and Sample Width

In this Section, the experiments conducted by Alzoubi, et al. [25] on 60 Ω per square sputter-deposited ITO on 127 -μm PET substrate, with substrate length of 10 cm and width of either 24 or 8 mm are presented. The effect of bending diameter, bending frequency and sample width on the percentage change in electrical resistance during the cyclic bending condition was investigated. Optical microscope images were obtained to observe crack initiation and propagation in the ITO layer.

The cyclic bending fatigue tester setup is shown in Figure 1. As the experiment starts, the two bending dies start to reciprocate vertically causing the sample to be bent around the lower die in the upward stroke, as shown in Figure 1b and around the upper die in the downward stroke, as shown in Figure 1c. The frequency of the cyclic motion was either 100 or 30 cycles per minute and the diameter of the bending die was either 8.9 or 5 mm.

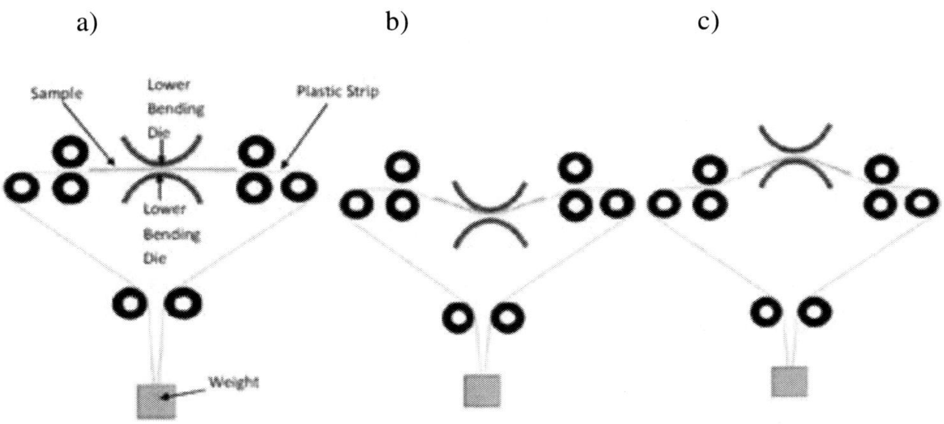

Figure 1. Cyclic bending tester setup (a- the sample is in neutral position, b- the sample is bent around the upper die, and c- the sample is bent around the lower die).

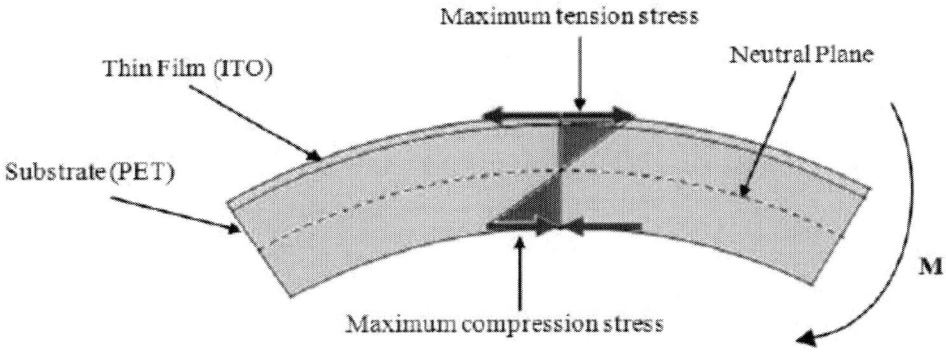

Figure 2. Bending schematic diagram.

When the sample is bent over the lower die in the upward stroke, the thin film at the upper-side of the sample is stretched, while the lower surface of the substrate is compressed (see Figure 2). The tension and compression are created by the momentum (M) that applied due to bending. In this situation, the maximum tension stress is developed at the thin film, while the maximum compression stress is developed at the lower surface of the substrate. When the sample is bent around the upper die in the downward stroke, the same but opposite way of stress can develop.

Thus, the maximum tension is developed on the lower surface of the substrate, and the maximum compression is developed on the thin film. With repetitive bending around upper and lower dies, the thin film will suffer from tension-compression cyclic loadings, which lead to fatigue and an ultimate failure. Optical images of the film surface are obtained to monitor the ITO surface uniformity. Electrical resistance of the ITO thin film is measured from both ends of the sample after a specified number of bending cycles while the sample is flat and compared to the initial resistance before starting the test. The percentage change in electrical resistance, $(\Delta R/R_o) \times 100$ %, is considered for all analysis, where R_o is the initial resistance and ΔR is the difference between current and initial resistance of the ITO film before testing.

To study the effect of bending diameter on the percentage change in electrical resistance, two different bending diameters were chosen, 5 mm and 8.9 mm. The experiments were performed using 8 mm-wide samples at room temperature and humidity and bending frequency of 100 cycles/minutes. The percentage change in electrical resistance is significantly greater at smaller mandrel diameters, as shown in Figure 3.

Tighter mandrels produce greater stresses in the thin film, cause cracks and then produce great increase in electrical resistance and ultimate failure. Moreover, the effect of sample width on the percentage change in electrical resistance was also studied, as shown in Figure 4.

The experiments were conducted at the following conditions: room temperature and relative humidity, bending diameter of 5 mm and bending frequency of 100 cycles/minute.

It was found that the width of the sample has an effect on percentage change in electrical resistance. This is due to the fact that the crack propagation in wider samples takes a longer time to cut the whole film, and therefore the percentage change in electrical resistance is lower than thinner samples tested under the same conditions. Similarly, the effect of bending frequency on the percentage change in electrical resistance was investigated considering two different bending frequencies of 30 and 100 cycles per minute under the following conditions: room temperature, room relative humidity, 8.9 mm bending diameter and 8 mm sample width. Figure 5 shows that the bending frequency has a negligible effect. This property depends on other mechanical properties like hardness, ductility, etc.

Figure 6 shows optical images for crack development on the ITO surface during the bending fatigue experiment. As can be noted, the cracks are initiated as early as 10 bending cycles. However, the density of the cracks increases with the bending cycles. As reported in the literature, cracks are initiated from surface imperfections [29], and then the neighbor micro cracks start to join each other to form bigger ones. For further explanation, the design of experiment (DOE) tool was used to study the significance of the effect of number of cycles, sample width, bending frequency and bending diameter on the percentage change in electrical resistance.

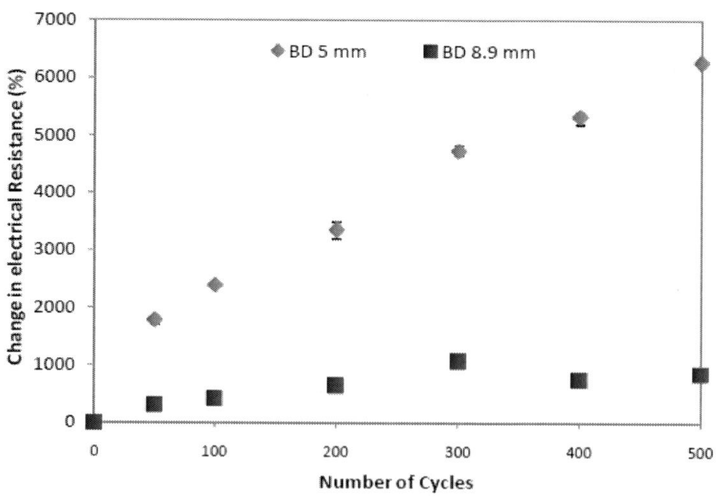

Figure 3. Effect of bending diameter on the percent change in electrical resistance; the experiments were conducted on an 8 mm-wide sample at a room temperature, room relative humidity and bending frequency of 100 cycles/minutes.

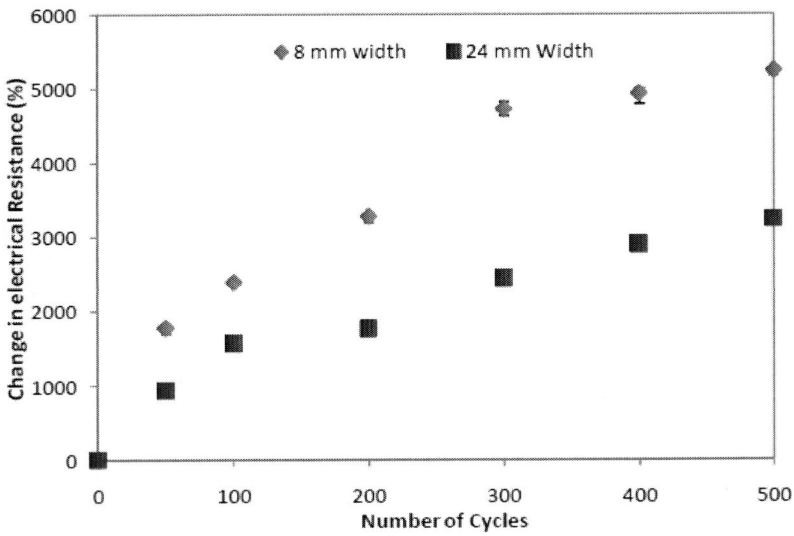

Figure 4. Effect of sample width on the percent change in electrical resistance; the experiments were conducted at a room temperature, room relative humidity, 5 mm bending diameter and 100 cycles/minute bending frequency.

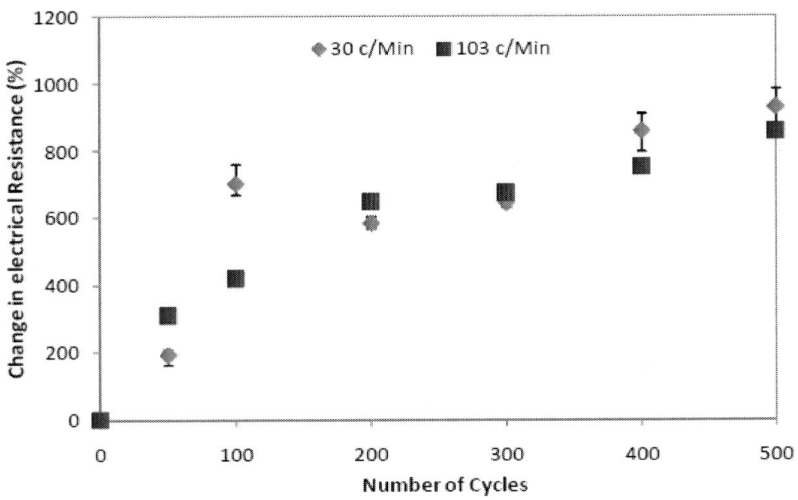

Figure 5. Effect of bending frequency on the percent change in electrical resistance; the experiments were conducted at a room temperature, room relative humidity, 8.9 mm bending diameter and 8 mm sample width.

Two levels were chosen for each factor except number of cycles, where seven levels were chosen, as shown in Table 1. The response of the DOE analysis was the percentage change in electrical resistance.

In Table 2, analysis of variance (ANOVA) results are presented. According to the statistics basics, the term (factor or the interaction between factors) is significant only if the P-value is lower than the significant level, α, which is selected to be 0.05 in this case.

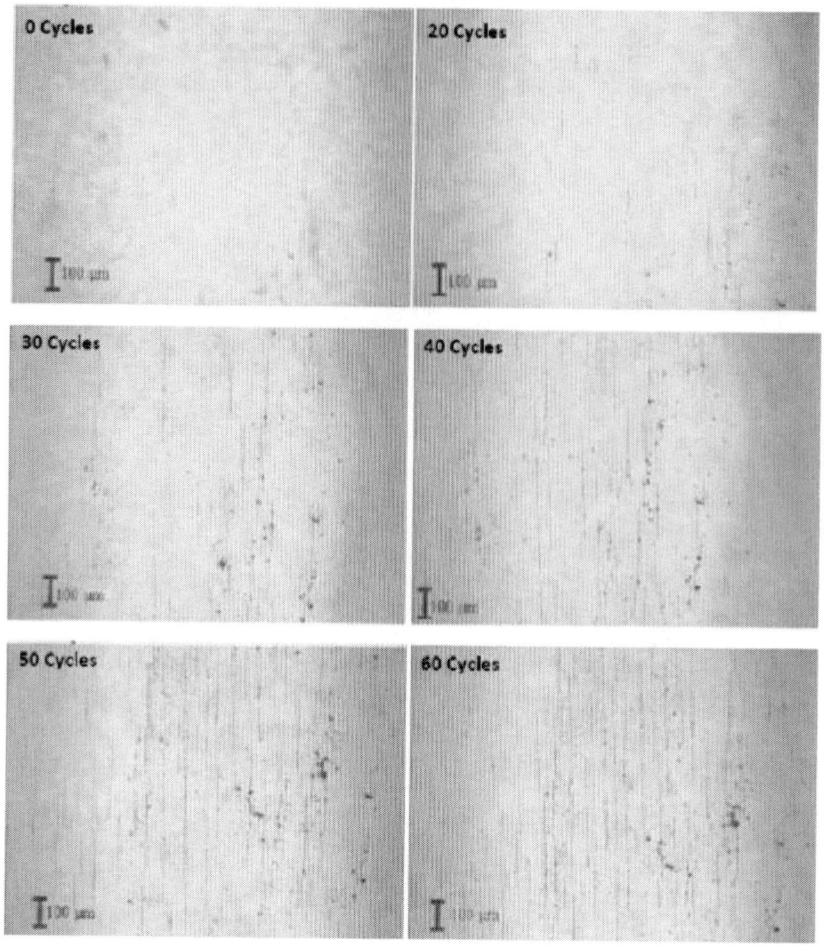

Figure 6. Cracks development during cyclic bending fatigue experiment; the experiments were conducted at a room temperature, room relative humidity, 5 mm bending diameter, 8 mm sample width, and 103 cycles/min bending frequency.

Table 1. The selected DOE factors and levels for the effect of bending diameter, frequency and sample width

Factor	Level 1	Level 2	Level 3	Level 4	Level 5	Level 6	Level 7
Number of cycles (NOC)	0	50	100	200	300	400	500
Sample width (W) (mm)	24	8	--	--	--	--	--
Bending Frequency (F) (cycles/min)	100	30	--	--	--	--	--
Bending Diameter (BD) (mm)	8.9	5	--	--	--	--	--

Table 2. ANOVA table for the DOE experiments

Source	DF	Seq SS	Adj SS	Adj MS	F	P
Nomber of cycles	6	80002535	52431487	8738581	6.08	0.000
Bending diameter	1	52111471	40021083	40021083	27.83	0.000
Sample width	1	13206189	14292620	14292620	9.94	0.003
Bending fequency	1	1798108	1618380	1618380	1.13	0.293
Nomber of cycles × Bending diameter	6	25328275	25328275	4221379	2.94	0.014
Bending diameter × Sample width	1	7565989	10102785	10102785	7.03	0.010
Sample width × Bending fequency	1	11974163	11974163	11974163	8.33	0.005
Error	59	84846339	84846339	1438074		
Total	76	276833069				

S = 1199.20; R-Sq = 69.35%; R-Sq(adj) = 60.52%.

In this ANOVA, only the significant terms were considered to save more degree of freedom for the error term, which increases the accuracy of the results. It is clearly seen that number of cycles, sample width and bending diameter are significant factors. However, the bending frequency is not significant, but it could not be removed from the ANOVA because its interaction with the sample width is significant. Furthermore, the following two factor interactions are significant: number of cycle – bending diameter, bending diameter – sample width and sample with – bending frequency.

3.1.2. Effect of Bending Fatigue, Temperature and Relative Humidity

In this Section, the effect of temperature and relative humidity on the electrical resistance under cyclic bending, conducted by Hamasha et al. [26] is reviewed with a further discussion and critique. In their study, two sets of experiments were conducted on a 60 Ω per square sputter-deposited ITO on 127 -μm PET substrate. The first set of experiments was designed to study the effect of temperature and relative humidity on resistance change of the thin film during cyclic bending. Eight samples were exposed to different temperatures and relative humidity while being subjected to cyclic bending condition at a constant bending frequency of 75 cycles per minute and a bending diameter of 76.2 mm. 20 °C and 70 °C were selected as levels of the temperature and 20% and 80% were selected as relative humidity levels. Microscopic images were obtained to characterize the damage on the thin film after 10,000 bending cycles at each combination of temperature and relative humidity, and the percentage change in electrical resistance for the ITO film was estimated for each sample after every 2,000 cycles.

In the second set of experiments, another eight samples were exposed to the same combinations of temperature and relative humidity, but without cyclic bending. The aim of this set of experiments is to check if the film-damage in the first set of experiments were due to the temperature and relative humidity alone or due to the cyclic bending. The exposure time was 133 minutes and 20 seconds, which equivalents to the time required to perform 10,000 cycles at the same frequency with the first set of experiments. The percentage change in electrical resistance was estimated for each sample at every 20 minutes and 40 seconds.

Moreover, the DOE tool was used to study the effects of temperature, relative humidity, bending fatigue and the interaction between them on the percentage change of electrical resistance for the ITO film.

Figure 7 shows the failure mode on the film's surface after 10,000 cycles of cyclic bending at different temperatures and humidity.

As shown in Figure 7(a), there is no sign of change on the film surface at the combination of a temperature of 20 °C and relative humidity of 20%. However, as the relative humidity increases during the cyclic bending, moisture damage (spot-like structures) starts to appear in the form of spots all around the surface, as shown in Figure 7(b). At a high temperature of 70 °C and a low relative humidity of 20%, some cracks (line-like structures) appear as shown in Figure 7(c). Figure 7(d) shows a catastrophic surface damage at a high temperature of 70 °C and a high relative humidity of 80%, where both cracks and spots appeared with a significant increase in both of them.

Figure 8 shows a higher magnification image of a selected moisture damage spot. The picture was captured at the following conditions: bending fatigue cycles of 10,000, temperature of 20 °C and relative humidity of 80 %.

Other optical images were obtained for the samples that were exposed to harsh environments without cyclic bending fatigue, as shown in Figure 9. In order to compare Figure 7 and Figure 9, images with the same magnifications were taken for both cases.

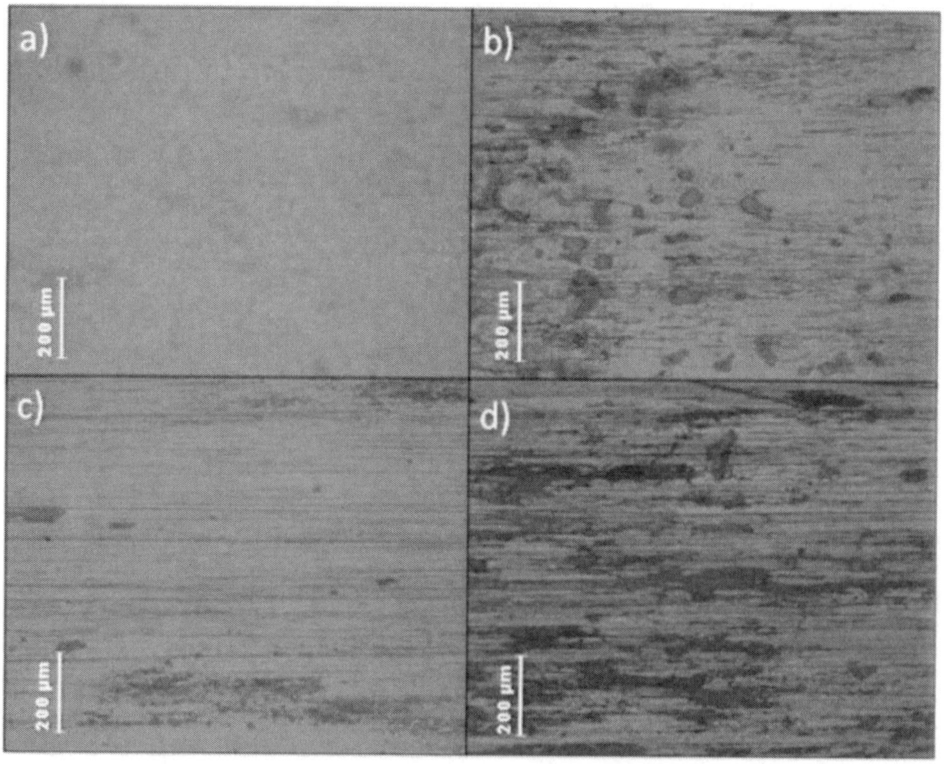

Figure 7. High magnification optical microscopic images after 10,000 bending cycles. a) Temperature: 20 °C, Relative humidity: 20%, b) Temperature: 20 °C, Relative humidity: 80 %, c) Temperature 70 °C, Relative humidity 20 %, and d) Temperature: 70 °C, Relative humidity: 80%.

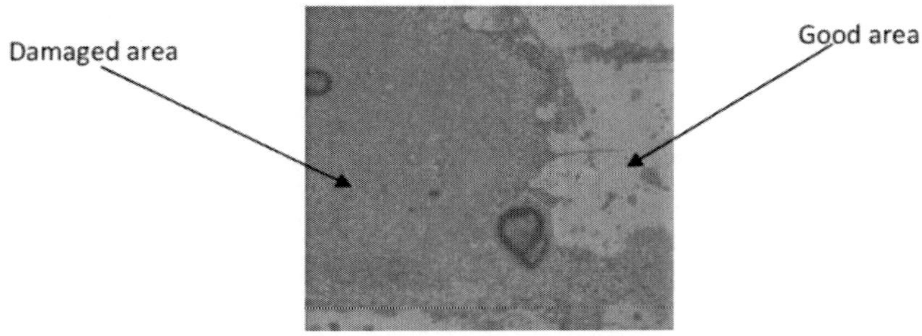

Figure 8. Optical microscopic image showing one selected moisture damage spot.

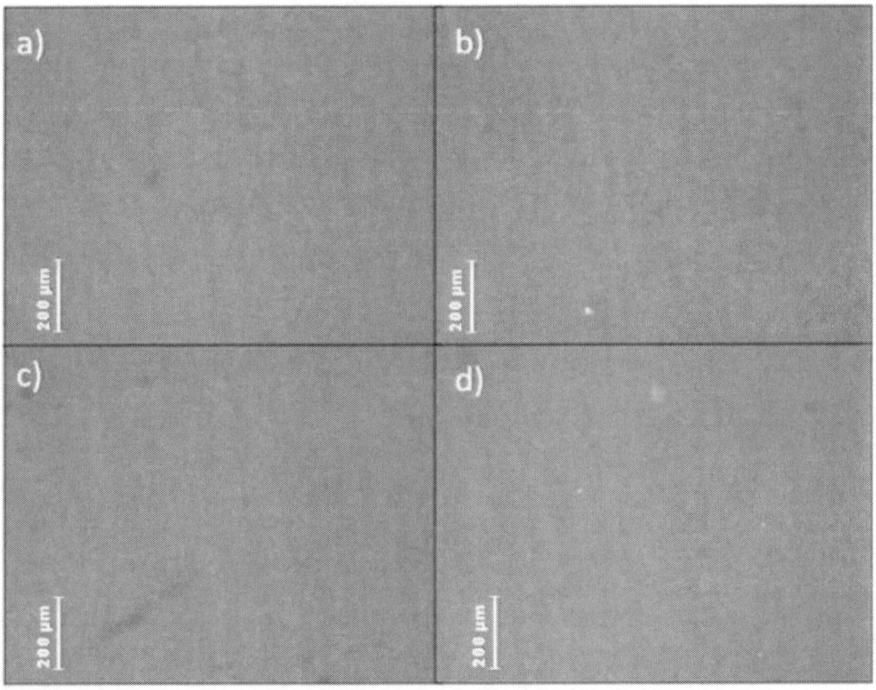

Figure 9. High magnification optical microscopic images after 133 minute and 20 seconds for ITO on PET under different temperature and relative humidity without involving of bending. a) Temperature: 20 °C, Relative humidity: 20%, b) Temperature: 20 °C, Humidity: 80%, c) Temperature: 70 °C, Relative humidity: 20%, and d) Temperature: 70 °C, Relative humidity: 80%.

However, there are no signs of cracks or spot-like damages found when the cyclic bending is not included. This means that cyclic bending is a significant factor in altering the surface integrity of the film, especially under high temperatures and relative humidity.

The relationship between the percentage change in electrical resistance and the number of cycles at different combinations of temperature and relative humidity is shown in Figure 10. The resistance did not exceed 1% of the original resistance for the samples cycled at a temperature of 20 °C and relative humidity of 20%. However, it exceeded 2% for the samples that cycled at a temperature of 20 °C and relative humidity of 80%, and the samples that cycled at 70 °C and relative humidity of 20%. However, there was a significant increase in the percentage change in electrical resistance of about 17% for the samples that cycled at a high temperature of 70 °C and relative humidity of 80%. There was no significant change in the percentage change in electrical resistance at any combination of temperature and relative humidity when the cyclic bending is not evolved, as shown in Figure 11. However, there was a slight increase in electrical resistance at high temperatures. DOE tools were used to identify the significant factors and factors' interactions. Two levels were selected for all factors except the time, where six levels were selected, as shown in Table 3. The DOE response was the percentage change in electrical resistance.

ANOVA results, as shown in Table 4, were obtained. Considering the confidence level of 0.05, temperature, bending, and relative humidity are significant factors. However, although the time is not a significant factor, the interaction between time and temperature is significant so it cannot be withdrawn from the ANOVA.

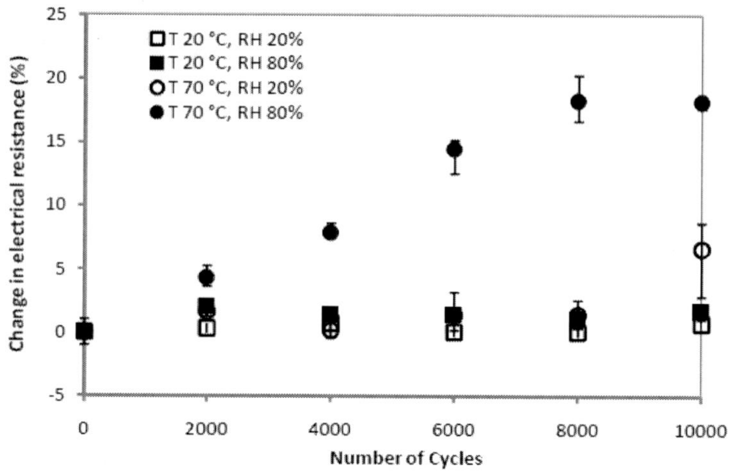

Figure 10. Increase in electrical resistance of ITO films with the number of bending cycles measured at two temperatures and relative humidity.

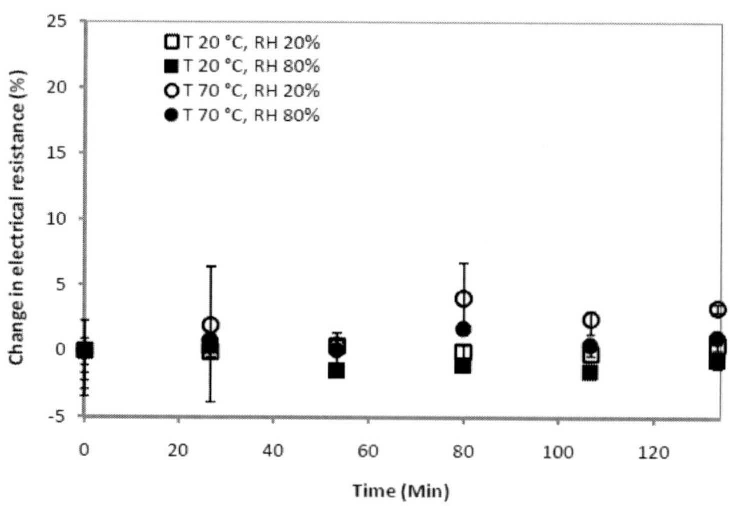

Figure 11. Increase in electrical resistance of ITO films with time at two temperatures and relative humidity.

Table 3. DOE factors and levels for the effect of time, bending fatigue, temperature and relative humidity of ITO/PET on the percentage change of electrical resistance

Factor	Level 1	Level 2	Level 3	Level 4	Level 5	Level 6
Time (Min)	0	26.67	53.33	80	106.67	133.33
Bending Fatigue	Yes	No	--	--	--	--
Temperature (°C)	20	70	--	--	--	--
Relative Humidity (%)	20	80	--	--	--	--

Table 4. ANOVA table for effect of bending fatigue, temperature and relative humidity of ITO thin film on the percentage change of electrical resistance

Source	DF	SS	MS	F	P
Time	5	51.565	10.313	1.82	0.146
Bending	1	196.928	196.928	34.66	0.000
Temperature	1	293.675	293.675	51.69	0.000
Relative humidity	1	61.176	61.176	10.77	0.003
Time×Bending	5	68.041	13.608	2.40	0.066
Time×Temperature	5	118.886	23.777	4.19	0.007
Bending× Temperature	1	6.412	6.412	1.13	0.298
Bending× Relative humidity	1	102.853	102.853	18.10	0.000
Temperature× Relative humidity	1	24.822	24.822	4.37	0.047
Bending× Temperature × Relative humidity	1	53.863	53.863	9.48	0.005
Error	25	142.025	5.681		
Total	47	1120.246			

S = 1199.20; R-Sq = 69.35%;R-Sq(adj) = 60.52%.

All two-factor interactions are significant except bending–temperature and time–bending. However, bending–temperature–relative humidity is the only significant three factor interactions.

Figure 12 shows the main effect plot for the considered factors. It can be seen that the percentage change in electrical resistance increases with temperature, relative humidity and number of cycles. Because of the formation of the cracks and spots, the main effect plot shows a high percentage change in electrical resistance for the bending cyclic samples. Figure 13 shows the interaction effect of the two-factor interactions.

The two-factor interaction is not significant in terms of affecting the response if the differences in the response between two factors at the upper and lower levels are the same, in other words, if the two curves representing the responses for two different factors are parallel in the two-factor interactions plot.

Recalling the ANOVA (table 4), all two-factor interactions are significant except bending–temperature and time–bending interaction. This conclusion is supported by the two-factor interaction as well.

3.2. Behavior of ITO Thin Film on PET Substrate under Stretching

In this subsection, the study on the behavior of ITO thin film grown on PET substrate under stretching by Hamasha et al. [27] is reviewed, discussed and critiqued. In their study, two ITO films grown on PET substrate by sputtering method with sheet resistances of 100 and 60 Ω per square were stretched up to 15% (strain) of the original length (60 mm) under three different strain rates, 0.01, 0.1 and 1.0 min^{-1}.

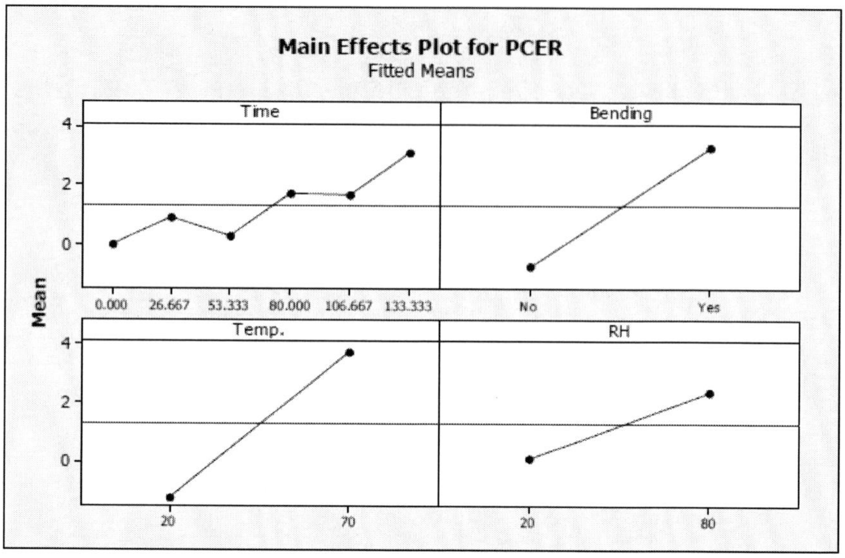

Figure 12. Main effect plot for the effect of time, bending fatigue, temperature and relative humidity of ITO/PET on the percentage change of electrical resistance.

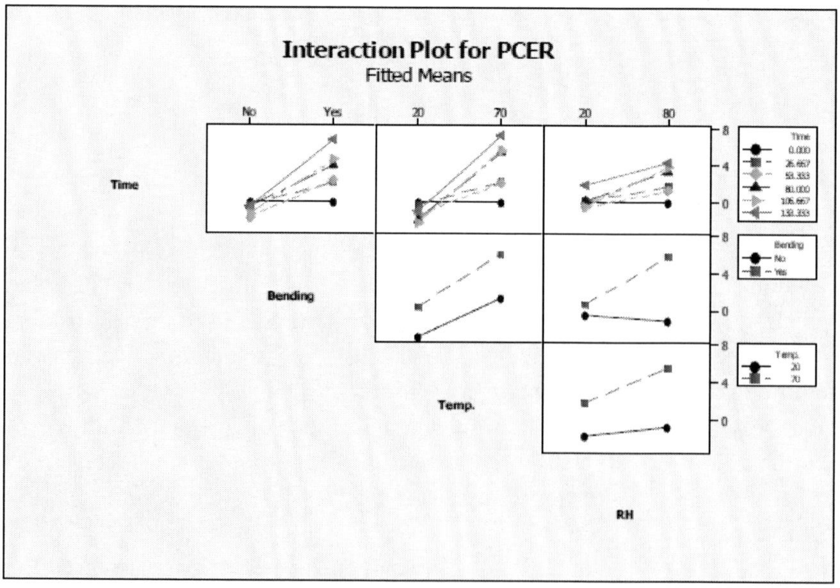

Figure 13. Interaction plot for the effect of time, bending fatigue, temperature and relative humidity of ITO thin film on the percentage change of electrical resistance.

Equations 1 and 2 refer to strain and strain rate, respectively.

$$Strain = \frac{\Delta L}{L_0} \tag{1}$$

$$Strain\ rate\ \frac{\Delta L}{L_0\, T} \tag{2}$$

where L_o is the initial distance between grips, ΔL is the increase in sample length and T is the stretching time in minutes. The initial length, L_o was 60 mm. Also the initial sample's width and thickness (including substrate) were 10 mm and 127 μm, respectively. The sample was gripped form both sides by Instron jaws before allowing both jaws to move away from each other, as shown in Figure 14, causing an increase in the sample's length and decrease in its width and thickness. The cracks propagation on the film after every 2% of strain was observed using an optical microscope. These observations showed two types of cracks: in the first type, the cracks were observed at 4% strain and more for both sheet resistances. The cracks in this type initiated perpendicularly to the tensile load and propagated towards the edge of the sample. In the second type, cracks initiated from the former cracks and propagated perpendicularly to them, and then stopped at the adjacent original crack (Figure 18 and 19). Hamasha et al. [27] called the first type of cracks "original cracks" and the second type "lateral cracks".

In order to measure the electrical resistance during stretching, ohmmeter terminals were connected to the thin film 20 mm away from the grips, as shown in Figure 14. In other words, the initial distance between the probe points was 40 mm. Under stretching, the strain gradually increases on the thin film causing crack initiation and propagation. Figure 15 shows the crack development at different strain levels under a strain rate of 0.01 min^{-1} for the film with 100 Ω per square sheet resistance.

It can be noticed that the width and intensity of the cracks greatly depends on the strain level. At 2% strain, there were no observed cracks; however, at 4% strain, many fine cracks were observed. At 8% strain, the cracks developed furthermore, and the density of the cracks increased. At this level of strain, the lateral cracks were observed to develop at different locations on the original cracks and propagated perpendicularly toward the next original crack. At 10% strain, similar to the original cracks, the lateral cracks became wider, and more intense.

Figure 16 shows the crack initiation and propagation under a strain rate of 0.01 min^{-1} for the film with 60 Ω per square sheet resistance.

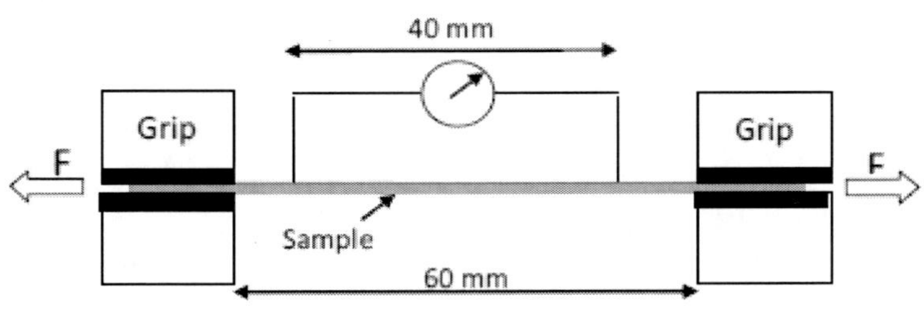

Figure 14. Schematic diagram of the specimen and the grips.

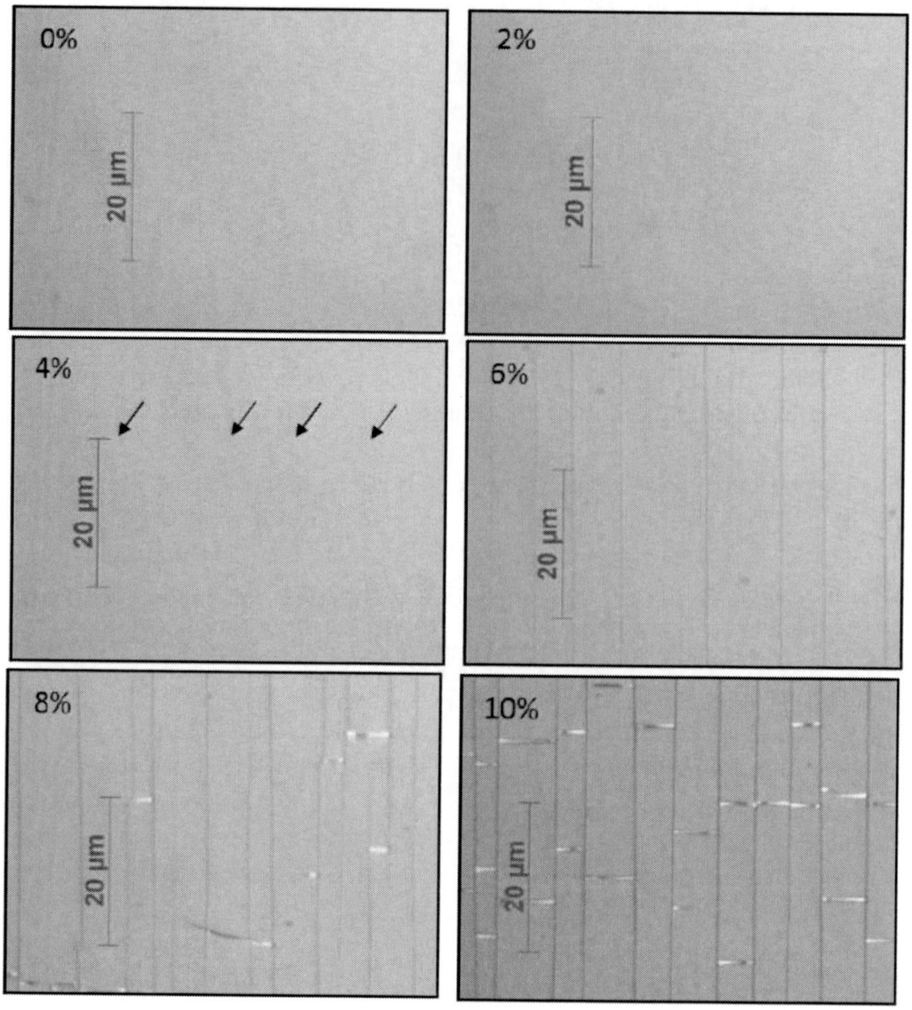

Figure 15. Crack development under different strain level for 100 Ω per square sheet resistance of ITO thin films on PET substrate.

The same crack initiation and propagation that was observed for the film with 100 Ω per square (Figure 15) was observed in this case too. The only difference was that the crack intensity is higher in the 100 Ω per square films. It should be noted here that the film with lower sheet resistance is thicker. Because the structure and composition of the ITO films were the same, it can be concluded that the thicker film has a wider average distance between the cracks after stretching.

The electrical resistance of the samples under stretching was monitored. The relative electrical resistance ($\Delta R/R_0$) was drawn versus strain at all considered strain rates sheet resistances. Figure 17 shows the effect of strain rate on $\Delta R/R_0$ for the film with a sheet resistance of 100 Ω per square.

The figure shows three distinguished regions. Initially, the resistance was constant ($\Delta R/R_0$ was zero) for all strain rates. This indicated that there were no cracks initiating in this period. The resistance then started to increase and $\Delta R/R_0$ curves were concave up to a certain strain. After that, the curve started to become convex.

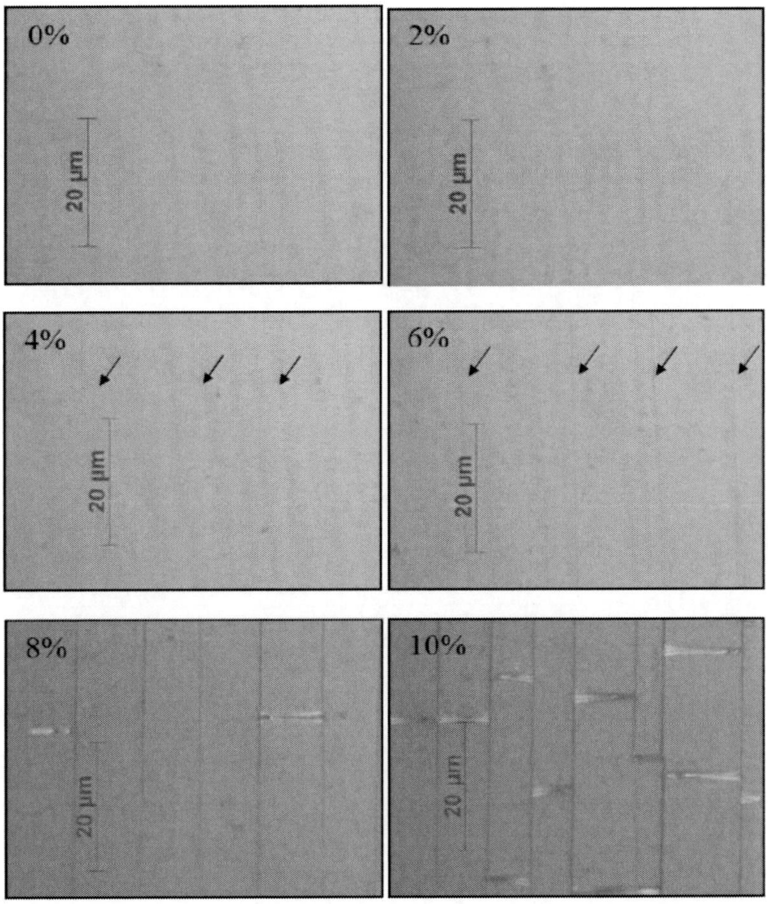

Figure 16. Crack development under different strain level for 60 Ω per square sheet resistance of ITO thin films on PET substrate.

The samples under a lower strain rate were more conductive at the beginning of stretching, but after a certain strain value, thier resistance accelerated up and became nonconductive, while the samples under a higher strain rate were still in the process of becoming nonconductive.

In other words, $\Delta R/R_0$ was higher at a higher strain rate at the beginning of the stretching process, but after a certain strain, $\Delta R/R_0$ at a lower strain rate increased quickly and exceeded $\Delta R/R_0$ for the higher strain rate. For the highest strain rate (1.0/min), a plateau was seen between 5 to 10% strain. $\Delta R/R_0$ was the largest for the highest strain rate below 5% strain, but beyond a strain of 12%, it showed the lowest $\Delta R/R_0$. The behavior of 60 Ω per square sheet resistance is the same as the behavior that explained in Figure 17, as shown in Figure 18.

$\Delta R/R_0$ – strain curves explain the relationship between the relative resistance change and strain, with a possibility to compare the effect of strain rate and sheet resistance on $\Delta R/R_0$. However, to investigate the level of significance of these factors on the $\Delta R/R_0$, a DOE with 2^3 factorial experimental design was considered. The response used was $\Delta R/R_0$ with three designed factors as follows: sheet resistance, strain rate and strain. The levels were as follows: sheet resistance levels are 60 and 100 Ω per square, strain rate levels are 0.01 and 1.0 min^{-1} and strain levels are 6% and 10%. The ANOVA results were obtained, as shown in Table 5.

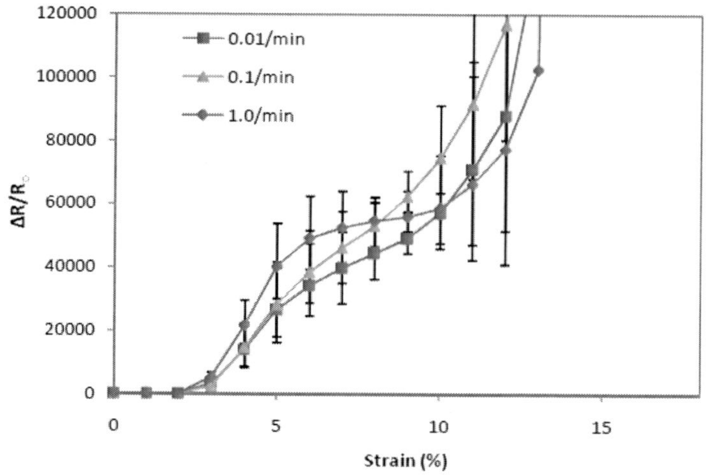

Figure 17. Strain rate effect on $\Delta R/R_0$ as a function of strain for 100 Ω per square sheet resistance. ITO thin film on PET substrate.

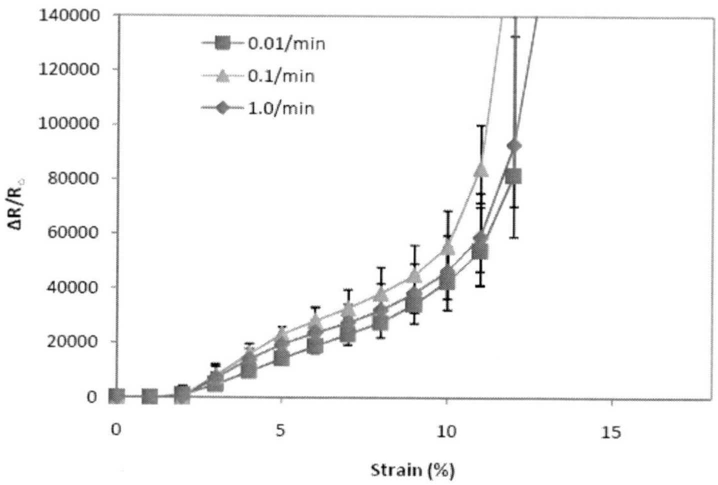

Figure 18. Strain rate effect on $\Delta R/R_0$ as a function of strain for 60 Ω per square sheet resistance ITO thin film on PET substrate.

Considering a confidence level of 0.05, it can be seen that the sheet resistance and strain are significant factors. Although the higher strain rate level was 100 times the lower strain rate level, the strain rate was still not significant. This is a good indication that during the roll-to- roll manufacturing, the speed of the roll is not a factor to alter the ITO thin film performance. Furthermore, no significant factor interactions were observed.

Figure 19 shows the main effects plot for the study factors on the relative electrical resistance change. The relative electrical resistance change was higher for the higher sheet resistance. Furthermore, the relative electrical resistance change was higher for the higher strain and strain rate. The strain rate has the minimum effect on the response between the three considered factors. However, the sheet resistance had the maximum effect in this study. Figure 20, two-factor interactions plot, shows no significant interaction between any two factors in terms of affecting the relative electrical resistance change significantly.

Table 5. ANOVA table for the response $\Delta R/R_0$

Source	DF	SS	MS	F	P
Sheet Resistance	1	3.38E+11	3.38019E+11	100.	0.000
Strain Rate	1	7.13 E+9	7135843690	2.13	0.155
Strain	1	4.8 E+10	4.8 E+10	14.3	0.001
Sheet Resistance× Strain Rate	1	2.66 E+9	2.66 E+9	0.79	0.380
Sheet Resistance× Strain	1	2.27 E+9	2.27 E+9	0.68	0.417
Strain Rate× Strain	1	3.15 E+9	3.15 E+9	0.94	0.340
Sheet Resistance× Strain Rate× Strain	1	2.8 E+9	2.81 E+9	0.84	0.367
Error	32	1.07E+11	3.36 E+9		
Total	39	5.1E+11			

S = 57938.2; R-Sq = 79.01%; R-Sq(adj) = 74.41%.

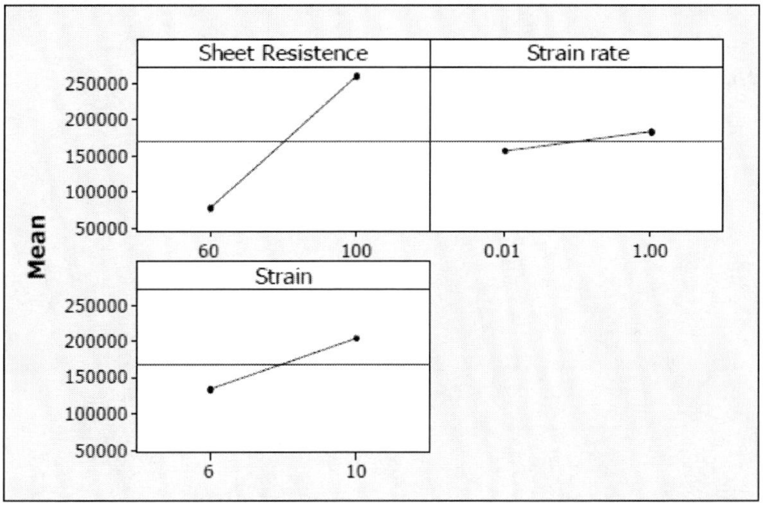

Figure 19. Main effect plot on the effect of sheet resistance, strain rate and strain on the relative electrical resistance change.

3.3. Behavior of ITO Thin Film on PET Substrate under Thermal and Isothermal Conditions

ITO is sensitive to thermal and isothermal conditions and the film electrical resistance increases gradually under such conditions. In addition, the thermal expansion coefficient mismatch between the ITO film and the substrate creates stress on the film when the film device is under high temperatures, such as operational condition and this stress gradually decreases when the device is turned off. The fluctuating temperature that is applied on the thin film between the turning on and off cycle could create micro-cracks as a consequence of fatigue. In this chapter, the experiments and results published by Hamasha, et al. [28] are reviewed, with further discussion and critique. In their work, two sets of experiments were conducted on a 60 Ω per square of sputter-deposited ITO on PET substrate.

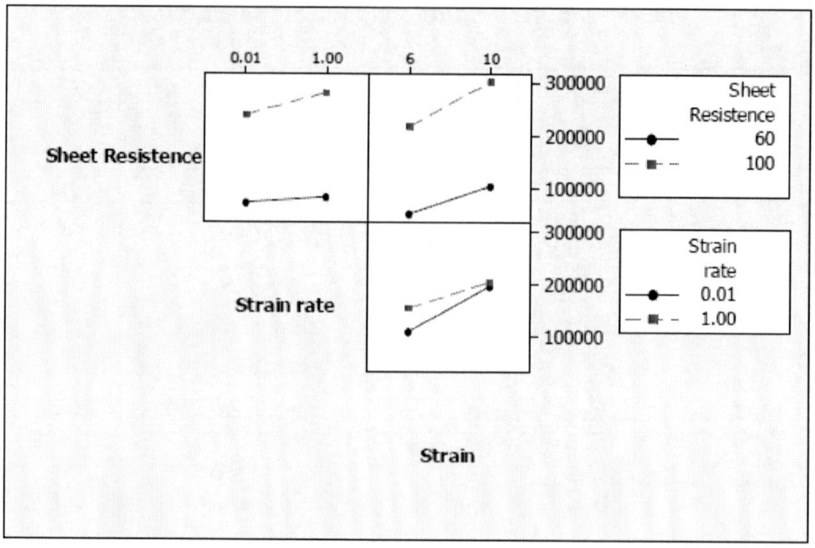

Figure 20. Two factor interactions plot on the effect of sheet resistance, strain rate and strain on the on the relative electrical resistance change.

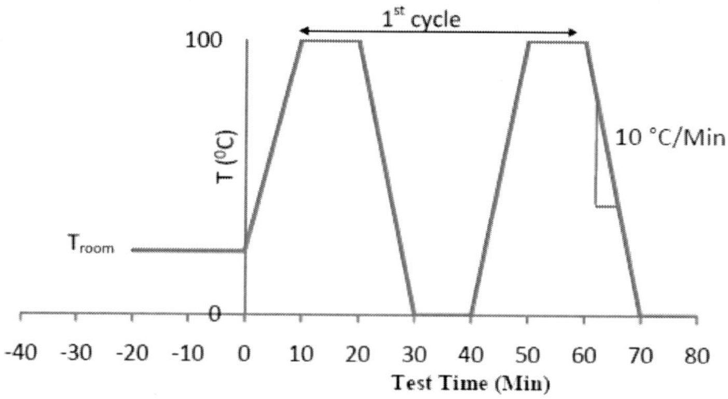

Figure 21. Thermal Cycling temperature profile.

The first set of experiments included four samples which were isothermally aged at 100 °C for 30 days and the other set of experiments included another four samples which were thermally cycled for 900 cycles. The thermal profile used for the thermal cycling experiment consisted of the following parameters: a high temperature of 100 °C, a low temperature of 0 °C, dwell time of 10 minutes, and ramp rate of 10 °C/min, as shown in Figure 21. The initial results showed that the electrical resistance of ITO thin film gradually increases under isothermal aging until the end of the 30 days. However, only very low degradation happened during the thermal cycling.

The isothermal aging was performed using an open environment oven. The percentage change in electrical resistance for each sample was obtained every 24 hours. The thermal cycling, which is a process of heating and cooling the samples periodically for a certain number of cycles, was also conducted on the thermal cycling chamber without controlling the relative humidity. In every cycle, the sample temperature is elevated to 100 °C for 10 minutes before cooling down at a cooling rate of 10 °C/minute to 0 °C, where it is kept for another 10 minutes before being heated up again to 100 °C at a heating rate of 10 °C/minute to start a new cycle (see Figure 21).

Figure 22 shows the relationship between the percentage change in electrical resistance and the number of cycle for each one of the thermally cycled samples. There was no appreciable change in the electrical resistance, and the percent change in electrical resistance did not exceed 30% over the test period. This slight increase may be due to the increased rate of surface reactions at high temperatures. The electrical resistance was measured every day during the isothermal aging test for all the four samples. The percentage change in electrical resistance increased steadily until about 10 days. After 10 days, it stabilized as shown in Figure 23. The percentage change in electrical resistance was about 130% at the tenth day.

At a magnification level of 50 KX, there was no observed difference in the surface morphology between as-produced, isothermally aged and thermally cycled samples (Figure 24).

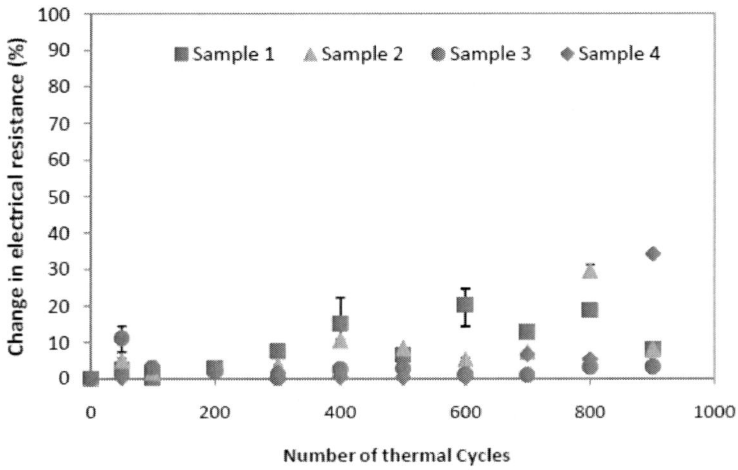

Figure 22. Electrical resistance change for ITO on PET samples after thermal cycling.

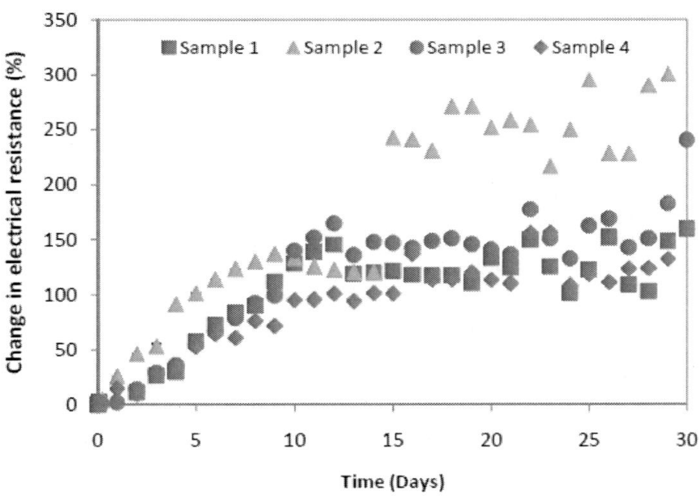

Figure 23. Electrical resistance change for ITO on PET samples after thermal aging.

a)

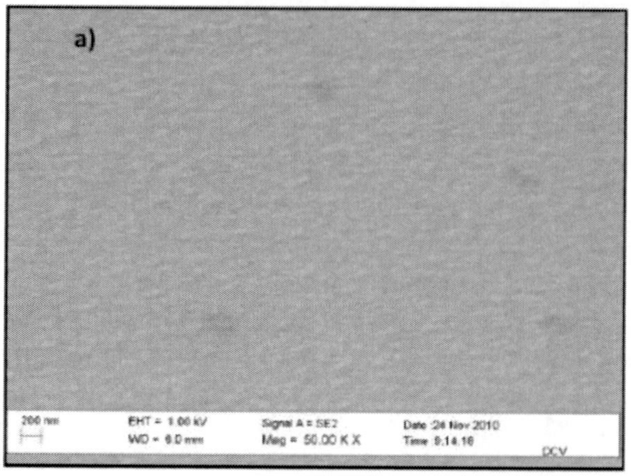

b)

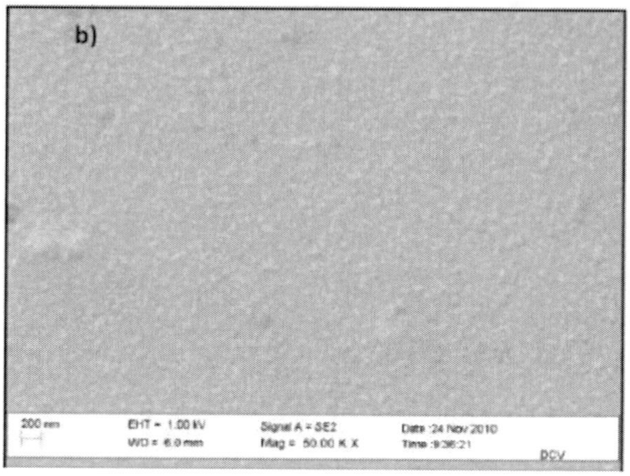

c)

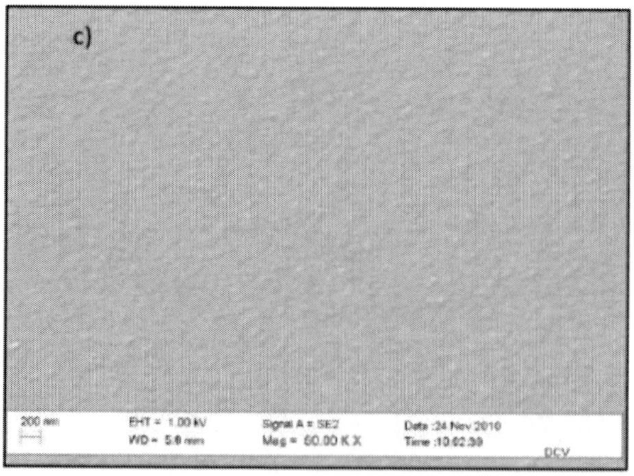

Figure 24. SEM images of ITO coated on PET film. a) as-produced, b) thermal-cycled and c) isothermally aged films.

The Energy-dispersive X-ray spectroscopy (EDX) spectra obtained for the three different samples (as produced, thermally cycled and thermally aged) also showed no significant change in oxygen, indium and Sn content (Figure 25). The X-ray diffraction (XRD) spectra of all the films subjected to three different conditions was also obtained, as shown in Figure 26. The as-grown film is mostly amorphous except for a relatively weak (222) peak of cubic I_2O_3.

In addition to the enhancement of the (222) peak, few other I_2O_3 peaks start to appear on thermally cycled film. The isothermally aged film has all the characteristic I_2O_3 peaks. It appears that the thermally aged film is annealed in air because the film is exposed to high temperatures over a longer time span.

In the case of thermal cycling in which the temperature is varied between 100 °C and 0 °C every 10 minutes, the XRD spectrum looks similar to the as grown film. The thermally cycled film could not get crystalized since the continuous exposure to high temperatures was not long enough.

The isothermally aged samples absorbed some oxygen during the annealing. The conductivity of the ITO film also depends on the oxygen vacancy. Oxygen vacancies give rise to the dangling indium bonds, which increase the conductivity. The increased oxygen content in the film reduces the oxygen vacancy and consequently decreases the conductivity of the material. For this reason, the electrical resistance increased for the isothermally aged samples.

Figure 27 shows a comparison of transmittance of the isothermally aged, and thermally cycled films to the as-produced film. For comparison purpose, the transmittance of bared PET is also given.

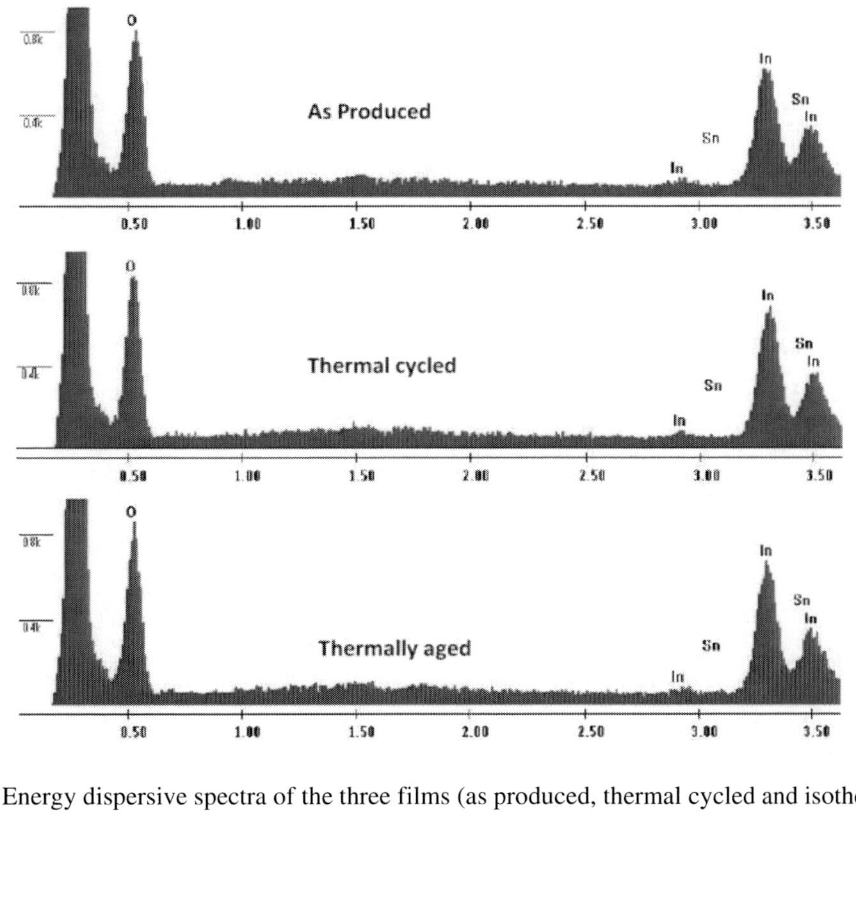

Figure 25. Energy dispersive spectra of the three films (as produced, thermal cycled and isothermally aged).

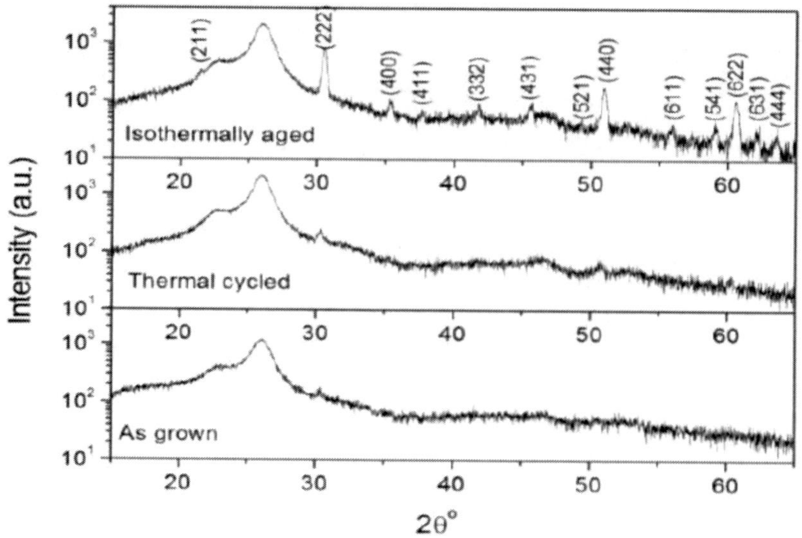

Figure 26. XRD spectra of the films. The isothermally aged film shows all the characteristic peaks of cubic I2O3. The hump around 25 °C is due to the amorphousity in the film or substrate.

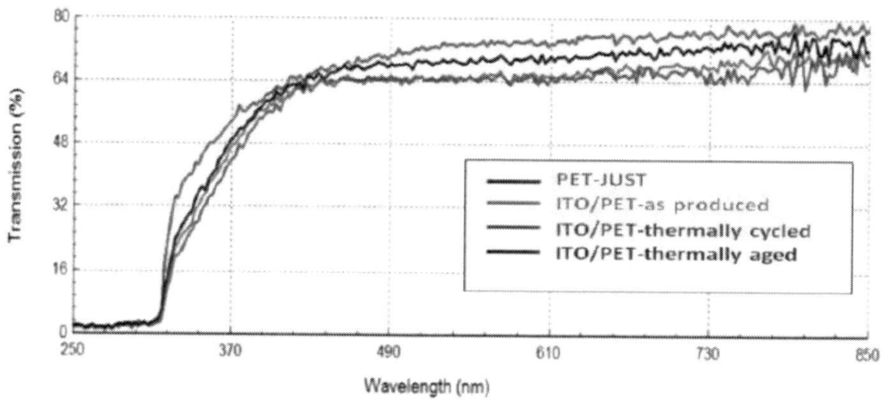

Figure 27. Transmittance of three films and PET.

The improved crystallinity and the increase in oxygen content in the thermally aged film are consistent with the observed increase in transmittance. It is clearly seen that the increased oxygen content and annealing improved the In_2O_3 bixbyite structure, thus enhancing the optical transmission.

Conclusion

This chapter mainly focuses on the reliability of ITO thin films on flexible substrates under different mechanical and/or environmental loads.

To estimate the reliability, four different experiments are reviewed, namely cyclic bending, stretching, isothermal aging, and thermal cycling experiments. Each experiment has a primary goal as follows.

The bending repetition of flexible electronic device can develop micro-cracks within the thin films and lead to degradation of the whole module. Therefore, the primary goal of the cyclic bending fatigue experiment is to see the effect of different factors on the crack development and the percentage change in electrical resistance for thin films on flexible substrate under cyclic bending conditions. Thin films can be subjected to stretching during manufacturing (e.g., roll to roll technology) of flexible electronic devices, leading to cracks and ultimately failure. Therefore, the primary goal of the stretching test is to give better understanding of the acceptable level of ITO film stretching during the manufacturing process, especially roll to roll technology.

During the usage of the film device, the film can be subjected to isothermal stresses in some situations, such as continuous operation. Heat is expected to change the properties of the thin film material over time for a wide range of materials. Therefore, the primary goal of the isothermal aging experiments is to characterize the effect of high temperatures on the ITO film over time. Moreover, fluctuation of temperature between the switching on and off of the film device can fatigue and crack the film due to straining-unstraining cycling stress. Therefore, the primary goal of the thermal cycling experiment is to characterize the effect of temperature fluctuating on the ITO thin films.

ITO film's electrical properties change is monitored in all experiments. In addition, other quality measures; such as cracks development while conducting cyclic bending fatigue and stretching experiments as well as surface morphology, material composition, crystallinity and transmittance while conducting thermal cycling and thermal aging are monitored.

The cyclic bending experiments reveal that the main effects of bending diameter, sample width, and the number of bending cycles have the greatest influence on the response percent change in electrical resistance. The resistance change is higher at higher bending cycles and smaller bending diameters. In a typical roll-to-roll process of flexible substrates, it is necessary to consider these factors through running the web over large diameter rolls with a minimum number of web travels over the rollers. Although bending frequency has little influence on the electrical resistance change, its interaction with the sample width is significant. Moreover, under a cyclic bending fatigue loading, the increasing humidity promoted the forming of moisture-damage spots and the increasing temperature enhanced the crack initiation and propagation. The worst case is the condition of cyclic bending fatigue loading with a high temperature and high humidity combination. The response of some factors highly depends on the level (high or low) of other factors, and this is a very important note to consider in the parameter optimization during the design and manufacturing processes.

Two types of crack form due to stretching. The first type of cracks initiates perpendicularly to the stretching direction and propagates towards the sample edge. The second type of cracks initiates from different points of the first type of crack and propagates perpendicular to them. There are distinguished regions in the relative electrical resistance change-strain curves. Initially, the resistance is constant ($\Delta R/R_0$ was zero) at all strain rates. This indicates that there were no cracks initiating in this period. The resistance then started to increase and $\Delta R/R_0$ curves were concave up to a certain strain. After that, the curve started to become convex. It can be concluded that the relative electrical resistance change under the high strain rate was higher at the beginning of the stretching process, but after a certain strain, the relative electrical resistance change accelerated up until the film became nonconductive while the relative electrical resistance change under high strain rate was in an acceleration process to achieve nonconductive film. ANOVA results showed that the strain and the sheet resistance are significant factors in the relative electrical resistance change considering α equals 0.05 while the strain rate was not a significant factor.

ITO is sensitive to environmental conditions and the electrical conductivity degrades as a consequence of harsh environmental condition. It looks like the ITO thin film that was exposed to the thermal cycling has relatively same electrical and optical properties as the original film, except for a nominal degradation in electrical properties. However, the electrical resistivity of the isothermally aged film gradually increases. It is attributed to the increase in oxygen content in the film. The isothermally aged film appears annealed as observed by the improved bixbyite In_2O_3 structure. The improved crystallinity resulted in a higher transmission as compared to as-produced and thermally cycled film.

REFERENCES

[1] P. Frach, H. Bartzsch, D. Glöß, M. Fahland, F. Hände, "Electrically insulating Al2O3 and SiO2 films for sensor and photovoltaic applications deposited by reactive pulse magnetron sputtering, hollow cathode arc activated deposition and magnetron-PECVD," *Surface and Coatings Technology,* Vol. 202, no. 22-23, pp. 5680–568, 2008.

[2] M. Choi, Y. Kim, C. Ha, "Polymers for flexible displays: From material selection to device applications, " *Progress in Polymer Science,* Vol. 33, no. 6, pp. 581–630, 2008.

[3] T. Hanada, T. Negishi, I. Shiroishi, T. Shiro, "Plastic substrate with gas barrier layer and transparent conductive oxide thin film for flexible displays, " *Thin Solid Films,* Vol. 518, no. 11, pp. 3089–3092, 2010.

[4] D. R. Cairns, G. P. Crawford, "Electromechanical properties of transparent conducting substrates for flexible electronic displays," *Proceeding of the IEEE,* Vol. 93, no. 8, pp. 1451- 1458, 2005.

[5] G. P. Crawford *Flexible Flat Panel Displays*, New York, John Wiley and Sons, Ltd., 2005.

[6] T. Zhang, M. Zhao, "Equilibrium depth and spacing of cracks in a tensile residual stressed thin film deposited on a brittle substrate," *Engineering Fracture Mechanics,* Vol. 69, no. 5, 589–596, 2002.

[7] Y. Leterrier, "Durability of nanosized oxygen-barrier coatings on polymers," *Progress in Materials Science*, Vol. 48, no.1, pp. 1–55, 2003.

[8] K. Alzoubi, S. Lu, B. Sammakia, M. Poliks, "Experimental Study of the High Cyclic Bending Fatigue of Thin Film Metal on Polyethylene Terephthalate for Flexible Electronics Applications, " *Proceedings of the ASME 2009 Inter. PACK Conference IPACK 2009*, July 19-23, 2009, San Francisco, California.

[9] K. L. Chopra, S. Major, D. K. Pandya, "Transparent Conductors—A Status Review" *Thin Solid Films* Vol. 102, no. 1, pp. 1-46, 1983.

[10] S. H. Mohamed, F. M. El-Hossary, G. A. Gamal, M. M. Kahlid, "Properties of Indium Tin Oxide Thin Films Deposited on Polymer Substrates", *Acta Physica Polonica A*, Vol. 115, no. 3, pp. 704-708, 2009.

[11] J. Zhou, "Indium Tin Oxide (ITO) Deposition, pattering and Schottky Contact Fabrication", *Master Thesis*, Rochester Institute of Technology, 2005.

[12] J. C. C. Fan, F. J. Bachner, G. H. Foley, "Effect of Oxygen Partial Pressure During Deposition on Properties of RF Sputtered Sn-Doped In2O3 Films", *Applied Physics Letters*, Vol. 111, no. 3, pp. 773 – 775, 1977.

[13] C.-H. Yang, S.-C. Lee, T.-C. Lin, S.-C. Chen, "Electrical and Optical Properties of Indium Tin Oxide Films Prepared on Plastic Substrates by Radio Frequency Magnetron Sputtering", *Thin Solid Films* Vol. 516, no. 8, pp. 1984-1991, 2008.

[14] K. A. Sierros, D. S. Hecht, D. A. Banerjee, N. J. Morris, L. Hu, G. C. Irvin, R. S. Lee, D. R. Cairns "Durable transparent carbon nanotube films for flexible device components" *Thin Solid Films*, Vol. no. 23, 518, pp. 6977–6981, 2010.

[15] P. O. Renault, P. Villain, C. Coupeau, P. Goudeau, K. F. Badawi, "Damage mode tensile testing of thin gold films on polyimide substrates by X-ray diffraction and atomic force microscopy" *Thin Solid Films* Vol. 424, no. 2, pp. 267–273, 2003.

[16] T. Tsuchiya, M. Hirata, N. Chiba, "Young's modulus, fracture strain, and tensile strength of sputtered titanium thin films" *Thin Solid Films* Vol. 484, no. 1-2, pp. 245–250, 2005

[17] D. R. Cairns, G. P. Crawford, "Electromechanical properties of transparent conducting substrates for flexible electronic displays," *Proceeding of the IEEE*, V. 93, 2005, pp. 1451.

[18] G. P. *Crawford Flexible Flat Panel Displays*, New York, John Wiley and Sons, Ltd., 2005.

[19] Yeong-shyung Chou, Jeffry W. Stevenson, John Hardy, Prabhakar Singh, "Material Degradation During Isothermal Ageing And Thermal Cycling of Hybrid Mica Seals Under Solid Oxide Fuel Cell Exposure Conditions", *Journal of Power Sources*, Vol. 157, pp 260–270, 2006

[20] M. J. Rizvi, C. Bailey, H. Lu, "Failure Mechanisms of AFC Joints Under Isothermal Ageing", *Microelectronics Journal* Vol. 39, pp 1101–1107, 2008

[21] P. J. Sebastain, "aging studies on CdSe thin films under ambient conditions", *Journal of physics and chemistry of solids*, Volume 52, Issue 8, pp963-968, 1998

[22] A. J. Tosser, C. R. Tellier, J. Launey, "2.25 Thickness dependent effects of thermal ageing in thin conductive Films", *Vacuum*, Vol. 27, Issue 4, pp 335-338, 1977

[23] B. Kaouache, P. Gergaud, O. Thomas, O. Bostrom, M. Legros, "I mpact of thermal cycling on the evolution of grain, precipitate and dislocation structure in Al, 0.5% Cu, 1% Si thin films" *Microelectronic Engineering* Vol. 70, pp 447–454, 2003

[24] T. K. Schmidt, T. J. Balk, G. Dehm, E. Arzt, "Influence of tantalum and silver interlayers on thermal stress evolution in copper thin films on silicon substrates" *Scripta Materialia*, Vol. 50, pp733–737, 2004

[25] Alzoubi, K., Hamasha, M. M., Lu, S., Samakia, B., "Bending Fatigue Study of Sputtered ITO on Flexible Substrate", *IEEE Journal of Display Technology*. Vol. 7, No. 11, pp. 593-600, 2011.

[26] Hamasha, M. M., Alzoubi, K., Lu, S., Desu, S. B., "Durability Study on Sputtered Indium Tin Oxide Thin Film on Poly Ethylene Terephthalate Substrate", *Thin Solid Films* Vol. 519, No. 18, pp. 6042-6047, 2011.

[27] Hamasha, M. M., Alzoubi, K., Lu, S., "Behavior of ITO Thin Film on PET Substrate under Stretching", *IEEE Journal of Display Technology*. Vol. 7, No. 8, pp. 426-433, 2011.

[28] Hamasha, M. M., Dhakal, T., Alzoubi, K., Albahri, S., Qasaimeh, A., Lu, S., Westgate, C. R. "Stability of ITO Thin Film on Flexible Substrate under Thermal Aging and Thermal Cycling Conditions", Invited paper, In press: *IEEE Journal of Display Technology*, 2012

[29] T. Li, Z. Suo, Deformability of thin metal films on elastomer substrates, *Int. J. Solids and Struct*. 43 (2006) 2351.

In: Indium
Editors: Hsaio G. Woo and Huang Tsai Choi
ISBN: 978-1-62257-696-8

Chapter 6

INDIUM(III)-COMPLEX OR INDIUM(III)-SALT AS EFFICIENT AND SUSTAINABLE CATALYST IN ASYMMETRIC ORGANIC SYNTHETIC TRANSFORMATIONS

Yi Li and Ming-Hua Xu*
Shanghai Institute of Materia Medica, Chinese Academy of Sciences, Shanghai, China

ABSTRACT

Recently, the utility of chiral indium(III) complexes or indium(III) salts as Lewis acids in asymmetric organic synthesis has received a great deal of interest due to their highly catalytic activity and excellent stereoselectivity, along with relatively low toxicity, good stability in air and water, and recyclability. There are a great number of examples involving indium(III) reagents in organic synthesis appeared in the last few years. A highlight of current developments on the use of chiral indium(III) complexes and indium(III) salts in various functional group transformations such as stereselective Friedel-Crafts alkylation, allylation reaction, Diels-Alder cycloaddition, Mukaiyama aldol condensation, carbonyl ene reaction, alkynylation reaction, Mannich-type reaction, Prins cyclization, ring opening of *meso*-epoxides, Petasis reaction, cycloaddition, *etc* is presented here, with particular emphasis on reaction diastetreoselectivity and enantioselectivity.

Keywords: Chiral indium(III) complex, indium(III) reagent, Lewis acid, asymmetric catalysis

* Corresponding author: Ming-Hua Xu. Shanghai Institute of Materia Medica, Chinese Academy of Sciences, 555 Zuchongzhi Road, Shanghai 201203, China, E-mail: xumh@mail. shcnc.ac.cn.

INTRODUCTION

In the past several years, the indium chemistry has undergone tremendous growth, and roles of indium have witnessed a wide variety of applications in organic synthesis. A diverse range of reactions using indium complexes or indium salts as Lewis acid catalysts have been developed. Among these applications, the utility of chiral indium(III) complexes in asymmetric organic synthesis has received a considerable amount of interest due to the high catalytic activity and excellent enantioselectivity.

On the other hand, simple indium(III) salts have shown as mild and effective Lewis acids, imparting high stereoselectivity in many organic transformations.

Moreover, indium(III) reagents often exhibit very low toxicity, good stability in air and water, and potential recyclability. Because of their particular importance in topical fields of research, a great number of reported examples involving indium(III) reagents in organic synthesis have appeared previously. [1-9] Although existing reviews have covered a broad scope of reactions on chiral indium(III) reagents, the high speed of publication on indium chemistry still demands an outlook on recent developments that are important to the synthetic community.

Thus, in this article, we highlight some of the general developments on the use of chiral indium(III) complexes or indium(III) salts in various asymmetric functional group transformations such as stereselective Friedel-Crafts alkylation, allylation reaction, Diels-Alder cycloaddition, Mukaiyama aldol condensation, carbonyl ene reaction, alkynylation reaction, Mannich-type reaction, Prins cyclization, ring opening of *meso*-epoxides, Petasis reaction, cycloaddition, *etc.* in the last few years.

We cover here works published between the beginning of 2005 and the end of 2011 and focus on new achievements in carbon-carbon bond-forming asymmetric reactions. We hope that this paper will provide non-specialists with some insight into this field and serve as a helpful tool in designing synthetic strategies in their research.

ASYMMETRIC FRIEDEL-CRAFTS ALKYLATION REACTION

Asymmetric Friedel-Crafts alkylation reaction has been proven to be a powerful tool for C-C bond formation in organic synthesis. [10]

Recently, Franz and co-workers reported the the first example of direct addition of indoles and electron-rich arenes to isatin electrophiles catalyzed by either the scandium(III)-inda-Pybox or indium(III)–inda-Pybox complex [11] (Scheme 1). The reaction proceeded efficiently to afford the monoaddtion product in high yield with excellent enantioselectivity (99% ee). Notably, the competing formation of the 3, 3'-bisindolyl oxindole was completely suppressed at a low temperature (-20°C).

To explain the stereochemistry, the authors proposed an octahedral or pentagonal-bipyramidal chelation model.

As shown in Figure 1, the amide carbonyl group of the isatin is bound in the apical position, and the nucleophile added from the less hindered *Si* face [12] with great stereocontrol, affording products with the shown absolute configuration.

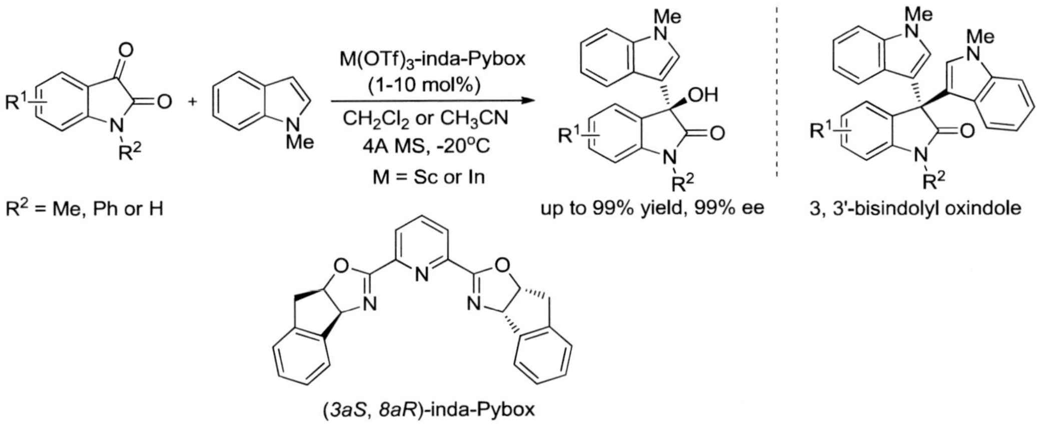

Scheme 1. Addition of Indoles to Isatins under the Catalysis of $Sc(OTf)_3$ or $In(OTf)_3$-Inda-Pybox.

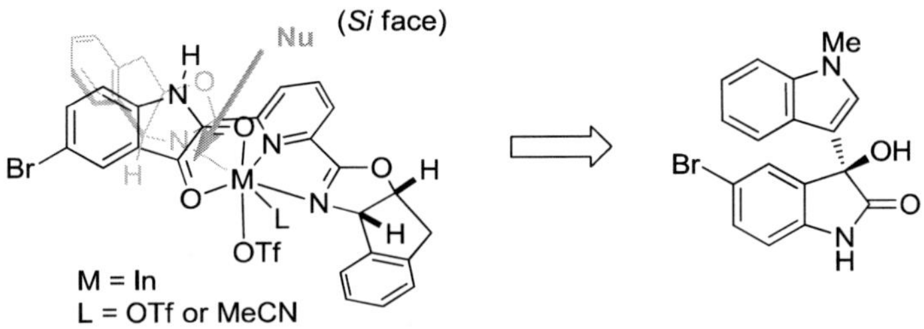

Figure 1. Stereochemical Model for the Asymmetric Addition.

An extension on the above indium(III) catalyst system was successfully applied to the enantioselective and regioselective addition of pyrroles to isatins. [13] Unlike indoles, pyrroles often show divergent reactivity in terms of the known C2 and C3 regioselectivity. With indium(III) triflate catalyst, interestingly, the reaction regioselectivity was found to be largely influenced by temperature and solvent. When 10 mol% of $In(OTf)_3$-inda-Pybox complex was used as a catalyst in acetonitrile at -20°C, a series of various *N*-alkylated isatin adducts were obtained in high yields with excellent enantioselectivities (94~>99% ee) and regioselectivities (90:10~>99:1) in favor of C2-regiosomers (Scheme 2).

It was noted that unprotected NH isatins are also effective reaction substrates, affording products with equal excellent enantioselectivities and regioselectivities (Scheme 3). Therefore, the indium(III)-catalyst exhibits great catalytic activation of isatin ketone moiety while also controlling the regio- and enantioselectivity for the addition. This catalytic asymmetric addition of pyrroles to isatins catalyzed by an indium(III)-Pybox complex represents an unprecedented method for preparation of highly optically active 3-hydroxy-2-oxindoles.

The asymmetric Friedel–Crafts alkylation of indoles with α, β-unsaturated aldehydes and ketones is an effective and direct method for synthesizing enantioenriched indole alkaloids. [14] Recently, Luo and co-workers [15] disclosed an asymmetric binary acid catalyst system that enables achievement of 1,2- and 1,4-additions in both a regio- and enantioselective manner in Friedel–Crafts alkylation of indoles with α, β-unsaturated α-keto esters. Interestingly, the simple combination of a chiral phosphoric acid with an indium(III) halide salt constitutes a highly efficient synergistic catalyst.

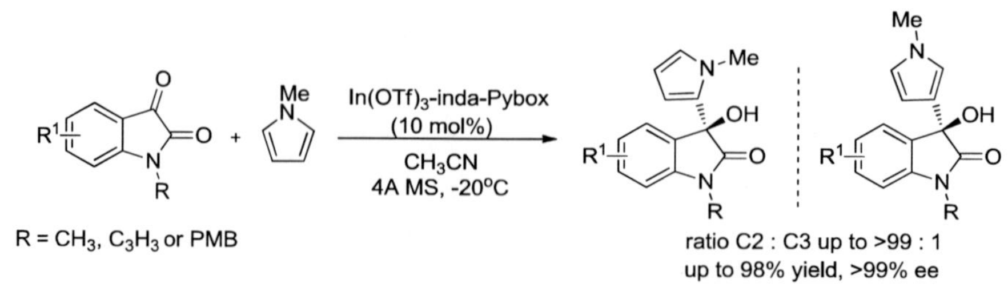

Scheme 2. Indium(III)-Catalyzed Addition of Pyrroles to *N*-Alkylated Isatins.

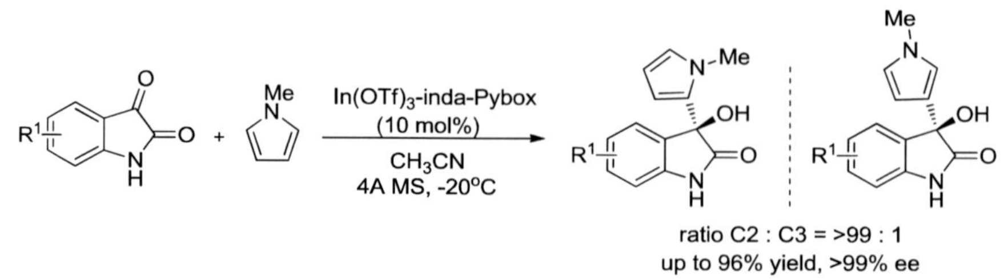

Scheme 3. Regio- and Enantioselective Indium(III)-Catalyzed Addition of Pyrroles to NH Isatins.

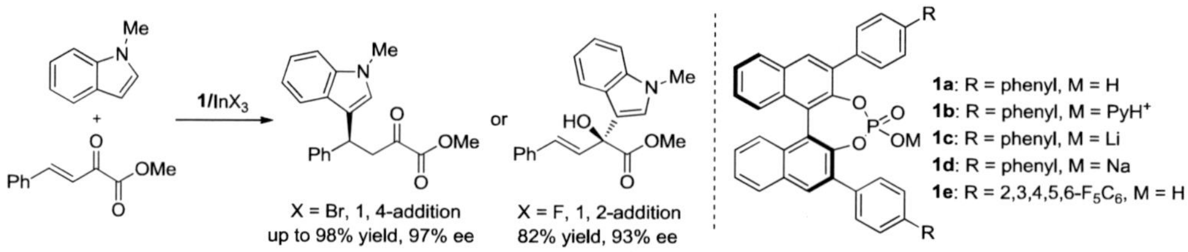

Scheme 4. Regioselective and Enantioselective 1, 2- and 1, 4-Addition Reactions.

It was found that use of different Lewis acids have a dramatic impact on the reaction regioselectivity. A simple swap of counteranions of indium(III) from fluoride to bromide or chloride switches the regioselectivity from 1,2- to 1,4-addition with excellent reactivity and enantioselectivity (Scheme 4).

By using the combination of **1a** and InF_3 as catalyst, a variety of α, β-unsaturated α-keto esters were subjected to the reaction with *N*-methyl indole and gave predominantly the 1,2-addition products in good yields and with up to 99% ee (Scheme 5).

In all cases, the reactions proceeded cleanly with only trace amounts of 1,4-adducts being detected and no bisindole products were observed. On the other hand, when the catalyst combination change to **1c**/$InBr_3$ or **1e**/$InBr_3$ binary systems, the reactions also work well but gave exclusively the 1,4-addition products in good yields and with excellent regioselectivities and enantioselectivities (Scheme 6).

According to the ESI-MS and IR experiments and known reports [16], the halides may serve as ligands that fine tune the indium(III) metal center in the catalytic systems to afford specific electronic and steric properties. However, the exact roles of indium halides on regioselective shift between 1,2- and 1,4-additions with excellent stereocontrol remains unclear.

$InCl_3$ has been reported to specifically catalyze intramolecular Friedel-Crafts cyclization of allyl chlorides tethered to an *N*-aryl moiety in a highly stereoselective manner. [17]

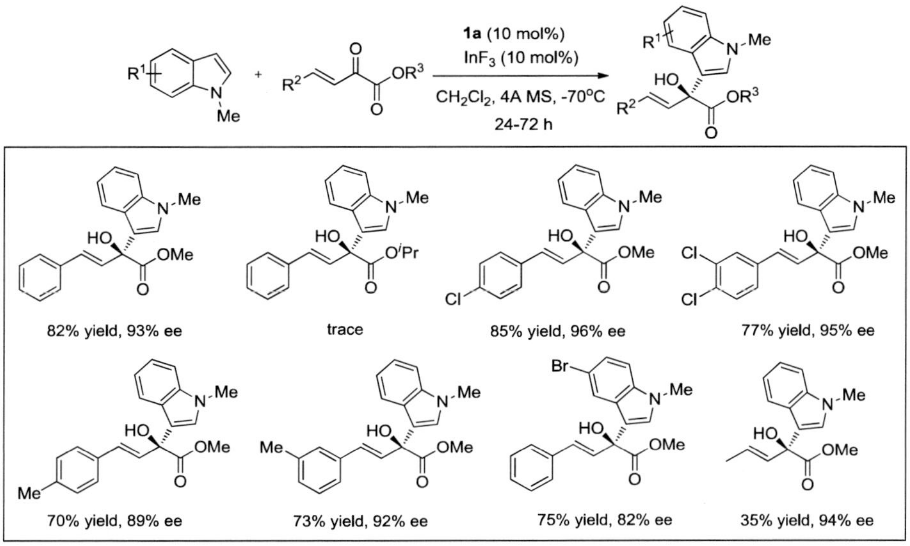

Scheme 5. Enantioselective 1,2-Addition of *N*-methyl Indoles with Various α-Keto Esters.

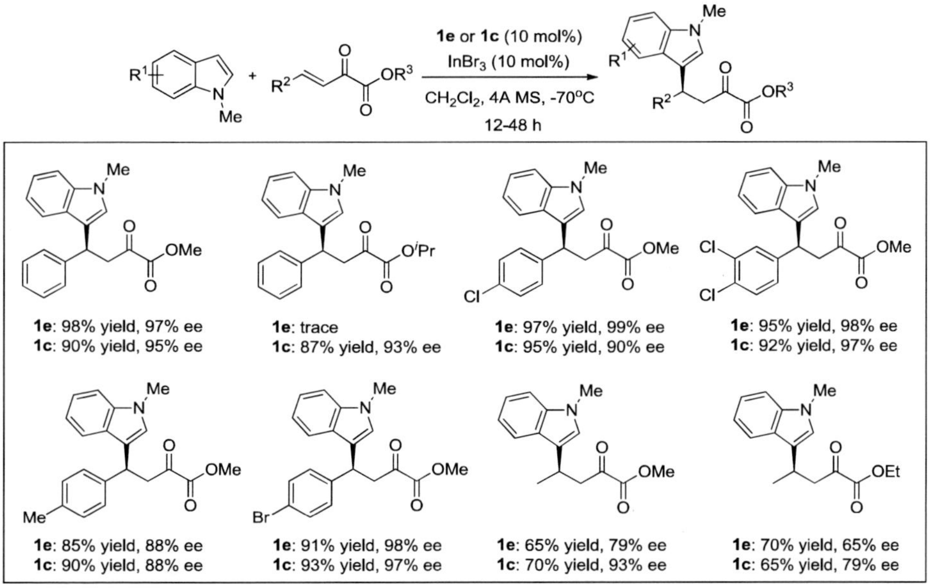

Scheme 6. Enantioselective 1, 4-Addition of *N*-Protected Indoles with Various α-Keto Esters.

With 10 mol% of $InCl_3$ and powdered 4 Å molecular sieves in 1,2-dichloroethane, a broad range of chiral allyl chlorides underwent intramolecular cyclization smoothly to furnish the corresponding 1,2,3,4-tetrahydroquinolines in good yields with up to >25:1 diastereoselectivity in favor of the 2,4-*cis*-diastereomer (Scheme 7). The catalytic activity of $InCl_3$ in this reaction was assumed to rely on its halophilic nature by activating the allyl system via chloride abstraction.

Very recently, Xu and co-workers reported a highly diastereoselective synthesis of α-arylglycines via asymmetric Friedel-Crafts alkylation of arenes with a chiral *N-tert*-butanesulfinylimino ester. [18]

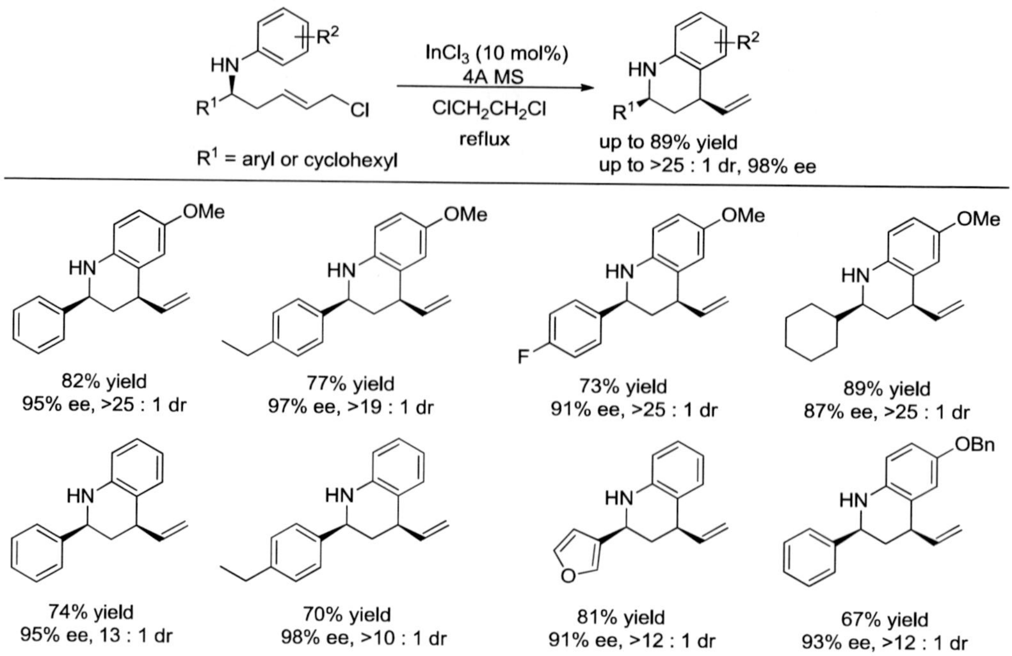

Scheme 7. $InCl_3$-Catalyzed Intramolecular Friedel-Crafts Reaction.

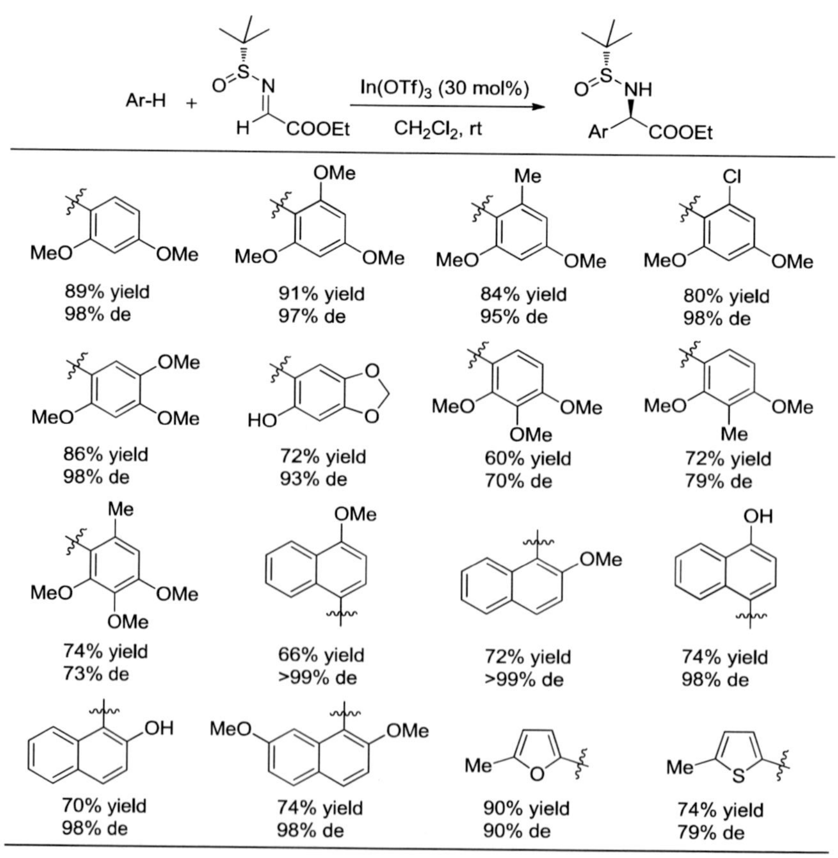

Scheme 8. Diastereoselective Friedel-Crafts Reaction.

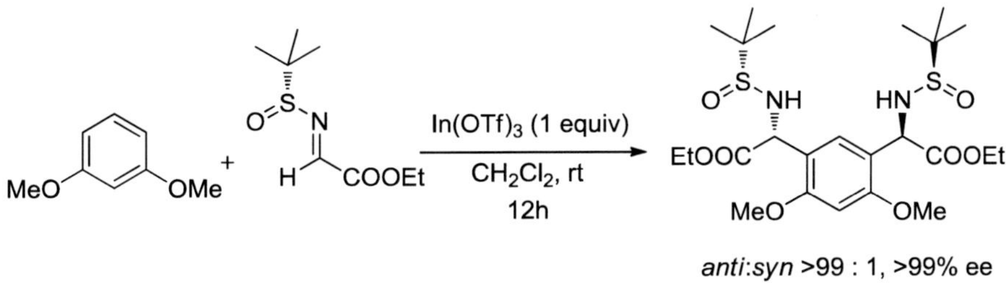

Scheme 9. Stereoselective Double Friedel-Crafts Alkylation.

Among the transition-metal Lewis acids screened, $In(OTf)_3$ showed superb catalytic activity and stereoselectivity. The reaction could be accomplished with ease in the presence of a catalytic amount of indium(III) triflate at room temperature. It allows exceptionally mild, efficient and ready access to various enantiomerically enriched α-arylglycines in good yields and with excellent diastereoselectivities (up to 99% de) (Scheme 8).

By this protocol, the potential of more challenging stereoselective double Friedel-Crafts alkylation was explored. As expected, when 1 equiv of $In(OTf)_3$ was employed, the reaction of dimethoxybenzene with (*R*)-*N*-tert-butanesulfinylimino ester was promoted and successfully afforded the corresponding dialkylation product in a relatively good yield (43%) with an extremely high diastereo- and enantioselectivity (*anti* : *syn* > 99 : 1, >99% ee) (Scheme 9). To rationalize the observed diastereofacial selectivity, a five-membered chelation transition state rather than four-membered chelation transition state [19] is proposed (Figure 2). The coordination of the Lewis acidic indium to both the imine nitrogen and carbonyl oxygen significantly activates the imino ester carbonyl. As the uncoordinated *N*-sulfinyl group adopts an approximately synperiplanar configuration, [20] in the case of (*R*)-*N-tert*-butanesulfinyl imino ester, the bulky *tert*-butyl group is positioned at the *si*-face of imine molecule. To avoid the steric repulsion, the arene adds preferentially from the sterically unblocked *re*-face of the imine C=N bond, thereby facilitating (*R*)-α-arylglycine product formation.

Asymmetric Allylations

The asymmetric addition of allylic nucleophiles to the carbonyl compounds or the carbonyl imino derivatives has been established as a powerful method for the synthesis of chiral homoallylic alcohols [21] or chiral homoallylic amines [22] which are important precursors to numerous organic compounds. Despite the development of a large array of chiral catalyst, enantioselective allylation using chiral indium(III) catalyst has been rarely explored. Loh *et al.* discovered that chiral indium(III)-Pybox complexes prepared from indium triflate and C_2-symmetric chiral Pybox ligands were suitable and effective catalysts for the enantioselective allylation of a variety of carbonyl compounds with allyltributylstannane. [23-24] In their early studies, an indium(III)-(3*aS*-8*aR*)-Pybox complex was employed in catalyzing the enantioselective allylation reaction of allyltributylstannane with aldehydes. The reactions proceeded smoothly to afford the homoallylic alcohols with moderate to high enantioselectivities (60–93% ee) in dichloromethane at –60 °C in the presence of powdered activated 4 Å molecular sieves (Scheme 10). [23]

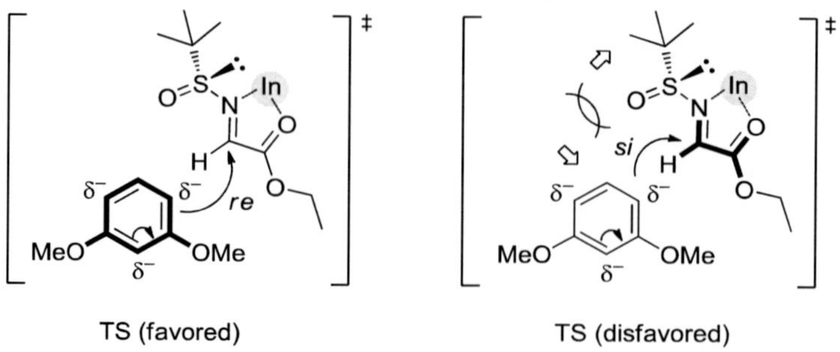

Figure 2. Mechanistic Proposals for Stereocontrol.

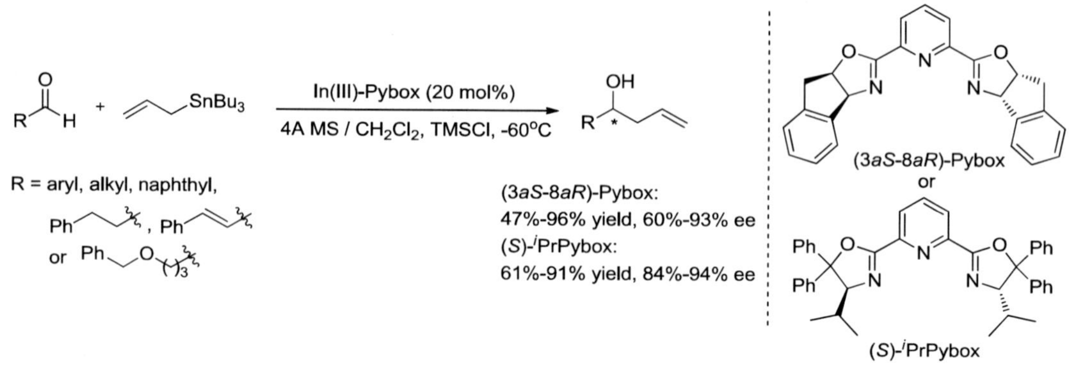

Scheme 10. In(III)–Pybox-Catalyzed Allylation of Aldehyde.

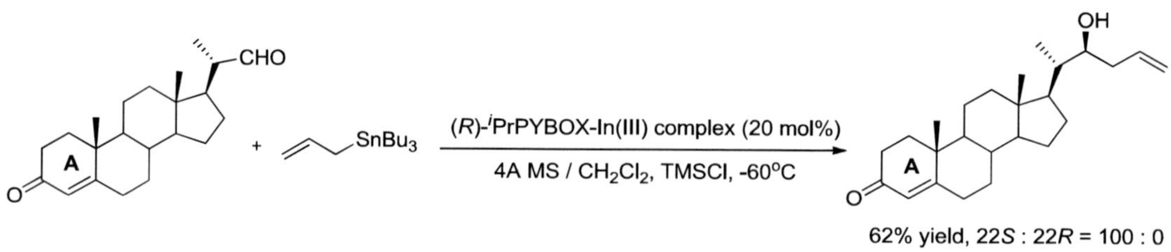

Scheme 11. Application to the Steroid Side-Chain Synthesis.

Soon after, the same authors demonstrated that the use of tetraphenyl-substituted (*S*)-^{i}PrPybox as a chiral ligand could further improve the reaction enantioselectivities (84–94% ee) and gave also better yields (Scheme 10). [24]

Using either of the chiral indium(III)-Pybox complexes, highly efficient allylation of steroidal aldehyde with allyltributylstannane can be carried out, leading to the corresponding allylic alcohol with excellent diastereoselectivity (22*S*:22*R*=100:0) (Scheme 11). Notably, chiral Pybox ligand could be easily recovered from the reaction mixture in good yield after the usual workup, and could be reused in this enantioselective allylation reaction without loss of enantioselectivity.

The same catalytic system was further applied to the enantioselective allylation of less reactive ketones. With chiral In(III)-^{i}PrPybox complex, tertiary homoallylic alcohols with moderate to high enantiomeric excesses (54–95% ee) were obtained (Scheme 12). In the case of 1-indanone, the best result of 95% ee was observed. [25]

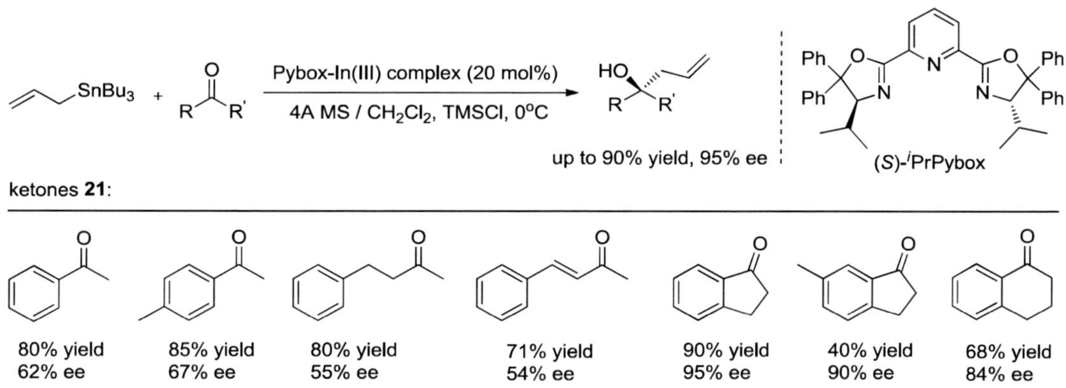

Scheme 12. In(III)-^{i}PrPybox-Catalyzed Allylation of Ketones.

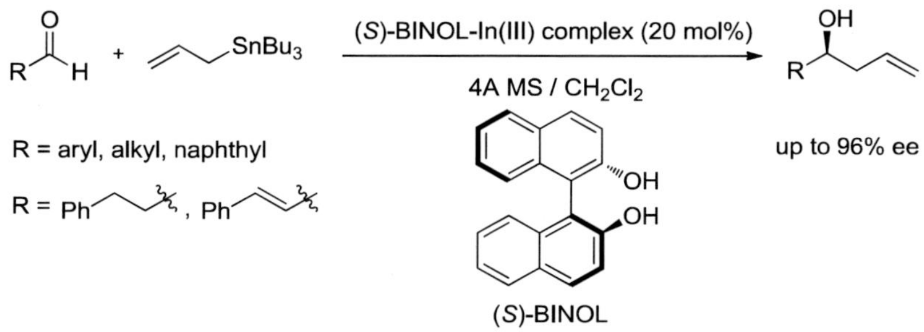

Scheme 13. In(III)–BINOL Catalyzed Allylation of Aldehydes.

Other than using the tridentate chiral Pybox ligands, Loh and co-workers also discovered that the asymmetric allylation reaction of allyltributyl stannane to aldehydes can be catalyzed by chiral In(III)-BINOL complex (Scheme 13). [26]

In the presence of 20 mol% of chiral catalytic complex prepared from $InCl_3$ and (*S*)-BINOL, allyltributylstannane reacted with both aromatic and aliphatic aldehydes to afford the corresponding homoallylic alcohols in good yields and high enantioselectivities (up to 96% ee). According to the ^{1}H NMR studies, it was considered that a new indium allylic species is initially generated by transmetallation of $InCl_3$ with allyltributyl stannane, which complexes with BINOL to form a monomeric active (*S*)-BINOL-In-allyl Lewis acid catalyst. Importantly, this active catalytic indium species is sensitive to the water.

By simply changing $InCl_3$ to $InBr_3$, the catalytic asymmetric allylation of ketones with allyltributylstannane in the presence of (*R*)-BINOL ligand can be easily accomplished to afford tertiary homoallylic alcohols in good yields and high enantioselectivities (up to 92% ee) (Scheme 14). [27] A transition state model of the stereochemical course of allylation catalyzed by the chiral (*R*)-BINOL-$InBr_3$ complex was proposed as depicted in Figure 3. The aromatic rings of the (*R*)-BINOL effectively blocked the *si* face of the complexed ketone from attack by allyltributylstannanes. As such, allyl attack from the *re* face of the ketone would be preferred, leading to the formation of (*R*)-enantiomer.

The application of chiral *N*-oxides in transition-metal catalysis has received much attention. [28] Feng and co-workers reported an interesting chiral indium complex prepared from $InBr_3$ and (*S*)-pipecolic acid-derived *N*, *N'*-dioxide (1:2 ratio) for the enantioselective allylation of ketones with tetraallyltin as the allylating agent. Under the optimized conditions, a range of aromatic ketones underwent successful allylation to furnish the corresponding homoallylic alcohols in moderate to high yields with good enantioselectivites (up to 83% ee) (Scheme 15). [29]

In this reaction system, the indium(III) bromide is considered to function as a Lewis acid to activate the ketone while chiral *N*, *N'*-dioxide acts as a Lewis base for the activation of tetraallyltin.

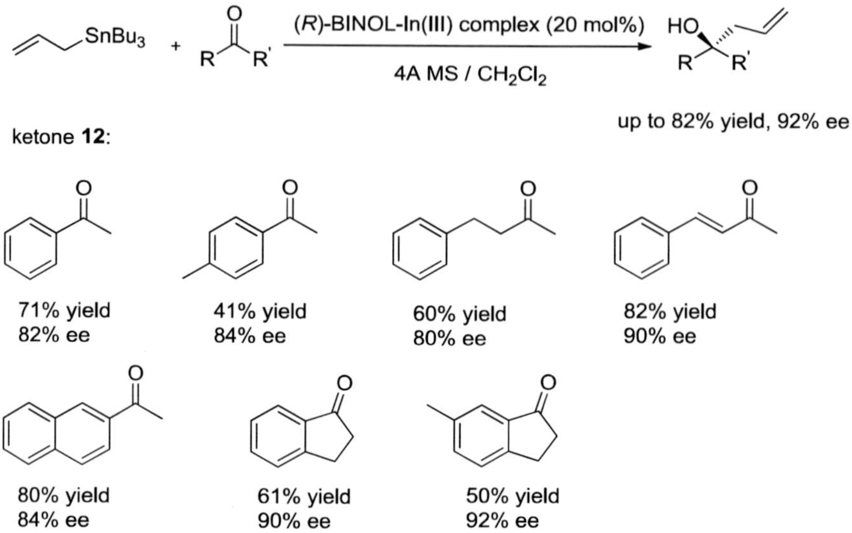

Scheme 14. In(III)–BINOL Catalyzed Allylation of Ketones.

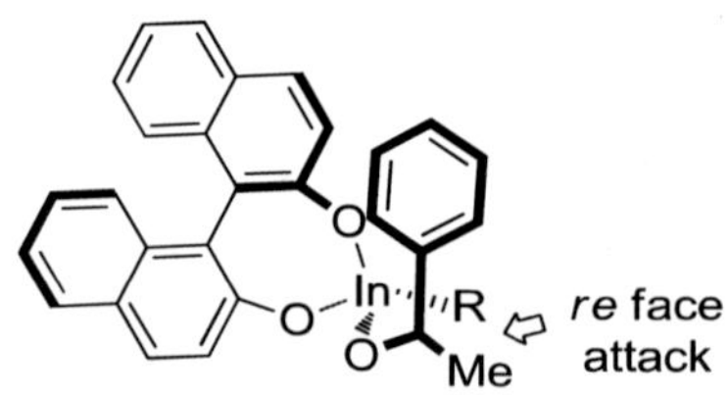

Figure 3. Proposed BINOL-In(III)-Ketone Transition State.

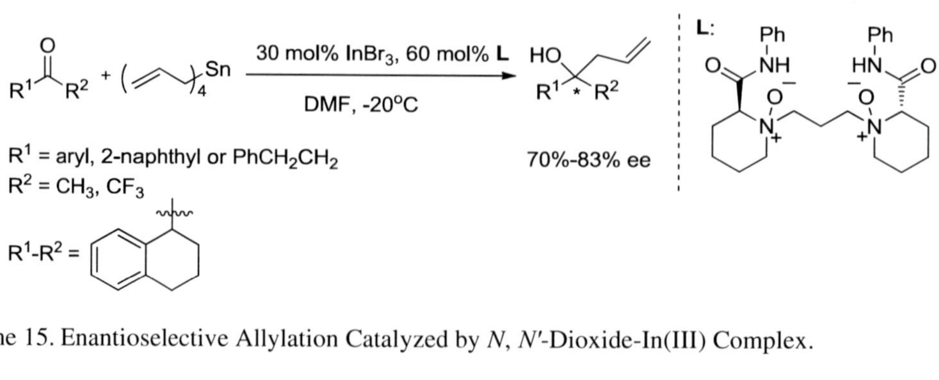

Scheme 15. Enantioselective Allylation Catalyzed by *N*, *N'*-Dioxide-In(III) Complex.

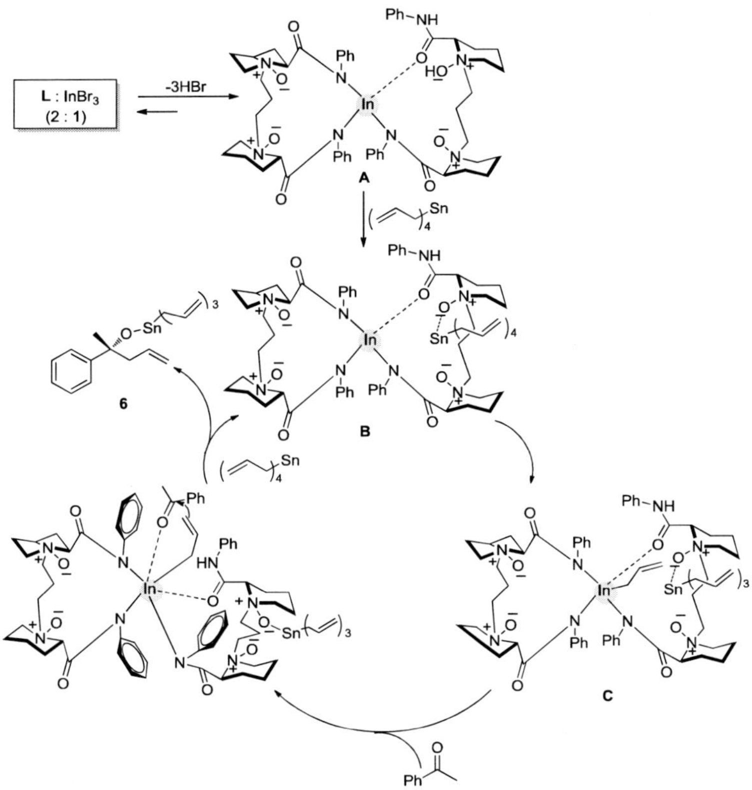

Figure 4. Plausible Catalytic Cycle.

As illustrated in Figure 4, the catalytic cycle of allylation was proposed. Initially, the In(III) complex A was generated by coordination of two molecules of *N*, *N'*-dioxide with $InBr_3$. [30] In the presence of tetraallyltin reagent, the *N*-oxide subsequently coordinates to the tin atom to form a possible hypervalent tin species (complex B). [31] The following transmetallation occurs to further form an allylic indium complex C. [32] As a result, the indium allyl attacks the activated ketone from the less sterically hindered *Si* face to give chiral homoallylic alcohol.

The asymmetric allylation of imines to furnish optically active homoallylic amines has been of considerable interest to organic chemists. [22] Among the strategies developed, diastereoselective allyl addition to chiral auxiliary derived imines has received major attention. [22b, 33] In most studies, a six-membered chair-like transition-state model was considered, in which the allylmetal generally acts as a Lewis acid to activate the imine. [33f-g] Interestingly, Xu and Lin *et al.* proposed an acyclic transition state in the presence of an additional rather strong Lewis acid.

As such, Zn-mediated allylation to chiral *N-tert*-butanesulfinyl imines was investigated. [34] When (*R*)-*N-tert*-butanesulfinyl imine is employed, in contrast to the closed cyclic transition state (TS-1) in the absence of other Lewis acid, an acyclic transition state (TS-2) might be engaged if the chelation of Lewis acid with both nitrogen and sulfinyl oxygen of imine is favored, thus directing allyl attack selectively from the sterically unblocked *si*-face to give (*S*)-amine product (Scheme 16).

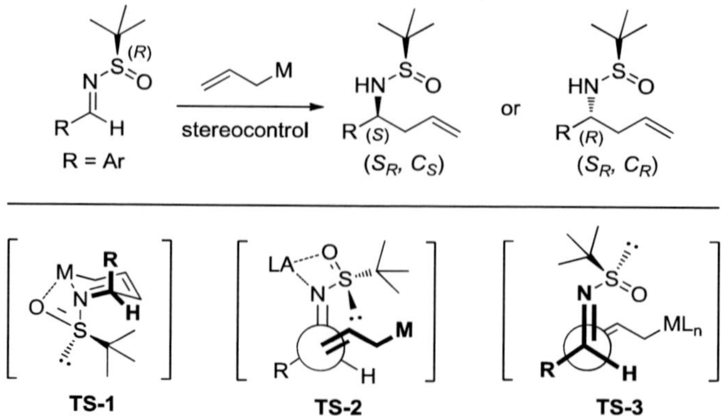

Scheme 16. Diastereoselective Allylation to (*R*)-*N*-*tert*-Butanesulfinyl Aldimines.

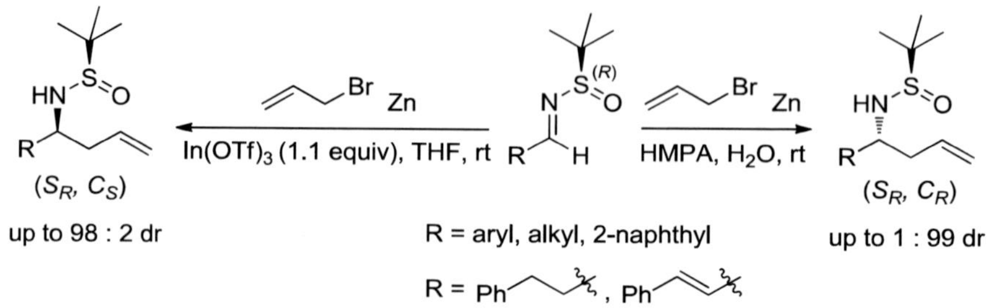

Scheme 17. Diastereoselective Allylation to (*R*)-*N*-*tert*-Butanesulfinyl Aldimines with Two Different Reaction Systems.

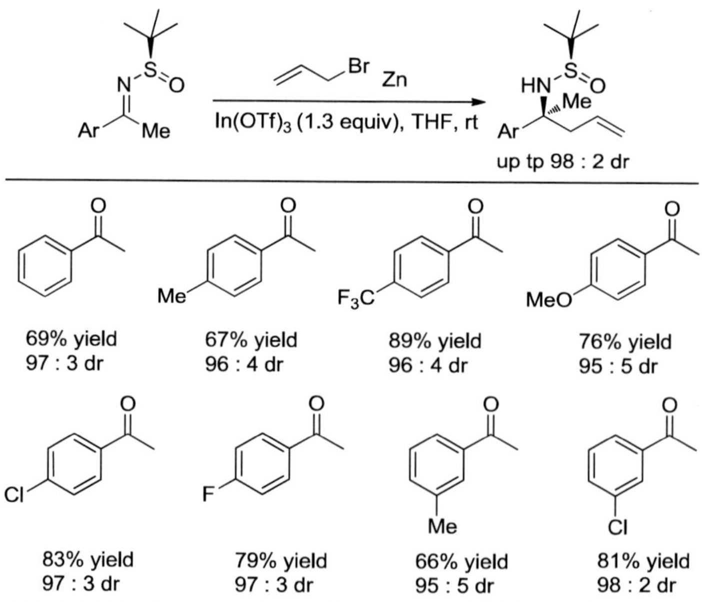

Scheme 18. Diastereoselective Allylation to (*R*)-*N*-*tert*-Butanesulfinyl Ketimines.

On the other hand, the authors assumed another type of acyclic transition state model TS-3, in which the allylmetal is coordinated to additional Lewis bases in the reaction system rather than the sufinyl oxygen. As the uncoordinated *N*-sulfinyl group adopts an approximate synperiplanar configuration, [20] the allyl addition to the less hindered *re*-face of imine would be preferred, thereby facilitating (*R*)-amine formation.

To carry out the idea, a series of Lewis acids such as $Zn(OTf)_2$, $Mg(OTf)_2$, $Cu(OTf)_2$, $In(OTf)_3$, $Yb(OTf)_3$, $La(OTf)_3$, $SnCl_4$ was examined. [35] Among them, $In(OTf)_3$ was found to be the unique, a best result of diastereoselectivity up to 98:2 could be obtained in THF. Fortunately, the complete opposite stereocontrol was achieved as expected when HMPA was used as solvent as well as Lewis base, giving (*R*)-product in 97% yield and with 98% de (1:99 dr) in the presence of proper amount of H_2O (Scheme 17). *N-tert*-Butanesulfinyl imines of various aromatic aldehydes proved to be excellent substrates in both systems and gave excellent drs in most cases.

Gratifyingly, less reactive *N-tert*-Butanesufinyl ketimines were also examined as substrates to produce the corresponding quaternary carboncontaining chiral homoallylic amines (Scheme 18). In the presence of indium(III) triflate, ketimines work equally well in THF, giving the corresponding enantiopure homoallylic amines in moderate yields and with excellent diastereomeric ratios. Remarkably, >95:5 dr was observed in all cases. As the allylation of ketimines did not take place under the optimized HMPA conditions, it was assumed that the added Lewis acid $In(OTf)_3$ significantly increases the reactivity of ketimines.

Asymmetric Diels-Alder Reaction

Asymmetric Diels-Alder cycloaddition reaction is one of the most important methods used for the synthesis of chiral cyclic molecules. Having successfully employed the chiral In(III)-BINOL complex in asymmetric allylation, Loh and co-workers further explored the catalytic ability of such complex in other organic transformations. During the study, they then discovered that the use of a (*S*)-BINOL-In(III) complex as precatalyst and allytributylstannane as activator could lead to an active chiral BINOL-In-allyl Lewis acid capable of catalyzing enantioselective Diels-Alder reaction. [36] Typically, this chiral indium(III) catalyst can be prepared from (*S*)-BINOL (22 mol%), $InCl_3$ (20 mol%), and allyltributylstannane (60 mol%) in the presence of activated 4 Å MS. It exhibited a broad applicability for the reactions of 2-methacrolein and 2-bromoacrolein with a variety of dienes including both cyclic and open-chain dienes, affording the respective Diels-Alder adducts with good yields and high enantioselectivities (up to 98% ee, *exo*/*endo* 99:1) (Scheme 19). Notably, the chiral indium(III) catalyst was found to be water-tolerant, thus the reaction can be carried out in aqueous media.

However, the sequence of water addition was critical for the chiral indium complex to function in aqueous media. Similar as in the asymmetric allylation, a plausible transition-state model is illustrated in Figure 5. The *re* face of the complexed *s-trans*-α, β-enal is effectively blocked by the aromatic rings of the (*S*)-BINOL. Therefore, diene addition from the non-hindered *si* face of the α, β-double bond take place to give the corresponding enantiomer.

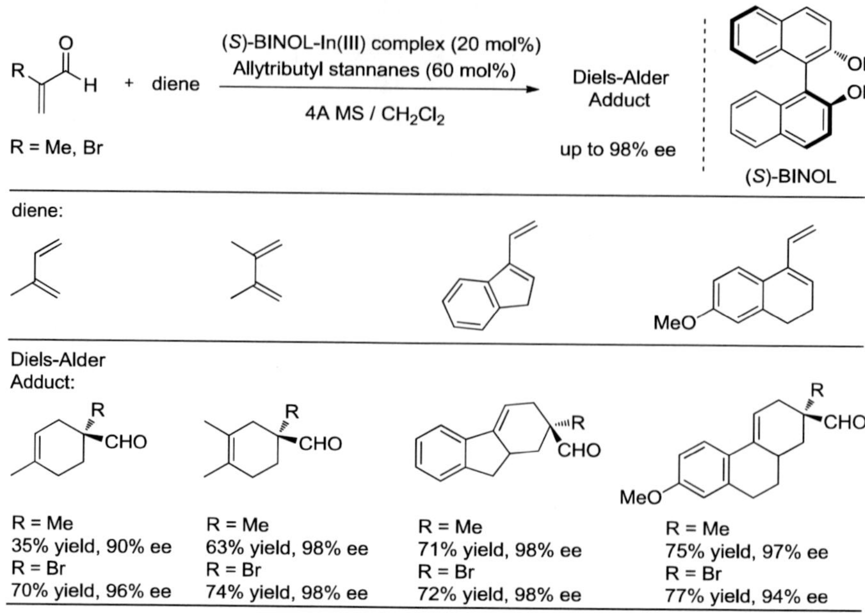

Scheme 19. In(III)-BINOL Complex-Catalyzed Aymmetric Diels-Alder Reactions.

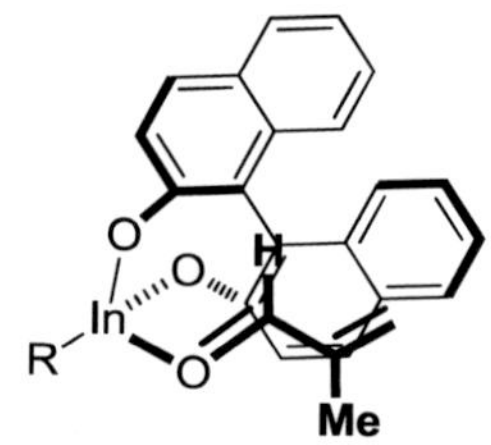

Figure 5. Pre-Transition-State Assembly of Catalyst and Dienophile.

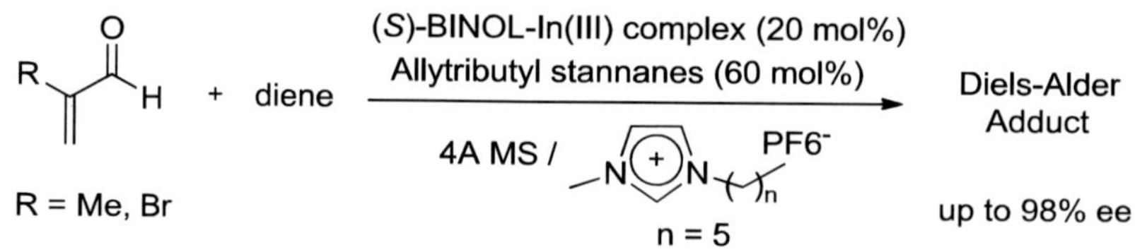

Scheme 20. In(III)-BINOL Complex-Catalyzed Aymmetric Diels-Alder Reaction in Ionic Liquid.

Extension of the chiral BINOL-In(III) catalytic system in [hmim][PF_6^-] ionic liquids was also successful and provides both improved yield (up to 92%) and stereocontrol (up to 98% ee) of Diels-Alder reaction of cyclic and open-chained dienes to 2-methacrolein and 2-bromoacrolein) (Scheme 20). [37] Most importantly, the chiral indium complex catalyst can be recycled up to seven successive cycles with comparable yields and enantioselectivities without loss of catalytic activity. In the previous study, Feng and co-workers have demonstrated that chiral *N*, *N'*-dioxide-In(III) complex can efficiently catalyze the enantioselective allylation of ketones with tetraallyltin. In an continuous work, they disclosed that these chiral indium(III) complexes can also be applied in the asymmetric hetero-Diels–Alder (HDA) reaction of Danishefsky's dienes with carbonyl compounds to afford highly substituted chiral dihydropyranones. [38] Among the chiral *N,N'*-dioxide ligands evaluated, the (*S*)-pipecolic acid derived *N,N'*-dioxides were found to be superior ligands. The catalyst can be easily prepared on a gram scale by mixing the *N,N'*-dioxide/In(OTf)$_3$ with 4 Å molecular sieves in THF and remains no loss of activity within 6 months. Under the optimized conditions, a wide range of aromatic, aliphatic, and heterocyclic aldehydes were investigated and various trisubstituted dihydropyranones were obtained with high diastereoselectivity and excellent enantioselectivity (up to 99%ee) (Scheme 21). [39]

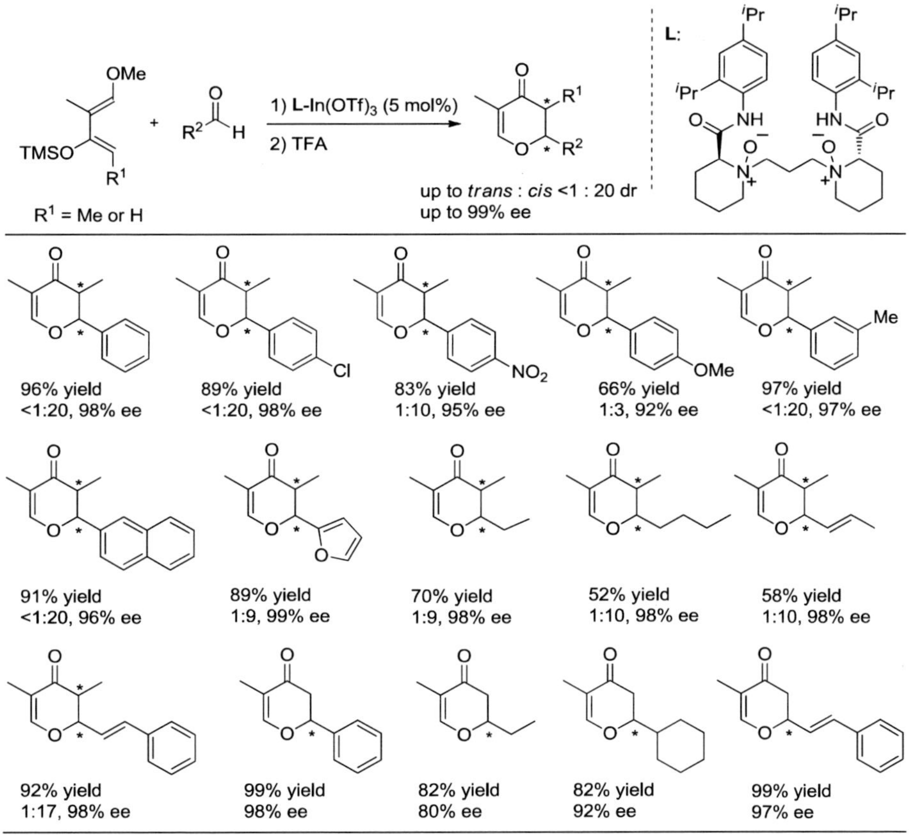

Scheme 21. Chiral *N,N'*-dioxide/In(OTf)$_3$-Catalyzed Asymmetric HDA Reaction with Diene.

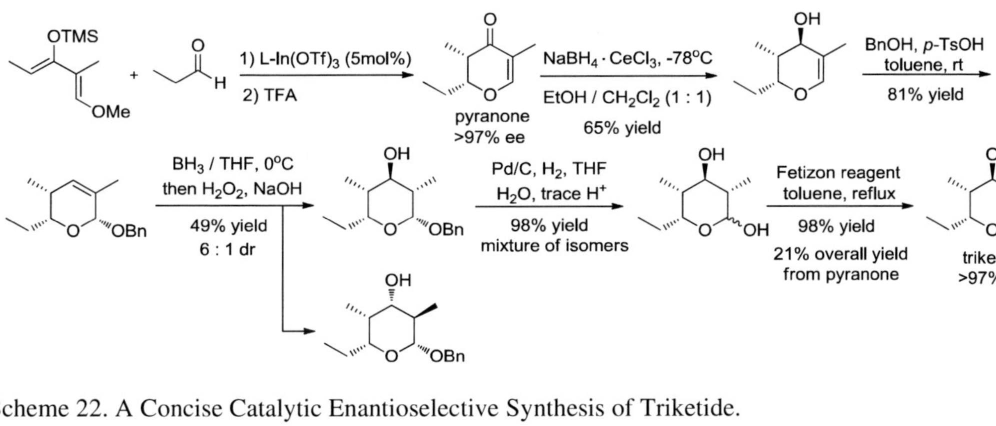

Scheme 22. A Concise Catalytic Enantioselective Synthesis of Triketide.

This indium-catalyzed highly stereoselective HAD reaction proves to be of great utility in complex molecule synthesis. For example, it has been successfully applied on a sub-gram scale to the practical synthesis of triketide. By utilizing the resulting cycloadduct pyranone as a key intermediate, the corresponding triketide was successfully accessed in overall yield of 21% with 97% ee (Scheme 22).

ASYMMETRIC MUKAIYAMA ALDOL REACTION

The asymmetric Mukaiyama aldol [40] reaction between enolsilane derivatives and aldehydes is one of the most important and widely used methods for the preparation of optically active β-hydroxy carbonyl compounds which are useful synthetic intermediates in organic synthesis and medicinal chemistry. Over the decades, much work has been focused on the development of efficient chiral Lewis acid catalysts. As an extension of the chiral In(III)-Pybox catalytic system, Loh *et al.* explored the Mukai-yama aldol reaction of aldehydes with (1-methoxy-2-methyl-propenyloxy)- trimethylsilane. [41] Indeed, the chiral indium complexes prepared by reacting $In(OTf)_3$ and various Pybox ligands in dichloromethane at room temperature in the presence of 4 Å MS can promote the enantioselective Mukaiyama aldol reaction, while bidentate bis-oxazoline ligands were ineffective. Ligand screening revealed that the (*S*, *S*)-^{i}PrPybox-In(III) complex was the optimal catalyst. However, the use of other indium salts such as InF_3, $InCl_3$ and $InBr_3$ would result in low yields and enantioselectivities. With the established best conditions, the reaction generally furnished β-hydroxyesters in moderate to good yields and enantioselectivities (51-92% ee) (Scheme 23). When 4-nitro-benzaldehyde was employed, the best enantiomeric excess of 92% was obtained. Most recently, Loh and co-workers achieved a highly efficient enantioselective Mukaiyama aldol reaction. Using a catalytic amount of in situ formed In(III)-Pybox complex, polymeric or hydrated glyoxylate underwent aldol reaction with enolsilanes derived from aryl ketones to give enantioenriched β-hydroxy ketones in excellent yields (up to 94% yield) and enantioselectivities (up to 98% ee) (Scheme 24). [42] Notably, the best catalytic system was determined by combination of 5 mol% of $InBr_3$, 5 mol% of $AgSbF_6$ and 6 mol% of Pybox in CH_3CN in the presence of 4 Å MS. The reaction was found to be sensitive to the steric hindrance of the reactants. Increasing steric bulk of both enolsilanes and glyoxylate esters will lead to increase in the enantioselectivity. Arylenolsilanes containing both electron-withdrawing and donating groups on the aromatic ring all proceeded smoothly under mild conditions with excellent stereocontrol.

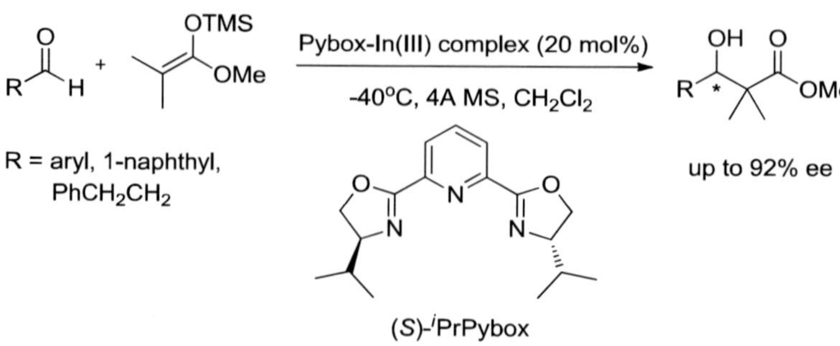

Scheme 23. Chiral In(III)-Pybox-Catalyzed Enantioselective Mukaiyama Aldol Reaction.

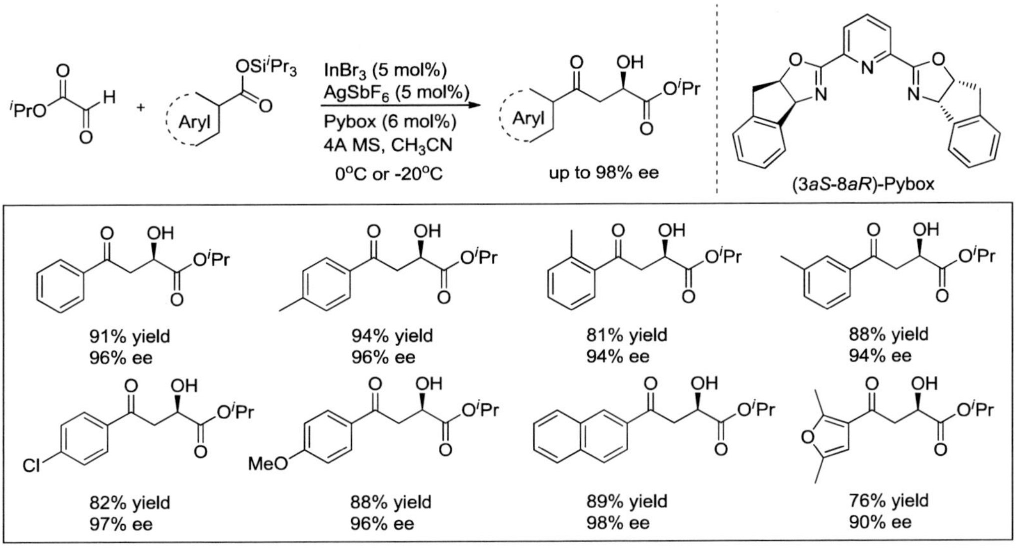

Scheme 24. Chiral In(III)-Pybox-Catalyzed Aldol Reaction with Enolsilanes.

The efficiency and unique stereoselectivity of the reaction was further demonstrated by employing α-substituted aryl ketone-derived enolsilane as reactant.

Under the standard reaction conditions, the reaction with isopropyl glyoxylate provided the *syn* aldol product with excellent diastereoselectivity (97 : 3 dr) and enantioselectivity (98% ee) (Scheme 25).

Asymmetric Ene Reaction

The carbonyl-ene reaction has attracted much attention [43] because it provides an efficient way to construct versatile useful building blocks. [44] Based on the finding of chiral indium(III) complex in asymmetric catalysis, Loh and co-workers reported an efficient asymmetric carbonyl-ene reaction using a catalytic combination of $In(OTf)_3$ and (3*aS*-8*aR*)-Pybox ligand (Scheme 26). [45]

Under the optimized conditions, the commercially available ethyl glyoxylate reacted highly enantioselectively with various 1,1-disubstituted and trisubstituted alkenes in the presence of 5 mol% of indium(III)-Pybox complex to give the expected homoallylic alcohol products.

In particular, a wide range of alkenes including aromatic, cyclic and acyclic aliphatic ones all underwent the carbonyl-ene reaction effectively to furnish the desired products in good to excellent yields and excellent enantioselectivities (up to 99% ee). The high enantioselectivity may be attributed to the glyoxylate coordinating to the cationic In(III) centre to form a facially discriminated complex.

This work represented an attractive strategy for the synthesis of chiral homoallylic alcohols containing a α-ester function, though the reactions generally took very long time (4-6 days).

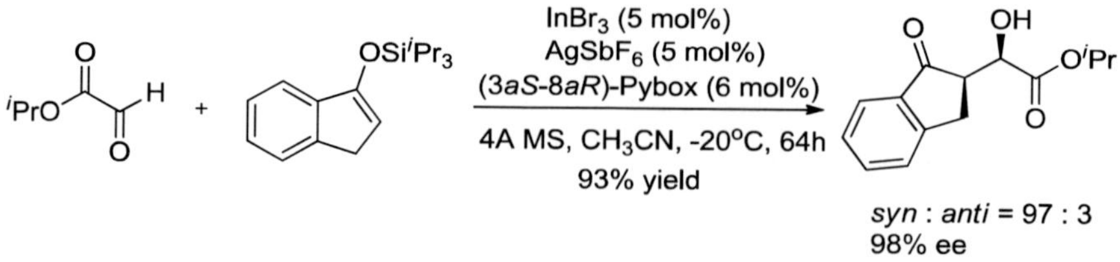

Scheme 25. Chiral In(III)-Pybox-Catalyzed Aldol Reaction with α-Substituted Enolsilanes.

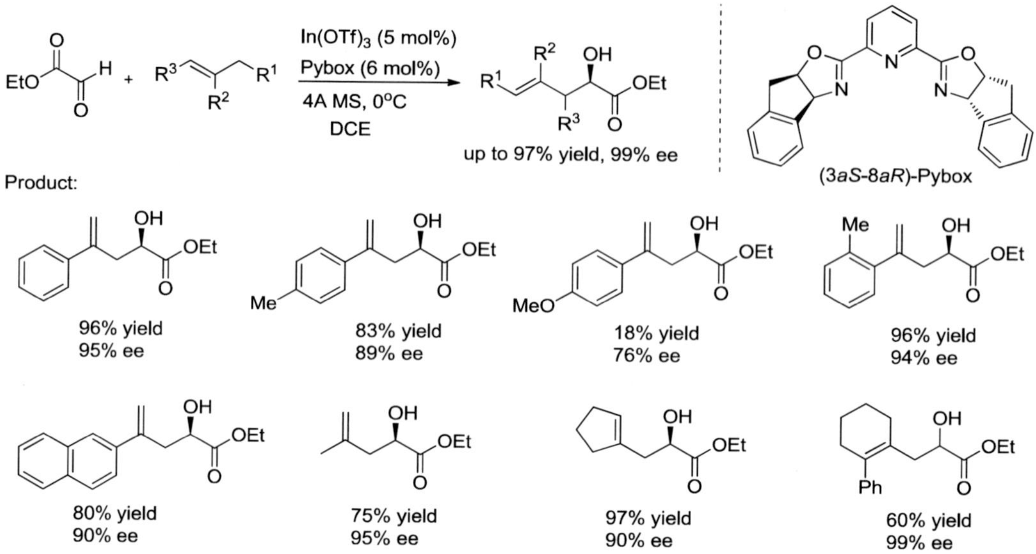

Scheme 26. In(III)-Pybox Complex-Catalyzed Aymmetric Ene Reactions.

ASYMMETRIC ALKYNYLATION

Direct catalytic asymmetric addition of terminal alkynes to aldehydes and ketones is a highly desired transformation for straightforward synthesis of valuable propargylic alcohols. In recent years, several highly enantioselective approaches have been reported using stoichiometric amounts of corresponding metal reagents, while asymmetric alkynylations using a catalytic amount of metal are rare. [46] In an early study, Shibasaki and co-workers developed an interesting new catalytic alkynylation of carbonyl compounds promoted by the combination of indium(III) salts ($InBr_3$ or $In(OTf)_3$) and $^iPr_2NEt$. As confirmed by *in situ* IR and NMR spectroscopic studies, the success of this catalytic reaction can be attributed to the dual activation of both alkyne [47] and carbonyl compound in the presence of a catalytic amount of indium(III) salt and amine base (Scheme 27). [48] Encouraged by these results, further development of asymmetric reaction conditions to produce versatile optically active propargylic alcohols was carried out. After careful examination, a highly enantioselective catalytic process was achieved by employing the In(III)/BINOL complex and Cy_2NMe. Notably, this catalytic system has broad generality for both aromatic and aliphatic aldehydes including very easily enolizable ones, as well as phenylacetylene, alkenylacetylenes, and alkylacetylenes, providing the desired propargylic alcohols in good yields and excellent enantioselectivities (up to 99% ee) (Scheme 28). [49] Mechanistically, two indium metals are likely involved in the catalytic cycle in terms of the observed rather strong positive nonlinear effects [50].

ASYMMETRIC MANNICH-TYPE REACTION

The enantioselective Mannich reaction is one of the most conceptually simple and atom-economical approaches to the construction of β-amino carbonyl compounds, which are valuable synthons in the preparation of natural products and biologically active molecules. [51-52]

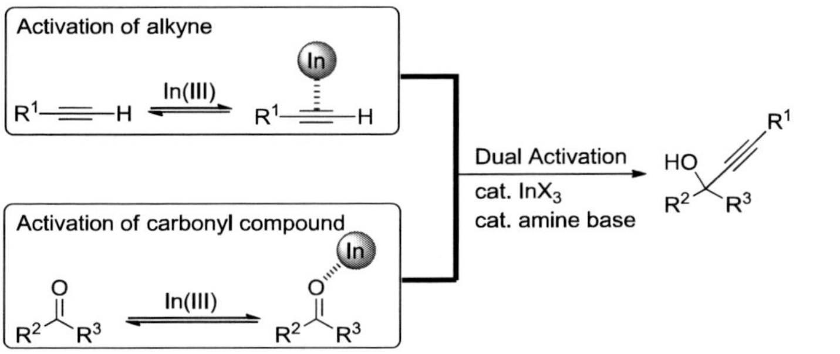

Scheme 27. Alkynylation via Dual Activation with Indium(III).

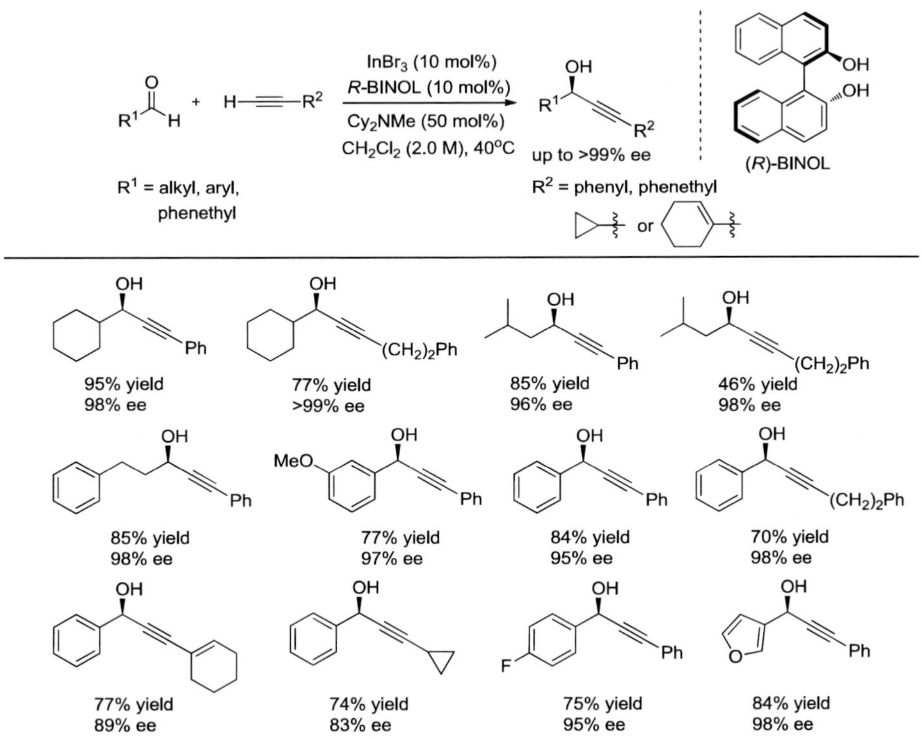

Scheme 28. $InBr_3$/BINOL Complex-Catalyzed Asymmetric Alkynylation of Aldehydes.

Although there have been several examples of $InCl_3$-catalyzed diastereoselective Mannich-type reaction of chiral amines, benzaldehyde, and 1-methoxy-2-methyl-1-trimethylsilyloxypropene, [53-54] less progress has been made in the application of chiral indium(III) complexes in catalytic enantioselective Mannich reaction. In 2005, Matsunaga and Shibasaki successfully demonstrated that a 2:1 complex of $In(O^iPr)_3$ and (*S,S*)-linked-binol was effective in generating the indium enolate in situ from *N*-(2-hydroxyacetyl)pyrrole, facilitating the asymmetric Mannich-type reaction with tosyl imines to give β-amino-α-hydroxy carboxylic acid derivatives in high enantioselectivities (91–98%ee, major diastereomer) and good yields (65–98%) (Scheme 29). [55]

In this catalytic system, the difference of R substituent of imines had significant effect on the *syn*/*anti* selectivity of the reaction. Alkenyl imines afforded *syn* adducts in good diastereoselectivities (*syn*:*anti*, 88:12–91:9) and high enantioselectivities (93–97% ee, *syn*). On the other hand, imines containing *ortho*-substituted aromatic rings, and cyclopropyl imine gave products with *anti* selectivity (*syn*:*anti*, 37:63–14:86) in high ee (92–98% ee, *anti*). Unlike the most known Lewis acidic indium catalysis by $In(OTf)_3$, $InCl_3$, or $InBr_3$, a Brønsted basic indium chiral catalyst was most likely generated in the reaction.

It was assumed that either an indium alkoxide or an indium phenoxide functions as a Brønsted base and deprotonates *N*-acylpyrrole at the α position to form the chiral indium enolate in situ. To explain the diastereoselectivity, two acyclic *anti*-periplanar transition state models as shown in Figure 6 were proposed. For imines with a planar configuration, TS-1 is favored to minimize gauche interactions. In contrast, imines with bulky R group would favor the TS-2 to avoid the steric repulsion between R and the pyrrole ring, affording the *anti*-Mannich adducts.

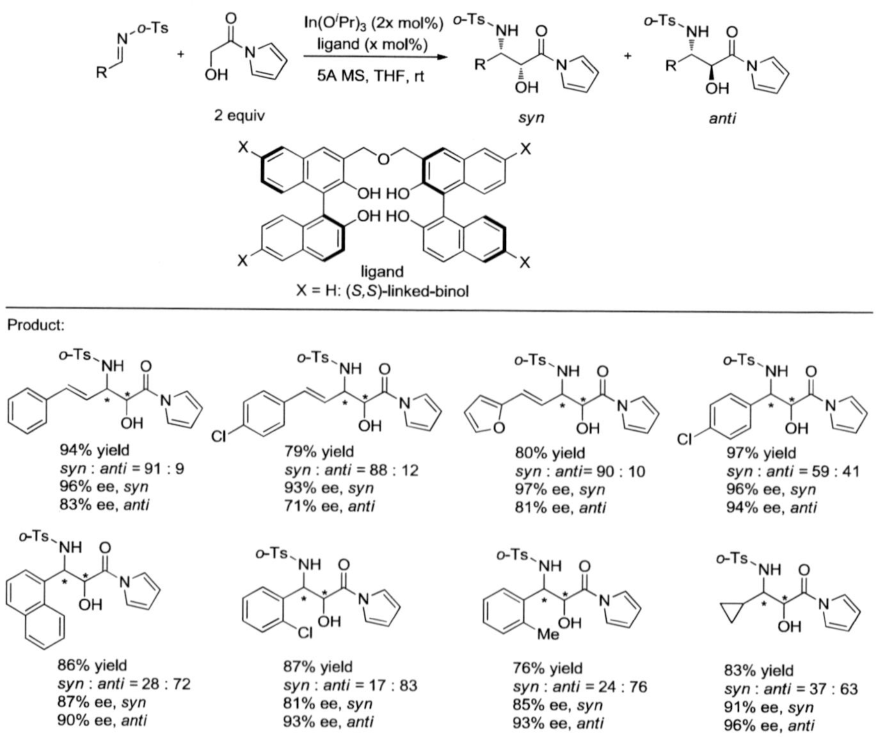

Scheme 29. $In(O^iPr)_3$/(*S,S*)-Linked-BINOL Complex Catalyzed Asymmetric Mannich Reaction.

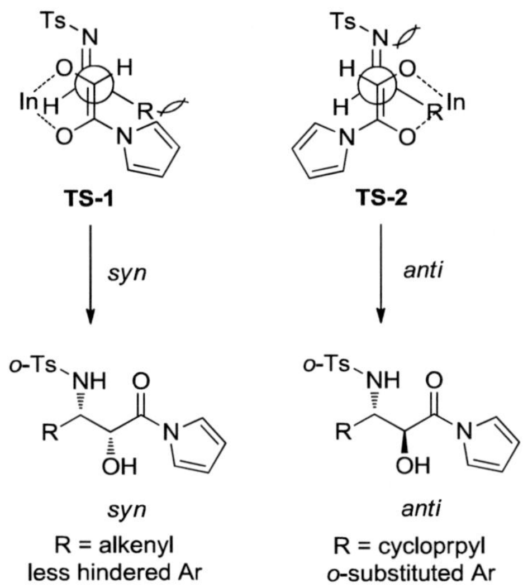

Figure 6. Proposed Transition-State Model of Mannich-Type Reactions.

ASYMMETRIC PRINS CYCLIZATION

Prins cyclization offers an useful olefin–aldehyde condensation method for the formation of diverse pyran rings, [56] which are commonly found in biologically or pharmaceutically active compounds. In an earlier work, Loh *et al.* found that a catalytic amount of an indium Lewis acid can catalyze a one-pot crossed Prins cyclization with high yield and *syn*-stereoselectivity. [57] Subsequently, this method was extended for the enantioselective synthesis of tetrahydropyran derivatives.

By employing optically active homoallylic alcohol as substrate, the $InCl_3$-mediated Prins cyclization reactions with aldehydes proceeded well in dichloromethane at a lower temperature to furnish the 2,4,6-trisubstituted tetrahydropyran products in moderate to good yields. In all cases, enantioselectivities were retained (Scheme 30). [58]

Furthermore, Loh and co-workers developed a versatile catalytic Prins cyclization using (*R*)-BINOL-In(III) complex as catalyst and trimethylsilyl halides as additives. All-*cis* configuration 4-halo-2,6-trisubstituted tetrahydropyran products were obtained in moderate to excellent yields. The efficiency and practicality of this method is demonstrated by the highly convergent enantioselective synthesis of (-)-centrolobine (Scheme 31). [59]

ASYMMETRIC RING OPENING OF *MESO*-EPOXIDES

The catalytic asymmetric ring-opening of *meso*-epoxides provides convenient access to enantioenriched 1, 2-difunctionalized compounds. [60] With chiral scandium(III)-bipyridine complex as catalyst, Schneider *et al.* reported a highly enantioselective alcoholysis and aminolysis of *meso*-epoxides. [61] Inspired by this work, the authors further extended this protocol to the enantioselective thiolysis of *meso*-epoxides.

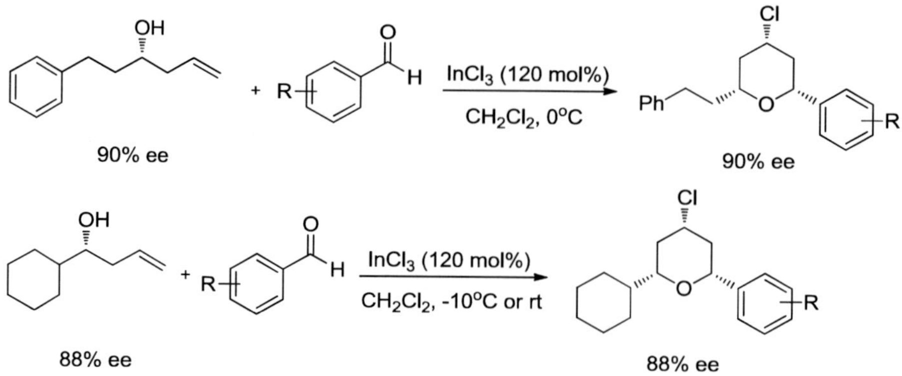

Scheme 30. In(III)-Mediated Asymmetric Prins Cyclization.

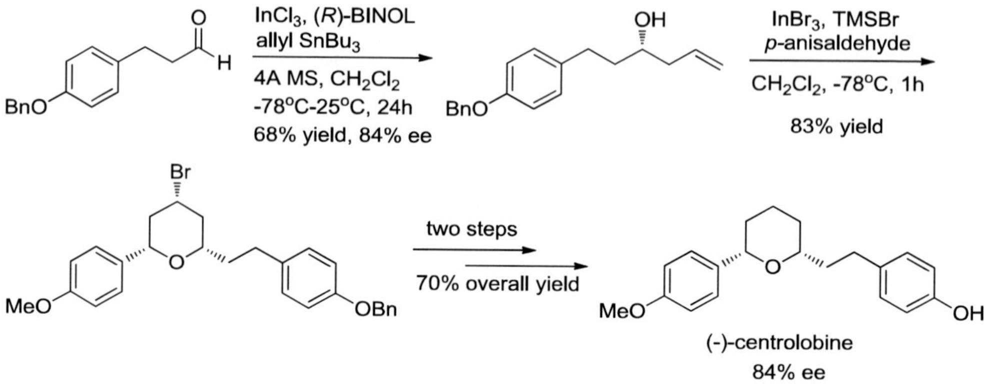

Scheme 31. The Synthesis of (-)-Centrolobine.

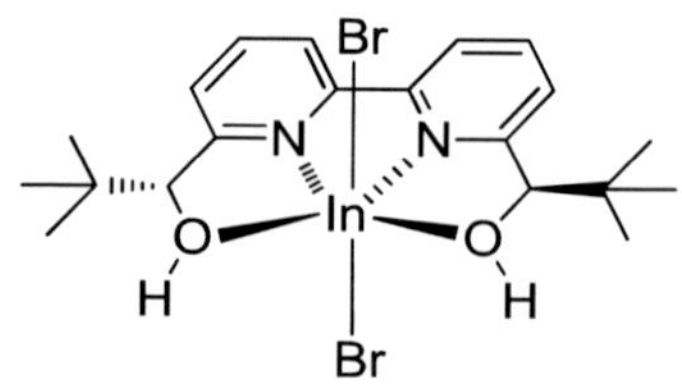

Figure 7. Chemical Structure of Chiral Catalyst.

By employing 10 mol% of $InBr_3$–bipyridine complex as catalyst, the ring-opening reaction of *cis*-stilbene oxides with various aromatic and aliphatic thiols was promoted to give 1,2-mercapto alcohols in high yields and excellent enantioselectivities (85-96% ee) (Scheme 32). [62] Especially noteworthy is the reaction appears to be no structural or electronic limitation in the thiol component; however, aliphatic *meso*-epoxides give rise to only low enantioselectivity in this catalytic system.

A crystal structure analysis of the chiral catalyst reveals a pentagonal-bipyramidal coordination geometry, where the indium is tetracoordinated by the bipyridine ligand through both nitrogen and oxygen atoms (Figure 7). It was considered that the chiral catalyst exhibits a combined Lewis acid–Brønsted acid activity, with the Lewis acidic indium center activating the epoxide.

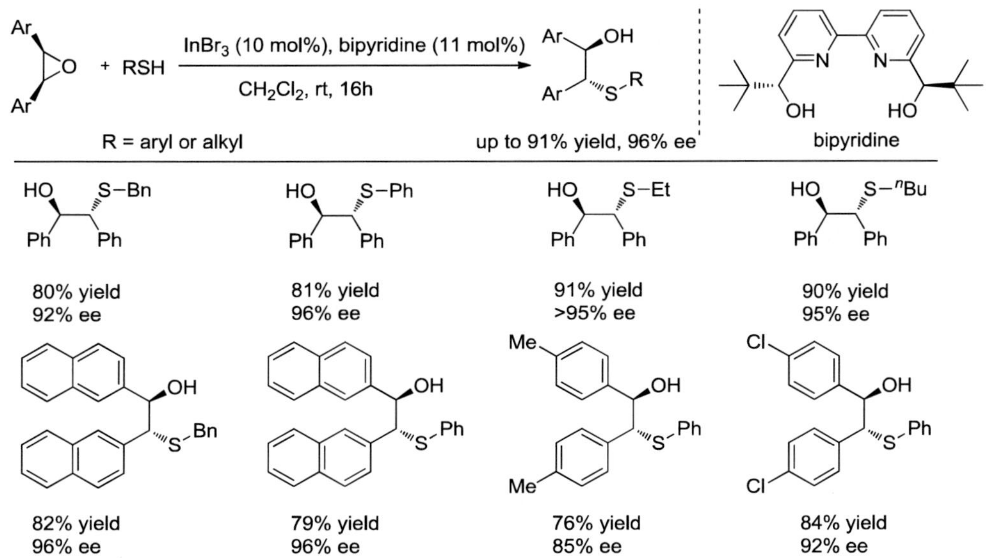

Scheme 32. Indium–Bipyridine-Catalyzed Ring Opening of *meso*-Epoxides with Thiols.

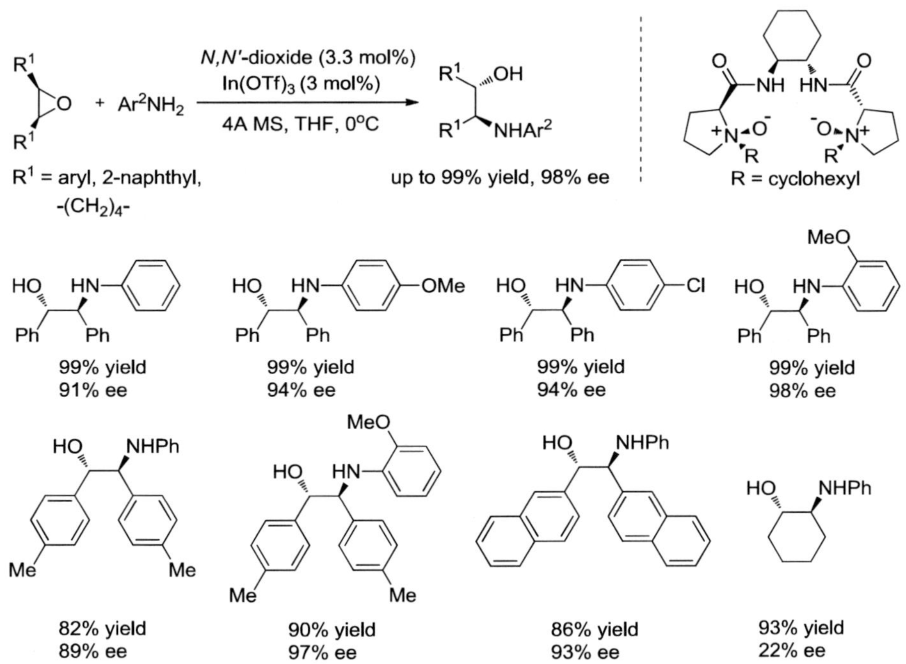

Scheme 33. *N,N'*-dioxide/In(OTf)$_3$ Complex Catalyzed Enantioselective Ring Opening of *meso*-Epoxides with Amines.

In addition, hydrogen-bonding between the hydroxyl protons and the incoming thiol nucleophile was assumed to activate and direct the incoming nucleophile to the indium-bound epoxide.

Recently, Feng and co-workers described the first example of proline-based *N, N'*-dioxide-In(OTf)$_3$ complex-catalyzed asymmetric ring-opening of *meso*-epoxides with aromatic amines (Scheme 33). [63] In general, the reaction can be completed in THF with 4 Å

MS at 0°C to furnish chiral β-amino alcohols in high yields (up to 99%) and excellent enantioselectivities (up to 99% ee). Unfortunately, aliphatic amines did not yield the desired products, presumably because stronger nucleophilicity made them coordinate irreversibly to the indium (III) complex. Although the mechanism is not clear, a plausible seven-coordinate transition state model [64,69] is proposed (Figure 8), in which indium is coordinated to the *N*-oxides and amide oxygens in a tetradentate manner. Due to the presence of two sterically bulky cyclohexyl substituents, the oxygen atom of the *meso*-stilbene oxide is bound to the indium center from the favorable less hindered backward side, meanwhile, the aniline was fixed by the hydrogen bond with the amide proton. Presumably, nucleophilic attack of the aniline nitrogen to the adjacent (*R*)-carbon of the epoxide is preferred.

ASYMMETRIC PETASIS REACTION

The Petasis boronic acid Mannich reaction is a versatile multicomponet reaction of boronic acids, amines and aldehydes that generates highly functionalized α-amino acids and β-amino alcohols. [65] For the stereoselective preparation of β, γ-unsaturated α-amino acids, asymmetric Petasis reaction represents a powerful and most direct approach. [66] Very recently, Xu *et al.* reported an In(III) Lewis acid-promoted highly diastereoselective Petasis Borono-Mannich reaction for the efficient synthesis of optically active β, γ-unsturated α-amino acids (Scheme 34). [67] Interestingly, the reaction can be promoted by a series of different transition metal Lewis acid catalysts including $In(OTf)_3$, $Yb(OTf)_3$, $Sc(OTf)_3$, $Zn(OTf)_2$, $Cu(OTf)_2$, $InBr_3$, $InCl_3$, $FeCl_3$, $FeBr_3$ and $FeCl_2$, but the best result was achieved with $InBr_3$ in terms of both yield and diastereoselectivity. In the presence of 10 mol% of $InBr_3$, a wide range of vinylboronic acids bearing different substituents were successfully reacted with glyoxylic acid and (*R*)-*tert*-butylsulfinamide, giving the corresponding β, γ-unsaturated α-amino acid products with excellent diastereoselectivities (93-99% de) in all cases. Notably, benzofuran-2-boronic acid and benzothiophen-2-boronic acid are also found to be suitable substrates, and they exhibited high reactivity and led to the formation of highly optically active α-(2-benzofuranyl)glycine (99% de) and α-(2-benzothiophenyl)glycine (96% de).

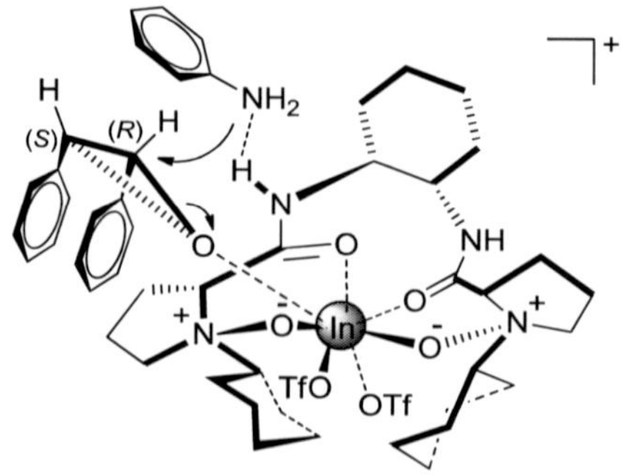

Figure 8. Proposed Transition State Model.

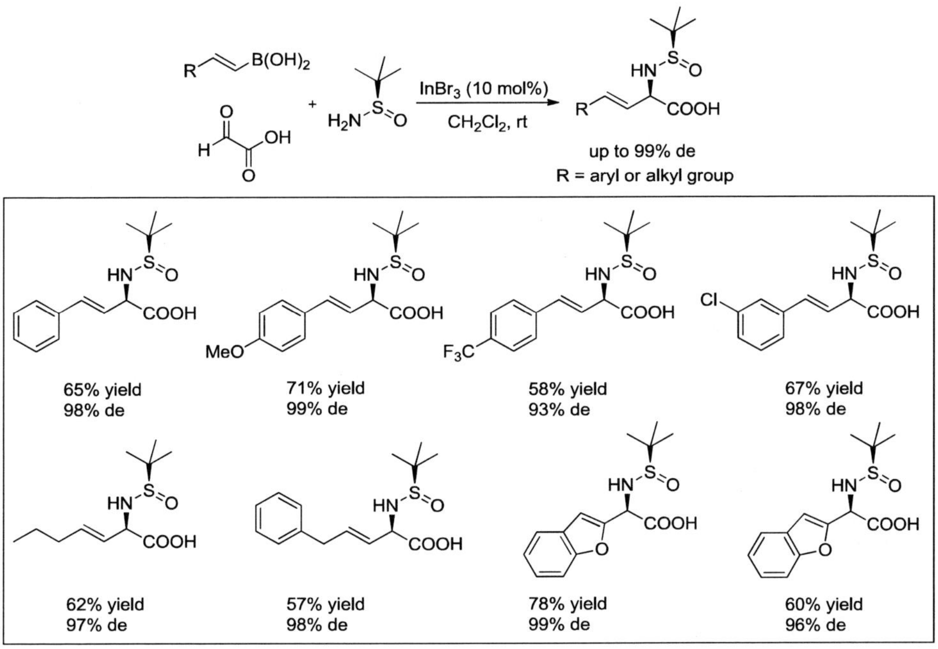

Scheme 34. $InBr_3$-Promoted Diastereoselective Petasis Reaction.

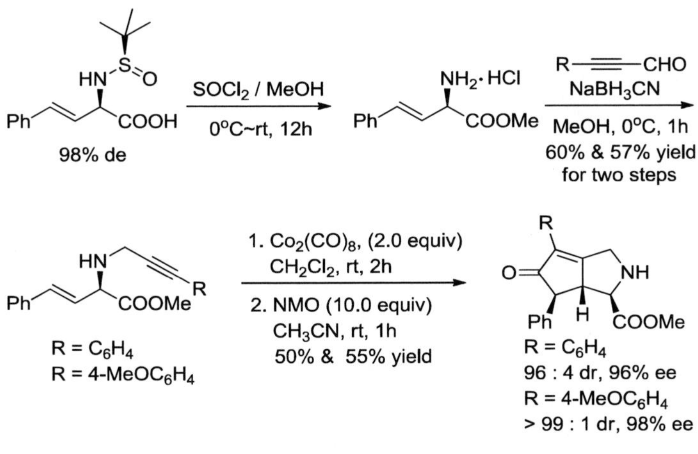

Scheme 35. Synthesis of Cyclopenta[*c*]proline Derivatives.

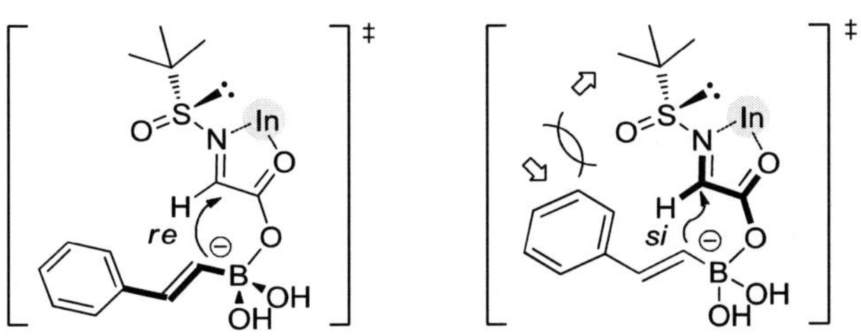

TS-1 (favored) **TS-2 (disfavored)**

Figure 9. Mechanistic Proposals for Stereocontrol.

The synthetic utility of the above methodology was subsequently demonstrated in the highly stereoselective construction of important bicyclio 3-azabicyclo[3,3,0]octane skeleton (Scheme 35). Removal of the sulfinyl group followed by concomitant esterification under $SOCl_2$/MeOH conditions furnished methyl ester. Reductive amination of methyl ester with 3-phenyl-2-propynal or 3-(4-methoxyphenyl)-2-propynal provided the key intermediate. Finally, under a typical Pauson-Khand reaction conditions, the corresponding cyclopenta[*c*]proline derivatives were obtained in moderate yield with high diastereoselectivities (96 : 4 dr and >99 : 1 dr).

A proposed transition state model for the observed reaction diastereofacial selectivity is shown in Figure 9.

In the presence of Lewis acid, coordination of imine nitrogen and carbonyl oxygen of the quaternary boronate complex to Lewis acidic metal is favored, which would form a stable five-membered ring chelation. Since the uncoordinated *N*-sufinyl group adopts an approximately synperiplanar configuration, [20] the styryl group preferred to migrate from the *Re* face rather than the *Si* face of the imine C=N bond because the latter is strongly shielded by the bulky *tert*-butyl group, thus results in the *R*-configured product.

Asymmetric Cycloaddition

An efficient bifunctionally catalyzed cycloaddition process was presented by Lectka *et al* for the synthesis of highly optically active *β*-lactams (Scheme 36). [68] The catalyst system involves the combined use of both a Lewis acid and a chiral Lewis base. By using In(III) triflate in combination with chiral cinchona alkaloid nucleophiles, cycloaddition of acetyl chloride and imino ester proceeded smoothly to give desired *β*-lactams in high yields (92-98% yield) as well as excellent diastereoselectivities (up to 60 : 1 dr) and enantioselectivities (95-98% ee).

During the examination of various Lewis acid, In(III) was found to be the most effective metal cocatalyst. Through rigorous investigation, the authors proposed a mechanism for the bifunctional catalyst system (Scheme 37). [69] Initially, a non-metal-coordinated zwitterionic enolate was formed by the reaction of complex In(III)-BQ with phenylketene. Imino ester then coordinates to the salicylate bound indium to provide a ternary complex. Intramolecular nucleophilic addition of enolate to the indium activated imine moiety occurs as the rate-determining step, leading to the formation of the new C-C bond. Finally, transacylation followed by catalyst regeneration allow for additional catalytic cycles.

Asymmetric Synthesis of *C*-(Alkynyl)-Pseudoglycals

Yadav *et al.* employed $InBr_3$ as a mild Lewis acid catalyst to effectively promote the stereoselective reaction of simple δ-hydroxy-α,β-unsaturated aldehydes with alkynyl silanes and give *C*-(alkynyl)-pseudoglycals in high yields (Scheme 38). [70] A possible reaction mechanism is illustrated in Scheme 39. First, activation of the aldehyde by indium(III) bromide occurs.

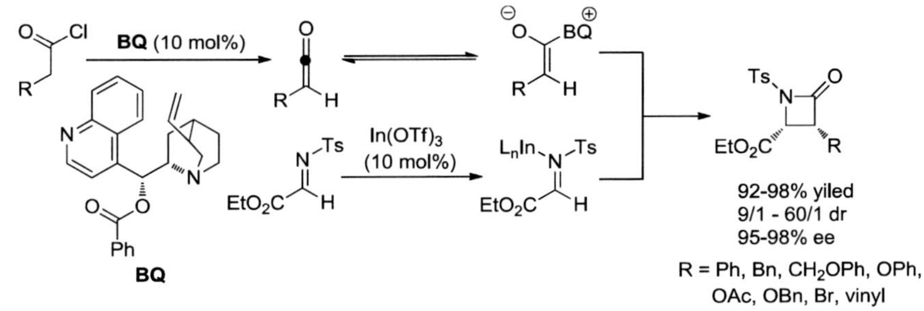

Scheme 36. β-Lactams Synthesis Catalyzed by In(OTf)$_3$/BQ.

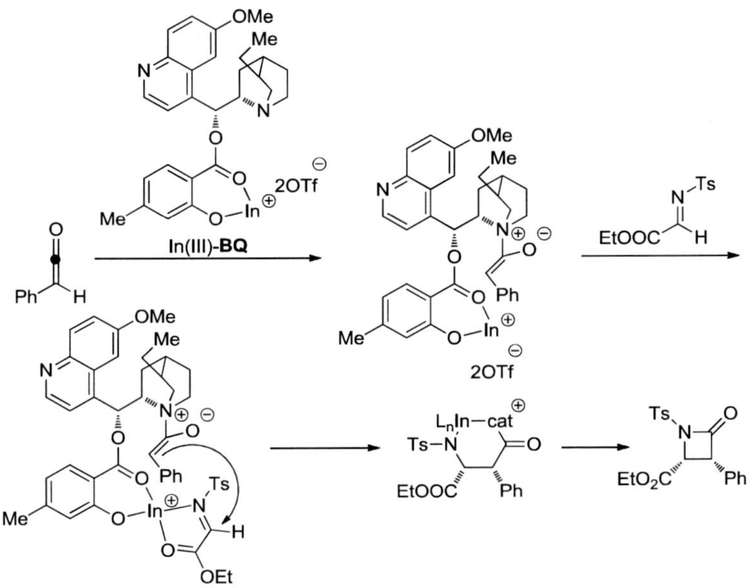

Scheme 37. Proposed Mechanism with Metal Complex In(III)-BQ.

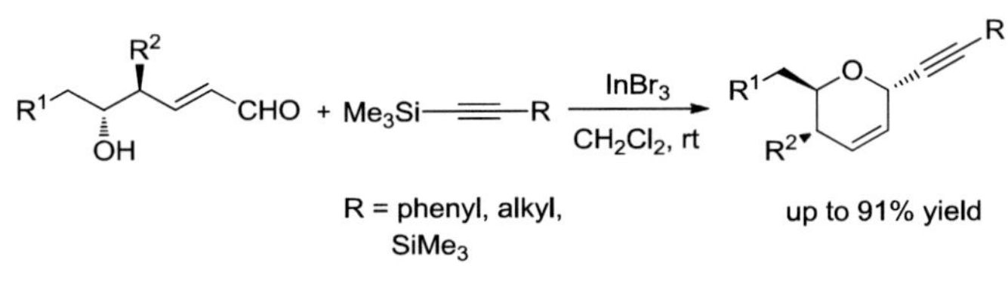

Scheme 38. InBr$_3$-Promoted Stereoselective Formation of *C*-(Alkynyl)-Pseudoglycals.

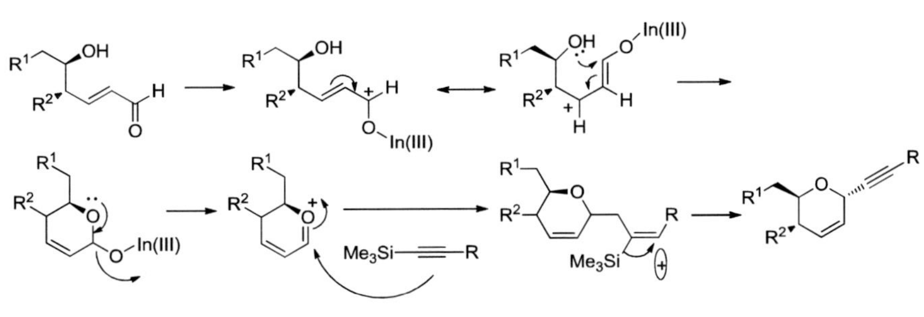

Scheme 39. Possible Mechanism for the Reaction.

Then, proton transfer followed by hydroxyl oxygen addition led to the formation of an oxonium intermediate. A further stereoselective nucleophilic attack by the silyl acetylene onto the oxonium gives the cationically stable intermediate because of β-silicon effect. [71] Finally, elimination of the trimethylsilyl group results in the formation of the product.

CONCLUSION

In recent years, indium reagents [72] have gained widespread application as effective Lewis acid catalysts in many fundamental organic reactions. This article summarizes the most recent developments of indium(III) chemistry in asymmetric catalysis since 2005. The utility of chiral indium(III) complexes and indium(III) salts as Lewis acids in catalytic asymmetric synthesis is demonstrated by the high levels of stereocontrol induced in a wide range of functional group transformations including asymmetric Friedel-Crafts alkylations, allylation reactions, Diels-Alder cycloadditions, Mukaiyama aldol condensation, carbonyl ene, alkynylation reactions, Mannich-type reaction, Prins cyclization, ring opening of *meso*-epoxides, Petasis reaction, cycloaddition, *etc.* Compared to other Lewis acids, indium(III) reagents have advantages of low toxicity, good stability and recyclability. Although this review shows extensive research in over the past seven years on the utility of chiral indium(III) complexes and indium(III) salts as Lewis acids, this area is far from being exhausted, further developments are certain to produce new class of elegant catalysts. It is hoped that this highlight will stimulate the development of new chiral indium reagents and their application in newly developed asymmetric reactions.

ACKNOWLEDGMENT

Financial support from the National Natural Science Foundation of China, and National Science and Technology Major Project is greatly acknowledged.

REFERENCES

[1] Li, C.-J., Chan, T.-H. *Tetrahedron* 1999, 55, 11149.
[2] Ranu, B. C. *Eur. J. Org. Chem.* 2000, 2347.
[3] Chauhan, K. K., Frost, C. G. *J. Chem. Soc., Perkin Trans.* 1 2000, 3015.
[4] Fringuelli, F., Piermatti, O., Pizzo, F., Vaccaro, L. *Curr. Org. Chem.* 2003, 7, 1661.
[5] Frost, C. G., Hartley, J. P. *Mini-Rev. Org. Chem.* 2004, 1, 1.
[6] Nair, V., Ros, S., Jayan, C. N., Pillai, B. S. *Tetrahedron* 2004, 60, 1959.
[7] Loh, T.-P., Chua, G.-L. *Chem. Commun.* 2006, 2739.
[8] Augé, J., Nadège, L.-G., Uziel, J. *Synthesis* 2007, 12, 1739 and references therein.
[9] Yadav, J. S., Antony, A., George, J., Reddy, B. V. S. *Eur. J. Org. Chem.* 2010, 591.
[10] Olah, G. A. Wiley-Interscience: New York, 1973.
[11] Hanhan, N. V., Sahin, A. H., Chang, T. W., Fettinger, J. C., Franz, A. K. *Angew. Chem. Int. Ed.* 2010, 49, 744.

[12] Evans, D. A., Fandrick, K. R., Song, H.-J., Scheidt, K. A., Xu, R. *J. Am. Chem. Soc.* 2007, 129, 10029. (b) Desimoni, G., Faita, G., Toscanini, M., Boiocchi, M. *Chem. Eur. J.* 2008, 14, 3630. (c) Zhao, J. F., Tsui, H. Y., Wu, P. J., Lu, J., Loh, T. P. *J. Am. Chem. Soc.* 2008, 130, 16492.

[13] Gutierrez, E. G., Wong, C. J., Sahin. A. H., Franz, A. K. *Org. Lett.* 2011, 13, 5744.

[14] For reviews on asymmetric Friedel–Crafts reactions, see: (a) You, S.-L., Cai, Q., Zeng, M. *Chem. Soc. Rev.* 2009, 38, 2190. (b) Poulsen, T. B., Jógensen, K. A. *Chem. Rev.* 2008, 108, 2903. (c) Bandini, M., Melloni, A., Umani-Ronchi, A. *Angew. Chem. Int. Ed.* 2004, 43, 550. (d) M. Bandini, M., Eichholzer, A. *Angew. Chem. Int. Ed.* 2009, 48, 9608. (e) Terrasson, V., de Figueiredo, R. M., Campagne, J. M. *Eur. J. Org. Chem.* 2010, 2635.

[15] Lv, J., Zhang, L., Zhou, Y., Nie, Z., Luo, S., Cheng, J.-P. *Angew. Chem. Int. Ed.* 2011, 50, 6610.

[16] Pagenkopf, B. L., Carreira, E. M. *Chem. Eur. J.* 1999, 5, 3437. (b) Denmark, S. E., Beutner, G. L., Wynn, T., Eastgate, M. D. *J. Am. Chem. Soc.* 2005, 127, 3774.

[17] Giera, D. S., Schneider, C. *Org. Lett.* 2010, 12, 4884.

[18] Li, Y., Ji, D.-M., Xu, M.-H. *Org. Biomol. Chem.* 2011, 9, 8452.

[19] Ji, D.-M., Xu, M.-H. *Chem. Commun.* 2010, 46, 1550.

[20] For related crystal structures and theoretical calculations: (*a*) Owens, T. D., Hollander, F. J., Oliver, A. G., Ellman, J. A. *J. Am. Chem. Soc.* 2001, 123, 1539. (*b*) Owens, T. D., Souers, A. J. Ellman, J. A. *J. Org. Chem.* 2003, 68, 3. (c) Schenkel, L. B., Ellman, J. A. *Org. Lett.* 2003, 5, 545. (*d*) Tietze, L. F., Schuffenhauer, A. *Eur. J. Org. Chem.* 1998, 1629. (*e*) Bharatam, P. V., Uppal, P., Kaur, A., Kaur, D. *J. Chem. Soc. Perkin Trans.* 2 2000, 43.

[21] Denmark, S. C., Fu, J.-P. *Chem. Rev.* 2003, 103, 2752.

[22] Yamamoto, Y., Asao, N. *Chem. Rev.* 1993, 93, 2207. (b) Ding, H., Friestad, G. K. *Synthesis* 2005, 2815.

[23] Lu, J., Hong, M.-L., Ji, S.-J., Loh, T.-P. *Chem. Commun.* 2005, 1010.

[24] Lu, J., Ji, S.-J., Teo, Y.-C., Loh, T.-P. *Org. Lett.* 2005, 7, 159.

[25] Lu, J., Hong, M.-L., Ji, S.-J., Teo, Y.-C., Loh, T.-P. *Chem. Commun.* 2005, 4217.

[26] Teo, Y.-C., Tan, K.-Y., T.-P. *Chem. Commun.* 2005, 1318.

[27] Teo, Y.-C., Goh, J.-D., Loh, T.-P. *Org. Lett.* 2005, 7, 2743.

[28] Chelucci, G., Murineddu, G., Pinna, G. A. *Tetrahdron: Asymmetry* 2004, 15, 1373. (b) Malkov, A. V., Kočovský, P. *Eur. J. Org. Chem.* 2007, 29. (c) Liu, X., Lin, L., Feng, X. *Acc. Chem. Res.* 2011, 44, 574.

[29] Zhang, X., Chen, D., Liu, X., Feng, X. *J. Org. Chem.* 2007, 72, 5227.

[30] Delpech, F., Guzei, I. A., Jordan, R. F. *Organometallics* 2005, 21, 1167.

[31] Denmark, S. E., Fu, J. *J. Am. Chem. Soc.* 2003, 125, 2208.

[32] Yasuda, M., Miyai, T., Shibata, I., Baba, A. *Tetrahedron Lett.* 1995, 36, 9497. (b) Shibata, I., Kato, H., Ishida, T., Yasuda, M., Baba, A. *Angew. Chem., Int. Ed.* 2004, 43, 711. (c) Takita, R., Fukuta, Y., Tsuji, R., Ohshima, T., Shibasaki, M. *Org. Lett.* 2005, 7, 1363.

[33] For examples, see: (a) Ken Lee, C.-L., Ling, H. Y., Loh, T.-P. *J. Org. Chem.* 2004, 69, 7787. (b) van der Sluis, M., Dalmolen, J., de Lange, B., Kaptein, B., Kellogg, R. M., Broxterman, Q. B. *Org. Lett.* 2001, 3, 3943. (c) Friestad, G. K., Ding, H. *Angew. Chem. Int. Ed.* 2001, 40, 4491. (d) Cook, G. R., Maity, B. C., Kargbo, R. *Org. Lett.* 2004, 6,

1741. (e) Vilaivan, T.,Winotapan, C., Banphavichit, V., Shinada, T., Ohfune, Y. *J. Org. Chem.* 2005, 70, 3464.

[34] Sun, X.-W., Xu, M.-H., Lin, G.-Q. *Org. Lett.* 2006, 8, 4979.

[35] Sun, X.-W., Ph.D. Thesis, Shanghai Institute of Organic Chemistry Chinese Academy of Sciences, 2007.

[36] Teo, Y.-C., Loh, T.-P. *Org. Lett.* 2005, 7, 2539.

[37] Fu, F., Teo, Y.-C., Loh, T.-P. *Org. Lett.* 2006, 8, 5999.

[38] Reviews: (a) Danishefsky, S. J., DeNinno, M. P. *Angew. Chem. Int. Ed.* 1987, 26, 15. (b) Corey, E. J., Guzman-Perez, A. *Angew. Chem. Int. Ed.* 1998, 37, 388. (c) Jórgensen, K. A. *Angew. Chem. Int. Ed.* 2000, 39, 3558. (d) Lin, L. L., Liu, X. H., Feng, X. M. *Synlett* 2007, 2147, and references therein.

[39] Yu, Z., Liu, X., Dong, Z., Xie, M., Feng, X. *Angew. Chem. Int. Ed.* 2008, 47, 1308.

[40] Kobayashi, S., Fujishita, B. Y., Mukaiyama, T. *Chem. Lett.*1990, 1455.

[41] Fu, F., Teo, Y.-C., Loh, T.-P. *Tetrahedron* 2006, 47, 4267.

[42] Zhao, J.-F., Tan, B.-H., Loh, T.-P. *Chem. Sci.* 2011, 2, 349.

[43] For a general review of the ene reaction, see: (a) Mikami, K., Shimizu, M. *Chem. Rev.* 1992, 92, 1021. (b) Berrisford, D. J., Bolm, C. *Angew. Chem., Int. Ed.* 1995, 34, 1717. (c) Mikami, K., Terada, M., Narisawa, S., Nakai, T. *Synlett* 1992, 255. (d) Mikami, K. *Pure Appl. Chem.* 1996, 68, 639.

[44] Nicolaou, K. C., Kim, D. W., Baati, R. *Angew. Chem. Int. Ed.* 2002, 41, 3701. (b) Hornberger, K. R., Hamblet, C. L., Leighton, J. L. *J. Am. Chem. Soc.* 2000, 122, 12894. (c) Felpin, F. X., Lebreton, J. *J. Org. Chem.* 2002, 67, 9192.

[45] Zhao, J.-F., Tsui, H.-Y., Wu, P.-J., Lu, J., Loh, T.-P. *J. Am. Chem. Soc.* 2008, 130, 16492.

[46] Lin, P. *Tetrahedron* 2003, 59, 9873. (b) Cozzi, P. G., Hilgraf, R., Zimmermann, N. *Eur. J. Org. Chem.* 2004, 4095. (c) Frantz, D. E., Fässler, R., Tomooka, C. S., Carreira, E. M. *Acc. Chem. Res.* 2000, 33, 373. (d) Tan, L., Chen, C.-Y., Tillyer, R. D., Grabowski, E. J. J., Reider, P. J. *Angew. Chem. Int. Ed.* 1999, 5, 38. (e) Boyall, D., López, F., Sasaki, H., Frantz, D., Carreira, E. M. *Org. Lett.* 2000, 2, 4233. (f) Frantz, D. E., Fässler, R., Carreira, E. M. *J. Am. Chem. Soc.* 2000, 122, 1806. (g) Anand, N. K., Carreira, E. M. *J. Am. Chem. Soc.* 2001, 123, 9687. (h) Fässler, R., Tomooka, C. S., Frantz, D. E., Carreira, E. M. *Proc. Natl. Acad. Sci. US* 2004, 101, 5843.

[47] For examples of catalytic activation of alkynes by In(III), see: (a) Tsuchimoto, T., Maeda, T., Shirakawa, E., Kawakami, Y. *Chem. Commun.* 2000, 1573. (b) Tsuchimoto, T., Hatanaka, K., Shirakawa, E., Kawakami, Y. *Chem. Commun.* 2003, 2454. (c) Nakamura, M., Endo, K., Nakamura, E. *J. Am. Chem. Soc.* 2003, 125, 13002. (d) Sakai, N., Annaka, K., Konakahara, T. *Org. Lett.* 2004, 6, 1527. (e) Tsuchimoto, T., Matsubayashi, H., Kaneko, M., Shirakawa, E., Kawakami, Y. *Angew. Chem. Int. Ed.* 2005, 44, 1336.

[48] Takita, R., Fukuta, Y., Tsuji, R., Ohshima, T., Shibasaki, M. *Org. Lett.* 2005, 7, 1363.

[49] Takita, R., Yakura, K., Ohshima, T., Shibasaki, M. *J. Am. Chem. Soc.* 2005, 127, 13760.

[50] Girard, C., Kagan, H. B. *Angew. Chem. Int. Ed.* 1998, 37, 2922. (b) Blackmond, D. G. *Acc. Chem. Res.* 2000, 33, 402. (c) Kagan, H. B. *Adv. Synth. Catal.* 2001, 343, 227. (d) Kagan, H. B. *Synlett* 2001, 888.

[51] For comprehensive reviews: (a) Risch, N., Arend, M., Westermann, B. *Angew. Chem.*

Int. Ed. 1998, 37, 1044, and references therein. (b)Tramontini, M., Angiolini, L. In: *Mannich Bases, Chemistry and Uses*; CRC: Boca Raton, FL, 1994, and references therein. (c) Volkmann, R. A. In: *Comprehensive Organic Synthesis*; Trost, B. M., Fleming, I., Eds.; Pergamon: Oxford, UK, 1991; Vol. 1, p 355 and references therein.

[52] *Organic Synthesis in Water*; Grieco, P. A., Ed.; Blacky Academic and Professional: London, UK, 1998. (b) Li, C. J. *Chem. Rev.* 1993, 93, 2023. (c) Kobayashi, S., Hamada, T., Manabe, K. *J. Am. Chem. Soc.* 2002, 124, 5640. (d) Akiyama, T., Onuma, Y. *J. Chem. Soc., Perkin Trans.* 1 2002, 1157. (e) Kobayashi, S., Manabe, K. *Acc. Chem. Res.* 2002, 35, 209.

[53] Loh, T.-P., Chen, S.-L. *Org. Lett.* 2002, 4, 3647.

[54] Chen, S.-L., Ji, S.-J., Loh, T.-P. *Tetrahedron Lett.* 2003, 44, 2405.

[55] Harada, S., Handa, S., Matsunage, S., Shibasaki, M. *Angew. Chem. Int. Ed.* 2005, 44, 4365.

[56] Adams, D. R., Bhatnagar, S. P. Synthesis 1977, 661. (b) Arundale, E., Mikeska, L. A. *Chem. Rev.* 1952, 51, 505.

[57] Chan, K.-P., Loh, T.-P. *Tetrahedron Lett.* 2004, 45, 8387.

[58] Lee, C.-H. A., Loh, T.-P. *Tetrahedron Lett.* 2006, 47, 1641.

[59] Chan, K.-P., Loh, T.-P. *Org. Lett.* 2005, 7, 4491.

[60] Reviews: (a) Schneider, C. *Synthesis*, 2006, 3919. (b) Pastor, I. M., Yus, M. *Curr. Org. Chem.* 2005, 9, 1. (c) Jacobsen, E. N., Wu, M. H. In: Comprehensive Asymmetric Catalysis, ed. Jacobsen, E. N., Pfaltz, A., Yamamoto, H. Springer, Berlin, 1999, vol. 3, p. 1309.

[61] Schneider, C., Sreekanth, A. R., Mai, E. *Angew. Chem., Int. Ed.* 2004, 43, 5691. (b) Bolm, C., Zehnder, M., Bur, D. *Angew. Chem., Int. Ed.* 1990, 29, 191.

[62] Nandakumar, M., Tschöp, A., Krautscheid, H., Schneider, C. *Chem. Commun.* 2007, 2756.

[63] Gao, B., Wen, Y., Yang, Z., Huang, X., Liu, X., Feng, X. *Adv. Synth. Catal.* 2008, 350, 385.

[64] Nandakumar, M. V., Tschçp, A., Krautscheid, H., Schneider, C. *Chem. Commun.* 2007, 2756.

[65] Candeias, N. R., Montalbano, F., Cal, P. M. S. D., Gois, P. M. P. *Chem. Rev.* 2010, 110, 6169.

[66] Petasis, N. A., Zavialov, I. A. *J. Am. Chem. Soc.* 1997, 119, 445. (b) Churches, Q. I., White, J. M., Hutton, C. A. *Org. Lett.* 2011, 13, 2900. (c) Jiang, B., Yang, C. –G., Gu, X. -H. *Tetrahedron Letters* 2001, 42, 2545. (d) Lou, S., Schaus, S. E. *J. Am. Chem. Soc.* 2008, 130, 6922.

[67] Li, Y., Xu, M.-H. *Org. Lett.* 2012, 14, 2062.

[68] France, S., Wack, H., Hafez, A. M., Taggi, A. E., Witsil, D. R., Lectka, T. *Org. Lett.* 2002, 4, 1603.

[69] France, S., Shah, M. H., Weatherwax, A., Wack, H., Roth, J. P., Lectka, T. *J. Am. Chem. Soc.* 2005, 127, 1206.

[70] Yadav, J. S., Raju, A. K., Sunitha, V. *Tetrahedron Lett.* 2006, 47, 5269.

[71] Takahiro, T., Isobe, M. *Tetrahedron Lett.* 1992, 33, 7911, and references cited therein.

[72] For a recent account on indium(I) catalysis, see: Schneider, U., Kobayashi, S. *Acc. Chem. Res.* 2012, DOI: 10.1021/ar300008t and references therein.

In: Indium
Editors: Hsaio G. Woo and Huang Tsai Choi
ISBN: 978-1-62257-696-8

Chapter 7

INDIUM: MILD AND USEFUL REAGENT FOR CHEMICAL TRANSFORMATIONS AND FURTHER APPLICATIONS

Tomoko Mineno[1,*] and Hitoshi Yoshimitsu[2]
[1]Faculty of Pharmacy, Takasaki University of Health and Welfare, Takasaki, Gunma, Japan
[2]Faculty of Pharmaceutical Sciences, Sojo University, Ikeda, Kumamoto, Japan

ABSTRACT

Indium (In) is classified as one of the post-transition metals and is located in group 13, similar to boron and aluminum, on the periodic table of the elements. Interest in the use of indium and indium-related reagents is growing. Many recent investigations in synthetic chemistry have focused on indium or indium-related reagents, and have shown distinct progress. A variety of applications have been explored, and there have been reports detailing organic reactions that utilized indium or indium-related reagents to realize chemical transformations, bond-formations, and bond-extensions. This chapter covers the valence of indium metal, highlights the chemistry of indium or indium-related reagents, such as chemoselective protection and deprotection methodologies, and includes applications of indium chemistry. These applicable examples demonstrate further prospects and suggest an expansive potential for the field.

INTRODUCTION

Indium is a metallic chemical element with atomic number 49. Indium is classified as one of the post-transition metals located in group 13, like boron and aluminum, in the periodic table of the elements. The electron shell configuration of indium is $1s^2\ 2s^2\ 2p^6\ 3s^2\ 3p^6\ 3d^{10}\ 4s^2\ 4p^6\ 4d^{10}\ 5s^2\ 5p^1$. The element is silvery-white, soft, and has very high malleability. Due to its

[*] Corresponding Author: mineno@takasaki-u.ac.jp.

relative rarity, this element tends to require efficient use in industry applications world-wide. Indium was discovered in 1863 by the German metallurgists Ferdinand Reich and Hieronymus Theodor Richter [1]. Indium was found in the samples of zinc ores, and was named for the characteristic bright indigo-blue line in its atomic spectrum.

Despite the discovery of indium, it was not until the early 1990s that scientists in the field of synthetic organic chemistry made certain attempts to investigate its role or that of indium-related reagents in organic synthesis. However, recent investigations into synthetic chemistry that focused on indium or indium-related reagents were prominently remarkable and showed a distinct contribution to the progress of indium chemistry.

As a result, a variety of applications have since been explored, and reports have focused on organic reactions utilizing indium or indium-related reagents in chemical transformations, bond-formations, and bond-extensions. This chapter focuses on the valence of indium metal, highlights the chemistry of indium and indium-related reagents, such as chemoselective protection and deprotection methodologies, and includes examples of the application of indium chemistry, as listed below. These applications clearly demonstrate the potential for further development and suggest the prospects for the field.

1. Chemistry of utilizing indium metal
    1-1. Reactions of amines
    1-2. Reactions of alcohols
    1-3. Reactions of carboxylic acids
2. Chemistry of utilizing trivalent indium reagents
    2-1. Trivalent indium salts of triflate
    2-2. Trivalent indium salts of halogens
3. Application of indium chemistry to glycoside chemistry

1. Chemistry of Utilizing Indium Metal

Chemoselective reactions at the intended reactive sites are a vital point for the success of an implemented sequence of chemical reactions. For this purpose, a variety of protecting groups and many strategic reaction methodologies have been developed. It is advantageous that indium metal is stable in water and air. Therefore, a reaction can generally be started without activating the indium metal, and without requiring anhydrous solvents. A low value for the first ionization potential of indium metal is the foundation of its propensity for application. This section describes the chemoselective reaction methods that utilize indium metal, including the protecting and deprotecting reactions of functional groups, as well as other useful reactions.

1-1. Reactions of Amines

1-1-1. Deprotection of Trichloroethyl Carbamates

Protecting groups such as the 2,2,2-trichloroethoxycarbonyl (Troc) moiety are efficiently employed as masking units for hydroxy and amino groups. The Troc group has been exceedingly useful in both multi-step and total syntheses. Although several methods such as zinc metal [2], electrolysis [3,4], zinc-lead coupling [5], and cadmium dust [6] are available for the cleavage of the Troc group from carbamates, an environmentally and ecologically safe and mild reaction method may be preferred. Indium metal-mediated reactions have been introduced as a mild and convenient method for the effective cleavage of the Troc group from carbamates [7]. The reaction proceeds in the presence of indium powder, and the optimization of the reaction conditions has been carried out, as listed in Table 1. The reaction was improved when the solvent was changed from MeOH-H_2O (5:1) to EtOH-H_2O (3:2), presumably due to the higher boiling point of EtOH (bp 78 °C) compared with that of MeOH (bp 65 °C). The most efficient deprotection of a Troc moiety was observed in the case of Troc-protected 4-aminobenzonitrile, requiring only 2 equiv of indium powder and 3 equiv of ammonium chloride (Table 1, entry 5).

Based on these optimized deprotection conditions, various aromatic and aliphatic carbamates were subjected to deprotection, as shown in Table 2. The method is discussed as an alternative chemoselective transformation for compounds accommodating acid-labile functionalities, and it has furnished corresponding free amines in good to excellent yields. Halogens, ketones, nitriles, and methoxy groups were not affected despite the reductive properties of indium metal. Ketal, however, is reported to have been cleaved. Selective deprotection of the Troc group has been discussed as triisopropylsilyl (TIPS), methoxyethoxymethyl (MEM), and *t*-butyldimethylsilyl (TBDMS) groups were resistant to these reaction conditions.

Table 1. The optimization of the reaction conditions[a]

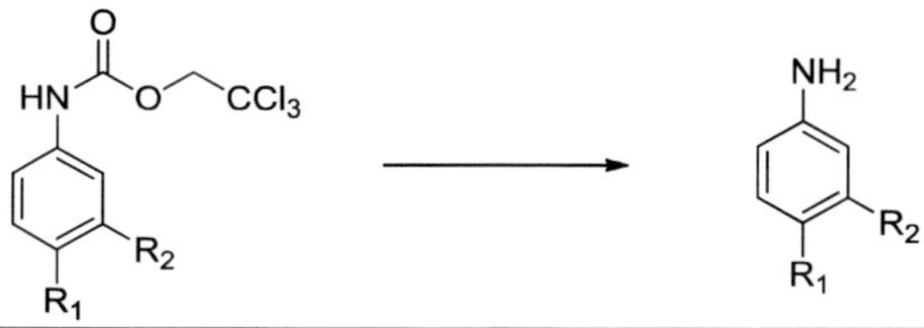

Entry	R_1	R_2	Solvent	In (eq.)	NH_4Cl (eq.)	Yield[b] (%)
1[c]	CN	H	MeOH/H_2O (5:1)	2	5	49
2	CN	H	MeOH/H_2O (5:1)	5	10	84
3	Ac	H	MeOH/H_2O (5:1)	5	10	22
4	H	Ac	MeOH/H_2O (5:1)	5	10	68
5	CN	H	EtOH/H_2O (3:2)	2	3	98

a. All reactions were performed at reflux. b. Isolated yields. c. Starting material (33%) was also recovered.

1-1-2. Synthesis of Carbamates as Selective Protection

Organic carbamates not only play an important role in synthetic chemistry but also are often the components of biologically and pharmacologically active compounds. Therefore, simple and efficient synthetic methods for the preparation of carbamates are valuable. Recent developments involving chemical reactions using catalytic amounts of indium metal have shown unique properties such as negligible toxicity and inertness against air and water, and have introduced synthetic methods to use carbamates in selective protection [8]. A number of proceeding reactions previously existed for the formation of carbamates [9,10], but a simple, efficient, and selective method for preparation of carbamates using a catalytic amount of indium has now been introduced. The reaction was first examined using 1.0 equiv of cyclohexylamine and 1.0 equiv of methyl chloroformate as the starting material, and it was possible to reduce the amount of indium to 0.1 equiv as the catalyst (Table 3). A structurally wide range of amines were then transformed to carbamates under optimal reaction conditions, as shown in Table 4. The applicability was verified for both aliphatic and aromatic amines, displaying good to high isolated yields. Also, the efficiency of the reaction for both primary and secondary amines was emphasized. However, an aromatic amine possessing a strong electron-withdrawing functionality, such as a nitro group, gave a lower yield (Table 4, entry 5). The reaction proceeded with not only methyl chloroformate but also with the chloroformates of other alkyl groups, such as vinyl and benzyl.

Table 2. Representative entries for the deprotection reactions of carbamates[a]

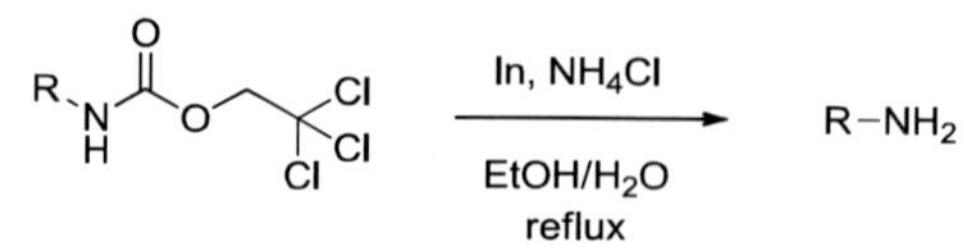

Entry	Starting Material	Product	Time (h)	Yield[b] (%)
1	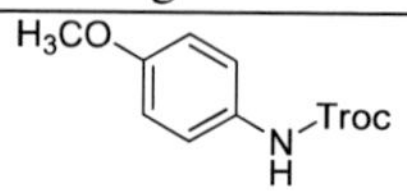	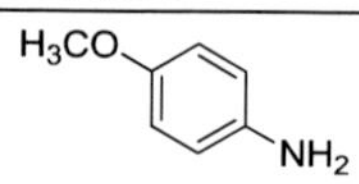	4	83
2	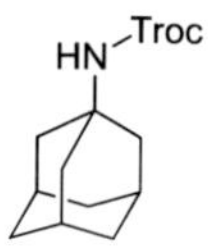	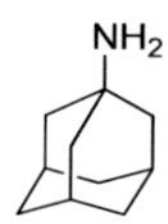	6	91
3	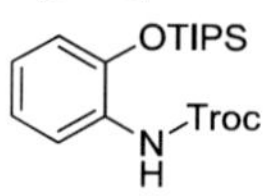	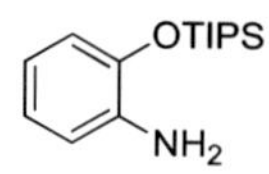	1.5	92
4	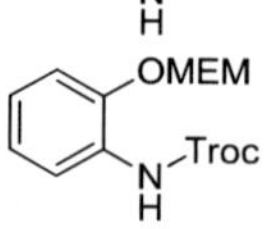	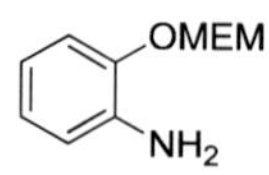	2	91
5	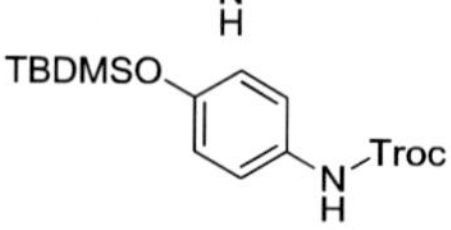	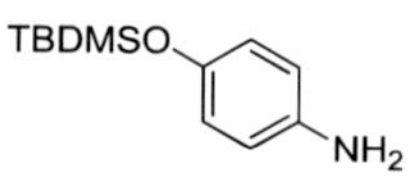	1.5	93

a. All reactions were performed at reflux. b. Isolated yields.

1-1-3. Facile Cleavage of the t-Butoxycarbonyl Group from di-t-Butylimidodicarbonate

Among the many protecting groups currently developed, the *t*-butoxycarbonyl (Boc) group is one of the most commonly used for amines. Generally, the Boc group shows chemical stability to basic conditions, and removal of the Boc group is conducted in acidic conditions. Often, di-*t*-butylimidodicarbonates, which are the primary amines being protected by the 2 Boc groups on the same nit-rogen atom, play an important role in organic synthesis. The Boc group is generally deprotected through organic or inorganic acids such as HCl [11], HBr [12], and trifluoroacetic acid [13]. However, these acids sometimes are incompatible with other acid-sensitive functionalities. Besides, despite the intention of eliminating one Boc group, occasionally both Boc groups are eliminated. Therefore, the methodology used to accomplish the facile cleavage of one Boc group from di-*t*-butylimidodicarbonate is quite valuable, and this was achieved through indium metal, along with zinc metal [14]. The removal of the Boc group from di-*t*-butylimidodicarbonate was effectively carried out by reflux in the presence of 2 equiv of indium powder or zinc powder using methanol as the solvent. The reaction unmasked only one Boc group, leaving the other Boc group intact, as shown in Table 5. Mostly, the yields were slightly higher for conversions using indium metal than for those using zinc metal. The reaction conditions were almost neutral as the pH of the reaction media was approximately 7.2. The reaction conditions were mild, and it was advantageous that the acid labile protecting groups such as tetrahydropyran (THP), TBDMS, and *t*-butyldiphenylsilyl (TBDPS) groups, as well as sulfonamide, were not affected.

Table 3. Reaction of cyclohexylamine with methyl chloroformate in the presence of indium

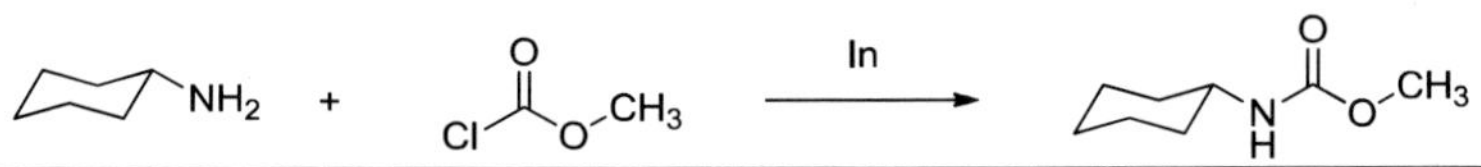

Entry	In (equiv)	Solvent	Temp.	Time (h)	Isolated Yield (%)
1	1.0	CH_3CN	r.t.	5	89
2	0.1	CH_3CN	r.t.	10	90
3	0.1	CH_3CN	reflux	3	92
5	0	CH_3CN	r.t.	10	40

Table 4. Representative entries for the synthesis of carbamates from amines

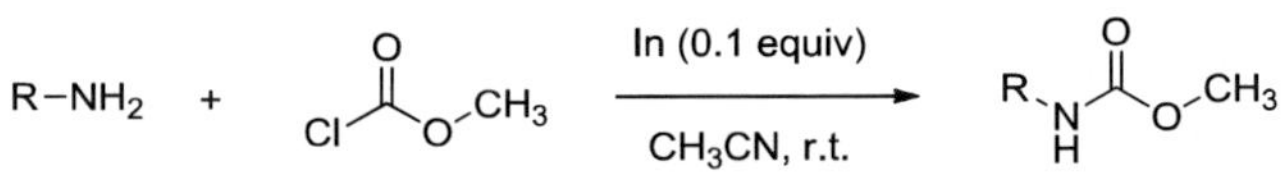

Entry	Starting Amine	Product	Time (h)	Isolated Yield (%)
1[a]	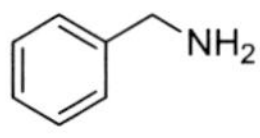	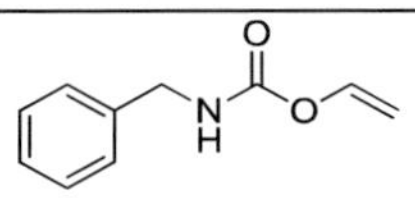	8	87
2[b]	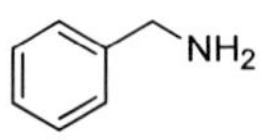	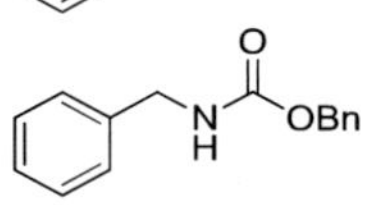	8	90
3	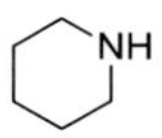	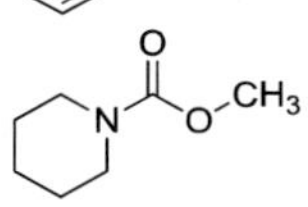	10	86
4	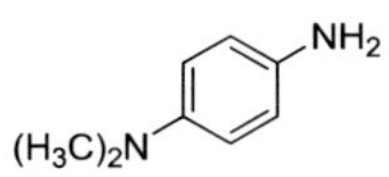	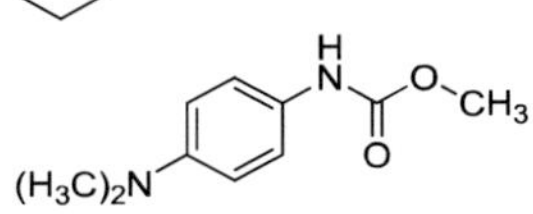	7	91
5	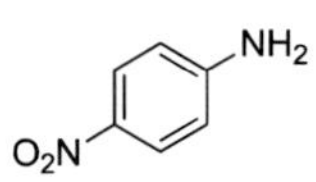	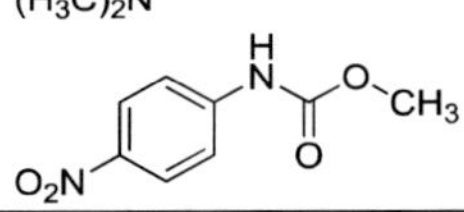	12	77

a. Vinyl chloroformate was used. b. Benzyl chloroformate was used.

Table 5. Representative entries for indium- and zinc-mediated conversion

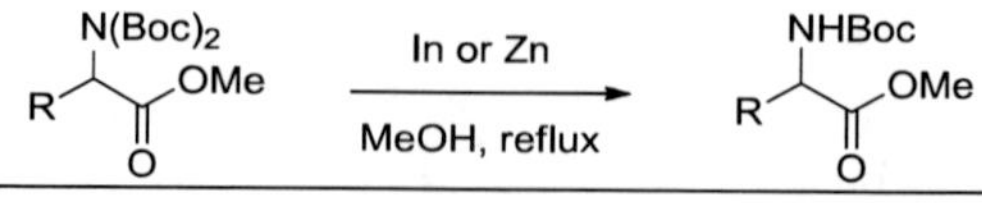

Entry	Substrate	Product	Zinc		Indium	
			Time (h)	Yield (%)	Time (h)	Yield (%)
1	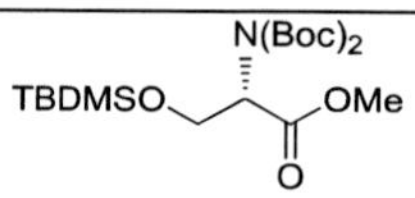	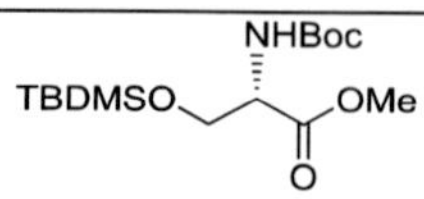	24	87	24	90
2	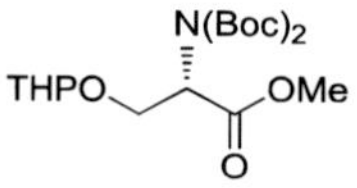	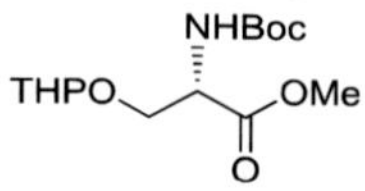	21	80	22	83
3	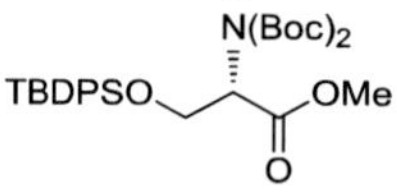	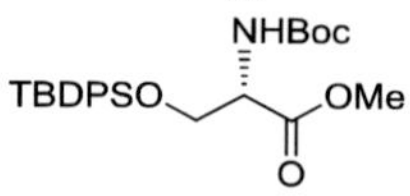	20	89	18	87
4[a]	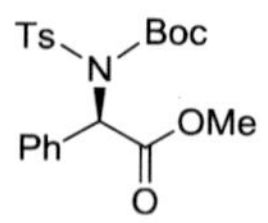	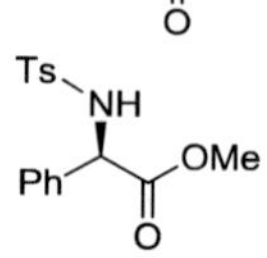	18	87	20	90

a. The substrate is protected by the tosyl and Boc groups.

1-2. Reactions of Alcohols

1-2-1. Deprotection of Troc and Trichloroacetyl Groups

The hydroxyl group is also a functional group that occasionally requires protection during the processes of multi-step and total syntheses. Although Troc and trichloroactyl groups have been widely used as the protecting groups, sufficient deprotection methods were not available for their removal. In order to overcome the drawbacks of those traditional methods [15-21], a new, mild, and chemoselective method for the removal of Troc and trichloroactyl groups using indium metal has been developed [22]. The general applicability of this methodology was determined starting with various substrates of trichloroethoxycarbonates and trichloroacetates that possess a diverse functionality. When the protected alcohols were subjected at reflux in the presence of 2 equiv of indium powder and aqueous ammonium chloride in methanol, the intended products of the corresponding free alcohols were given in excellent yields, as shown in Table 6. Neither chiral racemization nor the migration of silyl groups was observed. The deprotection reaction effectively proceeded without affecting the olefin, alkyne, alkoxy, or halogen moieties. In addition, a high chemoselectivity for the reaction conditions was demonstrated as the other protecting groups such as benzoyl (Bz), TBDPS, ethyl ester, and amide were stable under these conditions.

1-2-2. Synthesis of Carbonates from Alcohols

The reaction introduced in section ***1-1-2*** was applicable to the synthesis of carbonates from alcohols, and the result was presented in the same report [8]. Being similar to the case of carbamates, there are a couple of methods for preparing carbonates [23,24], which include reactions employing isocyanates [10,25] and carbon dioxide [26-29]. Although the reaction had to be conducted in a boiling solvent, a stoichiometric amount of indium was unnecessary. The synthesis of carbonates from alcohols was explained using 1.0 equiv of alcohols and 1.5 equiv of chloroformate as the starting materials, and the reaction mixture was heated at reflux in the presence of 0.1 equiv of indium. Structurally diverse alcohols were efficiently transformed into the corresponding carbonates, as listed in Table 7. Similar to the case of 4-nitroaniline (Table 4, entry 5), the use of 4-nitorophenol resulted in a lower yield (Table 7, entry 3). When tertiary alcohol was the starting substrate, the reaction did not proceed to completion and ended with only 41% yield despite an extended reaction time, due to the untoward effect of the steric bulkiness on the rate of the reaction (Table 7, entry 4). Consequently, the chemoselectivity of the method was examined by applying these reaction conditions to several amino alcohols (Table 8). The selective protection on amino functional groups in high yields was solely observed without the formation of carbonates or cyclization.

Table 6. Representative entries for the deprotection of Troc and trichloroactyl groups

$$\text{R-OP} \xrightarrow[\text{MeOH, reflux}]{\text{In, aq. } NH_4Cl} \text{R-OH} \qquad \text{P} = Cl_3CH_2OCO\text{- or } Cl_3CCO\text{-}$$

Entry	Starting Material	Product	Time[a] (h)	Yield[a] (%)	Time[b] (h)	Yield[b] (%)
1	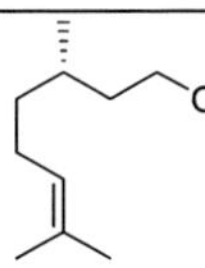	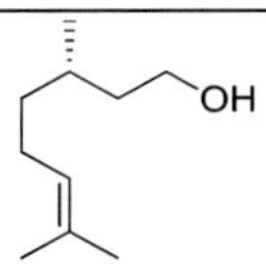	1	98	0.5	95
2	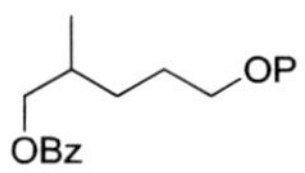	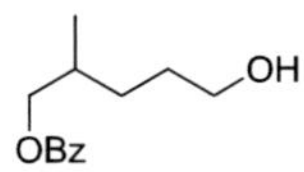	1.5	94	0.5	98
3	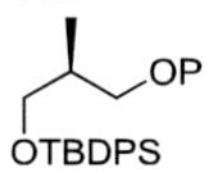	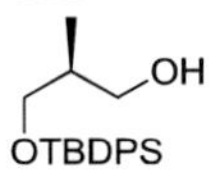	1	95	0.5	94
4	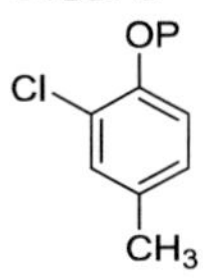	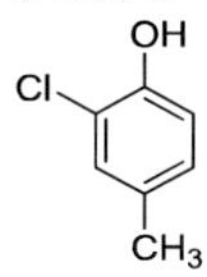	1.5	84	1	82
5	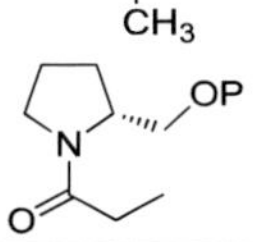	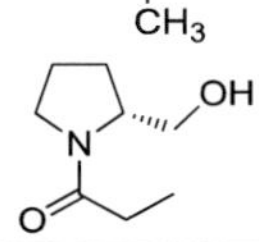	3	82	1.5	85

a. Troc deprotection. b. Trichloroacetyl deprotection.

1-2-3. Synthesis of Chiral Allylic Alchohols Using Sonication

Chiral allylic alcohols often serve as versatile intermediates during the course of multi-synthetic steps. Various methods are used for the synthesis of chiral allylic alcohols: chemical reactions through kinetic resolution [30], asymmetric reduction [31,32], enantioselective addition [33], and reductive elimination [34-38]. However, due to the drawbacks derived from reagents or conditions, these methods have often resulted in unsatisfactory yields with insufficient selectivity. A novel and practical method for the synthesis of chiral allylic alcohols using indium metal has recently been introduced [39]. Upon treating 2-iodomethyl-

O-isopropylidine acetals with 2 equiv of indium metal in methanol under sonication, reductive elimination smoothly furnished the corresponding chiral allylic alcohols in an efficient manner, and the general reaction path is illustrated in Figure 1. A couple of substrates were subjected to this transformation, and some examples are listed in Table 9. The compatibility with sensitive functional groups — acetal, ester, benzyloxy, and TBDMS groups — was confirmed. Additionally, the effectiveness of indium metal, in terms of yields and reaction time, is reportedly an improvement over other metals such as zinc, samarium, and tin. As for the use of sonication, the sonic waves could cut down the reaction time.

Table 7. Representative entries for the synthesis of carbonates from alcohols

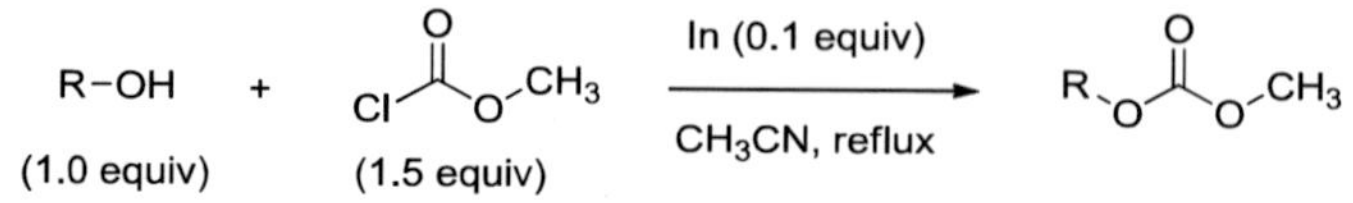

Entry	Substrate	Product	Time (h)	Isolated Yield (%)
1	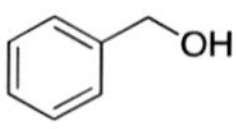	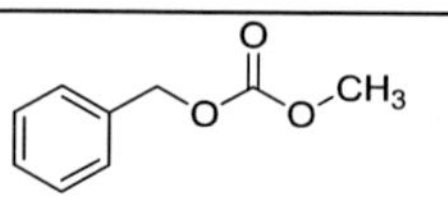	12	80
2	$CH_3(CH_2)_6CH_2OH$	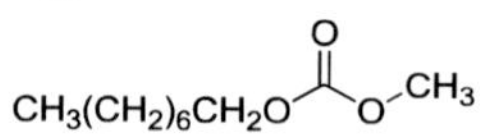	10	85
3	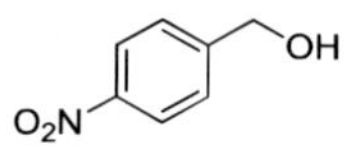	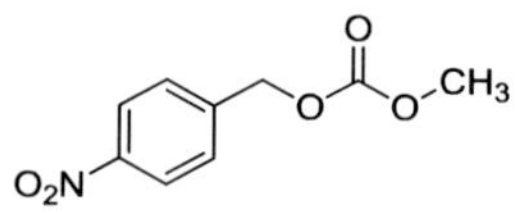	16	72
4	^{t}Bu–OH	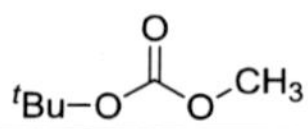	24	41

Table 8. Representative entries for the chemoselective protection of amino groups at room temperature in the presence of indium (0.1 equiv)

Entry	Starting Substrate	Product	Time (h)	Isolated Yield (%)
1	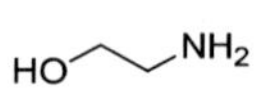	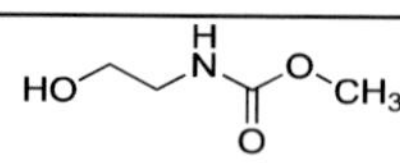	10	87
2	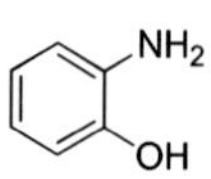	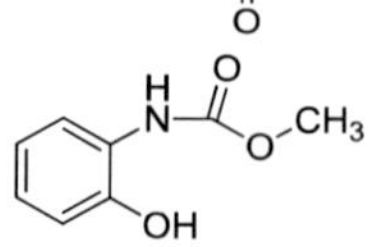	10	89

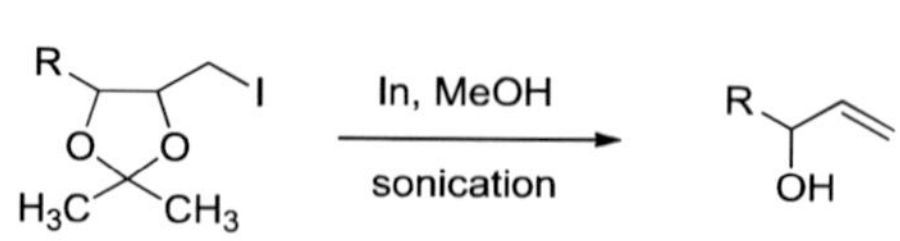

Figure 1. General reaction path for the synthesis of chiral allylic alcohols.

Table 9. Examples of the efficient synthesis of chiral allylic alcohols

Entry	Substrate	Product	Time (h)	Yield[a] (%)
1	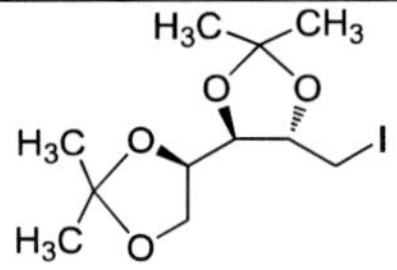	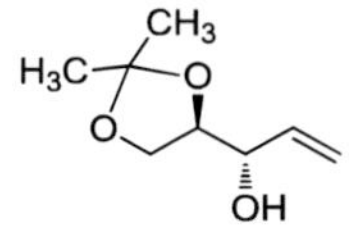	5.5	92
2	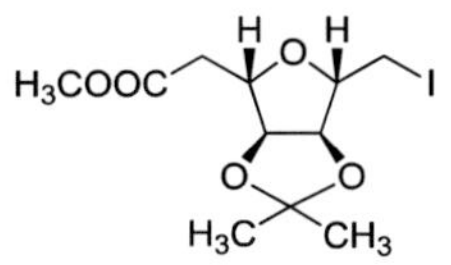	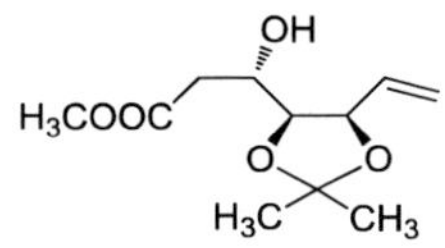	6.0	86
3	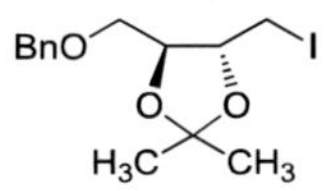	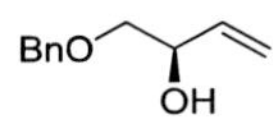	6.5	94
4	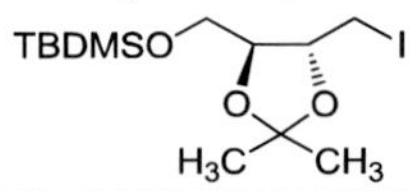	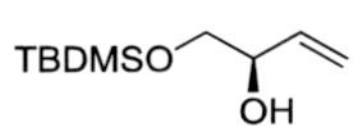	6.0	87

a. Isolated and unoptimized yields.

1-2-4. Deprotection of 4-Nitrobenzyl Ethers and Esters

The 4-nitorobenzyl protecting group has been used in the protection of functionalities due to its resistance to acidic conditions. For example, 4-nitorobenzyl esters are more stable to acidic hydrolysis than the other benzyl esters and have been frequently utilized in the field of organic synthesis. For the deprotection of a 4-nitorobenzyl group, the substrate is simply heated in the presence of indium powder and ammonium chloride in aqueous ethanol [40]. The indium metal works as an agent to reduce the nitro groups [41], then the concomitant cleavage of the benzylic C-O bond occurs to regenerate the free alcohols with 4-toluidine as the by-product. As seen in Table 10, which shows some of the report results, all substrates were cleaved in good yields. The selectivity is demonstrated as the benzyl ethers remain intact, suggesting the further utility of indium metal in protecting group manipulation.

Table 10. Deprotection of 4-nitorobenzyl ethers and esters

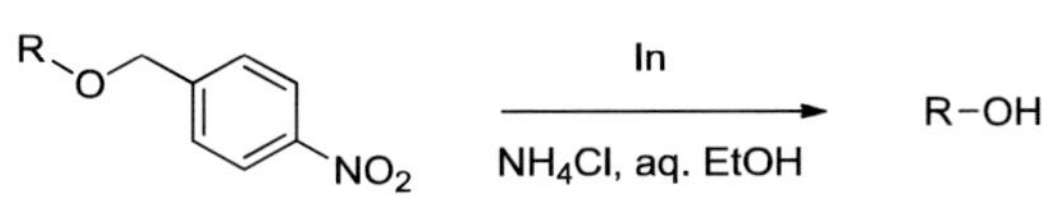

Entry	R	Yield (%)
1	$PhCH_2CH_2$	87
2	$PhOCH_2CH_2$	98
3	$PhCH_2OCH_2CH_2$	100

1-3. Reactions of Carboxylic Acids

1-3-1. Deprotection of Trichloroethyl Esters

Likewise, in the case of amines and alcohols, the 2,2,2-trichloroethyl moiety often serves as a convenient masking unit for carboxylic acids as well as for phosphorous compounds. The trichloroethyl moiety can be routinely cleaved from its esters using zinc/acetic acid [42], electrolysis [43], samarium(II) iodide [44], selenium/sodium borohydride [45], and cadmium/acetic acid [46]. An alternative indium-mediated methodology for the removal of a trichloroethyl group under mild conditions has been developed [47]. Optimal reaction conditions were initially explored using trichloroethyl benzoate, and indium metal was determined to be sufficiently effective to pursue the deprotecting process. The reaction mixture was heated at reflux in the presence of 2 equiv of indium powder and 5 equiv of ammonium chloride with tetrahydrofuran/water (10:1) as the solvent to achieve an exclusive cleavage to benzoic acid. A variety of aromatic and aliphatic carboxylates were then treated under developed reaction conditions to furnish the corresponding carboxylic acids in good to excellent yields. Table 11 lists the results of the deprotection. In both cases involving the electron-withdrawing and electron-donating functionalities of benzoate derivatives, the substrates were smoothly deprotected to the corresponding acids. A wide range for the applicability of the deprotecting reaction was indicated for both aromatic and aliphatic esters.

Discussion must include the chemical structure, in particular the precise length of the tethered alkyl chain. As shown in Table 12, as for the reaction using indium metal that was started with trichloroethyl carboxylates with a benzylic methylene moiety such as phenyl acetate and ω-phenylalkanoates, the structural outcome of the product is different, and results in an exclusive formation of 2,2-dichloroethyl esters with little amount of the deprotected forms. Meanwhile, in the case of carboxylates having an elongated tethered alkyl chain such as 4-phenyl butyrate (Table 12, entry 3) and 6-phenylhexanoate (Table 12, entry 4), when subjected to the reaction with indium metal, a mixture of 2,2-dichloroethyl esters and free carboxylic acids was generated. There are reports of dichloromethyl radicals and dichloromethyl anions through cleavage and reductive demonochlorination, as postulated from conventional methods. The reductive stability of radicals/anions can be a factor, as with the previous report on the dichotomy between radical and anionic dechlorination [48].

As before, the deprotection reaction has been incomplete for the substrates wherein a benzylic methylene moiety exists; therefore, it was a continuous challenge to overcome this shortcoming. Much thoughtful consideration of this limitation led to the discovery of a new reaction condition that could successfully cleave trichloroethyl esters with a benzylic methylene moiety and furnish the deprotected forms of the corresponding free carboxylic acids [49]. While reconsidering the reductive stability of radicals/anions [47,48,50], an alteration in the deuterated solvents was envisioned. When dissociation of a C-H bond is involved in the demonochlorination of trichloroethyl esters, the rate of the reaction can be slowed by replacing tetrahydrofuran with tetrahydrofuran-d_8, due to the kinetic isotope effect of the C-D bond [51,52]. A couple of substrates of trichloroethyl esters having a benzylic methylene unit as well as representative functional groups have been heated at reflux in the presence of 2 equiv of indium powder, using 5 equiv of deuterated ammonium chloride and tetrahydrofuran-d_8/deuterium oxide (10:1) as the solvent. The deprotecting condition is mild, and shows dramatic improvement while effectively forming the corresponding carboxylic acids. The results are summarized in Table 13. The electron-donating groups likely aided the high yields of carboxylates, while electron-withdrawing groups likely promoted the medium yields. Halogen and silyl groups were resistant to the reaction conditions. Although a small amount of 2-deuterio-2,2-dichloroethyl esters were formed, the deprotection of trichloroethyl esters then became a simpler process.

Table 11. Representative entries for the deprotection of trichloroethyl esters

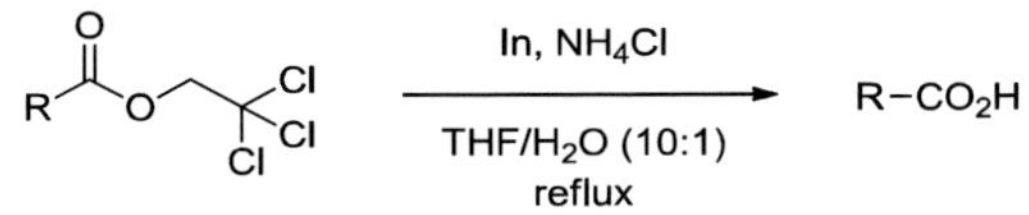

Entry	Substrate	Product	Time (h)	Isolated Yield (%)
1	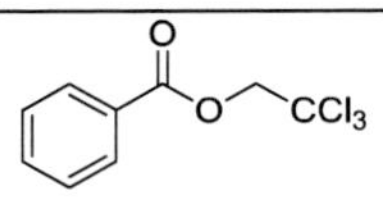	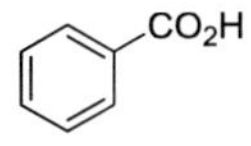	7	92
2	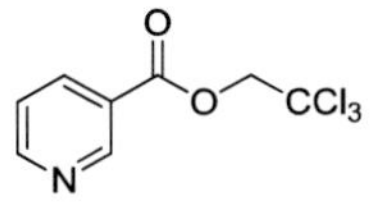	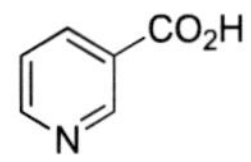	11	95
3	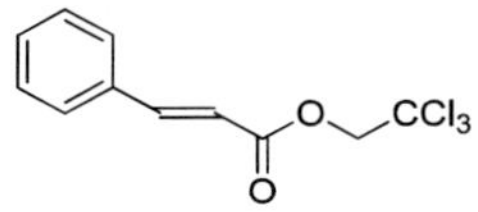	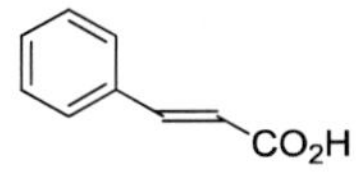	15	76
4	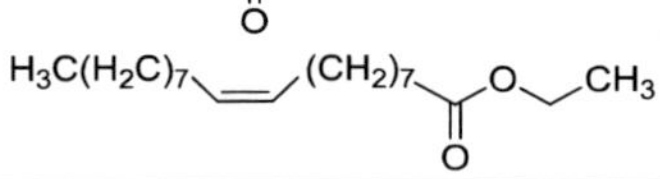	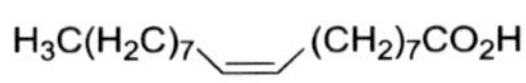	48	88

Table 12. Reductive demonochlorination by indium-mediated reactions

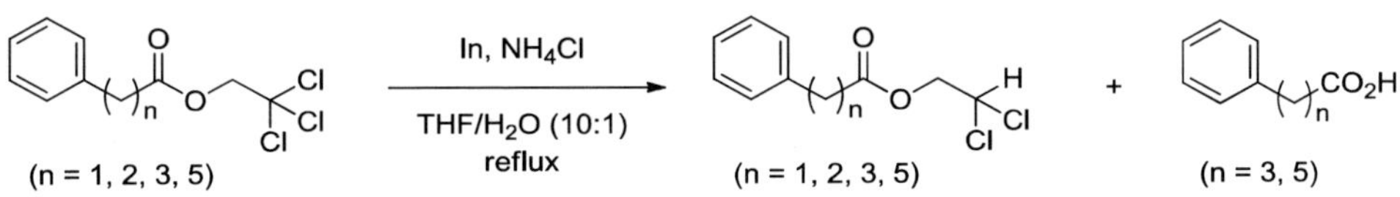

Entry	Substrate	Product	Time (h)	Isolated Yield (%)
1	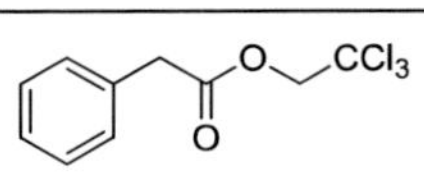	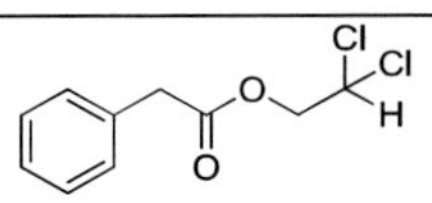	20	54
2	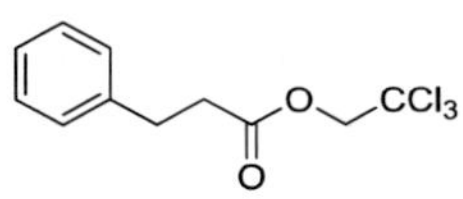	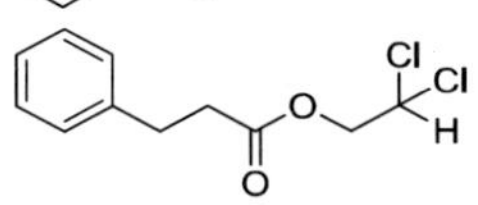	18	85
3	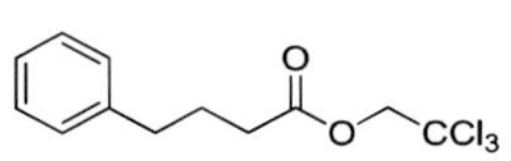	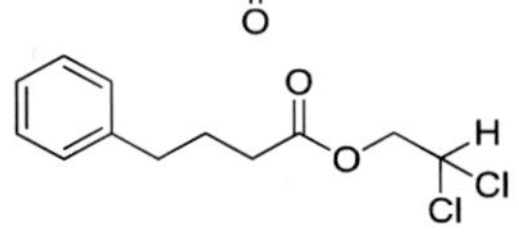	24	57 (+27)[a]
4	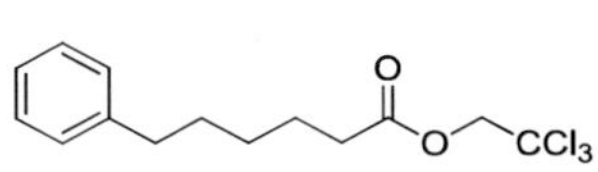	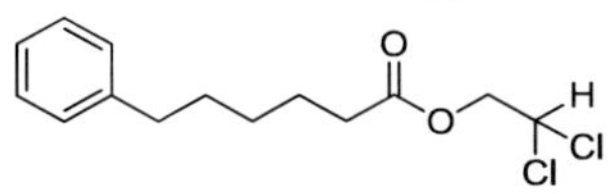	24	60 (+27)[a]

a. In parentheses are the yields of the corresponding free carboxylic acids.

Table 13. Reaction using THF-d_8/D_2O (10:1) as the solvent

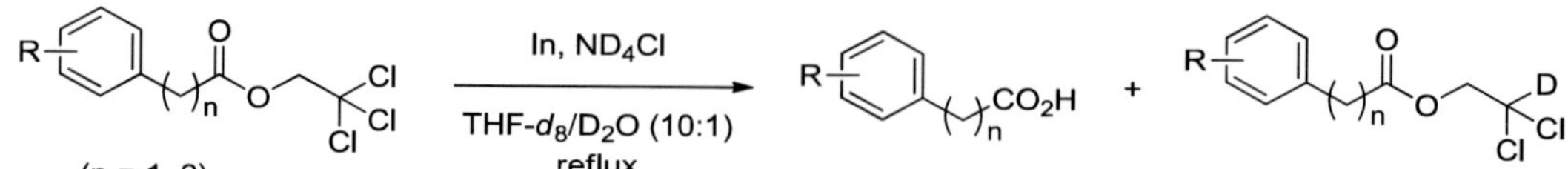

Entry	Substrate	Products and Yields[a] (%)				Time (h)
1	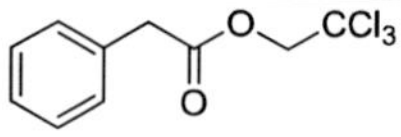	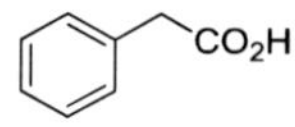	80	-	ND	40
2	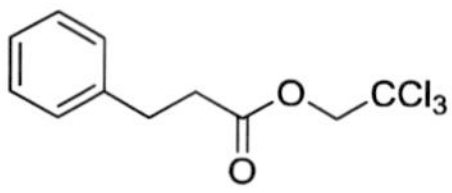	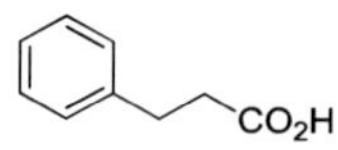	72	-	ND	40
3	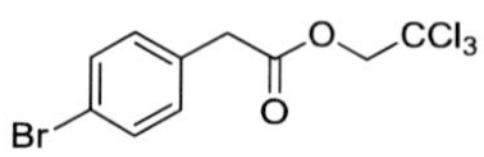	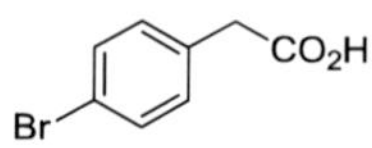	54	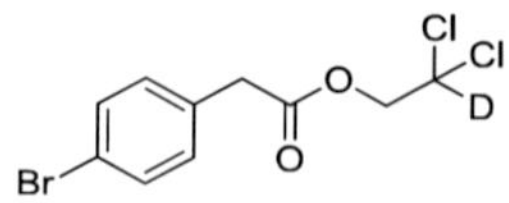	3	40
4	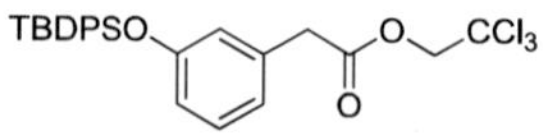	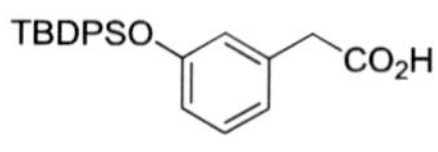	70	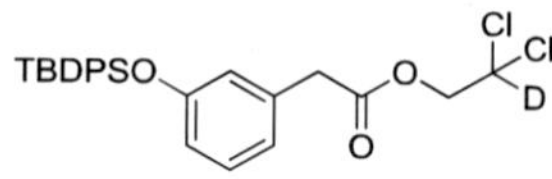	28	40
5	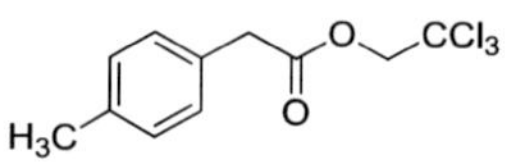	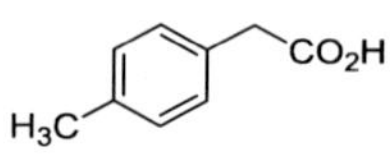	85	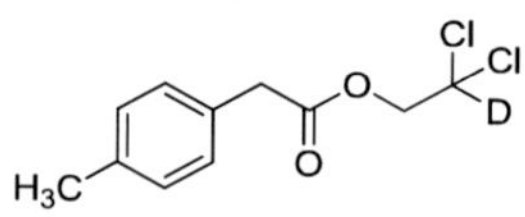	6	40

a. Isolated yields and the free carboxylic acids were indicated after silica gel separation.

2. Chemistry of Utilizing Trivalent Indium Reagents

Among indium reagents, trivalent indium salts have gained attention due to the moderate reactivity attained under mild conditions. The properties are quite unique and different from reagents of aluminum or boron in the same 13 group. The trivalent indium salts are useful as Lewis acids due to the stability in water and the compatibility with oxygen and nitrogen. There is increased interest in using them, particularly in catalytic processes developed utilizing their properties in organic synthesis.

This section begins with effective and useful chemical transformations, including the tetrahydropyranylation and depyranylation of alcohols, which are smoothly catalyzed by indium(III) triflate. Following are esterification methods using indium(III) halides, such as the high-yielding methyl esterification of carboxylic acids catalyzed by indium(III) chloride. The esterification of long-chain substrates is also exemplified by the effective usage of indium(III) chloride.

2-1. Trivalent Indium Salts of Triflate

2-1-1. The Tetrahydropyranylation and Depyranylation of Alcohols

THP is an attractive protecting group and is commonly used in order to protect the functionality of alcohols. THP-protected alcohols are normally stable under basic conditions and are compatible with metal halides, alkyl lithiums, Grignard reagents, and catalytic hydrogenation. Among the many known methods for the tetrahydropyranylation and depyranylation of alcohols, the most commonly used reagents that catalyze both transformations are *p*-toluenesulfonic acid (PTSA) [53,54], pyridinium *p*-toluenesulfonate (PPTS) [55], ammonium chloride [56], zirconium(IV) chloride [57], iodine [58,59], potassium dodecatangestocobaltate trihydrate [60], lithium bromide [61], acetonyltriphenylphosphonium bromide [62], and tetrabutylammonium tribromide [63]. Despite these precedents, and in order to improve their drawbacks, an efficient, mild, and practical method for these transformations is desired, and that was attained via the use of a trivalent indium reagent [64]. Indium(III) triflate can act as a Lewis acid and in recent years, its roles have been constantly elucidated in various synthetic transformations. It was discovered that indium(III) triflate could affect the tetrahydropyranylation and depyranylation of alcohols under different reaction conditions. The tetrahydropyranylation of various primary and secondary alcohols was carried out using a catalytic amount of 0.005 equiv of indium(III) triflate in dichloromethane at 0 °C, with stirring for 30-50 minutes (Figure 2, Eq. 1), and examples are listed in Table 14. The tetrahydropyranylation was successful over a range of functional groups, furnishing the protected alcohols in good yields. The method was investigated using alcohols that also possess other functional groups such as Bz, TBDPS, and TBDMS, in order to find successful THP protection and provide satisfactory yields without affecting the other protected moieties. However, the trichloroethanol derivative yielded the product in only 33% yield (Table 14, entry 2), which could possibly be ascribed to the inductive electron-withdrawing effect of the trichloromethyl moiety.

Subsequently, the depyranylation of the THP ethers was implemented to obtain the corresponding alcohols in good to excellent yields, using the same catalyst but a different molar ratio in aqueous methanol at room temperature (Figure 2, Eq. 2), as shown in Table 15. Compared to the amounts of indium(III) triflate required for protection, 0.05 equiv, or 10-fold, was required for deprotection reactions along with an extended reaction time for completion. Selective deprotection proceeded in the case of the TBDPS-protected alcohol unit, but did not for the TBDMS-protected unit. The TBDPS group would be the favored version to manipulate by protection and deprotection in this case.

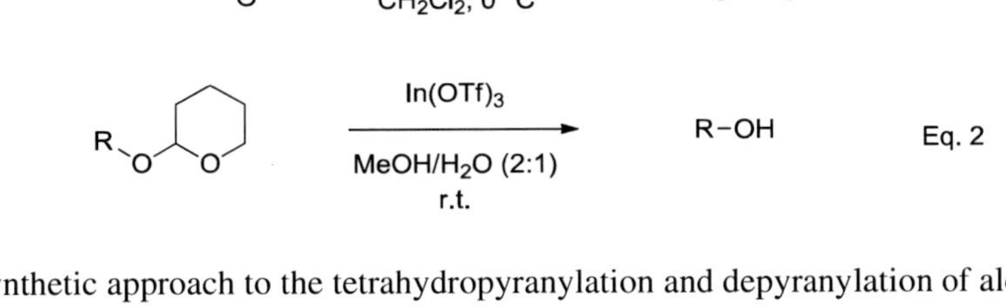

Figure 2. The synthetic approach to the tetrahydropyranylation and depyranylation of alcohols.

Table 14. Examples of the tetrahydropyranylation of alcohols[a]

Entry	Starting Material	Product	Time (min.)	Yield[b] (%)
1	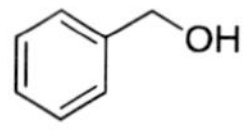	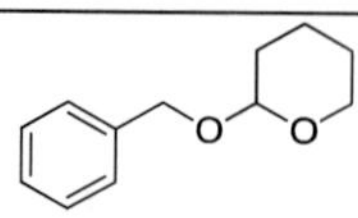	30	85
2	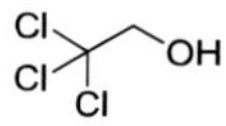	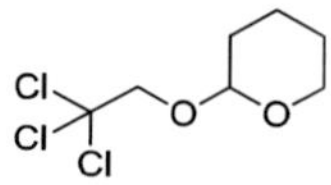	50	33
3	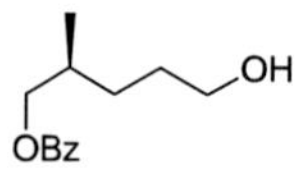	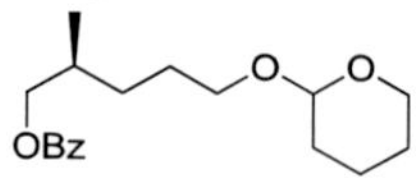	30	65
4	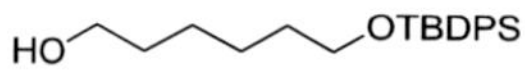	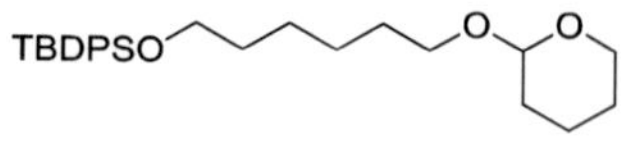	30	82
5	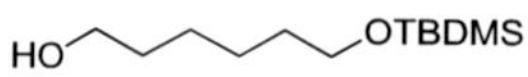	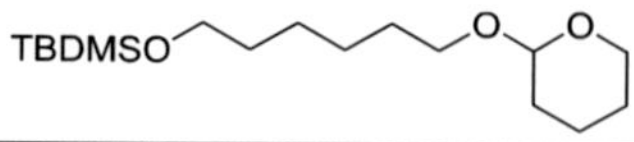	30	78

a. All reactions were performed at 0 °C using 0.005 equiv of $In(OTf)_3$. b. Isolated yields.

Table 15. Examples of the depyranylation of THP ethers[a]

Entry	Starting Material	Product	Time (h)	Yield[b] (%)
1	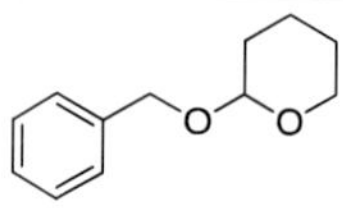	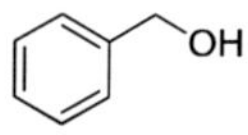	6	91
2	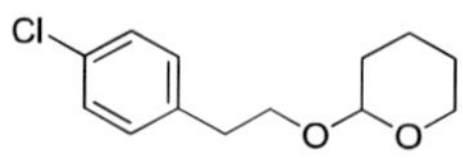	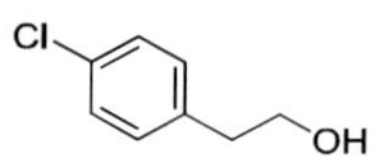	6	85
3	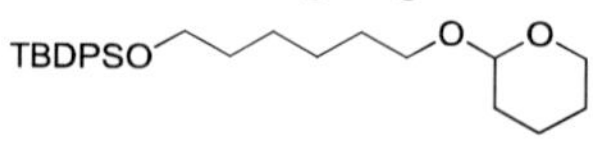	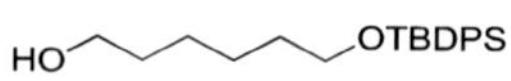	12	81
4	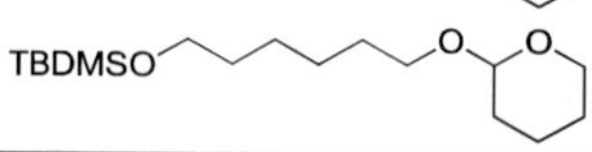	N. R.	-	-

a. All reactions were performed at r.t. using 0.05 equiv of $In(OTf)_3$. b. Isolated yields.

2-1-2. The Acetylation of THP Ethers

In addition to these results, it became apparent that indium(III) triflate promotes the acetylation of THP ethers in the presence of an acetic anhydride reactant (Figure 3, Eq. 1) [64]. The method reported is fast and carried out at room temperature and involves no special reaction conditions or reagents other than 0.05 equiv of indium(III) triflate. Besides, this one-step transformation enables a change from the acid-labile THP protecting group into a base-labile acetyl protecting group. A couple of representative examples of this efficient acetylation of THP ethers are illustrated in Table 16. Furthermore, based on similar reaction conditions using 0.05 equiv of indium(III) triflate, reactants other than acetic anhydride were investigated. The satisfactory results are explained here [65].

Earlier attempts to convert THP ethers into acetate were accomplished by heating the reaction mixture at reflux in the presence of a catalytic amount, 20 mol%, of indium(III) iodide which is generated *in situ* by the reaction of indium metal and iodine [66,67]. Although the reaction periods were longer and required air-sensitive reagents, the method introduced applied not only to the acetylation of THP ethers but also to the acetylation of methoxymethyl (MOM) ethers (Figure 4). In both cases, THP ethers and MOM ethers of primary alcohols were smoothly converted to acetates while those of secondary alcohols and phenols resulted in a simple deprotection to the parent hydroxyl compounds. The reaction conditions are tolerable to various functionalities including hydroxyl-protecting TBDMS and benzyl groups, and are very clean and high-yielding. Other than trivalent indium reagents, a catalytic amount of copper(II) triflate was introduced for the conversion of THP ethers into the corresponding acetates [68].

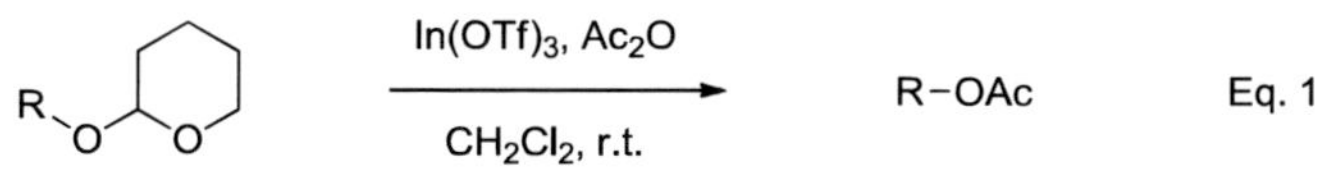

Figure 3. A synthetic approach to the acetylation of THP ethers.

Table 16. Examples of the acetylation of THP ethers[a]

Entry	Starting Material	Product	Time (h)	Yield[b] (%)
1	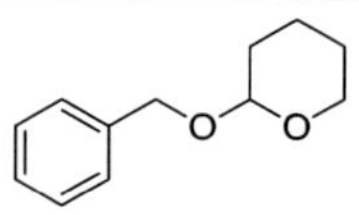	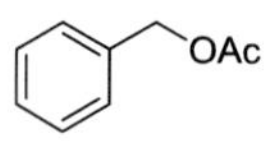	1	93
2	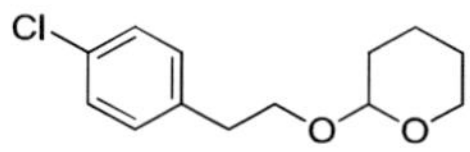	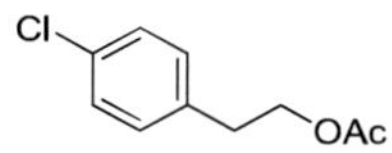	2	90
3	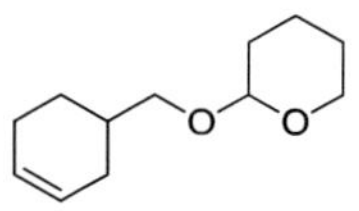	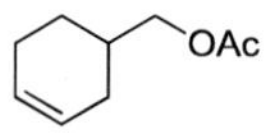	1	85

a. All reactions were performed at r.t. using 0.05 equiv of $In(OTf)_3$. b. Isolated yields.

$$RCH_2OTHP \text{ and } RCH_2OMOM \xrightarrow[\text{EtOAc, reflux}]{In,\ I_2} RCH_2OAc$$

Figure 4. The acetylation of THP ethers and MOM ethers.

2-2. Trivalent Indium Salts of Halogens

2-2-1. The Methyl Esterification Reaction

Among many chemical reactions, methyl esterification is one of the most basic, and a number of methods have been proposed. The most general method is the treatment of *N*-methyl-*N*-nitrosourea under basic conditions in order to generate diazomethane *in situ* [69]. However, this procedure may not be satisfactorily secure because of the carcinogenic properties. The methyl esterification of carboxylic acids proceeded in methanol in the presence of a catalytic amount of indium(III) chloride [70]. The direct condensation of carboxylic acids with alcohols using a catalyst from the group 4 transition elements has also been reported [71].

Unlike the reaction conditions of a methanol/trimethylsilyl chloride combination that releases a proton during the course *in situ*, indium(III) chloride can serve as an efficient and mild Lewis acid and does not release a proton. An array of carboxylic acids in methanol was treated with 0.2 equiv of indium(III) chloride at reflux, or under sonication conditions at room temperature, and gave a good to excellent yield of the corresponding methyl esters. The general applicability of this method was evaluated using various compounds with diverse substituents or protecting groups, and the method worked for both aliphatic and aromatic carboxylic acids.

The reaction conditions were very mild, and functionalities such as the 9-fluorenylmethoxycarbonyl (Fmoc) group and the olefin unit of α,β-unsaturated carboxylates were not affected (Table 17). In addition, it was reported afterward that esterification proceeded either via sonication at room temperature or with a catalytic amount of indium(III) chloride tetrahydrate, instead of indium(III) chloride, to complete well in the presence of water molecules.

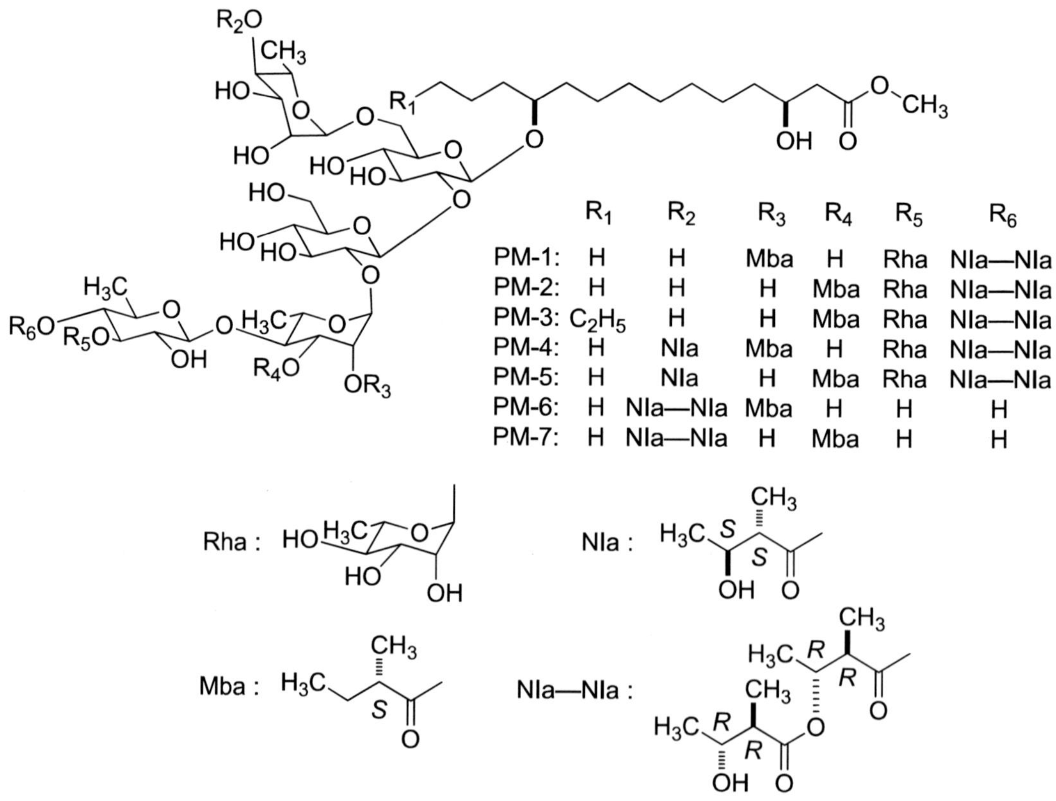

	R_1	R_2	R_3	R_4	R_5	R_6
PM-1:	H	H	Mba	H	Rha	Nla—Nla
PM-2:	H	H	H	Mba	Rha	Nla—Nla
PM-3:	C_2H_5	H	H	Mba	Rha	Nla—Nla
PM-4:	H	Nla	Mba	H	Rha	Nla—Nla
PM-5:	H	Nla	H	Mba	Rha	Nla—Nla
PM-6:	H	Nla—Nla	Mba	H	H	H
PM-7:	H	Nla—Nla	H	Mba	H	H

Figure 5. Structures of PM-1 to PM-7.

Table 17. Examples of methyl esterification[a]

$$R{-}CO_2H \xrightarrow[\text{MeOH}]{InCl_3 \text{ or } InCl_3{\cdot}4H_2O} R{-}CO_2Me$$

Entry	Substrate	Product	Catalyst	Temp.	Time (h)	Yield (%)
1	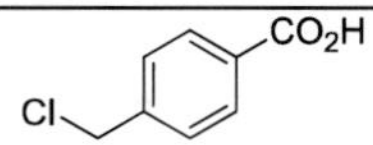	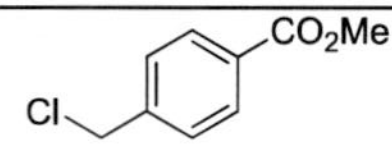	$InCl_3$	reflux	40	85[b]
2	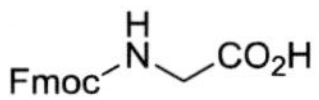	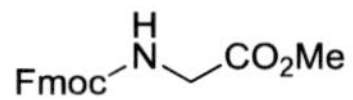	$InCl_3$	reflux	40	96
3	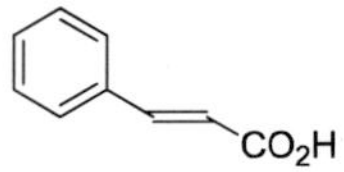	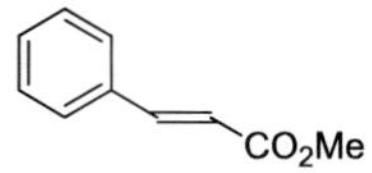	$InCl_3$	reflux	48	93
4	$CH_3(CH_2)_7CH{=}CH(CH_2)_7CO_2H$	$CH_3(CH_2)_7CH{=}CH(CH_2)_7CO_2Me$	$InCl_3$	r.t.[c]	18	86
5	$CH_3(CH_2)_7CH{=}CH(CH_2)_7CO_2H$	$CH_3(CH_2)_7CH{=}CH(CH_2)_7CO_2Me$	$InCl_3{\cdot}4H_2O$	reflux	21	98

a. All reactions were performed in MeOH using 0.2 equiv of $InCl_3$. b. Side product of methyl 4-(methoxymethyl)benzoate was also formed in 12% yields. c. The reaction was carried out employing sonication at room temperature.

Table 18. Esterification of palmitic acid with cetyl alcohol catalyzed by metal chloride

Entry	Metal salt	Yield (%)
1	$GaCl_3$	97.5
2	$InCl_3$	98.6
3	$ZnCl_2$	81.5

2-2-2. The Esterification of Long-chain Substrates

Esters with long chains are indispensable components, particularly in the fields of food, cosmetics, and agriculture industry. However, conventional long chain acids and long chain alcohols have been treated by esterification reactions catalyzed by corrosive mineral acids. There is a need to find a replacement catalyst, which is environmentally green, mild, and safe, but that can actively produce esters with long chains. Meanwhile, an investigation into the catalytic performance of some multi-valent metal salts in the esterification of saturated and unsaturated long-chain acids and long-chain alcohols has been reported [72]. Esterification of palmitic acid (hexadecanoic acid) with cetyl alcohol (1-hexadecanol) was observed employing metal chloride as the catalyst. Table 18 shows part of the summary for the report, and among the catalysts, indium(III) chloride showed a particularly high catalytic performance during the esterification, and resulted in a very high yield for cetyl palmitate. These results suggest that indium(III) chloride can be a highly effective catalyst not only for the production of cetyl palmitate but also for the synthesis of other esters with long chains.

3. Application of Indium Chemistry to Glycoside Chemistry

In general, glycoside can be isolated from a glycoside mixture via HPLC analysis [73-75]. This separation by HPLC analysis is popular in most laboratories. The advantages of HPLC analysis over other chromatographic procedures include sensitivity, rapidity, and purity. When an RI detector is coupled with the HPLC system, the HPLC system supplies an ideal procedure for separation of the glycoside mixtures in plants or animals. Normal HPLC peaks commonly show a sharp and symmetrical shape. However, if natural glycosides possess a carboxyl group, HPLC peaks mainly show a peak tailing or broadening. The phenomenon of the peak tailing or broadening indicates troublesome treatments in the separation of glycoside mixtures. It is our understanding that the methyl esterification of a carboxyl group is an available and effective method for solving this problem. Among conventional methods of methyl esterification, the most common is the use of diazomethane, whereby a solution of resin glycoside mixtures in methanol is treated with diazomethane at room temperature. However, in many cases, the outcome is that the reaction with diazomethane has not reached expectations, particularly with regard to yield [76]. This is because methyl esterification with diazomethane commonly uses a low polar solvent such as ether, dichloromethane, or chloroform, whereas resin glycoside mixtures use a high polar solvent like methanol in consideration of the solubility of the glycoside mixtures. Therefore, upon treatment with diazomethane, the use of a high polar solvent was not ideal and was predicted to produce a low yield. By comparison, when the glycoside mixtures were treated with indium(III) chloride in methanol, following the preceding reported procedure [70], a high yield could be envisioned. And in fact, the application to resin glycoside mixtures produced a high yield, as expected. Consequently, 7 resin glycoside methyl esters (PM-1—PM-7) were isolated from the resin glycoside mixtures of methyl esterification (Figure 5) [77]. Indium(III) chloride successfully catalyzed methyl esterification and is thus considered substantially useful for the investigation of glycosides possessing a glucuronic acid as a sugar unit.

Conclusion

This chapter covers recent developments in synthetic chemistry and focuses on indium and indium-related reagents. A variety of applications were explored and thoroughly elucidated, and a number of recent reports were cited concerning the progress of indium chemistry. However, some areas of indium chemistry remain unclear and require further investigation. This chapter provides useful, effective, and selective chemical alternatives as modern methodologies, particularly for functional group transformations. In addition, in the latter, an application of indium chemistry to glycoside chemistry was presented. It is our expectation that further extensive research on indium metal will proceed, and that this chapter may serve as a guide.

References

[1] G. Schwedt. *CLB Chemie in Labor und Biotechnik* 2009, *60*, 256-259.
[2] G. Just; K. Grozinger. *Synthesis* 1976, 457-458.
[3] L. V. Hijfte; R. D. Little. *J. Org. Chem.* 1985, *50*, 3940-3942.
[4] M. F. Semmelhack; G. E. Heinsohn. *J. Am. Chem. Soc.* 1972, *94*, 5139-5140.
[5] L. E. Overman; R. L. Freerks. *J. Org. Chem.* 1981, *46*, 2833-2835.
[6] G. Hancock; I. J. Galpin; B. A. Morgan. *Tetrahedron Lett.* 1982, *23*, 249-252.
[7] T. Mineno; S.-R. Choi; M. A. Avery. *Synlett* 2002, 883-886.
[8] J.-G. Kim; D.-O. Jang. *Tetrahedron Lett.* 2009, *50*, 2688-2692.
[9] P. Adams; F. A. Baron. *Chem. Rev.* 1965, *65*, 567-602.
[10] S. Ozaki. *Chem. Rev.* 1972, *72*, 457-496.
[11] L. A. Carpino. *J. Org. Chem.* 1964, *29*, 2820-2824.
[12] R. D. Allan; G. A. R. Johnston; R. Kazlauskas; H. W. Tran. *J. Chem. Soc., Perkin Trans. 1* 1983, 2983-2985.
[13] A. S. Ripca; R. S. Bohacek; D. H. Rich. *Bioorganic & Med. Chem. Lett.* 1998, *8*, 357-360.
[14] J. S. Yadav; B. V. S. Reddy; K. S. Reddy; K. B. Reddy. *Tetrahedron Lett.* 2002, *43*, 1549-1551.
[15] T. B. Windholz; D. B. R. Johnston. *Tetrahedron Lett.* 1967, *27*, 2555-2558.
[16] M. Imoto; N. Kusunose; S. Kusumoto; T. Shiba. *Tetrahedron Lett.* 1988, *29*, 2227-2230.
[17] S. Hanessian; R. Roy. *Can. J. Chem.* 1988, *63*, 163-165.
[18] A. F. Cook. *J. Org. Chem.* 1968, *33*, 3589-3593.

[19] M. F. Semmelhack; G. E. Heinsohn. *J. Am. Chem. Soc.* 1972, *94*, 5139-5141.
[20] R. Yamada; N. Negoro; K. Bessho; K. Yanada. *Synlett* 1995, 1261-1263.
[21] V. Schwarz. *Collect. Czech. Chem. Commun.* 1962, *27*, 2567-2569.
[22] M. Valluri; T. Mineno; R. M. Hindupur; M. A. Avery. *Tetrahedron Lett.* 2001, *42*, 7153-7154.
[23] A.-A. G. Shaikh; S. Sivaram. *Chem. Rev.* 1996, *96*, 951-976.
[24] J. P. Parrish; R. N. Salvatore; K. W. Jung. *Tetrahedron* 2000, *56*, 8207-8237.
[25] T. E. Waldman; W. D. McGhee. *J. Chem. Soc., Chem. Commun.* 1994, 957-958.
[26] D. J. Darensbourg; S. A. Niezgoda; J. D. Draper; J. H. Reibenspies. *J. Am. Chem. Soc.* 1998, *120*, 4690-4698.
[27] T. Sakakura; Y. Saito; M. Okada; J.-C. Choi; T. Sato. *J. Org. Chem.* 1998, *63*, 7095-7096.
[28] M. Yoshida; N. Hara; S. Okuyama. *Chem. Commun.* 2000, 151-152.
[29] R. N. Salvatore; F. Chu; A. S. Nagle; E. A. Kapxhiu; R. M. Cross; K. W. Jung. *Tetrahedron* 2002, *58*, 3329-3347.
[30] V. S. Martin; S. C. Woodard; T. Katsuki; Y. Yamada; M. Ikeda; K. B. Sharpless. *J. Am. Chem. Soc.* 1981, *103*, 6237-6240.
[31] I. Ojima; N. Clos; B. Cecilia. *Tetrahedron* 1989, *45*, 6901-6939.
[32] C. Fuganti; P. Grasselli; S. Servi; F. Spreafico; C. Zirotti; P. Casati. *J. Chem. Res. Synop.* 1984, 112-113.
[33] W. Oppolzer; R. N. Radinov. *Tetrahedron Lett.* 1991, *32*, 5777-5780.
[34] L. C. Swallen; C. E. Boord. *J. Am. Chem. Soc.* 1930, *52*, 651-660.
[35] K. C. Nicolaou; M. E. Duggan; T. Ladduwahetty. *Tetrahedron Lett.* 1984, *25*, 2069-2072.
[36] S. Chandrasekhar; S. Mohapatra; M. Takhi. *Synlett* 1996, 759-760.
[37] S. Chandrasekhar; M. Takhi; J. S. Yadav. *Tetrahedron Lett.* 1995, *36*, 5071-5074.
[38] A. Furstner. *Tetrahedron Lett.* 1990, *31*, 3735-3738.
[39] J. S. Yadav; B. V. S. Reddy; K. S. Reddy. *Tetrahedron* 2003, *59*, 5333-5336.
[40] C. J. Moody; M. R. Pitts. *Synlett* 1999, 1575-1576.
[41] C. J. Moody; M. R. Pitts. *Synlett* 1998, 1028.
[42] R. B. Woodward; K. Heusler; J. Gosteli; P. Naegeli; w. Oppolzer; R. Ramage; S. Ranganathan; H. Vorbruggen. *J. Am. Chem. Soc.* 1966, *88*, 852-853.
[43] M. F. Semmelhack; G. E. Heinsohn. *J. Am. Chem. Soc.* 1972, *94*, 5139-5140.
[44] A. J. Pearson; K. Lee. *J. Org. Chem.* 1994, *59*, 2304-2313.
[45] Z.-Z. Huang; X.-J. Zhou. *Synthesis* 1989, 693-694.
[46] Y. Genisson; P. C. Tyler; R. N. Young. *J. Am. Chem. Soc.* 1994, *116*, 759-760.
[47] T. Mineno; H. Kansui; T. Kunieda. *Tetrahedron Lett.* 2007, *48*, 5027-5030.
[48] T. A. Mitchell; D. Romo. *Heterocycles* 2005, *66*, 627-637.
[49] T. Mineno; H. Hirayama; K. Nakahara; M. Yamashita; H. Kansui; H. Moriwaki. *Tetrahedron Lett.* 2010, *51*, 6045-6048.
[50] N. Mitsuo; T. Kunieda; T. Takizawa. *J. Org. Chem.* 1973, *38*, 2255-2257.
[51] R. A. Lynch; S. P. Vincenti; Y. T. Lin; L. D. Smucker; S. C. Subba Rao. *J. Am. Chem. Soc.* 1972, *94*, 8351-8356.
[52] K. B. Wiberg; L. H. Slaugh. *J. Am. Chem. Soc.* 1958, *80*, 3033-3039.
[53] K. F. Bernady; M. B. Floyd; J. F. Poletto; M. J. Weiss. *J. Org. Chem.* 1979, *44*, 1438-1446.

[54] E. J. Corey; H. Niwa; J. Knolle. *J. Am. Chem. Soc.* 1978, *100*, 1942-1943.
[55] M. Miyashita; A. Yoshikoshi; P. A. Grieco. *J. Org. Chem.* 1977, *42*, 3772-3774.
[56] J. S. Yadav; D. Srinivas; G. S. Reddy. *Synth. Commun.* 1998, *28*, 1399-1404.
[57] N. Rezai; F. A. Meybodi; P. Salehi. *Synth. Commun.* 2000, *30*, 1799-1805.
[58] N. Deka; J. C. Sarma. *Synth. Commun.* 2000, *30*, 4435-4441.
[59] H. M. S. Kumar; B. V. S. Reddy; E. J. Reddy; J. S. Yadav. *Chem. Lett.* 1999, 857-858.
[60] M. H. Habibi; S. Tangestaninejad; I. Mohammadpoor-Baltork; V. Mirkhani; B. Yadollahi. *Tetrahedron Lett.* 2001, *42*, 2851-2853.
[61] M. A. Reddy; L. R. Reddy; N. Bhanumathi; K. R. Rao. *Synth. Commun.* 2000, *30*, 4323-4328.
[62] Y.-S. Hon; C.-F. Lee. *Tetrahedron Lett.* 1999, *40*, 2389-2392.
[63] S. Naik; R. Gopinath; B. K. Patel. *Tetrahedron Lett.* 2001, *42*, 7679-7681.
[64] T. Mineno. *Tetrahedron Lett.* 2002, *43*, 7975-7978.
[65] T. Mineno; N. Nikaido; H. Kansui. *Chem. Pharm. Bull.* 2009, *57*, 1167-1170.
[66] B. C. Ranu; A. Hajra. *J. Chem. Soc., Perkin Trans. 1* 2001, 355-357.
[67] B. C. Ranu; A. Hajra. *J. Chem. Soc., Perkin Trans. 1* 2001, 2262-2265.
[68] K. L. Chandra; P. Saravanan; V. K. Singh. *Tetrahedron Lett.* 2001, *42*, 5309-5311.
[69] S. M. Hecht; J. W. Kozarich. *Tetrahedron Lett.* 1973, *14*, 1397-1400.
[70] T. Mineno; H. Kansui. *Chem. Pharm. Bull.* 2006, *54*, 918-919.
[71] K. Ishihara; S. Ohara; H. Yamamoto. *Science* 2000, *290*, 1140-1142.
[72] K. Mantri; R. Nakamura; K. Komura; Y. Sugi. *Chem. Lett.* 2005, *34*, 1502- 1503.
[73] H. Yoshimitsu; M. Nishida; T. Nohara. *Tetrahedron* 2001, *57*, 10247-10252.
[74] H. Yoshimitsu; M. Nishida; T. Nohara. *Phytochemistry* 2003, *64*, 1361-1366.
[75] H. Yoshimitsu; H. Miyashita; M. Nishida; T. Mineno; T. Nohara. *Chem. Pharm. Bull.* 2010, *58*, 1043-1046.
[76] M. Ono; T. Kawasaki; K. Miyahara. *Chem. Pharm. Bull.* 1989, *37*, 3209-3213.
[77] M. Ono; A. Takigawa; T. Mineno; H. Yshimitsu; T. Nohara; T. Ikeda; E. Fukuda-Teramachi; N. Noda; K. Miyahara. *J. Nat. Prod.* 2010, *73*, 1846-1852.

In: Indium
Editors: Hsaio G. Woo and Huang Tsai Choi
ISBN: 978-1-62257-696-8

Chapter 8

EXPERIMENTAL STUDY OF INDIUM ATOM USING ELECTRON AND OPTICAL SPECTROSCOPY

Dragutin Sevic*, Maja S. Rabasovic, Vladimir Pejcev and Bratislav P. Marinkovic
Institute of Physics, University of Belgrade, Serbia
V. Pejcev is also with Faculty of Natural Sciences,
University of Kragujevac, Serbia

ABSTRACT

In this chapter we overview important characteristics of indium and present our experimental results regarding indium lines obtained by electron spectroscopy and laser induced breakdown spectroscopy (LIBS). In the first series of experiments high purity indium granules were melted in the Knudsen type oven in order to obtain the effusive atomic beam which was perpendicularly crossed by monoenergetic electron beam. In the second series of experiments investigated were both high purity indium granules, to get reference signals, and suspected indium traces within basis plate of liquid crystal display. The surface of sample was hit by laser beam in order to obtain the plasma plume that emits the spectral lines which are the fingerprints of the present atomic and ionic species. Optical emission spectrum has been acquired using streak camera. The most intense In (I) lines of the spectrum at 303.9 nm, 325.6 nm, 410 nm, 451 nm are detectible in both series of experiments, however with different resolution. The optical diagnostics of the indium lines could be useful for detecting indium in electric waste, because the detecting device could be made portable. In this chapter we also show similarities and differences between the measurements obtained by the electron spectroscopy and laser induced breakdown spectroscopy applied on indium as the target.

* Corresponding author: sevic@ipb.ac.rs.

1. INTRODUCTION

Indium is a soft, gray metallic element which belongs to the elements of group III of the Periodic Table. Its atomic number is 49 and the ionization energy IP = 5.7864 eV. Indium has an open shell structure. It was found and spectroscopically identified as a minor component in zinc ores.

Because of its low melting point, 429.75 K, it is technologically attractive especially in the semiconductor and optoelectronics. Its current primary application is in transparent electrodes made of indium tin oxide in liquid crystal displays (LCD). Indium Oxide is a transparent ceramic material, when doped with Tin (usually about 10%) it becomes conducting but stays transparent. Indium Tin Oxide (ITO) can be applied by vacuum coating to a wide range of glass and plastic materials to make transparent conducting panels, liquid crystal displays (LCD) and antistatic applications [1-3]. Uses of indium are rapidly increasing and there are many other electronic devices which contain indium. It is relatively rare and ranks 61st in crustal abundance. It is about three times as abundant as silver. Indium has two stable isotopes ^{115}In and ^{113}In with an abundance of 95.7% and 4.3%, respectively. The $^{115}In^+$ stored in a radio-frequency trap and laser cooled has been investigated as a possible optical frequency standard with high stability and accuracy. The narrow $5s^{2\,1}S_0$–$5s5p\,^3P_0$ transition in In^+ at a wavelength of 236.5 nm could be used as a reference clock transition [4,5].

The spectral lines of neutral and ionized indium are of interest in the plasma spectroscopy and determination of plasma parameters [6]. The solar indium abundance is also analyzed in the astrophysical investigations [7]. Many plasma diagnostics techniques require electron impact excitation cross sections. Another very important role of indium is the application of an indium single ion as a reference for an optical frequency standard of unprecedented stability and accuracy. In^+ is the only one with an alkaline-earth-like spectrum. A narrow transition of a laser cooled ion, stored in the radio-frequency trap serves as the clock transition [8].

Indium is a particularly suitable as a tracer for two line atomic fluorescence (TLAF) thermometry measurements [9-11]. This method works by using two diode lasers with wavelength of 410 nm and 451 nm to excite the $^2S_{1/2}$ resonance state of indium atoms seeded into a flame. Owing to the greater oscillator strength of atoms then molecules, the strong fluorescence signals can be obtained at low excitation. Advantage of the indium atoms is that the spin–orbit splitting in the *5p* ground state of indium leads to an energy spacing that is equal to kT in typical combustion environments (1000–3000K) [9].

Excitation cross sections of In (II) spectral lines are experimentally studied in [12]. In this chapter we present experimental results regarding indium lines obtained by electron spectroscopy and laser induced breakdown spectroscopy (LIBS) [13-17].

In the series of experiments using electron spectroscopy high purity indium granules were melted in the Knudsen type oven in order to obtain the effusive atomic beam which was perpendicularly crossed by monoenergetic electron beam.

The LIBS is a method of atomic emission spectroscopy in which a pulsed laser is used as an excitation source. The rapid heating of the material in the focal volume induces a laser ablation and the ejected material dissociates into excited ions and atomic species. At the end of the laser pulse, the plasma cools and excited ions and atoms emit characteristic optical radiation which could be used to analyze the spectral lines of the components in material [18-20].

The plasma plume formed at the surface of an indium target has been achieved in our experiments using the Nd:YAG laser and it's optical emission spectrum has been acquired using streak camera. The most intense In (I) lines of the spectrum at 303.9 nm, 325.6 nm, 410 nm, 451 nm are detectible by both methods, however with different resolution. We also show other similarities and differences between the measurements obtained by the electron spectroscopy and laser induced breakdown spectroscopy applied on indium as a target. The optical diagnostics of the indium lines could be useful for detecting indium in electric waste, because the detecting device could be made portable.

As the atom of 13th column (formely labeled as IIIa group) indium has the $5s^25p$ configuration resulting in the $5p\ ^2P_{1/2}$ ground state. The metastable ($5p\ ^2P_{3/2}$) and resonant ($6s\ ^2S_{1/2}$) levels are positioned at 0.274 eV and at 3.022eV from the ground ($5p\ ^2P_{1/2}$) level, respectively. The results that concern the elastic differential cross sections (DCS) and resonant excitations of indium atoms are obtained using the electron spectroscopy in our laboratory and published recently [13, 14]. The possibility of detection of optically forbidden transitions is an advantage of electron spectroscopy. After we had implemented the optical LIBS technique the first results regarding indium are presented in [15-17].

Indium atom is the first of a series of the $5p$ elements of the Periodic Table, which ends by xenon atom. The atoms of this series are characterized by the relatively small value of the dipole static polarizability α_d, ranging from 30.4 a_0^3 for In to 27.3 a_0^3 for Xe [21]. Recent experimental measurements and theoretical investigations of DCS for Ca [22] and Yb [23,24] atoms with large α_d values (~150 a_0^3) have demonstrated a considerable difference in the form of DCS angular dependence as compared to noble gases. For example, DCSs for the Ca atom have three minima at 10 eV and 20 eV and two minima at 40, 50, and 100 eV. In the case of the Yb atom at 80 eV no structure of the "minimum-maximum" type was noticed within the entire angular range, while at lower energies, i.e., at 10–60 eV, two minima were found. Regarding indium atom in this study, the energy dependence of the angular positions of DCS minima is also determined [13].

2. ESMA - ELECTRON SPECTROMETER FOR METAL ATOMS

For measurements presented in this chapter we have employed the electron spectrometer that has been described in more detail elsewhere [13, 14]. It consists of an oven for the production of the atomic beam, monochromator and analyzer. The electron beam is emitted from a hairpin cathode, with the current at the end of the hemispherical monochromator of the order of 1–10 nA. The monochromator is fixed and the analyzer can be positioned from −30° to 150° with respect to direction of the electron beam around the atomic beam. The electron spectrometer can be operated in three different modes: recording of electron energy loss spectra, scanning the impact energy and angular distribution of elastically and inelastically scattered electrons. Schematic overview of hemispherical electron spectrometer ESMA is shown in Figure 1. Both Analyzer and Monochromator consist of complex system of cylindrical electrodes, i.e. electrostatic lenses.

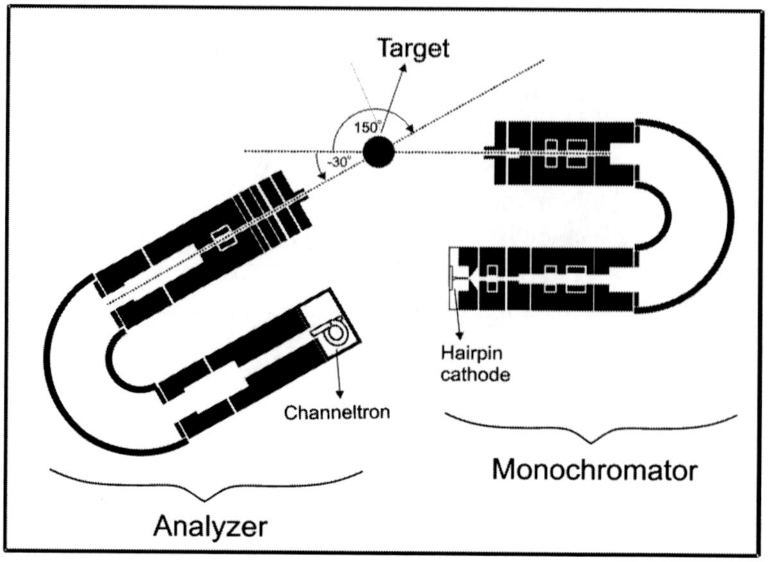

Figure 1. Schematic overview of hemispherical electron spectrometer ESMA.

With small modifications this experimental setup was used before for experiments with silver, magnesium, lead and other metal atoms [25-35].

Indium vapor beam was produced by heating the oven crucible containing In metal (99.9% purity). The oven had to be modified after experiments with other targets in order to achieve higher temperatures. The oven consists of two separate heaters with the same diameters, one for the top of the stainless steel crucible and nozzle, and the other for the body of the crucible, and their temperatures were controlled by two thermocouples. The heating currents of those heaters were about 5.3 and 3.4 A, respectively. They provided a variable temperature difference between the top and the bottom. The nozzle was maintained at approximately 100 K higher temperature than bottom of the oven. The atomic beam was effused through a 20 mm long channel in the cap of the oven crucible that has an inner diameter of 2.5 mm, so the aspect ratio was 0.125. Working temperature was approximately 1300 K and the metal-vapor pressure was about 10 Pa (or 0.07 Torr). Water cooling of oven, insulation ceramics, as well as several tantalum shields protected the channel electron multiplier from overheating. The background pressure was of the order of 10^{-5} Pa.

The energy scale was calibrated by measuring the position of the feature in the energy dependence of elastic scattering, attributed to the threshold energy of the 6s $^2S_{1/2}$ excitation of indium at 3.022 eV, as can be seen in Figure 2. The uncertainty of the energy scale is determined to be 100 meV, overall energy resolution of 120 meV, and the angular resolution was estimated to be 1.5°. The position of the real zero scattering angle was determined before each run by checking the symmetry of the inelastically (excitation of the $^2S_{1/2}$ state) scattered electron distribution around the mechanical zero at positive and negative angles (from −10° to +10°). The associated error in the angular scale determination was ±0.2°. Measurements at 0° were not reliable due to two reasons: first, there was some access noise originating from primary beam electrons although the energy selection of the analyzer was very efficient;

second, the present angular resolution was not sufficient to record the true intensity of strongly forward peaked angular distribution.

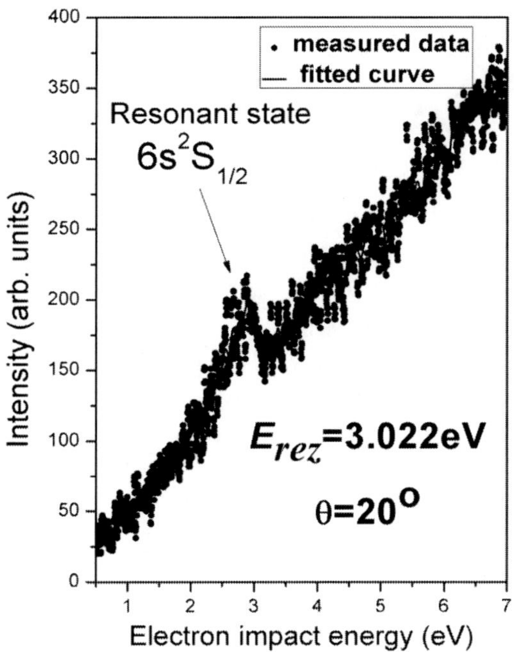

Figure 2. Position of the resonant state 6s $^{2}S_{1/2}$.

In the research presented in this chapter relative DCSs for the elastic scattering are obtained as follows. For a given incident energy E_0, the position of the analyzer was varied from 10° to 150° and the angular distribution of scattered electrons was measured. A channeltron was used for single electron counting. Thus obtained angular distribution was converted into relative DCSs by using the appropriate effective length correction factor [36]. We have applied the correction factors of Brinkmann and Trajmar [37] modified for our experimental conditions. The total error for relative DCS values arises from statistical errors, angular scale uncertainty, energy scale uncertainty, and estimation of the applied effective path-length correction. The statistical error was less than 0.10 (except at the DCS minima where it was less than 0.40). The contribution to the error from the angular uncertainty is up to 0.06 for small scattering angles and high energies. The contribution of the uncertainty in the energy scale is determined to be less than 0.05. The contribution to the error from the geometrical effective path-length correction factor is estimated to be 0.06.

3. Time Resolved Laser Induced Breakdown Spectroscopy

The laser induced breakdown spectroscopy (LIBS) is a sensitive tool for a rapid identification of the elemental composition of a sample regardless the aggregate states and with a little sample preparation.

Time resolved laser induced fluorescence (TR LIF) and Time resolved laser induced breakdown spectroscopy (TR LIBS) have similar hardware requirements. A detailed description and some of the preliminary results of our TR LIF/LIBS system are recently published [14-17,38-41]. The arrangement is based on the tunable Optical Parametric Oscillator (OPO) system (Vibrant 266, made by Opotek, Inc, pumped by Nd:YAG laser) to excite samples and on the detection part based on Hamamatsu streak camera (C4334-01) with high spatial and temporal resolution. The output of the OPO can be, for TR-LIF mode of operation, continuously tuned over a spectral range from 320 nm to 475 nm. The fluorescent properties of beet root juice were investigated in our laboratory using TR-LIF spectroscopy [38-39]. The detection of indium and silver lines were presented by us recently [15-17,41].

The system for LIBS measurements is shown schematically in Figure 3. Compared to the Time resolved laser induced fluorescence (TR-LIF) system presented in [38], it is easily noticed that focusing lens is added. Although LIBS measurements are possible when angle between the laser beam and optical axis of spectrograph is 90°, as in our TR-LIF measurements, better results are obtained when this angle is minimized, as shown in Figure 3.

Our laser is more than powerful enough (365 mJ at 1064 nm, variable OPO output >5mJ) for LIBS. The laser beam is focused on the surface of the samples (granule of pure indium or basis plate of LCD) using a convex lens with a 40 mm focal length to create a spark or breakdown in the sample. A portion of the plasma light is collected in two ways: by the Ocean Optics spectrograph and by the Hamamatsu streak camera. The Ocean Optics spectrograph detects wavelengths between 200 - 850 nm. The wavelength range of the Hamamatsu streak camera depends of the grating that is used in camera's spectrograph. For measurements presented in this chapter the grating of 50 lines/mm was used covering a 330 nm spectral range with the resolution of about 3.6 nm. The LIBS measurements are done in the air and at room temperature. To be on the safe side regarding Hamamatsu Streak Camera, which is very sensitive to the light, preliminary experiments were made using OPO output (5 mJ at 400 nm) and Ocean Optics spectrograph. It should be noted that using of streakscope enables time resolved diagnostics.

In most of our experiments, the plasma plume was formed at the surface of an indium target by using the OPO output (powerful enough for LIBS) of our tunable Vibrant laser system and plasma emission spectrum has been acquired by the streak camera. Namely, the laser induced fluorescence and breakdown spectra of the samples in our experiments are recorded using streakscope (Hamamatsu model C4334-01 with integrated video streak camera). The fundamental advantage of the streakscope is its two dimensional nature, that is especially important in measuring time-resolved LIBS and LIF spectra. The data are acquired and analyzed using High Performance Digital Temporal Analyzer (HPD-TA) software, provided by Hamamatsu. The basic package we obtained with the streak camera provides very advanced analysis of line profiles; however, image processing methods (2-D) are quite rudimentary. Using advanced image processing techniques we show how capabilities of basic software provided by Hamamatsu could be improved [16].

Finally, it should be pointed out that energy of OPO output is much smaller than energy of pumping Nd:YAG, which is as high as 365 mJ at 1064 nm, so we could use even more power for future LIBS experiments.

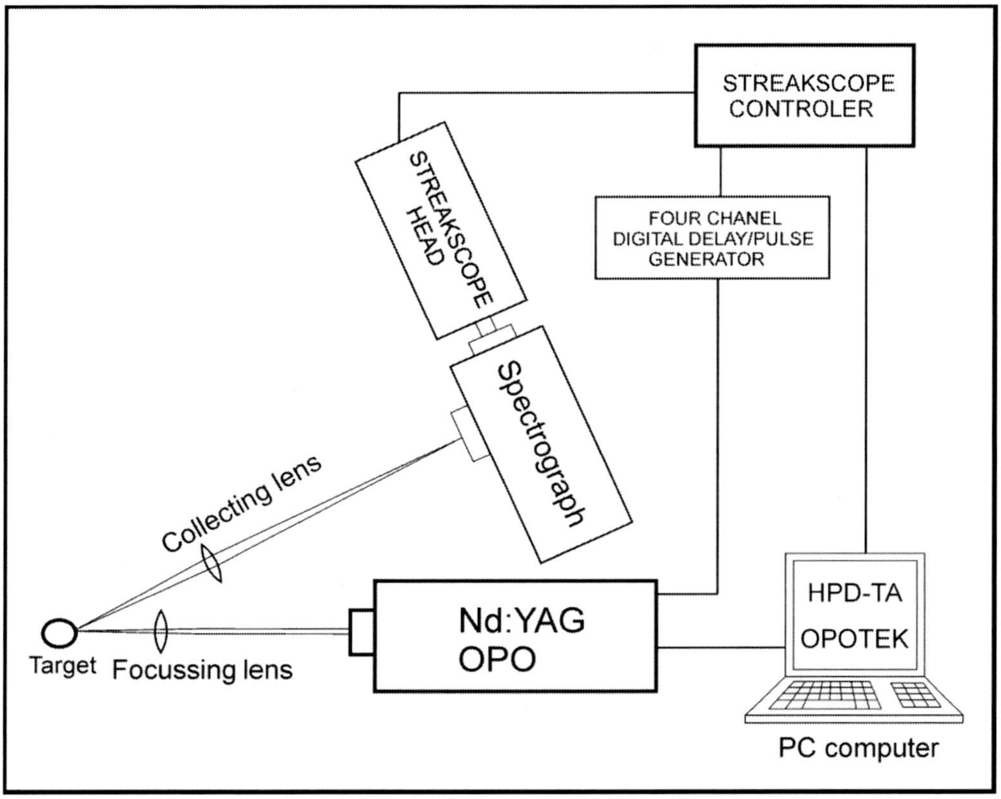

Figure 3. Our TR-LIF system modified for LIBS experiments.

4. Experimental Results Using ESMA Spectrometer

First, we present our experimental results regarding elastic electron–indium-atom scattering in the intermediate energy range. The results were presented and explained in detail in [13]. The measurements were performed at incident electron energies of E_0=10, 20, 40, 60, 80, and 100 eV within the scattering angles ranging from 10° to 150°. Theoretical calculations have been carried also in [13]. The relative values of experimental DCSs are normalized at 20° to the SEPASo theory with absorption potential VA_3, as explained and calculated in [13] are presented in Table 1 and Figure 4. As shown in [13], our experimental results agree well with theoretical calculations.

We have also measured the differential cross sections for the electron-impact excitation of the resonance transition $5p^2P_{1/2}$–$6s\ ^2S_{1/2}$ of In atom at small scattering angles [14]. The incident electron energies were E_0 = 10, 20, 40, 60, 80 and 100 eV, while the small scattering angles ranged from 1° to 10° in steps of 1°. The forward scattering function method has been used for normalizing the generalized oscillator strengths (GOS) to the known optical oscillator strength and obtaining the absolute DCS values. [42,43]. The absolute DCSs for the resonant excitation of indium are presented in Table 2 and Figure 5.

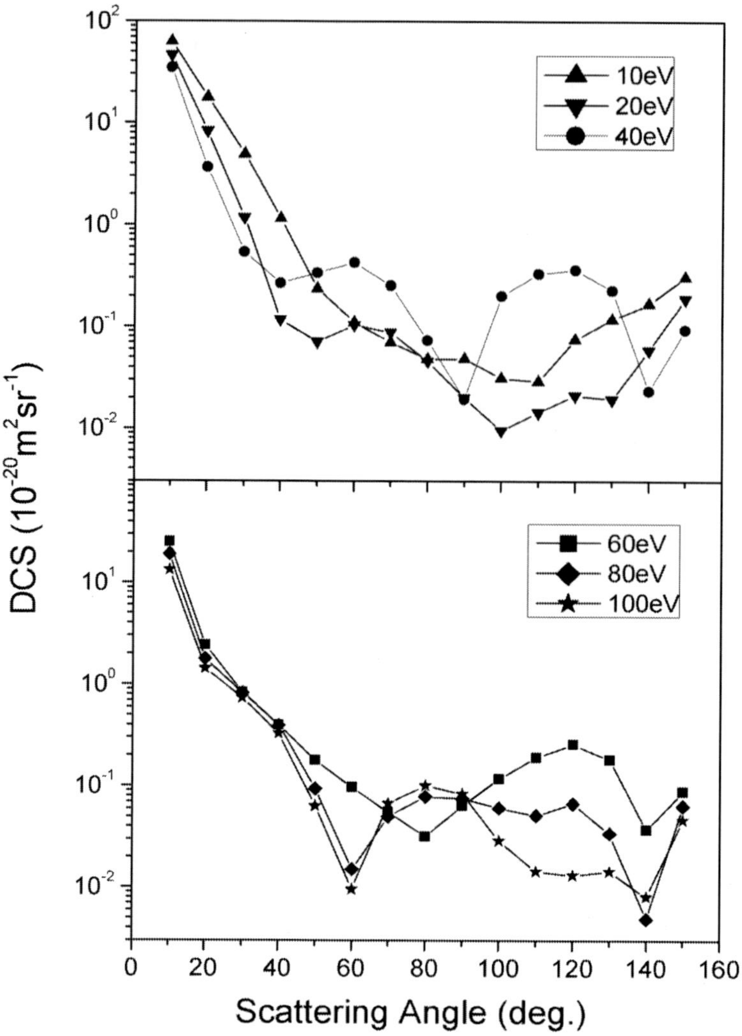

Figure 4. Differential elastic electron–indium-atom scattering cross sections in units of 10^{-20} m^2 sr^{-1} measured at 10 eV, 20 eV, 40 eV, 60 eV, 80 eV and 100 eV.

As one can see from Figure 5, DCSs decrease from 1° to 10° and the slope of DCSs increase with the increasing incident energy. As seen from Figure 5(a), DCS at 10 eV is about three times less relating to DCS at 20 eV at all measured angular region. Note also that DCS at 20 eV and 1° is 2.3 times larger, but at 10° is 1.4 times less than the DCS at 40 eV.

Figure 5(b) presents the results of DCSs at other three energies: 60 eV, 80 eV and 100 eV. So, at 100 eV the DCS value at 1° amounts 310.7 m^2 sr^{-1}, but at 10° DCS reaches the value 1.37 m^2 sr^{-1} and that is the largest variation in measured DCS values at small scattering angle range.

To compare In with other atoms studied in our laboratory it could be interesting to use the results for Pb [30] and Zn [44]. The similar FSF procedure is used for the normalization of relative experimental DCS values at small scattering angles for Zn atom at all measured impact energies except at 10 eV where the excitation energy for Zn (ω = 5.79 eV) does not satisfy the necessary condition $E_0 \geq 2.5\ \omega$. In that work a linear fit of the log(GOS) as a function of log(K^2) has been used for the obtaining the absolute GOS values. In the normalization procedure described in [42] it was pointed out that the linear variation of the GOS with K^2 is extensive for impact energy values close to the ω, but it covers fewer and fewer scattering angles as energy increases. In our recent published paper for Pb atom [30] it can be seen that the shape of the GOS values is similar to our results for In atom. For both of these atoms we used nonlinear fit instead of the linear one. This is supported by the relativistic distorted wave calculations [30]. The GOS values for Pb decrease fastly with the increase of the impact energy, while in the case of indium the GOS values show the maxima around 3° and tend to decrease both at smaller and larger scattering angles.

Table 1. Experimental differential elastic cross sections, in units of 10^{-20} m^2 sr^{-1}, for electron scattering by indium atom normalized at 20° to the *SEPASo* (with absorption potential VA_3), as calculated in [13]

Angle	Electron energy (eV)					
(deg)	10	20	40	60	80	100
10	62.7	45.6	34.6	25.1	19.1	13.25
20	17.8	8.26	3.66	2.41	1.78	1.429
30	4.91	1.174	0.540	0.834	0.824	0.734
40	1.143	0.116	0.266	0.395	0.395	0.327
50	0.236	0.071	0.337	0.177	0.0931	0.0632
60	0.110	0.1031	0.425	0.096	0.0151	0.0095
70	0.070	0.0876	0.252	0.055	0.0497	0.0668
80	0.048	0.0457	0.073	0.032	0.0781	0.1000
90	0.048	0.0199	0.0193	0.0636	0.0753	0.0828
100	0.031	0.0095	0.20	0.117	0.0607	0.0289
110	0.029	0.0143	0.330	0.189	0.0512	0.0144
120	0.074	0.0211	0.362	0.256	0.0670	0.0132
130	0.118	0.0192	0.227	0.180	0.0343	0.0143
140	0.167	0.0589	0.0234	0.037	0.0049	0.0081
150	0.307	0.189	0.093	0.088	0.0636	0.0464

Table 2. Differential cross sections in units of 10^{-20} m^2 sr^{-1}, for electron excitation of the 6s $^2S_{1/2}$ state of the indium atom

Angle	Electron energy					
(deg)	10 eV	20 eV	40 eV	60 eV	80eV	100 eV
1	13.9	47	108.7	155.4	233.4	310.7
2	12.5	41.0	86.8	117	169.1	216.6
3	10.9	34.7	65.4	80	107.8	123.9
4	9.2	28.7	46.8	51.5	62.1	61.9
5	7.59	23	32.2	32.2	33.5	29.0
6	6.10	18.0	21.4	20.0	17.5	13.6
7	4.80	13.7	13.9	12.5	9.1	6.74
8	3.73	10.3	9.0	8.0	4.88	3.64
9	2.91	7.7	5.98	5.14	2.76	2.17
10	2.32	5.88	4.20	3.29	1.88	1.37

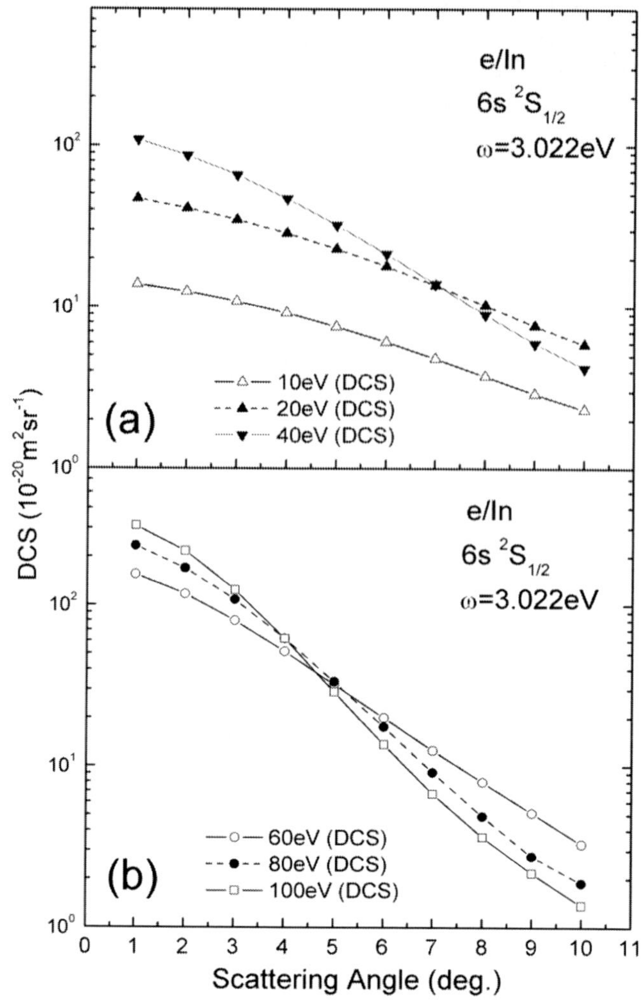

Figure 5. Differential cross sections for excitation of the resonant 6s $^2S_{1/2}$ state of the indium atom at: (a) 10 eV, 20 eV and 40 eV; (b) 60 eV, 80 eV and 100 eV electron-impact energies and at small scattering angles (in units of 10^{-20} m^2 sr^{-1}).

5. Experimental Results Using LIBS and Ocean Optics Spectrometer

Our first LIBS experiments were made using OPO output (5 mJ at 400 nm) and Ocean Optics spectrometer. Streak Camera is very sensitive to the light, so great care should be taken to avoid optical damage of the camera system.

LIBS spectra for almost pure indium sample (used in our ESMA experiments) and basis plate of LCD are presented in Figure 6 (a) and (b).

Samples were excited by OPO output (350 nm, 5 mJ). However, the OPO, the second harmonic (532 nm) and the fourth harmonic (266 nm) are not visible, or more or less not visible in Figure 6, because they were background subtracted.

Indium spectral lines are easily seen in Figure 6, however, lines of other elements could be noticed, as well. Sodium is of no surprise, because LCD basis plate is made of glass. However, aluminium is somehow surprising, because laser was focused on Indium Tin oxide electrode, not on other parts of handheld calculator.

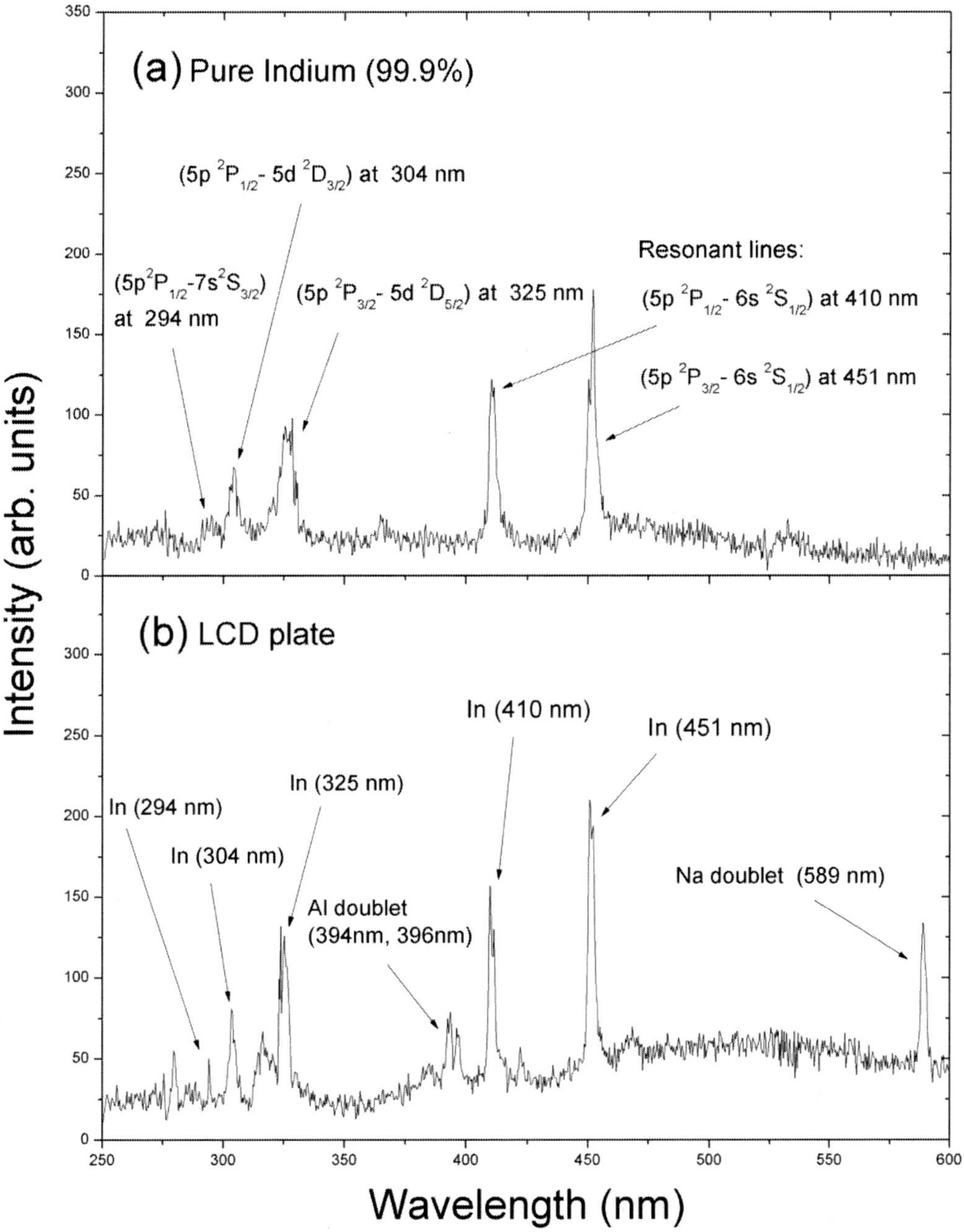

Figure 6. LIBS signal acquired from pure indium sample (a) and basis plate of LCD (b). Laser harmonics and OPO were background subtracted.

On both graphs in Figure 6, a LIBS spectrum with certain strong spectral features is standing out from the continuous background plasma light. Indium spectral lines seen in the Figure 6 are in good agreement with the data provided by NIST Atomic Spectra Database [45]. The two resonance emission lines from In (I) were observed at peak centers of 410 nm and 451 nm on both graphs in Figure 6. The principal mechanism of occupation of the excited states of indium atoms and ions in the laser generated plasma is the two-electron recombination [46]. The spectral lines of atomic sodium Na (I) (588.995 nm and 589.592 nm) and Al doublet lines (394 nm and 396 nm) can still be observed in Figure 6 (b).

6. Experimental Results Using LIBS and Hamamatsu Streak Camera

The streak image of In (I) and In (II) emission lines formed as a result of a laser generated plasma plume at the indium target surface is presented in Figure 7. The image has been acquired in photon counting mode operation using HPD-TA software with the time scale adjusted to 200 ns. The image is shown as acquired by a streak camera, without any processing by camera software. Because the deflection of streak sweep is not completely straight, there is a geometric distortion of the streak image in the sweep direction, visible here as a slightly, just noticeably, tilted spectrum. This distortion is easily removed by a curvature correction tool of the HPD-TA. The image is presented as grayscale, where different signal intensities are coded as different image intensities. Two black spots on top of the image (just under the wavelength scale) and light hellos around them correspond to camera saturation caused by OPO (350 nm) and laser second harmonic (532 nm) excitation. Although the streak camera has its hardware protection, this saturation should be avoided, and camera software warns operator about that, but it is difficult to prevent it in LIBS measuring. This is the reason why preliminary shots are recorded using Ocean Optics Spectrometer.

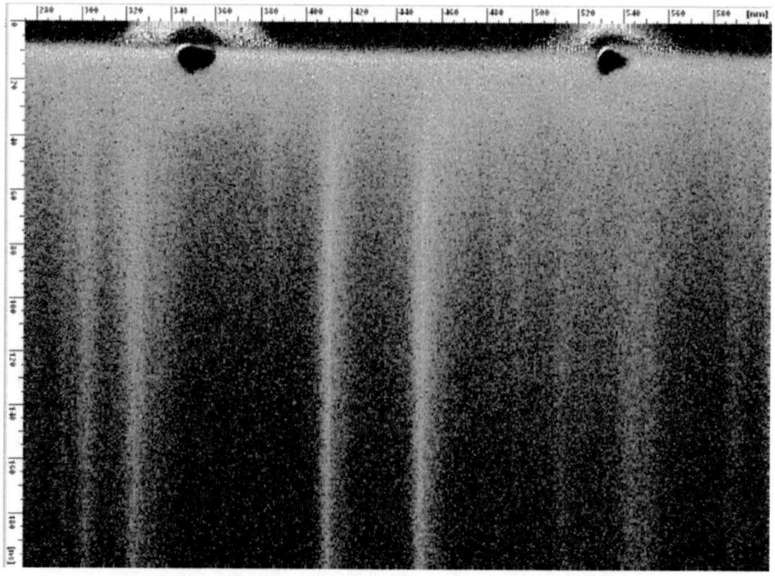

Figure 7. LIBS signal acquired from pure indium sample.

The narrow luminescence lines corresponding to the electron transitions in In (I) and In (II) are superimposed on a broad-band continuum. One dimensional profile of the streak image shown in Figure 7 is presented in Figure 8. The main spectral lines observed in Figures 7 and 8 are listed in the Table 3.

As can be seen from Figures 7 and 8, the most intensive lines of the spectrum are the atomic In (I) resonant emission lines at 410 nm and 451 nm. The other two atomic In (I) lines visible in streak image at 303.9 nm and 325.8 nm are also detectible in Figures 7 and 8. The radiation of In (II) ions from indium plasma was also observed in Figures 7 and 8, however 75 % of the intensity of line radiation was concentrated on these four most intensive In (I) lines. The two lines of In(II) at 289 nm and 294 nm are overlapped to the visible atomic In (I) line at 293.2 nm (see Table 3). Another group at 463.8 /465.6/ 468.1 nm of three In (II) lines with less intensity is also perceptible in Figures 7 and 8. Less intensive In (II) line is observed at 383.5 nm.

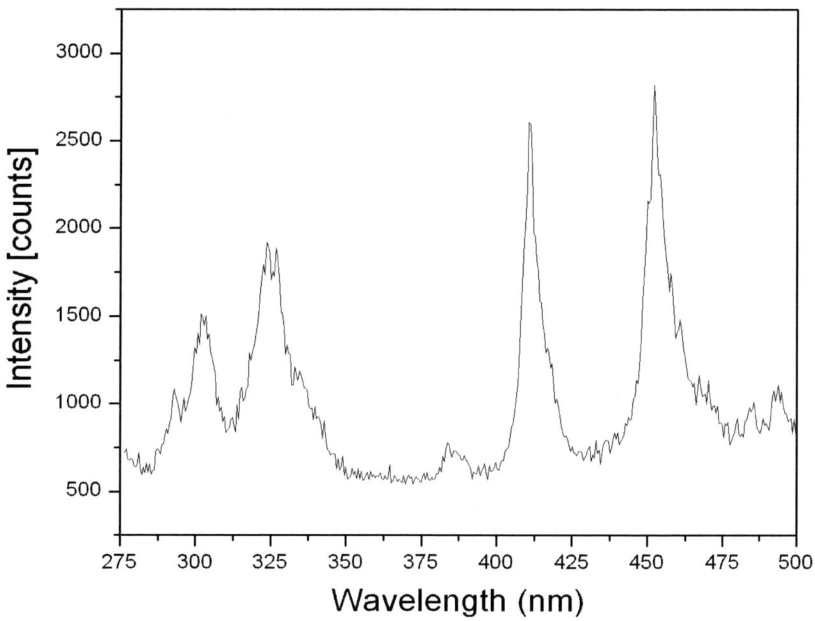

Figure 8. One dimensional profile of the streak image shown in Figure 7.

It should be pointed out that signals obtained with Ocean Optics spectrometer have no time resolution, in other words, they are integrated in time, so sometimes they look better compared to the line profiles from streak camera, where certain time window is selected for analysis.

7. Overall Analysis of Experimental Results

Now we will overview experimental results and show similarities and differences between the measurements obtained by the electron spectroscopy and laser induced breakdown spectroscopy.

Table 3. The most intense spectral lines of In (I) and In (II) in the emission spectrum of laser-generated indium plasma

λ (nm)	Transition	Atom, ion	E (eV)
468.1	$5s5d^3D-5s4f^3F^\circ$	In (II)	E_{high}-E_{low} = 2.647eV
465.6	$5s5d^3D-5s4f^3F^\circ$	In (II)	E_{high}-E_{low} = 2.663eV
463.8	$5s5d^3D-5s4f^3F^\circ$	In (II)	E_{high}-E_{low} = 2.672eV
451.1	$5p^2P^\circ_{3/2}-6s^2S_{1/2}$	In (I)	2.75 eV
410.2	$5p^2P^\circ_{1/2}-6s^2S_{1/2}$	In (I)	3.02 eV
383.5	$5s5d^1D-5s4f^1F^\circ$	In (II)	E_{high}-E_{low} = 3.233eV
325.6	$5p^2P^\circ_{3/2}-5d^2D_{5/2}$	In (I)	3.808 eV
303.9	$5p^2P^\circ_{1/2}-5d^2D_{3/2}$	In (I)	4.08 eV
294.1	$5s5p^1P^\circ-5s6s^1S$	In (II)	E_{high}-E_{low} = 4.215eV
293.2	$5p^2P^\circ_{3/2}-7s^2S_{1/2}$	In (I)	4.23 eV
289	$5s5p^1P^\circ-5s5d^1D$	In (II)	E_{high}-E_{low} = 4.289eV

In electron scattering experiment, an atomic beam of indium was created in a high vacuum system, by heating the oven. The indium atoms were excited from the ground $^2P_{1/2}$ state and thermally populated metastable $^2P_{3/2}$ state to the $^2S_{1/2}$ and $^2D_{3/2,5/2}$ states. The energy difference of 0.27 eV between $^2P_{1/2}$ and $^2P_{3/2}$ arises from the spin-orbit splitting of the two J states of the ^2P term. A schematic diagram of energy levels in indium atom is depicted in Figure 9.

The energy loss spectrum at 20 eV electron impact energy and scattering angle of 10° is illustrated in Figures 10 and 11. Figure 11 presents the same spectrum as the spectrum shown in Figure 10, redrawn with X scale calculated from energy loss in eV, so note that spectrum shown in Figure 11 is not optical emission spectrum. The X scale of Figure 11 is shown in nm, to make Figures 6, 7, 8 and 11 easy comparable. The energy loss scale was calibrated by measuring the position of the $5s5p^2\ ^2D_{3/2}$ excitation at 175.76 nm (or 7.055 eV) attributed to the position of the $6s\ ^2S_{1/2}$ excitation of indium at 410 nm (or 3.022 eV). The spectral lines of In I $(5p^2P_{1/2}-6s^2S_{1/2})$ at 410 nm and $(5p^2P_{3/2}-6s^2S_{1/2})$ at 451 nm correspond to the resonance doublet. The other two lines of the $5s^25d\ ^2D_{1/2,\ 3/2}$ states at 325.6 nm and 303.9 nm are resolved while $5s^26d\ ^2D_{1/2,\ 3/2}$ states are not resolved, we see only the line $(5p^2P_{1/2}-6d^2D_{3/2})$ at 256 nm. The line $(5p^2P_{1/2}-7s^2S_{1/2})$ at 275 nm is also apparent in energy loss spectra. We could also see the autoionizing states $5s5p^2\ ^2S_{1/2}$ at 175.76 nm (7.055eV) and inseparable doublet $5s5p^2\ ^2P_{1/2,3/2}$ at 165.66 nm (7.396 eV) and 164.98 nm (7.516 eV), respectively, in Figure 11. The excitation of that autoionizing state in regard to the ground $5p\ ^2P_{1/2}$ state of neutral In atom could be obtained by recording the energy loss spectra [25].

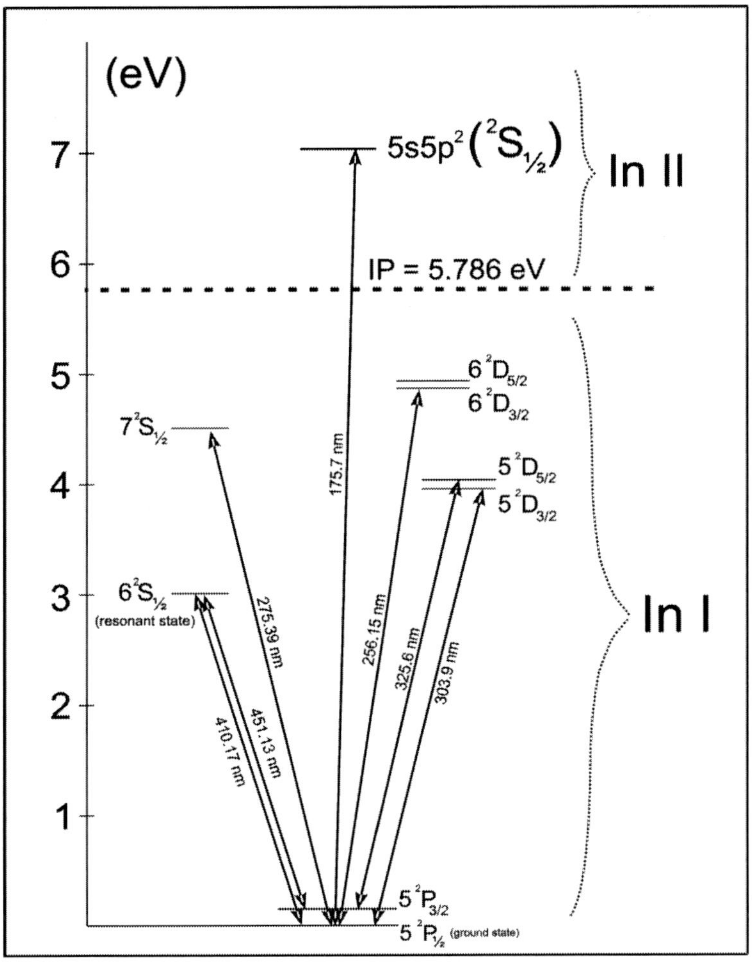

Figure 9. Diagram of energy levels in indium that have been observed in energy loss spectrum in electron scattering experiment.

Wavelength resolution of our LIBS could be determined using the usual full width half maximum (FWHM) method. For line at 451 nm, FWHM width is about 6 nm, corresponding to energy resolution of about 37 meV, which is better than 120 meV of electron spectrometer.

Electron spectroscopy has an advantage over the optical spectroscopy in possibility of detection of both optically allowed and forbidden transitions. However, the resolution of our electron spectroscopy apparatus is lower than resolution when using LIBS (120 meV vs. 37 meV). Another advantage of LIBS is the possibility of operation under normal temperature and air pressure, while electronic spectroscopy measurements are carried out only under vacuum conditions, and by heating the indium sample. Electron spectroscopy could be used for measurements that include a broad range of excitation energies. Our electron spectrometer is equipped with electron lenses that allow us to achieve the incident energy range from 10 eV to 100 eV.

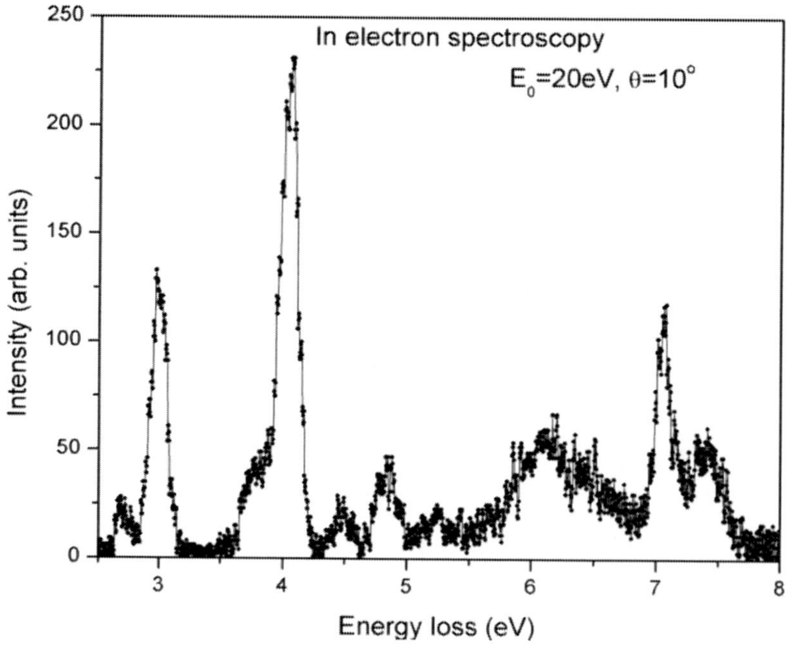

Figure 10. Energy loss spectrum of indium at 20 eV electron impact energy and θ=10° scattering angle.

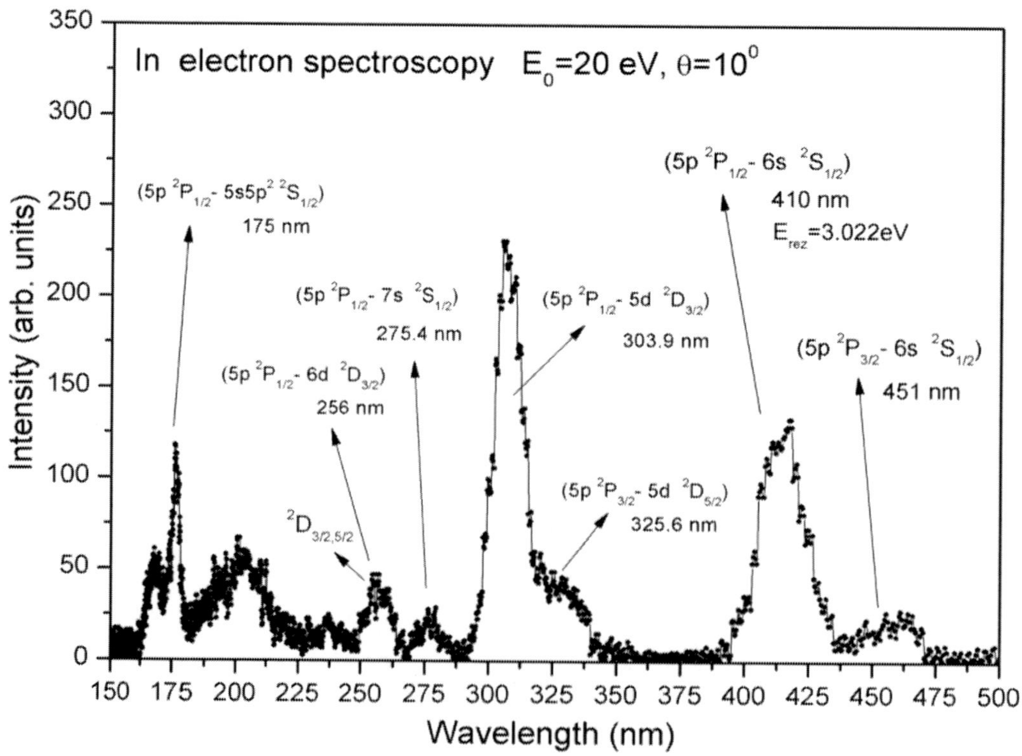

Figure 11. Energy loss spectrum of indium at 20 eV electron impact energy and θ=10° scattering angle. The spectrum is the same as the spectrum shown in Figure 10, redrawn with X scale calculated from energy loss in eV.

CONCLUSION

In this chapter we overviewed the important characteristics of indium and present our experimental results regarding indium lines obtained by electron spectroscopy and laser induced breakdown spectroscopy (LIBS). We analyzed and compared the results of both spectroscopy techniques. Possibility of detection of optically forbidden transitions is an advantage of electron spectroscopy. The most intense In (I) lines of the spectrum at 303.9 nm, 325.6 nm, 410 nm, 451 nm are detectible in both series of experiments, however with different resolution. Namely, the resolution of our electron spectroscopy apparatus is lower than resolution achieved when using LIBS.

We have obtained absolute values for generalized oscillator strengths and differential cross-sections for the excitation from the ground $5p^2P_{1/2}$ to the $6s^2S_{1/2}$ state of indium atom in the intermediate electron energy range from 10 to 100 eV, at small scattering angles up to 10°. So far as we know, there are no other available experimental data until our measurements. Moreover, differential cross sections for the elastic-electron scattering by indium atom in the intermediate energy region were studied for the first time both experimentally and theoretically. Experimental investigation of elastic-electron scattering by In atoms was performed using crossed beam technique for the energy range from 10 to 100 eV. The measurements were performed for a large angular range from 10° up to 150° and with a good angular resolution of 1.5°. The scattered electron intensity was measured as a function of the scattering angle. Angular distributions were multiplied by effective length correction factors in order to obtain relative DCS.

Many optical-based plasma diagnostic techniques require electron-impact excitation cross section, especially in astrophysics researches. The spectral lines obtained by electron spectroscopy can be combined with plasma emission measurements to select what emission lines to use in the analysis. In our measurements time resolved optical emission spectrum has been acquired using streak camera. Using of streak camera for measurements may improve the analysis approach, although good results were obtained by using simple spectrograph.

The optical diagnostics of the indium lines could be useful for detecting indium in electronic waste, because the detecting device could be made portable, operating under normal temperatures and air pressure. The study of indium lines presented in this paper could be widened on other trace element measurements in any kind of electric waste and for the other multi-elemental analysis. These results should be useful tool in environmental physics and ecology.

ACKNOWLEDGMENTS

The work has been done within the project 171020 financed by the Ministry of Education and Science, Republic of Serbia and ESF COST Action FA0906 UV4growth.

REFERENCES

[1] C. Calì, R. Macaluso, M. Mosca, *Spectrochim. Acta* Part B, 2001, 56, 743-751.

[2] A. De Bonis, A. Galasso, V. Marotta, S. Orlando, A. Santagata, R. Teghil, S. Veronesi, P. Villani, A. Giardini, *Appl. Surf. Sci.* 2006, 252 4632-4636.

[3] M. Kanehara, H. Koike, T. Yoshinaga, T. Teranishi, *J. Am. Chem. Soc.* 2009, 131, 17736-17737.

[4] J.A. Sherman, W. Trimble, S. Metz, W. Nagourney, N. Fortson, arxiv:/arXiv:physics/0504013.

[5] M. Eichenseer, A. Yu. Nevsky, Ch. Schwedes, J. von Zanthier, H. Walther, *J. Phys.B: At. Mol. Opt. Phys.* 2003, 36, 553.

[6] S. Djeniže, A. Srećković, and S. Bukvić, *Spectrochim. Acta,* Part B 61, 588 (2006)

[7] N. Vitas, I. Vince, M. Lugaro, O. Andriyenko, M. Gosic, and R. J. Rutten, *Mon. Not. R. Astron. Soc.* 2008,384, 370.

[8] Th. Becker J. v. Zanthier, A. Yu. Nevsky, Ch. Schwedes, M. N. Skvortsov, H. Walther and E. Peik, *Phys. Rev.* A, 2001 63, 051802.

[9] J. Hult, I.S Burns, C.F. Kaminski, *Proceeding of the Combustion Institute,* 2005, 30, 1535-1543.

[10] Q.N. Chan, P.R. Medwell, P.A.M. Kalt, Z.T. Alwahabi, B.B. Dally, G.J. Nathan, *Applied Optics,* 2010, 49,1257-1266.

[11] R.S.M. Chrystile, I.S. Burns, J. Hult, C.F. Kaminski, *Optics Letters,* 2009, 34, 2492-2494.

[12] Yu.M. Smirnov, *Opt. Spectrosc.* 2010, 34, 677-684.

[13] M.S. Rabasovic, V.I. Kelemen, S.D. Tosic, D. Sevic, M.M. Dovhanych, V. Pejcev, D.M. Filipovic, E. Yu. Remeta, B.P. Marinkovic, *Physical Review* A, 2008, 77, 062713-11.

[14] M.S. Rabasovic, S.D. Tosic, D. Sevic, V. Pejcev, D.M. Filipovic, B.P. Marinkovic, *Nucl. Instrum. Meth.* B, 2009, 267, 279-282.

[15] M.S. Rabasovic, D. Sevic, M. Terzic and B.P. Marinkovic, Proceedings of the 25th Symposium on Physics of Ionized Gases, *Publ. Astron. Obs. Belgrade,* 2010, 89, 229-232.

[16] D. Sevic, M.S. Rabasovic, B.P. Marinkovic, IEEE Trans. Plasma Sci.2011, 39(11), 2782-2783, *Special issue on images in plasma science.*

[17] M.S. Rabasovic, D. Sevic, V. Pejcev, B.P. Marinkovic, *Nucl. Instrum. Meth. B.* 2012, 279, 58-61. Special issue: Proceedings of the *Fifth International Conference on Elementary Processes in Atomic Systems*, Belgrade, Serbia, 21-25 June 2011, Edited by Bratislav Marinković and Károly Tőkési.

[18] A.K. Shuaibov, A.I. Dashchenko, I.V. Shevera, *Tech. Phys. Lett.* 2000, 26, 832-834.

[19] S. Pandhija, N.K. Rai, A.K. Rai, S.N. Thakur, *Appl. Phys.* B, 2010, 98, 231-241.

[20] A.K. Shuaibov, A.I. Dashchenko, I.V. Shevera, *High Temperature,* 2001, 39, 333- 337.

[21] T. M. Miller and B. Bederson, *Advances in Atomic and Molecular Physics*, edited by D. R. Bates and B. Bederson (Academic, New York, 1977) Vol. 13, 1–55.

[22] S. Milisavljevic, D. Sevic, R. K. Chauhan, V. Pejcev, D. M. Filipovic, R. Srivastava, and B. P. Marinkovic, *J. Phys.* B, 2005, 38, 2371.

[23] B. Predojević, D. Šević, V. Pejčev, B. P. Marinković, and D. M. Filipović, J. *Phys. B*, 2005, 38, 1329.

[24] V. I. Kelemen, M. M. Dovhanych, and E. Yu. Remeta,*J. Phys*. B, 2008, 41, 035204.

[25] B. Predojevic, D. Sevic, V. Pejcev, B.P. Marinkovic, D.M. Filipovic, J. Phys. B: At. *Mol. Opt. Phys.* 2004 36, 2371-2383.

[26] B. Predojević, V. Pejčev, D. M. Filipović, D. Šević, B. Tomčik and B. P. Marinković, *J. Phys. B: At. Mol. Opt. Phys.* 2011, 44, 055208.

[27] S. D. Tošić, M. S. Rabasović, D. Šević, V. Pejčev, D. M. Filipović, Lalita Sharma, A. N. Tripathi, Rajesh Srivastava, and B. P. Marinković, *Phys. Rev. A,* 2008, 77 012725.

[28] B. Predojević, V. Pejčev, D. M. Filipović, D. Šević and B. P. Marinković, *J. Phys. B: At. Mol. Opt. Phys.* 2008, 41, 015202.

[29] S. Milisavljević, M. S. Rabasović, D. Šević, V. Pejčev, D. M. Filipović, Lalita Sharma, Rajesh Srivastava, A. D. Stauffer, and B. P. Marinković, *Phys. Rev. A,* 2007, 76 022714.

[30] S. Milisavljević, M. S. Rabasović, D. Šević, V. Pejčev, D. M. Filipović, Lalita Sharma, Rajesh Srivastava, A. D. Stauffer, and B. P. Marinković, *Phys. Rev. A,* 2007, 75 052713.

[31] B. Predojević, V. Pejčev, D. M. Filipović, D. Šević and B. P. Marinković, *J. Phys. B: At. Mol. Opt. Phys.* 2007, 40(10) 1853-1861.

[32] S Milisavljević, D Šević, R K Chauhan, V Pejčev, D M Filipović, R Srivastava and B P Marinković, *J. Phys. B: At. Mol. Opt. Phys.* 2005, 38 (14) 2371 – 2384.

[33] S Milisavljević, D Šević, V Pejčev, D M Filipović and B P Marinković, *J. Phys. B: At. Mol. Opt. Phys.* 2004, 37 (18) 3571-3581.

[34] V. I. Kelemen, D. Šević, V. Pejčev, D. M. Filipović, E. Yu. Remeta and B. P. Marinković, *Nucl. Instrum. and Meth. in Phys. Res. B,* 2009, 267(2) 283 – 287.

[35] S. D. Tošić, V. I. Kelemen, D. Šević, V. Pejčev, D. M. Filipović, E. Yu. Remeta and B. P. Marinković, *Nucl. Instrum. and Meth. in Phys. Res. B,* 2009, 267(2) 283 – 287.

[36] M. S. Rabasović, S. D. Tošić, V. Pejčev, D. Šević, D. M. Filipović and B. P. Marinković, Facta Universitatis, *Series Phys. Chem. Technol.* 2008, 6(1) 119 – 125.

[37] R. T. Brinkmann and S. Trajmar, *J. Phys.* 1981, E 14, 245.

[38] M.S. Rabasovic, D. Sevic, M. Terzic, S. Savic-Sevic, B. Muric, D. Pantelic, B.P. Marinkovic, *Acta Physica Polonica A*, 2009, 116, 570-572.

[39] M. S. Rabasović, D. Šević, M. Terzić, and B. P. Marinković, *Phys. Scr.* 2012, T149 014076. Proc. Photonica 2011: 3rd *International School and Conference on Photonics*, 29 August–2 September 2011, Belgrade, Serbia, Edited by Jovana Petrović, Milutin Stepić and Ljupčo Hadžievski.

[40] Maja S. Rabasovic, Dragutin Sevic, Mira Terzic, and Bratislav P. Marinkovic, *Nucl. Instrum. Meth. B.* 2012, 279 16-19. Special issue: Proceedings of the *Fifth International Conference on Elementary Processes in Atomic Systems*, Belgrade, Serbia, 21-25 June 2011, Edited by Bratislav Marinković and Károly Tőkési.

[41] Maja Rabasović, Dragutin Šević, Mira Terzić, Bratislav P. Marinković, *Proc. XX European Conf. on the Atomic ad Molecular Physics of Ionized Gases* (XX ESCAMPIG), 13 – 17 July 2010, Novi Sad, Serbia, European Conference Abstracts Vol. 34B editors: Zoran Lj. Petrovic, Gordana Malovic and Dragana Maric, ISBN: 2-914771-63-0, Poster contribution P3.52.

[42] Z. Felfli, A. Z. Msezane, D. Bessis, *Phys. Rev. Lett.* 1998, 81, 963.

[43] N. B. Avdonina, Z. Felfli, A. Msezane, J. Phys. B: At. *Mol. Opt. Phys.* 1997, 30, 2591.
[44] R. Panajotovic, D. Sevic, V. Pejcev, D.M. Filipovic, B.P. Marinkovic, *Int. J. Mass. Spectrom.* 2004, 233, 253. 282
[45] Yu Ralchenko, A.E. Kramida, J Reader and NIST ASD Team (2010), *NIST Atomic Spectra Database (version 4.0.1)*, [Online]. Available: http://physics.nist.gov/asd [2011, May 30] National Institute of Standards and Technology, Gaithersburg, MD.
[46] A.K. Shuaibov, A.I. Dashchenko, I.V. Shevera, *Tech. Phys. Lett.* 2000, 26, 832-834.

In: Indium
Editors: Hsaio G. Woo and Huang Tsai Choi
ISBN: 978-1-62257-696-8

Chapter 9

NEW POTENTIALLY INDUSTRIAL TYPE OF INDIUM SULFIDE-MANGANESE ORE

V. I. Silaev*[1]*, A. V. Kokin*[2]*, D. V. Kiseleva*[3]*,
***N. N. Piskunova*[1] *and V. P. Lutoev*[1]**
[1]Institute of Geology, Komi Science Center,
Ural Branch of Russian Academy of Sciences, Syktyvkar, Russia
[2]Insitute North Caucasian Academy of Civil Service,
Rostov-on-Don, Russia
[3]Insitute of Geology and Geochemistry, Ural Branch of Russian Academy of Sciences,
Yekaterinburg, Russia

ABSTRACT

Indium is traditionally extracted from lead-zinc, tin-sulfide, copper- and copper-zinc-sulfide, skarn sulfide-magnetite deposits as an trace element at bulk content of some dozens ppm. The discovery in Russia in 1988 of sulfide-manganese ores with 30 to 300 ppm content of indium that had a considerable quantity of indium sulfides revealed the natural existence of a potentially industrial type of indium ore unknown before.

GEOLOGICAL SETTING AND GENERAL CHARACTERISTICS OF THE DEPOSIT

Vysokogornoye deposit is situated within Okhotsk volcanic belt, a part of high-yield Yuzhno-Verhoyansk metallogenic province, belonging to the eastern fold zone margins of Siberian Platform [7]. Its geological record reveals two structural levels: 1) the *lower* sediment formations of 800 m of thickness, composed of intensely dislocated and hydrothermally metamorphosed terrigene rocks of Verkhoyansk complex (Menkchensk and Chambinsk strata, P_2; 2) and the *upper* volcanogenic-sediment formations of 300-550 m, containing andesidacits, dacits and tuffs of average composition (Dustachansk series, K_{1-2})

and the above lying rhyolite-dacits and rhyolites (Kogarsky complex, K_2). In 25-30 km to the north of the deposit, there has developed a vein matter of lamprophyres and lamproites.

Prolific ore bodies of Vysokogornoye deposit are embedded mostly in hydrothermally metamorphosed Perm terrigene formations (Figure 1), represented by abundant lodes and intersecting crush zones of submeridional, north-west and rarely south-west extension, steep dipping west and east. Sulfide-manganese emplacement occurs in the following forms: 1) individual almost monoalabandine veins with aureoles of carbonate-manganese mineralization; 2) alabandine veins and veinlets in crush zones accompanied by quartz-polysulfide mineralization; 3) en echelon systems of alabandine and alabandine-pyrit-sphalerite-pyrrhotine veins in crush zones; 4) alabandine and alabandine-pyrit-sphalerite-pyrrhotine veins in metasomatites and hornfels in wallrocks; 5) small-scale veinlets and seams of carbonate-manganese, sometimes with alabandine shots.

Figure 1. Indium sulfide-manganese ores at Vysokogornoye deposit: a – beats of vein №6 (indicated by arrows), b – typical baring of massive ore.

Historically, there were three periods of ore formation at the deposit under consideration. The first period, coinciding with formation of granodiorite-porphyry dikes in volcanic rocks of dustachansk series, involved areal hydrothermal-metosomatic metamorphosis of terrigenous rocks of the lower structural level and volcanic rocks of the upper one, marked by berisitization, propylitization and argillization. Ore formation in the given period followed two mineralization stages: the early galena-sphalerite-pyrite-arsenic-pyrite one and the late galena-sphalerite-pyrite-pyrrhotine one with cassiterite and stannite. The second period, associated with Late Cretaceous corniferous granitoid intrusion, was marked by formation of sulfide-manganese (alabandine) and quartz-carbonate-manganese veins. Isotopic signature of carbonium and oxygen ($\delta^{13}C = -9.13 \pm 0.77$, $\delta^{18}O = 9.53 \pm 1.89$ ‰), testifies to the endogenic origin of the carbonate matter in the alabandine ores. Formation of the late quartz-carbonate-sulfate-antimonide mineralization, superimposed on alabandine ores, is considered to be the third period. The process is spacely associated with plagiogranites that terminated the formation of Kogarsk magmatic complex.

Our research leads us to conclude that in the given case we deal with the so called direct cycle of deposit formation: started with pyrite and arsenopyrite, followed by copper pyrite, sphalerite (here alabandine as well) and galena, and terminated by sulphosalts, antimonides and bismuthides. Based on the results of physical-chemical model study of such like mineralizations, we can assume the following thermodynamic parameters for the given emplacement: 1) formation of essential sulfide minerals at T = 500–300 °C, P > 2 kbar; 2) formation of later sulphasalt associations at T = 300–150 °C и P = 1.5–0.5 kbar. The replacement of sulfides by sulfidoantimonides and then by carbonates was conditioned by progressing reduction in sulphur activity at the time when the temperature of mineral formation was decreasing. For sulphasalt stage of mineralization, we can assume pH and Eh values of 3.5–4 and –200 mV correspondingly.

RESEARCH METHODS

Complex mineral-geochemical research of indium sulfide-managanese ores was conducted on the bases of Institute of Geology of Komi SC, Ural Branch of RAS. They consisted of a chemical test of atomic absorption recognition of some components; assessment of carbonic matter (C_{org}) content by coulometric titration method to test pH value using rapid-response analyzer An-7529M; diagnosing microelement content by MS-ICP method with quadrupole mass spectrometer Perkin Elmer ELAN 9000; thermal analysis done with automatic derivatograph Shimadzy DTG-60A/60AH; XD analysis on high-precision diffractometer Shimadzy XRD-6000; IR-spectroscopy of absorption on Fourier-transform spectrometer Lumex InfraLUM FT-02; Raman spectroscopy with the help of Horiba LabRam HR 800 system; Moessbauer spectroscopy performed with MS-1104Em; analytical scanning electron microscopy (SEM) performed with complex Jeol JSM-6400; atomic force microscopy (AFM) using NT-MDT Ntegra Prima; determination of fluid inclusions content with gas chromatograph "Zvet-800"; analysis of isotopic signature of carbon and oxygen in carbonates and carbon-bearing matter with the help of mass spectrometer Delta V Advantage and sampling systems Gas Bench II and Flash EA1112 correspondingly.

GENERAL CHARACTERISTICS OF INDIUM-MANGANESE ORES

Ores that are most prolific have a nearly hundred-per-cent alabandine composition, specimens varying in size from gigantic and coarse-crystalline to fine-medium-grained. The most important trace minerals are managanese carbonate, pyrite, pyrrhotine, sphalerite, galena, indium sulfates, sulfa-antimonides of various composition and cassiterite. The chemical make-up of the considered ores can be characterized as following (mass %): SiO_2 0.10–3.24; TiO_2 0.01–0.02; Al_2O_3 0.50–1.10; Fe_2O_3gen 1.81–10.85; MnO 71.04–81.47; MgO 0.01–0.10; CaO 0.10–0.92; ZnO 0–0.63; PbO 0–0.86; Na_2O 0.05–0.06; K_2O 0.02–0.06; P_2O_5 0.01–0.02; CO_2 1.59–3.93; S 32.36–35.36. The represented data show that these ores have 87–97 mass % alabandine composition, with 6.5–10.5 mass % content of trace elements. The content of carbonic matter recognized in them ranges from 0.05 to 0.85 mass %.

The obtained heating curves allow to detect small broadened exothermal effects at the interval of 350–550°C and a system of exothermal peaks in the period between 630 and

750°C with extremums at 685–690, 715–720, and 730–735°C. The first of the above mentioned intervals corresponds to the burn-off of carbonic matter. The obtained thermal data allow us to conclude that this matter is associated with lower kerite according to its degree of metamorphization. The higher temperature exothermal effects reflect the stepwise thermal dissociation of sulfides. The first peak results from oxidation of alabandine itself, and the second two correspond to thermal dissociation of sulfide inclusions. The pattern of the peaks on the studied heating curve clearly shows the dominant role of sphalerite among the other sulfide inclusions. The correlation of the weight change curve with the system of exothermal effects allowed to diagnose a sizable "bump" reflecting weight gain as a result of sulfide formation. At the interval of the highest temperatures on the thermal curves of alabandine ores, there is a clear endothermal effect with the extremum at 860 C, corresponding to a significant weight loss. This reflects the thermal dissociation of Ca-Mn monocarbonates.

The composition of the sufide-managanese ores was diagnosed to contain abundant trace elements, including rare earths whose contents, normalized to chondrite contents, demonstrate a negative trend, complicated by a strongly expressed europium maximum. The latter is a possible evidence of intense reducing conditions of ore formation. Thus, lanthanides in the studied sulfide-manganese ores can be classified as syngenetic inclusions, which is proved by their tendency to beneficate ores in the upward direction along the vein cut.

The results of special tests show the Au content in indium-manganese ores to reach 4 ppm and to have a rather uneven distribution. In addition, some samples had a high concentration of Pd, Rh and Re. It is important that the Co/Ni coefficients are very high, exceeding the average rate for the Earth crust as a whole and for most rocks and ores, except for basaltic derivatives. This geochemical characteristic suggests that sulfide-manganese mineralization is genetically related not to a granitoid, but to a deep basaltoid magma chamber.

To determine the composition of fluid inclusions in alabandine ores we carried out a chromatographic analysis of pyrolyzates obtained by heating to 300–400 °C, which is much lower than the temperature of thermal dissociating of early sulfides. The latter gives reason to believe that our findings do correspond to the composition of fluid inclusions. The content of the fluid phase in alabandine ores varies from 300 to 13000 mkg/g. Moreover, the sample of coarse-crystalline idiomorphic alabandine, taken at the top of the ore vein, has the maximum fluid saturation. The fluid inclusions consist predominantly of water (79–99.5 %) and the constant characteristic impurities of sulfurous components – H_2S и SO_2. The occurrence of the latter can be indicative of phase heterogeneity of the inclusions due not only to the sulfide, but also to the sulfate content.

The content of a whole number of trace elements in sulfide-manganese ores is much higher than the corresponding clarkes (Figure 2). The most important of these elements form the following sequence of relative benefication according to the size of concentration clarkes: [Mo>Rh,Pd (5–9)] < [Pb>Te>Se>Bi,As (10–50)] << [Sb>In>Sn>Zn>Re (350–900)] << [Cd>Ag (2000–7500)]. It is evident that the range of trace elements beneficating alabandine ores – Ag, Sn, Zn, Pb, Sb, As – closely matches the composition of the alabandine-accompanying cassiterite-polysulfide-sulphoantimonide mineralization. Besides, it's important to notice that the beneficiating alabandine ores group includes elements of sub-ore (Bi, Mo, Sn), as well as supra-ore (Sb, Pb, Ag) aureole, which is another indicator of a relatively low degree of geochemical differentiation of the studied deposit.

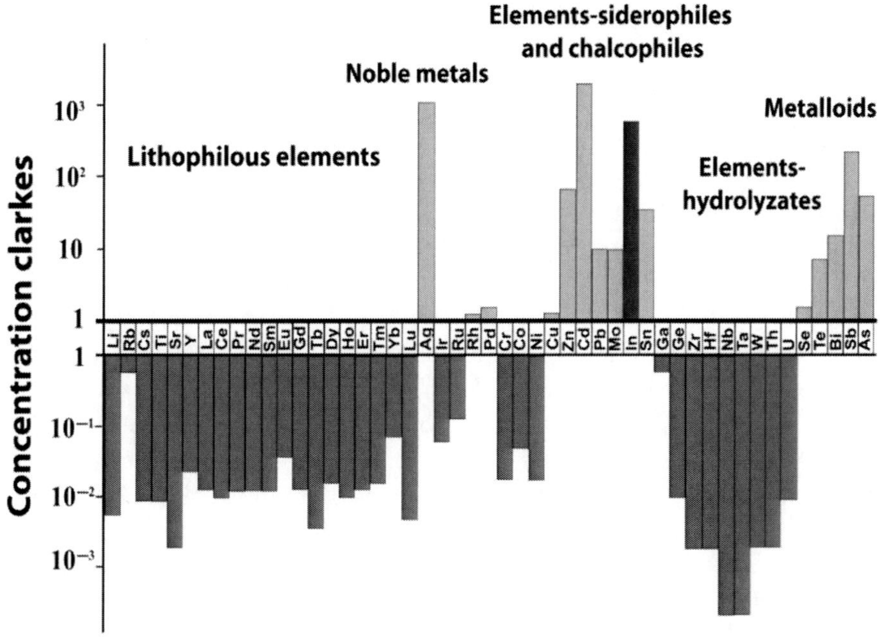

Figure 2. Relative concentrations of microelements in indium sulfide-manganese ores.

Among the accumulated in sulfide-manganese ores trace elements, indium takes a special place, rating among the most important for the modern economy rare metals. Our findings show that the bulk content of indium in the considered ores corresponds to the level of its benefication of hub elements, for example cassiterite and sphalerite, in the deposits that are traditional sources of this element. Detailed studies using SEM analysis allowed us to detect over 90 minerals in sulfide-manganese ores, including 15 proper indium sulfides and 4 minerals with a considerable indium admixture. The summary mineral inventory of Vysokogornoye deposit at this date can be presented as following.

Silicates: 1) guartz; 2) muscovite; 3) manganoceladonite $K(Mn,Mg,Fe)(Al,Fe)[Si_4O_{10}](OH)_2$; 4) chlorite; 5) manganochlorite $(Mn,Mg,Fe)_5Al[AlSi_3O_{10}](OH)_8$; 6) epidote; 7) allanite $(Ca,REE)_2$ $(Al,Fe,Mn)_3[Si_3O_{12}](OH)$; 8) tinzenite $(CaMn,Fe)_3Al_2[BSi_4O_{15}](OH)$; 9) mangantinzenite $(CaMn,Fe)_3(Al,Mn)_2[BSi_4O_{15}](OH)$; 10) titanite $CaTiSiO_5$; 11) titanite-malayaite-vyuanyatite $CaTiSiO_5$-$CaSnSiO_5$-$CaAlSiO_4(OH)$. *Oxysalts:* 12) calcite; 13) dolomite; 14) calcite-Mn; 15) manganocalcite; 16) calciorodochrosite; 17) rodochrosite; 18) siderite; 19) manganosiderite; 20) hydroxyl-fluorapatite; 21) hydroxyl-fluorapatite-Mn; 22) hydroxyl-fluorapatite-Sr; 23) xenotime $Y[PO_4]$; 24) monazite $Ce[PO_4]$; 25) hydroxyl-phosphate Mn-REE; 26) skorodite $Fe[AsO_4]2H_2O$. *Oxides:* 27) cassiterite; 28) pyrophanite $MnTiO_3$; 29) goethite $FeO(OH)$; 30) manganite $MnO(OH)$. *Chalcogenides:* 31) alabandine MnS; 32) alabandine-In; 33) alabandine-In,Sn; 34) solid solution of argentite-Sn in alabandine; 35) troilite; 36) pyrrotite; 37) pyrite; 38) chalcopyrite; 39) stannite Cu_2FeSnS_4; 40) stannite-kesterite $Cu_2(Fe,Zn)SnS_4$; 41) solid solution of alabandine in stannite; 42) sphalerite-Mn,Fe ; 43) sphalerite-In,Sn; 44) sphalerite with band of Zn-In phase; 45) galena;

46) argentite; 47) lapharetite $AgInS_2$; 48) roquesite, $CuInS_2$; 49) manganoindite $(Mn,Fe)InS_2$; 50) solid solution of In_2S_3 in manganoindite; 51) In_2S_3 phase in alabandine; 52) sakuraiite, (Cu,Zn,Fe,In,Sn)S; 53) sakuraiite-Ag; 54) petrukite-like mineral $(Cu,Mn,Fe,In,Sn)_3S_4$; 55) petrukite-like mineral Ag-containing; 56) sulfides $(Fe,Ag)_{2-x}S$; 57) solid solution of Ag in sulfides $(Ag,Fe,Mn)_{2-x}S$; 58) sulfides $(In,Ag)_{2-x}S$; 59) sulfides $(Sn,Ag)_{2-x}S$; 60) Sn-Cu-Ag sulfide In-containing; 61) chalcosine-covellite; 62) sulfoarsenide Zn-Ag; 63) sulfoarsenide Ag-Fe-Zn-In-Sn; 64) arsenopyrite; 65) arsenopyrite Co-Ni- containing; 66) arsenopyrite Sn-containing; 67) arsenopyrite In,Sn-containing; 68) sulfoarsenide Zn; 69) sulfoarsenide In-Sn-Fe-Zn-Ag; 70) miargyrite; 71) pyrargyrite; 72) playfarite; 73) robinsonite; 74) fizelite; 75) launayite; 76) boulangerite; 77) tennantite-tetrahedrite; 78) freieslebenite-Cu; 79) sulfoantimonides Ag-Ni-Zn; 80 jamesonite; 81) sulfoantimonides In-Sn-Cu; 82) sulfoantimonides In-Sn-Pb; 83) sulfoantimonides In-Zn-Cu; 84) sulfoantimonides Zn-Pb-Cu; 85) sulfoantimonides Sn-Pb; 86) sulfoantimonides In-Pb-Zn; 87) stibnite Sb_2S_3; 88) tetradymite Bi_2Te_2S. *Elements:* 89) sulfur; 90) bismuth; 91) silver; 92) gold.

The most important science- and application-wise mineralogical characteristic of the studied sulfide-manganese ores is their rather even saturation with various inclusions of indium and indium-stannum sulfides. The size of these inclusions varies from 1 to 250 mkm, with 80 % of the inclusions not exceeding 10 mkm. The AFM data show that the size of individuals of indium phases is not limited to micron level, dropping down to at least meso-nanometer values. All the mentioned above minerals are intensely beneficated in manganese and are, sometimes, not stoichiometrically stable. It is possible that at least some of them are new kinds of minerals.

The most part of indium and indium-stannane sulfide phases is fixed in alabandine, where they form emulsionlike impregnation mainly consisting of subisometric grains of 1 to 10 mkm in size (Figure 3, a-d). Larger and more elongated formations of indium phases are associated with blind fractures. Their intergrowths are places of occurrence of sulfoarsenides and cassiterites. In areas of brittle-plastic deformations in alabandine near to its border with carbonates, indium phase grains reach the size of a few hundred micrometers, forming the characteristic structures of cementation (Figure 3, e). In carbonates, indium minerals are presented by rather small (up to 10 mkm), mostly elongated forms (Figure 3, f). The edges of such forms reveal, as a rule, the signature of resorbtion testifying to indium sulfides being replaced by carbonates. Many indium sulfides inclusions are characterized by mosaic and lamellar heterogeneity due to considerable variation of indium content.

The largest formations are presented by indium containing sphalerite-like phase, whose individuals are characterized by a clearly defined zonal structure. The core part in such individuals is built up of In-Sn-contaning iron-manganese sphalerite; the fringe is composed of zinc-indium phase identified before in some other deposits in the form of small homogeneous grains [15, 17]. The determined girdle fabric sphalerite-like formations are strictly attributed to alabandine. They are found in the form of irregular, often elongated grains 30–170 mkm in size. The proportions of the core – proper sphalerite areas of the grains – in relation to zinc-indium fringe range widely, sphalerite even sometimes remaining as the nucleolus. Empiric formula for minerals girdle fabric individuals: *core part* – $(Zn_{0.70-0.76}Mn_{0.12-0.17}Fe_{0.07-0.08}Cu_{0.01-0.10}In_{0.01-0.03})S$; *fringe* – $(Zn_{0.26-0.50}Mn_{0.07-0.19}Fe_{0.03-0.07}Cu_{0.17-0.32}In_{0.18-0.29})S$.

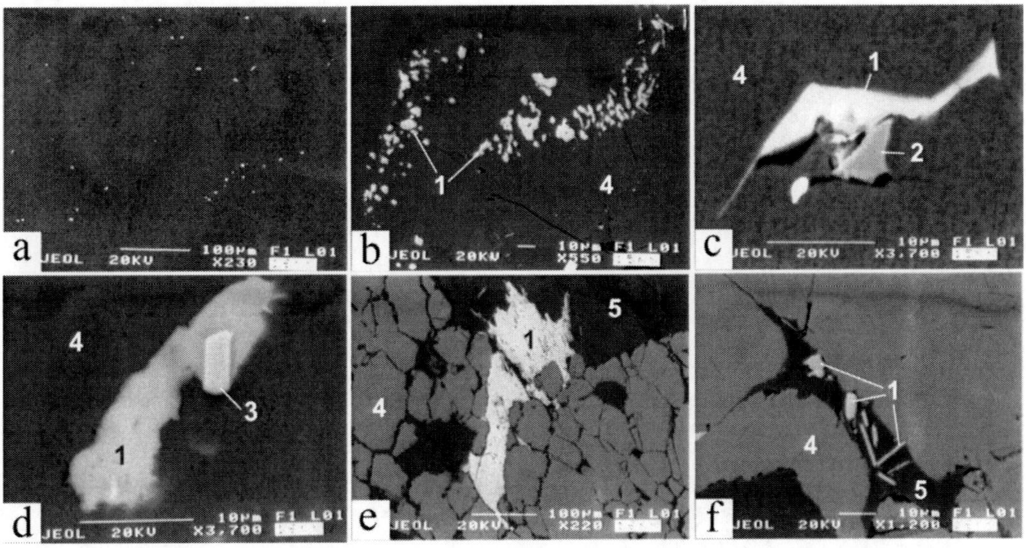

Figure 3. Indium sulfides inclusions in alabandine ores: a–c – in alabandine; d – at alabandine – carbonates boundary; E – in carbonates. Minerals: 1 – indium sulfides, 2 – arsenopyrite; 3 – cassiterite; 4 – alabandine; 5 – rhodochrosite. SEM-images in elastically reflected electrons view.

Most common among other indium minerals is copper-zinc sakuraiit-like monosulfide with empiric formula $(Mn_{0-0.25}Fe_{0-0.47}Cu_{0.02-0.46}Zn_{0.05-0.43}Ag_{0-0.13}In_{0.18-0.48}Sn_{0-0.17})S$. The analysis show that the content of this phase varies widely and almost constantly, the most zinc beneficated variations corresponding to the above described fringe zinc-indium phase. The latter supports the earlier expressed opinion about continuity of transitions between sakuraiit and zinc-indium minerals [15], and also about transitions between sakuraiit and minerals of stannite-kesterite family [6, 16, 17, 18].

Another common indium sulfide observed in the studied ores is petrukite-like phase with nominal stoichiometry of Me_3S_4. Its localization and size morphology are identical with the above presented sakuraiit. Analysis revealed that the detected phase corresponds to petrukite by its proportions between kaotines and sulphur, at the same time being consistently distinct from it by contaning twice as much indium and stannum. The empiric formula for this mineral can be presented as $(Mn_{0.11-0.70}Fe_{0-0.18}Cu_{0.66-0.93}In_{0.56-0.96}Sn_{0.66-1.22})_3[S_{3.98-4}\,Sb_{0-0.02}]$.

A somewhat rare trace element in alabandine ores is manganese-indium monosulfide, which is stoichiometrically similar to iron-indium sulfide – indite [4]. That's why we tentatively name our mineral manganoindiit, assigning to it formula $MnIn_2S_4$. In our case this sulfide occurs mainly in the form of carbonate inclusions as big as 10 µm. It is characterized by admixtures of iron, copper, zinc and stannum. Some manganoindiit formations are marked by indium oversaturated disrupting their stoichiometry. This is suggestive of the presence of dissolved admixture of sesquisulfide In_2S_3 in its content in the amount of as much as 5–25 mole per cent %. An even more rare element is roquesite, whose inclusions, unlike those of manganoindite, are localized mainly in alabandine. The roquesite detected in our study is characterized by admixtures of zink, manganese and iron, which is reflected in its empiric formula – $(Cu_{0.92}Mn_{0.06}Fe_{0.02})(I_{n0.88}Zn_{0.12})S_2$. Finally, the most rare mineral, which presents a rather interesting case, is an indium mineral stoichiometrically similar to lapharetite. In our study, this phase had a proportion of silver was considerably bigger than that of indium, which can be clearly seen in the calculated empiric formula – $Ag_{1.05-1.53}(In_{0.32-0.84}Sn_{0.11-0.15}Fe_{0-0.22})_{0.45-0.95}S_2$. This difference speaks of the alabandine ores having lapharetite with a dissolved in it solidphase solution of some other silver sulfide with a similar to lapharetite stoichiometry.

In close association with indium and stannum-indium sulfides, there have been detected numerous sulfoantimonide (Figure 4) and sulfoarsenide phases with some of them having a significant share of indium.

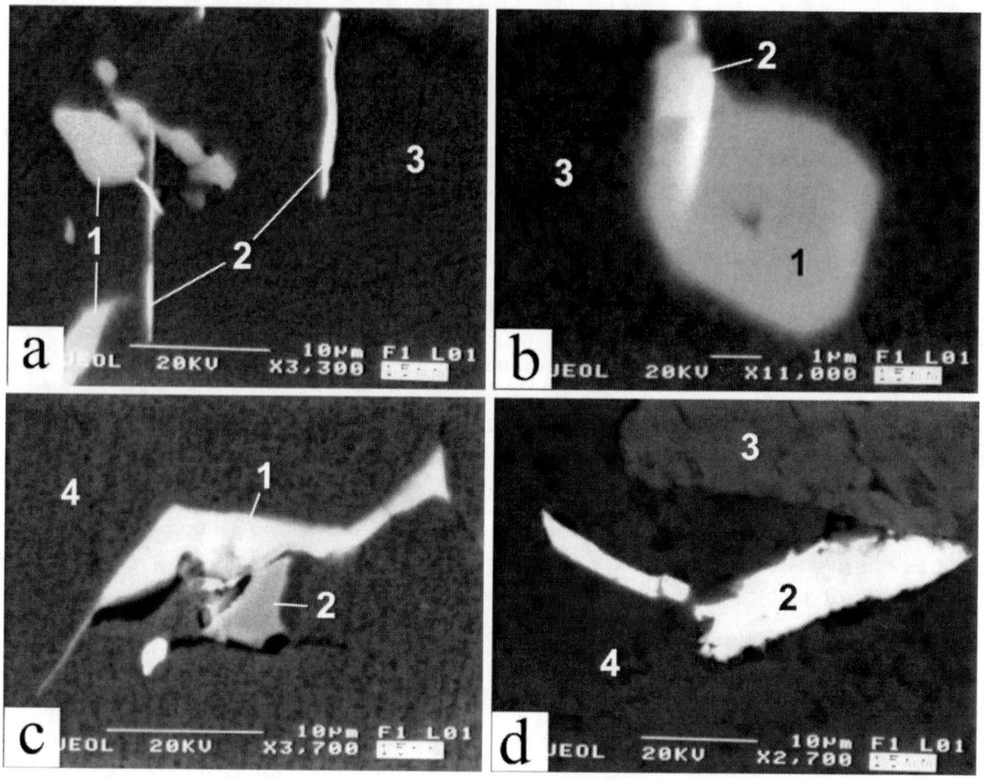

Figure 4. Association of indium sulfide and sulfoantimonides inclusions in alabantide (a–c) and carbonate (d). 1 – indium sulfides; 2 – sulfoantimonides; 3 – alabandine; 4 – rodochrosite. AFM-images in elastically reflected electron view.

Properties of Alabandine as Concentrator of Indium Microminerals

AFM images of the surface of alabandine individuals demonstrate typical rhombohedral-pyramidical etch pits that are characteristic of minerals of fluid-hydrothermal origin. Moreover, the polished specimens of alabandine reveal ultraporiness arising from a multitude of evenly distributed cavities of meso-nanometric size. Many areas of alabandine ores affected by brittle-plastic deformations give rise to development of small-scale cross-spur veinlets of manganese carbonates which create a picture of microcementation of crushed alabandine. Some alabandine specimens were analyzed with AFM. Subject to analysis were surfaces of recent shears, fractures, and surfaces of natural growth.

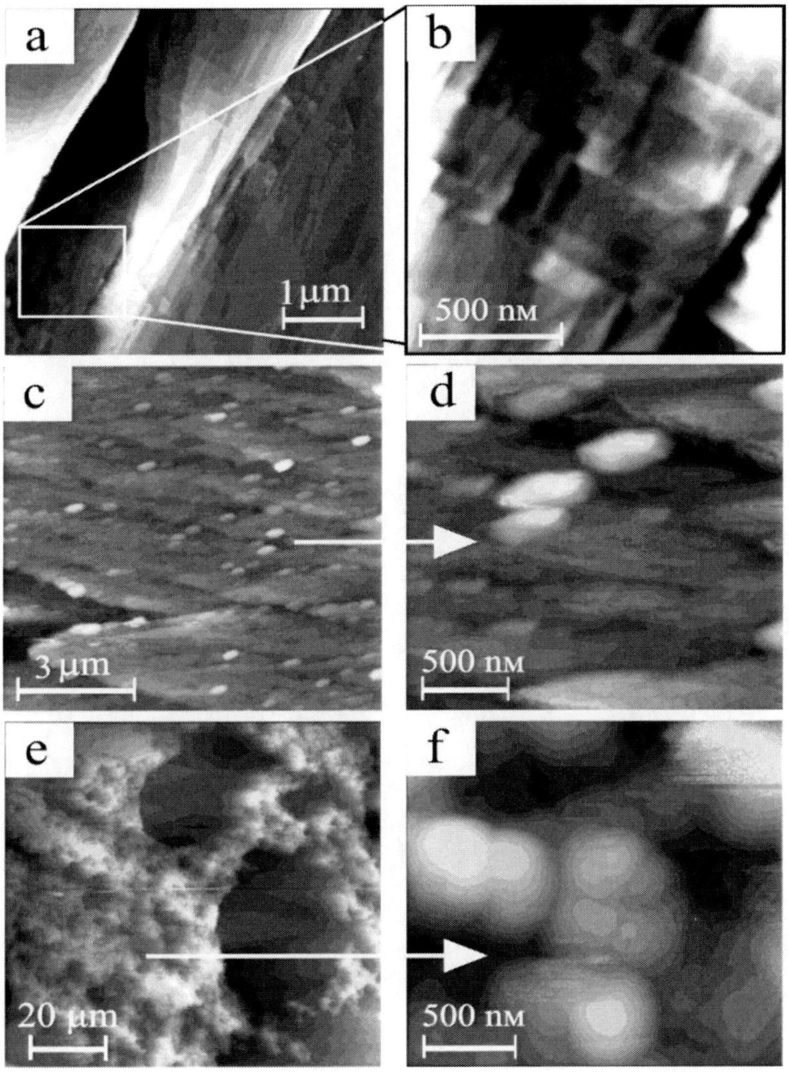

Figure 5. AFM-images of alabandine surfaces: a, b – step structure of crystal shear along cleavage; c, d –grains of micromineral inclusions on a sporadic shear surface; e – pores on a fracture surface; f – picture *e* zoomed, globular particles of alabandine oxidation products.

The surfaces of alabandine grain shears often have characteristic long steps (Figure 5, a). The systems of transversally crossing elementary steps on the images obtained with the help of high-resolution AFM show that even a natural crystal rich in inclusions usually splits into planes (001), associated with absolute cleavage of alabandine (Figure 5, b). On some surfaces of the shear there are uneven areas with broken steps a few nanometers high (Figure 5, c). According to some observations, these very areas are a place of concentration of oval-shaped micromineral inclusions with the diameter varying from 220 to 500 nm (Figure 5, d).

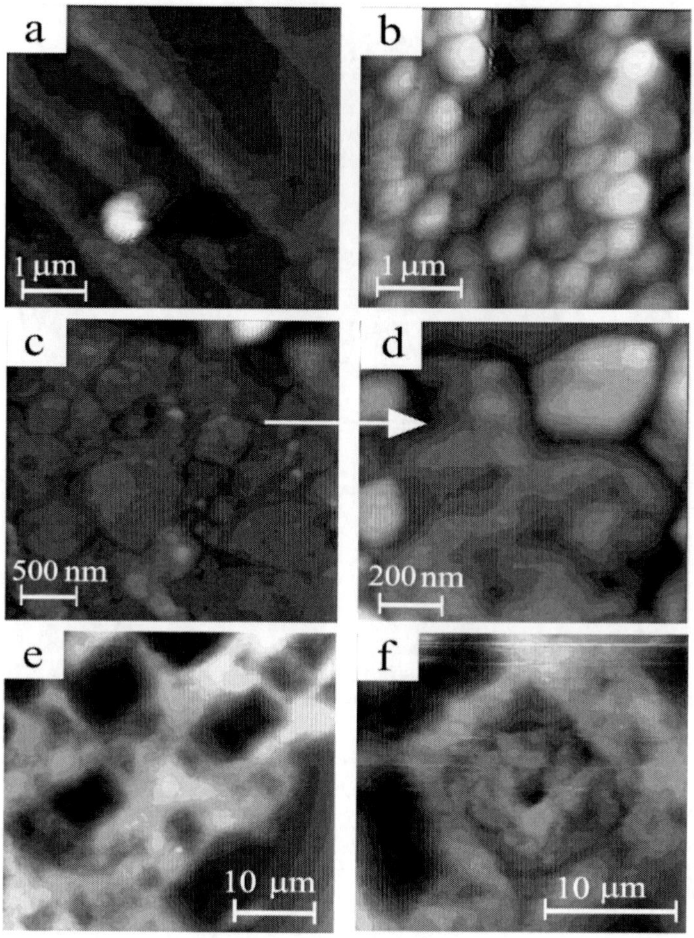

Figure 6. AFM-images of relief of an octahedral alabandine facet: a – 20 nm growth steps; b – surface globular skins of oxidation products; c, d – traces of decrepitation; e – cavities left after spalling of xenomineral inclusions; f – a particle of half-dissolved inclusion in one of the cavities.

Alabandine crystals are easily crushed along cross *fractures*, which are not visually detectable, but are diagnosed well by etching. The most characteristic feature of mantle rock on the walls of such microfractures are pores (Figure 5, e) of 300 nm to almost three micrometers in diameter. Besides, a detailed study of the walls revealed some thinnest epigenetic skins with meso-nanosize globular structure (Figure 5, f). Globule average size was 330 nm with average squared difference 100.9 nm, which is the evidence of its rather high competence (variation coefficient doesn't go much higher than 30 %). The presented below Raman-spectroscopy data show that globular skins consist of alabandine topochemical oxidation products.

It can be seen that the crystal faces have growth steps of different height, with terraces between them covered by skins of globular structure (Figure 6, a, b). Some areas of such surfaces experienced typical decrepitation resulting in grit fragments of 100 to 500 nm (Figure 6, c, d). This can be indicative of alabandine crystallization under fast and significant temperature reduction. Such temperature setting could also have caused partial dissolution of the surface, which can be concluded from the rectangular cavities left after spalled or dissolved xenomineral idiomorphic inclusions (Figure 6 e, f).

The most part of the studied alabandine contains only a small amount of iron, having the empiric formula $(Mn_{0.88–1}Fe_{0–0.12})S$. In addition, there are two rare (5–6 %) phases, one of which has an almost even content of manganese and iron – $(Mn_{0.33–0.59}Fe_{0.41–0.67})S$, and the other is characterized by admixtures of not only iron, but also of copper, indium and stannum – $(Mn_{0.86–0.96}Fe_{0.01–0.02}Cu_{0–0.04}In_{0.03–0.20}Sn_{0–0.06})S$. The analysis of alabandine content reveal that alabandine grains beneficated by trace elements are the least sulphur stoichiometric. The level of stoichiometry of alabandine increases with the appearance of micro-nano formations of other sulfides. All this can be explained by the presence and later decomposition of alabandine based solid solutions, which leads to formation of a more admixture-wise pure manganese sulfide. The aforementioned admixture beneficated alabandine specimens most probably present not affected or the least affected by exsolution primary phases.

Structural analysis of alabandine and phase diagnostics of finally dispersed inclusions and oxidation products was conducted by the X-ray diffraction and Raman-spectroscopy methods. The results showed that alabandine has four brightest and typical reflections (given in nm in the order of decrease of intensity, indices in brackets *hkl*): 0.2602–0.2609 [200] – 0.1841–0.1846 [220] – 0.1507 [222] – 0.3007–0.3015 [111]. Calculated with this data values for the parameter of unit cell range between 0.5215–0.5220 ± 0.0001 nm. We can conclude that the value of interplanar spaces and the parameter of unit cell, which directly depends on the first, inversely correlate with the iron content determined in both alabandine samples, coefficients of pair correlation ranging between 0.5–0.9. This speaks of a significant and strong minus relation between the considered parameters of the mineral constitution. Such correlation can be explained only by replacement of some part of larger (0.091 nm) ions Mn^{2+} by smaller (0.080 ем) ions Fe^{2+}.

The Raman-spectroscopy analysis of alabandine was done with the help of lasers 532 and 514 nm, the illumination power ranging from 1.3 to 130 mW. The received spectra were compared to the spectra in RRUFF data base.

The studied alabandine Raman-spectrum was discovered not to correlate with the spectrum assigned to the Mexican alabandine in RRUFF (Mina Preciosa, Puebla, Mexico). In our case, we registered only a whole-space line with weak peaks of Raman dispersion in the diapason of 100–600 sm^{-1} (Figure 7). This seems very logical, since the active vibrations in alabandine lattice with NaCl structure and Fm3m symmetry aren't excited in combination dispersion mode. This mineral can have only one optic mode of vibrations F_{1g} c $k = 0$, changing the dipolar moment, but not polarizing the lattice.

That is the reason why alabandine is active in IR spectrum (has a wide peak), but doesn't reveal any activity in the combinational light dispersion mode which is unable to provide electro-magnetic re-emission of light with a shift to vibration frequency. Therefore, the peaks of Raman dispersion in alabandine Raman-spectra can be caused only by admixture phases, in particular alabandine oxidation products. Our studies, conducted by Raman-spectroscopy method, did reveal existing on alabandine surfaces a multitude of nano-micrometer grains of pyrochroit $Mn(OH)_2$ and manganite MnO(OH), detected even at the lowest (1.3 mW) laser power (Figure 8). It was also discovered that manganite can be formed synthetically by a short-term exposure over alabandine emissions by a power of 13 mW and stronger. On the growth facets and fractures surfaces in the studied alabandine, there were detected numerous micro-areas with nano-thick skins of native sulphur (Figure 9), which are also, apparently, products of sulfide oxidation. Moreover, shear surfaces have numerous inclusions of primary metals, for example rhodochrosite and siderite, sometimes combined with the products of their own oxidation (Figure 10, 11).

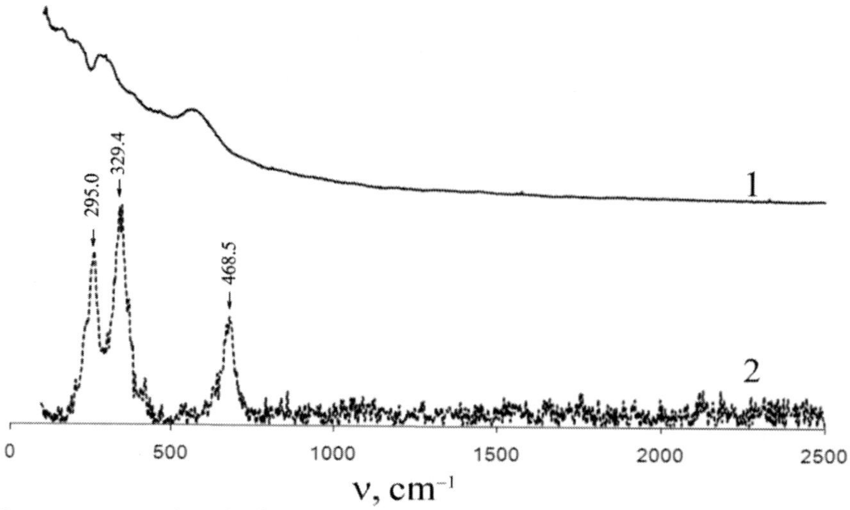

Figure 7. Raman-spectra: 1 – obtained on a growth facet of the studied alabandine (laser 514 nm, power 13 mW); 2 – assigned to Mexican alabandine from RRUFF data base (laser 532 nm).

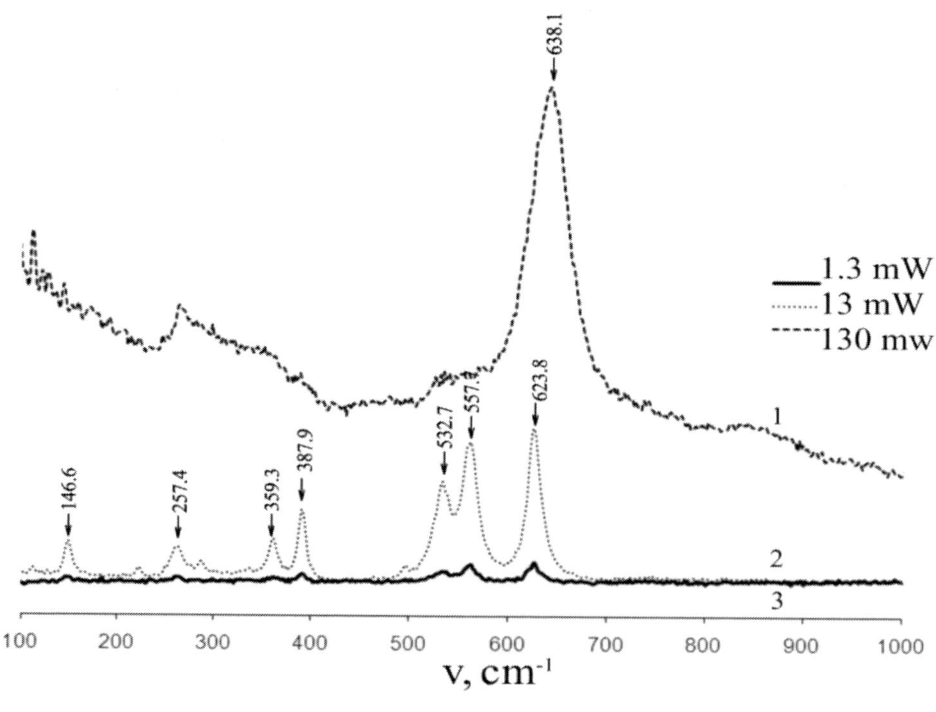

Figure 8. Raman-spectra associated with alabandine oxidation products, received at a different laser radiation power: 1 – pyrochroit, 130 mW; 2 – manganite, 13 mW; 3 – manganite traces, 1.3 mW.

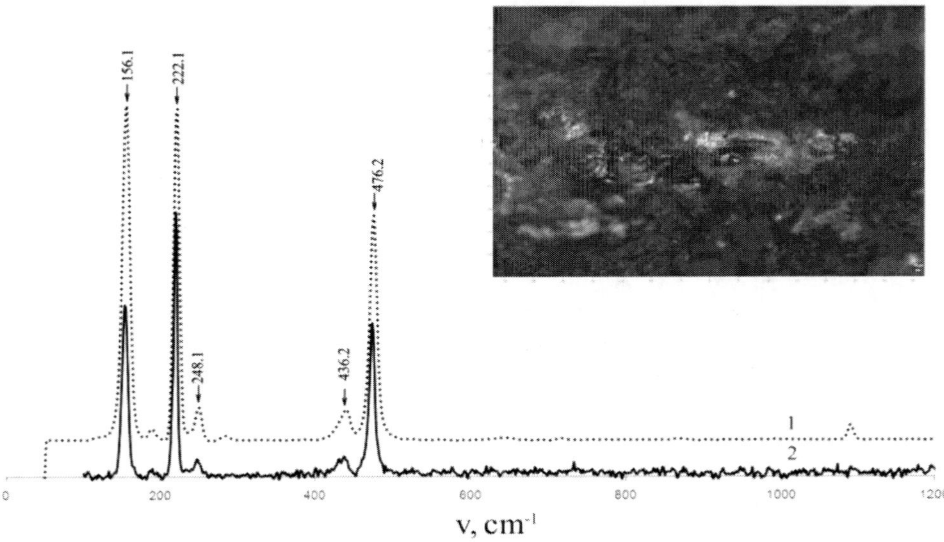

Figure 9. Raman-sectrum for native sulphur from RRUFF data base (1) and the spectrum received on the oxidated surface of the studied alabandine with skins of native sulphur (2).

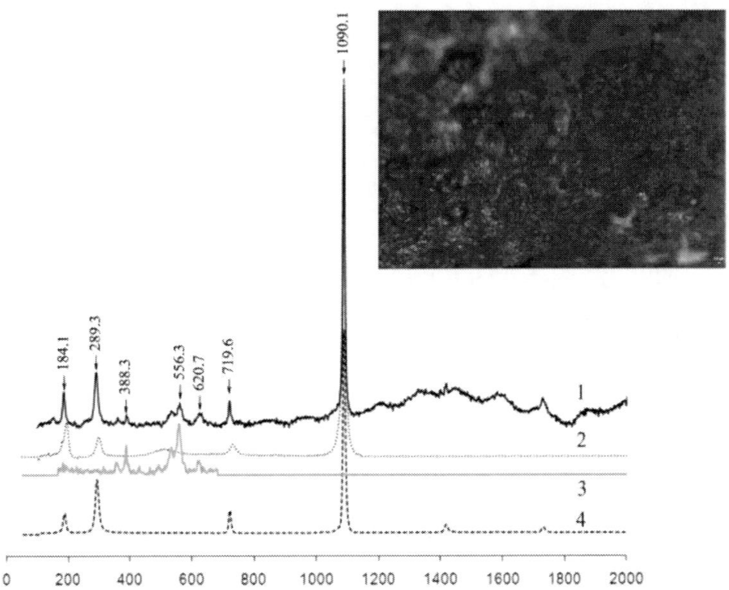

Figure 10. Raman-spectrum received on alabandine shear surface from carbonate inclusions and manganese products of their oxidation (1), in combination with spectra of siderite (2), manganite (3) and rhodochrosite (4) taken from RRUFF data base.

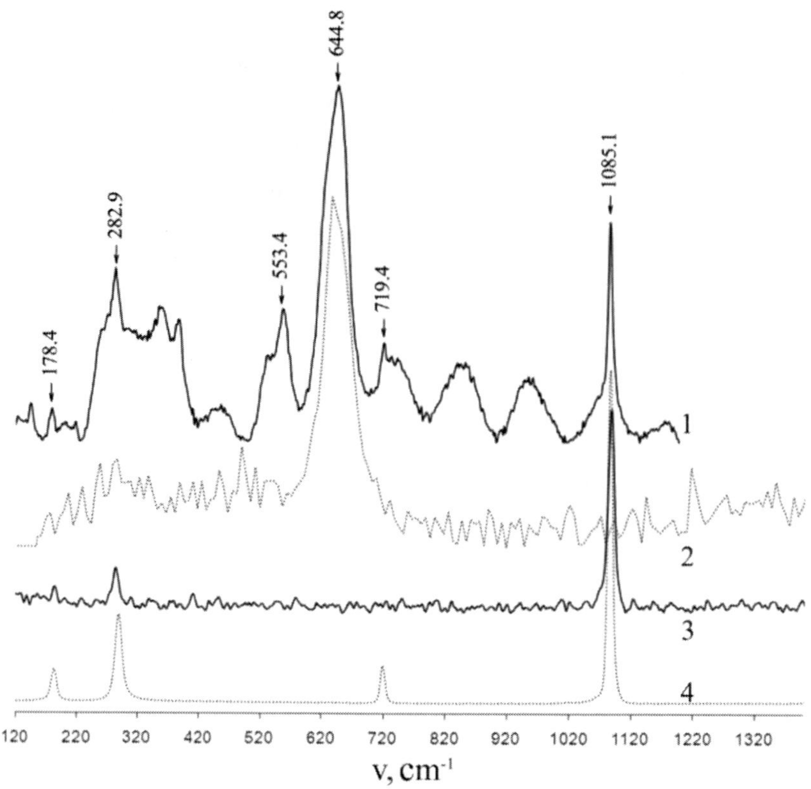

Figure 11. Raman-spectrum received on alabandine surface with admixtures of carbonates and iron-bearing products of their oxidation (1) together with spectra for hematite (2) siderite (3) and rhodochrosite (4) taken from RRUFF data base.

We attach particular importance to detection and reliable Raman-spectroscopy phase diagnostics of a idiogenous to alabandine carbonic matter (CM). In this case, we used laser radiation of a minimum (1.3 mW) power to prevent the process of technogenic smut formation. The received spectra (Figure 12) have two typical of CM wavelengths of IR-spectrum absorbtion, associated with A_{1g}- E_{2g2} vibration modes of carbonium atoms (at 1318.3 and 1573.9 sm^{-1} correspondingly).

These wavelengths occurred shifted far into the long-wavelength area compared to their location in the crystalline graphite spectrum [9]. It is also noteworthy that in the Raman-spectrum of the analyzed CM, the long-wave peak is more intensive than the short-wave peak in contrast to technogenic smut whose spectrum has a converse relationship of peaks intensities.

All this can be explained by amorphism and a rather low level of metamorphization of CM. According to the results of an isotopic geochemical study (Table 1), the considered CM compared to the CM in sedimentary Phanerozoic rocks and Vaygach-South-Novozemelsk stratiform polymetallic ore formations is characterized by 2-7 ‰ heavier carbon and 1-3 ‰ heavier nitrogen.

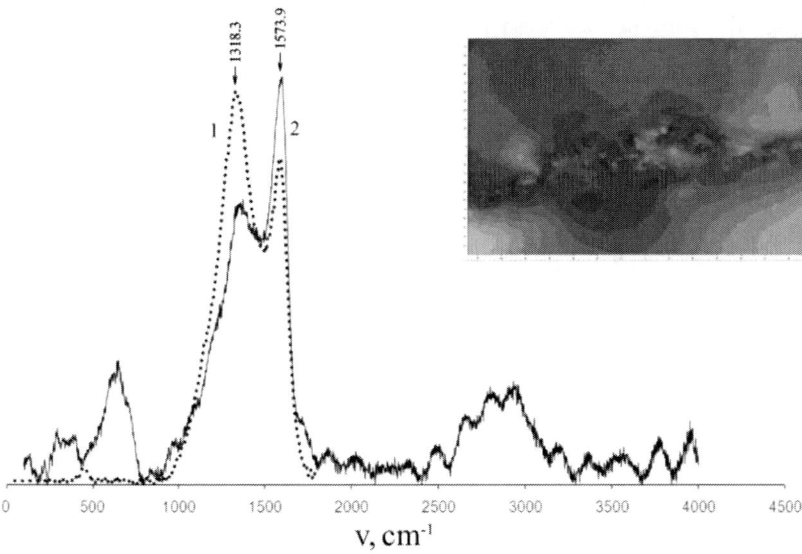

Figure 12. Raman-spectra on alabandine surface with carbonic matter inclusions: (1) and spectrum of one kind of technogenic smut from Sosnogorsk Gas Processing Plant (Russia) (2).

Compared to CM in Palaeoproterozoic rocks and sedimentary manganese ores (supergroup Transvaal, S. Africa), the studied CM is isotopically 1-4 ‰ heavier in carbon and 1–6 ‰ in nitrogen. Moreover, isotopic contents of carbon and nitrogen in CM indium ores are discovered to correlate directly, whereas in CM of sedimentary rocks and anthraxolite they have inverse correlation.

Table 1. Results of dual determination of isotopic content of carbon and nitrogen in carbonic matter from alabandine ores

Order number	Sample weight, mkg	$\delta^{13}C$,‰ PDB	$\delta^{15}N$, ‰ atm
1	1020	–23.14	0.91
2	672	–24.33	–1.26
3	985	–21.70	Not determined
4	882	–22.88	0.98

Note: measurement inaccuracy is ± 0.2 ‰ (1σ).

Structural Iron as an Indicator of Isomorphic Capacity of Alabandine

Alabandine at Vysokogornoye deposit presents interest not only as a source of manganese and indium, but also as an independent element, being a promising material for creating new semiconductor devices with colossal magnetoresistance effect. This is possible, because with the typical to it face-centered lattice of NaCl type, alabandine has a special order of the magnetic structure, determined by the number and spatial arrangement of admixture

Fe^{2+} ions. [1, 3, 11]. Up to the present day, all the studies of electronic and magnetic structure and the corresponding properties of manganese sulfide have been conducted exclusively on synthetic materials. We were the first to use Moessbauer spectroscopy to analyze structural states of ions of admixture iron in naturally occurring alabandine.

The objects used for our Moessbauer research were a series of typical alabandine samples that had a bulk admixture of iron ranging from 1.3 to 7.6 mass %. Moessbauer spectra were received at MS-1104Em spectrometer in speed diapason from –11 to +11 mm/sec at the ambient temperature of the samples. To avoid the texturing effect in the spectra, the preparations were made as frayed sampling materials. Isometric shift was determined in relation to α-Fe.

The Moessbauer spectra reveal three peaks of gamma-quantum absorption, that we interpreted as a singlet with an isometric shift of around 1mm/sec and a doublet with the splitting of around 0.6 mm/sec and isometric shift of about 0.3 mm/sec (Figure 13). The correlation of the spectra components varies a lot. The maximum contribution of the doublet is observed in the spectra received from low-iron content samples (Alb-4, 5; Fe = 2.7–2.9 mass %), and the minimum one – in the spectra received from high-iron content samples (Alb-1, 2; Fe = 3.8–7.6 mass %). A doublet with the same parameters was detected in the spectrum of pyrite, selected purely from indium sulfide-manganese ores (Alb-6P). The parameters of the spectrum answer the literature data on pyrite. The traces of pyrrhotine phase, discovered in sulfide-manganese ores by AFM method, were not detected.

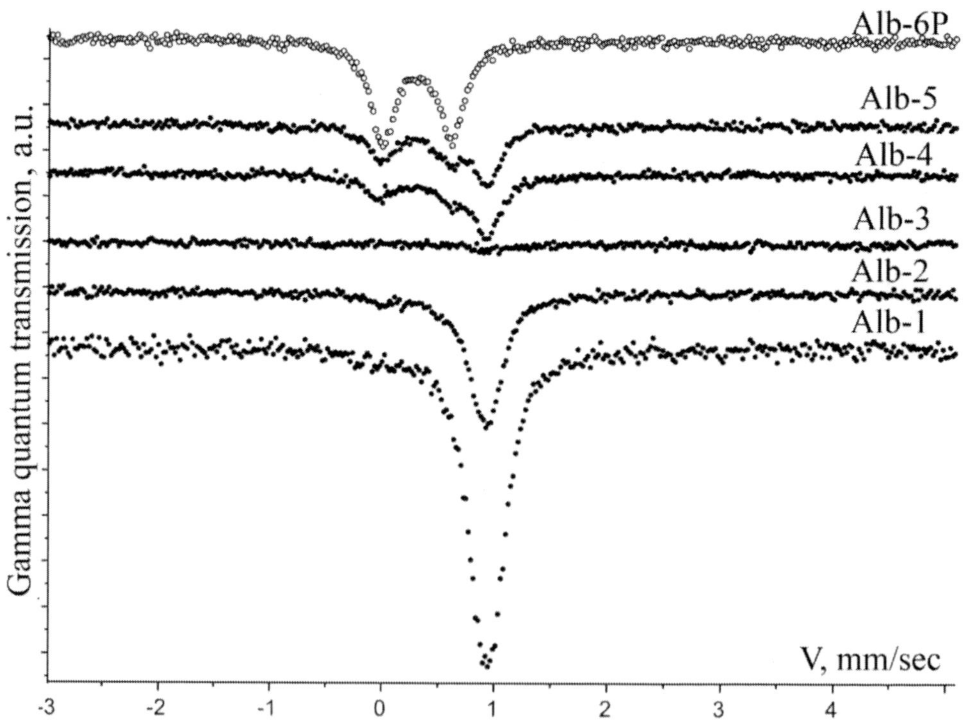

Figure 13. Panoramic Moessbauer spectra of indium sulfide-manganese ores (Alb-1 – Alb-5) and pyrite admixture in them (Alb-6P).

Fitting of the received spectra (Figure 14) confirms the adequacy of the dual-component model presented by a singlet with an isometric shift of about 0.95 mm/sec from isomorphic iron in alabandine itself and a doublet from pyrite inclusions. He results of the analysis and the parameters of Moessbauer spectra signals are presented in Table 2.

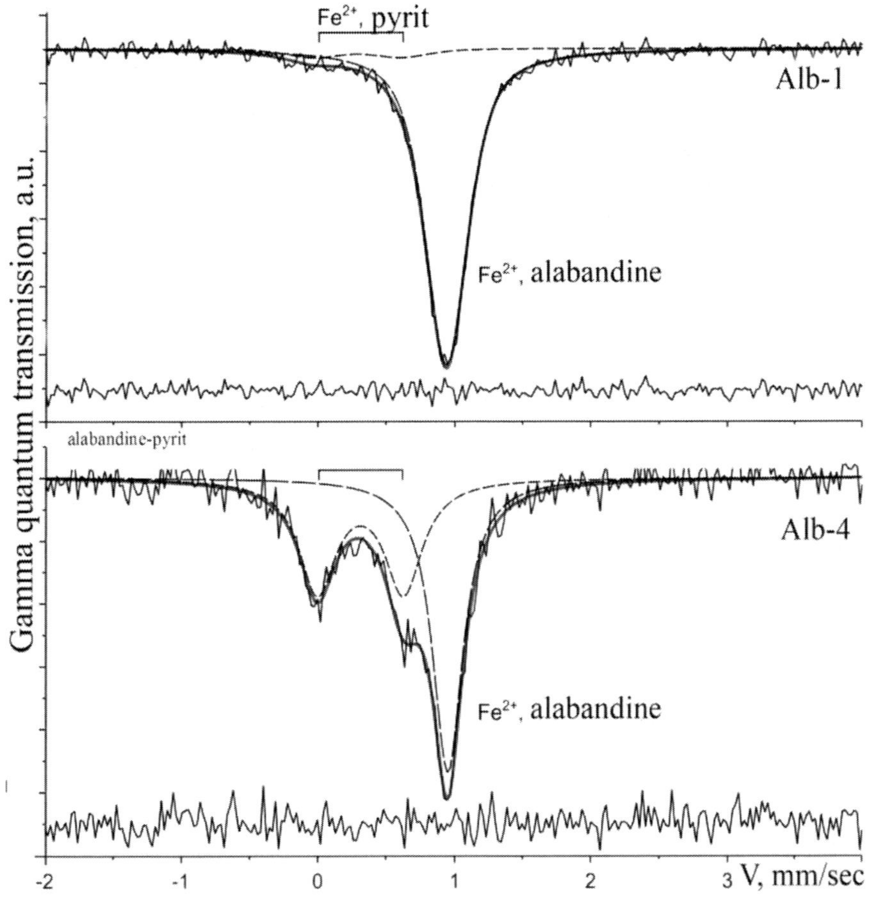

Figure 14. Results of fitting of Moessbauer spectra of alabandine with high (Alb-1) and low (Alb -4) iron content. The bold line – model spectrum. Dotted lines – detected components. Fine lines below the spectrum – the difference between the experimental and model spectrum.

Integral value of Moessbauer effect with absolute coefficient of correlation is proportionate to bulk iron content in sulfide-manganese ores (Figure 15, a). The content of divalent iron ions is proportional to integral intensity of absorption in alabandine singlet peak (Figure 15, b). The remainder of bulk iron content less of its divalent form in alabandine correlates well with the integral intensity of Moessbauer doublet of pyrite (Figure 15, c). The heterophase admixture in alabandine, mainly pyrite, contains from 5 to 60 % of total iron content.

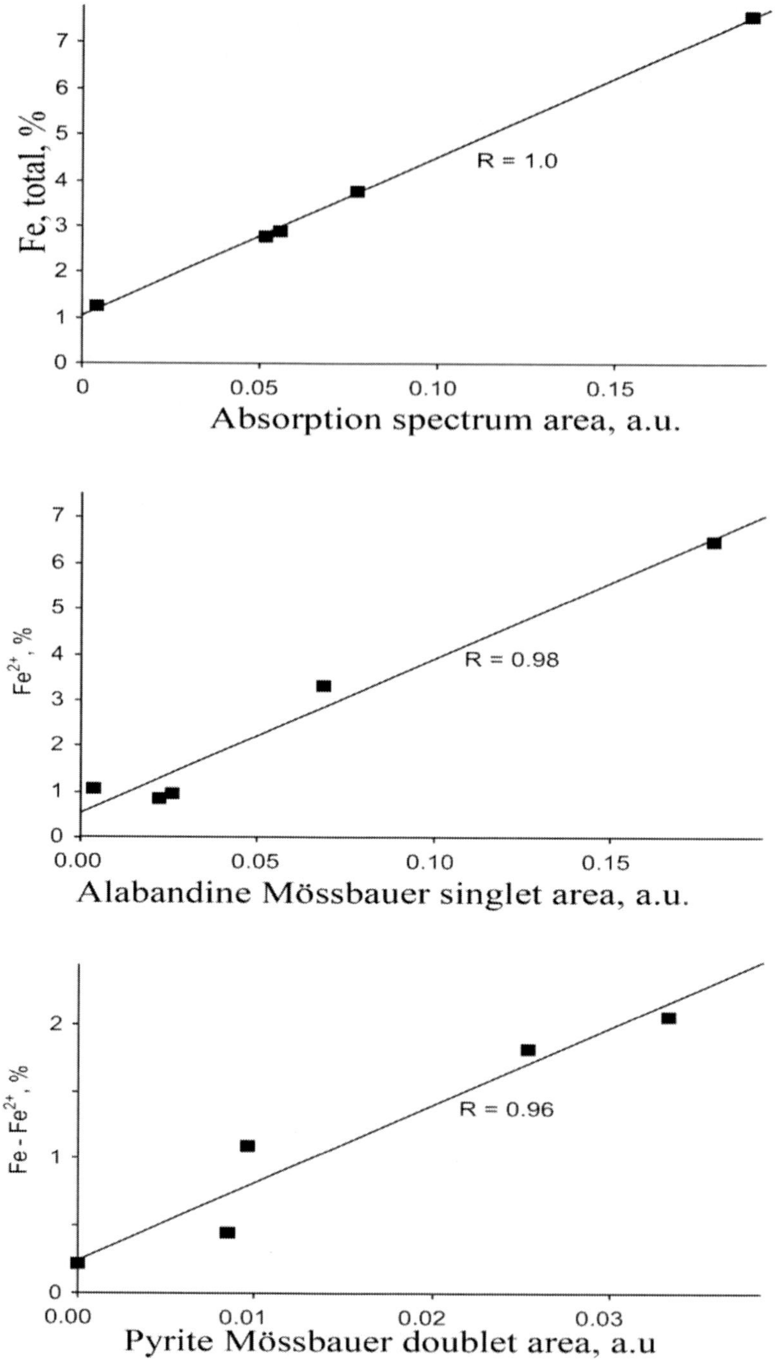

Figure 15. Correlation of Moessbauer effect value with iron concentration in indium sulfide-manganese samples: a – spectrum absolute area with bulk iron content; b – partial area under singlet corresponding to alabandine and Fe^{2+} contents in alabandine lattice; c – partial area under doublet corresponding to pyrite and differential of bulk iron content and Fe^{2+} content in alabandine.

Table 2. Parameters of Moessbauer spectra of indium sulfide-manganese ores with admixture of isomorphic and heterophase iron

no. of sample	Alabandine			Pyrite				Fe, mass%	
	IS, mm/sec	G, mm/sec	A, %	QS, mm/sec	IS, mm/sec	G, mm/sec	A%	Gross	Isomorphic (Fe^{2+})
Alb -1	0.940± 0.002	0.371± 0.004	94.9± 0.7	0.60	0.31	0.3	0.51± 0.8	7.59	6.50
Alb -2	0.948± 0.002	0.337± 0.005	89.0± 0.8	0.60	0.31	0.3	11.0± 1.0	3.76	3.30
Alb -3	0.97± 0.03	0.4± 0.1	100	–	–	–	–	1.27	1.05
Alb -4	0.953± 0.003	0.27± 0.01	51.1± 1.1	0.632± 0.012	0.314± 0.007	0.34± 0.02	48.9± 1.7	2.77	0.96
Alb -5	0.954± 0.003	0.26± 0.01	40.6± 0.9	0.603± 0.008	0.316± 0.004	0.32± 0.01	59.4± 1.4	2.90	0.83
Alb -6P	–	–	–	0.602	0.304	0.28	100		

Note: IS – isometric shift; QS – quadrupole splitting; G – half-width of the peaks, A – relative area of spectrum signal.

Table 3. Moessbauer parameters of ferromanganese alloy

Spectrum	Ferromanganese, Fe^0				Ferralumosilicit, Fe^0			
	QS, mm/sec	IS, mm/sec	G, mm/sec	A %	QS, mm/sec	IS, mm/sec	G, mm/sec	A %
Original	0.234 ± 0.005	–0.050 ± 0.002	0.35 ± 0.001	88 ± 1	0.74 ± 0.03	0.06 ± 0.02	0.35 ± 1	12 ± 2
n-spectrum	0.230 ± 0.002	–0.045 ± 0.001	0.224 ± 0.003	86 ± 1	0.75 ± 0.01	0.083 ± 0.006	0.224 ± 0.003	14 ± 1

Cubic symmetry (m3m) and ion type of chemical bonds in alabandine are known to be conditioned by coordination of every ion Mn^{2+} by six sulphur atoms at a rather symmetrical arrangement of 12 ions of manganese in the second coordination sphere. The determined singlet character of Moessbauer spectrum from isomorphic iron ions with zero quadrapole splitting (QS) testifies to the cubic context being retained even when Mn^{2+} ions are replaced by Fe^{2+} ions. However, a high concentration of admixture ions may stimulate superexchange interaction between equivalent Fe^{2+} ions connected by sulphur ions, which leads to quadrupole splitting of Moesbauer absorption peak. Moessbauer spectra of synthetic analogies of alabandine $Fe_xMn_{1-x}S$ with iron content of over 15 % are known to have a low resolution doublet with 0.2 mm/sec splitting and an isometric shift of 0.92 mm/sec. A close value of isometric shift was received for the natural occurring alabandine (Table 2). The row of samples with a progressing iron content increase (Alb-5 → Alb-4 → Alb-2 → Alb-1) has a stable trend of singlet (G) widening from 0.26 to 0.37 mm/sec. This can be due to an unresolved quadrapole splitting appearing as a result of the local decrease of lattice symmetry because of formation of segregations of isomorphic iron ions in alabandine structure.

The Product of Indium Sulfide-Manganese Ores Processing

At present, there are two technologies of sulfide-manganese ores conversion in Russia: 1) into ferromanganese (A.N. Seregin, Ferrous Metal Research Institute, Moscow) and 2) into ultrapure manganese metal (A. L. Baturin, Research Institute of Physical and Organic Chemistry of South Federal University, Rostov-on-Don). Both technologies were established before industrially potential indium admixture was discovered, that's why the developers weren't concerned with preserving this admixture in the products. The results of our research show that ferromanganese alloy derived from indium sulfide-manganese ores has the following average composition (mass %): Mn 78.38; Fe 4.88; Al 9.07; Si 5.75; S 1.23; V 0.28; Cr 0.24; Ni 0.17. The analysis revealed that even though technological conversion significantly changes the concentration of many admixture elements (Figure 16), 80 % of original indium, in spite of its high volatility, is preserved.

The main phase in converted products, according to the X-ray diffraction analysis record, is cubic modification of ferromanganese α-$FeMn_4$ (Figure 17). The X-ray diffraction pattern of the alloy has a rather intensive reflex of d/n = 0.261 nm, corresponding to the reflection from (200) in alabandine structure. Moessbauer spectrum of ferromanganese contains a wide, rather asymmetric singlet a detailed analysis of which is performed method of restricting peaks following the algorithm invented by G.V. Novikov [6]. As a result we received a so called n-spectrum having more restricted signals compared to the original spectrum. Table 3 and Figure18 present the results of fitting and nuclear gamma resonance parameters.

The contour of the spectrum consists of a pair of doublets, the doublet with narrow separation (0.23 mm/sec) and nearly zero negative isometric shift predominating. Only 10 % of iron are contained in the second doublet with a wide separation (0.74 mm/sec) and also nearly zero, but positive isometric shift. The main doublet (88 % of area) can be referred to the structure α-$FeMn_4$. The doublet with wide separation is referred to Fe^{2+} ions in low-symmetry iron-bearing phase. The chemical composition record and the obtained by

Moessbauer spectroscopy method data on distribution of iron ions allow us to assume that the alloy under consideration is mainly composed of the combination α-$(Fe_{0.22}Al_{0.76}Cr_{0.01}Ni_{0.01})Mn_4$. A small part of manganese (2.7 %) remained unrecovered. About 12 % of iron is contained in low-ordered ferrialuminosilicit of $(Fe_{0.19}Al_{0.71}V_{0.10})Si_{3.88}$ composition.

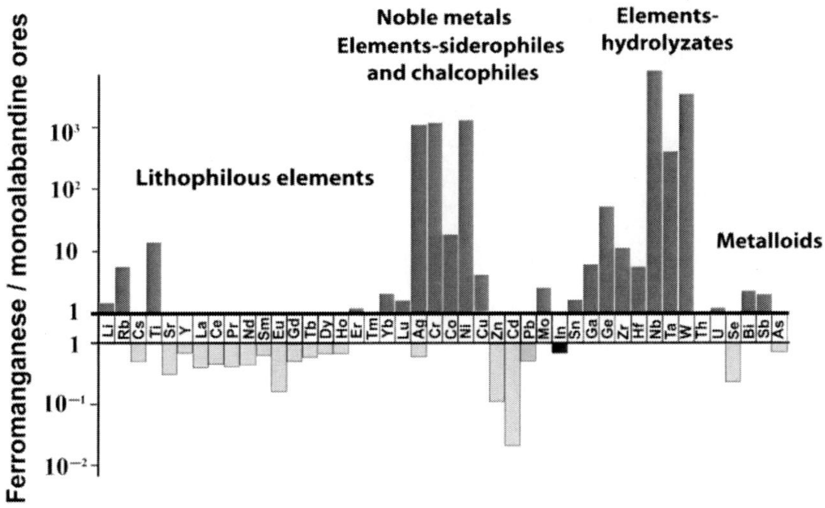

Figure 16. Concentration clarkes of the essential admixture elements in ferromanganese alloy derived from indium sulfide-manganese ores.

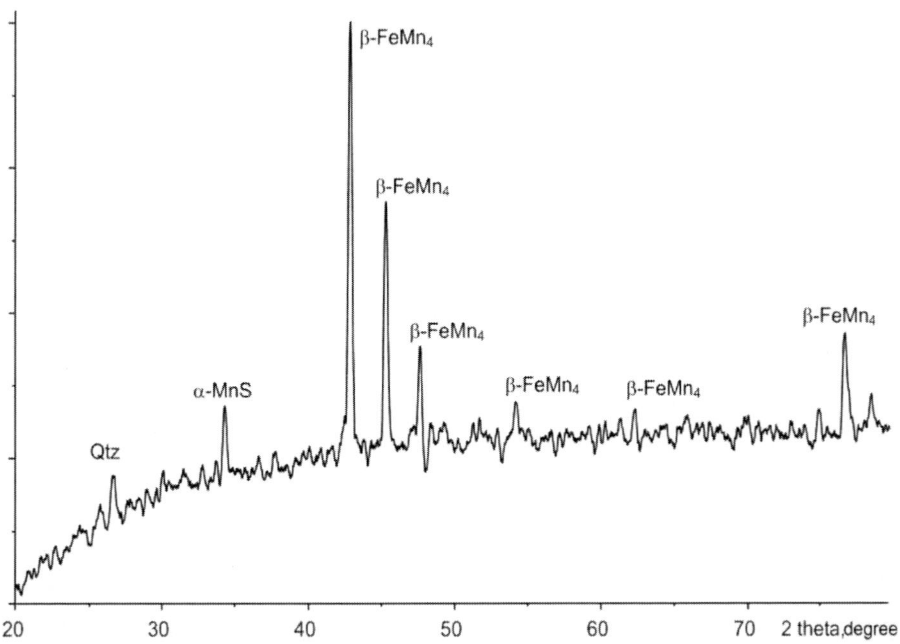

Figure 17. Typical X-ray diffraction pattern of ferromanganese alloy derived from indium sulfide-manganese ores.

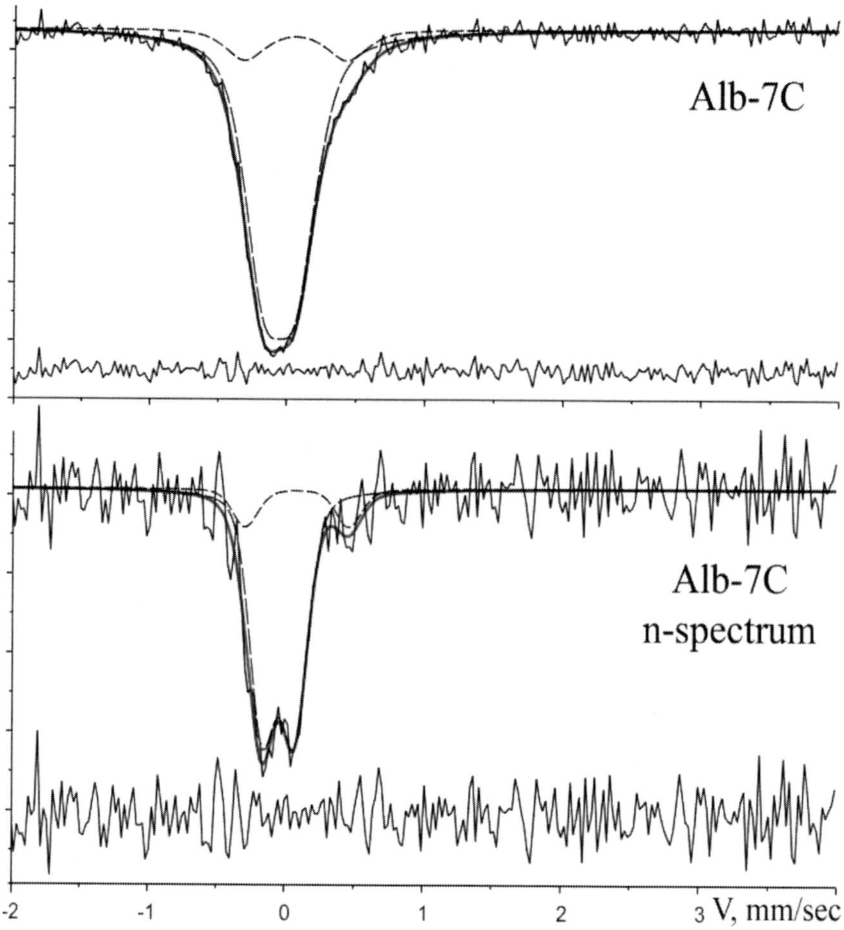

Figure 18. Fitting of the original and restricted (n-spectrum) Moessbauer spectra of ferromanganese alloy. The dotted lines show detected signals, the solid think line - summary simulated spectrum, at the bottom of the spectra – differential of the experimental and simulated spectra.

Conclusion

The results of the conducted research bring us to the conclusion that the discovered in Russia indium sulfide-manganese deposit presents a phenomenal natural object. The mineralization that has developed at this deposit is high-temperature (with crystallization of ore minerals at 460–500°C) and, according to the aggregative state of mineral-forming medium, pneumatolyth hydrothermal. However, a dramatic predominance of manganese over iron doesn't let it fit into the existing schemes of endogenic mineral-geochemical differentiation.

The deposit has received a unique in character and intensity indium benefication of sulfide-manganese ores, present not only in the form of numerous proper mineral, but also isomorphic admixture in various sulfides and sulfosalts of iron, copper, zinc, stannum, lead and silver. Another peculiarity is that the benefication was performed by a rather volatile metal, easily subject to subliming. This can be explained by a significant content of stannum admixture in sulfide-manganese ores, presented by all main types and classes of minerals –

chalcogenides, oxides, and silicates. The geological practice shows stannum-containing deposits to be the most indium-bearing and stannum-contaning minerals to be most common concentrators of this rare metal.

It is noteworthy that indium and a considerable, if not major, part of other economically promising microelements is presented in sulfide-manganese ores by minerals that have nano-micrometer size of individuals (from a few hundred nanometers to 100 mkm). Moreover, there is a reason to believe that the stated above dispersion level isn't extreme, reaching up to the first nanometers. The reliable experimental data show that the discovered in many ore minerals uniformity of microindividuals bulk distribution and low coefficients of their size variation can be indicative of indium sulfide-manganese ores forming in the conditions of a very high supersaturation.

Extensive signs of decomposition of high-temperature solid solutions at the deposit, conservation of many undecomposed solidphase mixtures or mixtures with incomplete exsolution in alabandine ores, non-stoichiometry of many phases speak, firstly, about crystallization of ore minerals in rather non-equilibrium conditions and in a specific geochemical medium, and secondly, about a rapid decrease of mineral formation temperature. It's quite possible that the conservation process of solid solutions was very much affected by the intensive invasion of manganese admixture into practically all minerals.

Lamprophyr and lamproite formations in the ore cluster, occurrence of minerals of axinite group in the ores, a rather high concentration of femic (basitophylic) microelements, rhenium benefication, which was already by academician A.E. Fersman claimed to spring from nearly the border between lower mantle and Earth core, high value of Co/Ni coefficient – all this makes us suppose that the considered mineralization process is related to the deep, perhaps mantle magma chamber.

ACKNOWLEDGMENT

The authors would like to acknowledge O.V. Lysyuk, G.N. Modyanova, S.T. Neverov, V.A. Radayev, I.V. Smoleva, and S.N. Shanina for their assistance in chemical part of the research.

REFERENCES

[1] Aplesnin S.S.; Ryabinkina L.I.; Abramova G.M. *Solid state physics* (on russ.). 2004, vol.46, № 11, pp. 2000–2005.

[2] Gavrilenko V.V.; Pogrebs N.A. Zapiski *VMO* (on russ.). 1992, vol. 121, № 2, pp. 41–47.

[3] Heikens H. H.; van Bruggen C. F.; Haas C. J. *J. Chem Solids.* 1978, vol. 39, pp. 833–840.

[4] Genkin A.D.; Muravieva I.V. Zapiski *VMO* (on russ.). 1963, vol. 92, № 4, pp. 445–457.

[5] Ishihara S.; Endo Yu. *Bull. Geolog. Japan.* 2007, vol. 58 (1/2), pp. 7–22.

[6] Kissin S. A.; Owens D. R. *Canad. Mineral.* 1989, vol. 27, № 4, pp. 673–688.

[7] Kokin A.V.; Silaev V.I.; Baturin D.V.; Filippov V.N. Yakutia alabandine – a new mineral type of industrial manganese mineralization; Rostov-on-Don: *"Rostizdat"*, 2011, 208 p.
[8] Kokin A.V.; Silaev V.I.; Kiseleva D.V.; Filippov V.N. Doklady *RAN* (on russ.). 2010, vol. 430, № 3, pp. 359–364.
[9] Kholodkevish S.V.; Berezkin V.I.; Davydov V.Yu. *Solid state physics* (on russ.). 1999, vol. 41, № 8, pp. 1412–1415.
[10] Komarova G.N.; Novorossova L.E. Geokhimia (on russ.). 1959, №8, pp.861-867.
[11] Loseva G.V.; Ryabinkina L.I.; Ovchinnikov S.G. *Solid state physics* (on russ.). 1991, vol. 33, № 11, pp. 3420–3422.
[12] Mormil S.I.; Amosov L.A. *Mineral raw materials* (on russ.). 2000, № 6, pp. 135–143.
[13] Murao S.; Deb M.; Furuno M. *Ore Geology Reviews.* 2008, vol. 33, pp. 490–504.
[14] Novikov G.V. (2006). *N-procedure.* URL: http://messbauer.iem. ac.ru/rus/n-procedura.html.
[15] Ohta E. Japan. *Mining Geology*. 1989, vol. 39, № 6, pp. 355–372.
[16] Petruk W. *Canad. Mineralogist.* 1973, vol. 12, № 1, pp. 46–54.
[17] Semenyak B.I.; Nedashkovich A.P.; Nikulin N.N. *Ore geology* (on russ.). 1994, vol. 36, № 3, pp. 230–236.
[18] Shimizu M., Kato A., Shiozawa T. *Canad. Mineral.* 1986, vol. 24, pp. 405–409.
[19] Sinclair W. D., Kooiman G. J. A., Martin D. A., Kjarsgaard I. M. *Ore Geology Reviews.* 2006, vol. 28, pp. 123–145.
[20] Wood S. A., Samson I. M. *Ore Geology Reviews.* 2006, vol. 28, pp. 57–102.

In: Indium
Editors: Hsaio G. Woo and Huang Tsai Choi

ISBN: 978-1-62257-696-8

Chapter 10

TIN-DOPED INDIUM OXIDE THIN FILMS FABRICATED BY THE SOL-GEL METHOD

Chu-Chi Ting
Graduate Institute of Opto-Mechatronics Engineering,
National Chung Cheng University, Chia-Yi, Taiwan, Republic of China

ABSTRACT

Tin-doped indium oxide (ITO) thin films were fabricated by the sol-gel spin-coating method with different indium sol-gel solutions synthesized from different indium precursors: $In(NO_3)_3$ or $InCl_3$ (denoted as N-ITO and Cl-ITO, respectively). For both N-ITO and Cl-ITO thin films, the increase of mobility and the reduction of carrier concentration/resistivity with increasing annealing temperatures from 400-700 °C are related to the increase of crystallization/densification and the annihilation of oxygen vacancies. The refractive index (1.84 at λ = 550 nm), packing density (0.83), resistivity (4.27×10^{-3} Ω-cm), and optical band gap (3.95 eV) of N-ITO thin films are higher than that of Cl-ITO thin films, which can be attributed to the higher densification, lower crystallinity, and more free charge carriers of N-ITO thin films. These properties make the indium nitrate-derived ITO thin films have better potential applications for some commercial products.

1. INTRODUCTION

Tin-doped indium oxide (ITO; Sn^{4+}-doped In_2O_3) thin films have attracted many researchers to study because of its good transparent and conductive properties for the application as the transparent electrodes of light-emitting diodes [1], liquid crystal displays [2], and thin-film solar cells [3].

Indium oxide possesses the cubic bixbyite structure with a lattice constant of 10.117 Å. (C-type, rare-earth sesquioxide structure: *Ia*31 space group, No. 206). It is well known that the conductive mechanism of ITO thin film is due to the formation of free charge carriers created by Sn^{4+} ions doping and oxygen vacancies in In_2O_3 lattice [20, 21]. The generation of one oxygen vacancy can create two free charge carriers [36]. In addition, two Sn^{4+} ions replacing the position of two In^{3+} ions in In_2O_3 lattice can also create two free charge carriers. Furthermore, 10 mol% Sn^{4+} ions can dissolve into In_2O_3 lattice to form the ITO solid solution with lowest resistivity [4].

There are many methods for the fabrication of ITO thin films such as evaporation [5, 6] and sputtering [7, 8]. However, above-mentioned technologies are correlated with the vacuum and high-temperature processes, resulting in the high manufacturing cost. The sol-gel spin-coating method is a very easy, economical, and nonvacuum technique to get high-quality ITO thin films.

There are two kinds of precursors (or starting materials) can be used to synthesized the sol-gel solution. One is inorganic metal salts, such as $M(NO_3)_x$ and MCl_x. The other is organic metal alkoxides, such as $M(OR)_x$ and $M(O_2CR)_x$. After proper hydrolysis and polycondensation reactions, the precursors can react with H_2O and chelate regents (eg., acetic acid and acetylacetone) to form the stable sol-gel solutions. Generally, indium chloride ($InCl_3$) [9, 10], indium nitrate [$In(NO_3)_3$] [11, 12], indium acetate [$In(O_2CCH_3)_3$] [13], and indium propoxide [$In(OC_3H_7)_3$] [14, 15] can be utilized to act as indium starting materials for the synthesis of the ITO sol-gel solutions. Moreover, indium acetate and indium propoxide are more expensive than indium chloride and indium nitrate, which inhibits their commercialized applications.

In this chapter, ITO thin films were fabricated by the sol-gel spin-coating method with different indium sol-gel solutions synthesized by different indium precursors (i.e., indium nitrate or indium chloride), which leads to the resulting ITO thin films with obviously different structural, optical, and electrical properties. We demonstrate that the refractive index, packing density, optical band gap, and resistivity of the ITO thin films synthesized from indium nitrate (denoted as N-ITO) are higher than that of the ITO thin films synthesized from indium chloride (denoted as Cl-ITO). These better optical and electrical properties make N-ITO thin films have better potential applications.

2. Fabrication of ITO Thin Films

2.1. Fabrication of ITO Sol-Gel Solution

Anhydrous indium chloride ($InCl_3$, 99.99%, Alfa Aesar) or anhydrous indium nitrate [$In(NO_3)_3$, 99.99%, Alfa Aesar] was added in the mixed solution of acetic acid (HAc, 99.5%, Acros) and 2-methoxyethanol (2-MOE, 99.5%, Merck) with the molar ratio of $In/H_2O/HAc/2\text{-}MOE=1/1/20/12$. Figure 1 shows the schematic flowchart of the synthesis of indium chloride-derived ITO sol-gel solution.

To dissolve anhydrous indium chloride, the mixed solution was refluxed at 80 °C for 3 h for the formation of clear indium solution. Anhydrous tin chloride ($SnCl_4$, 99%, Acros) dissolving in ethanol (EtOH, 99.9%, JT-Baker) was then dropped into the refluxed indium solution to form ITO solution. The doping concentration of Sn^{4+} ions was fixed at 10 mol% (Sn/In). Finally, the ITO solution was followed by stirring for 10 h at 50 °C for further homogeneous hydrolysis and polymerization reaction to acquire the indium chloride-derived ITO sol-gel solution.

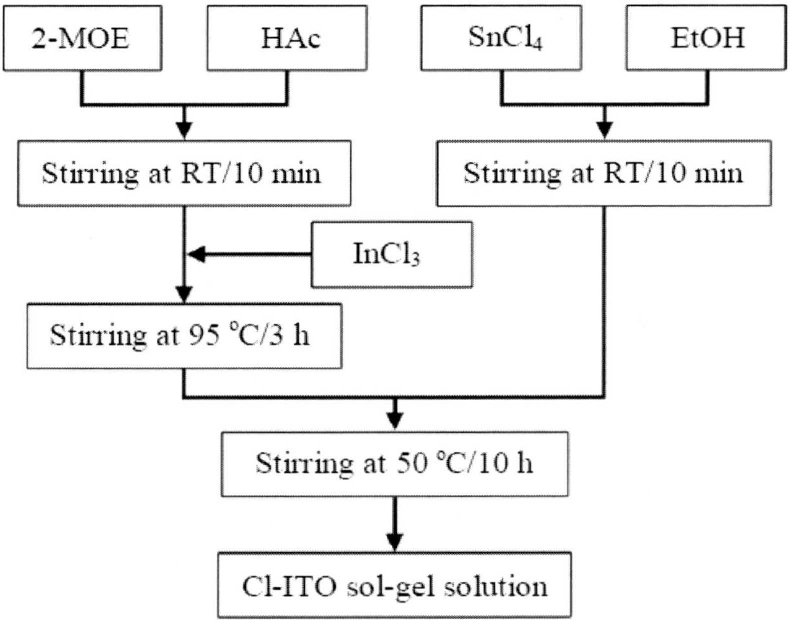

Figure 1. Schematic flowchart of the synthesis of indium chloride-derived ITO sol-gel solution.

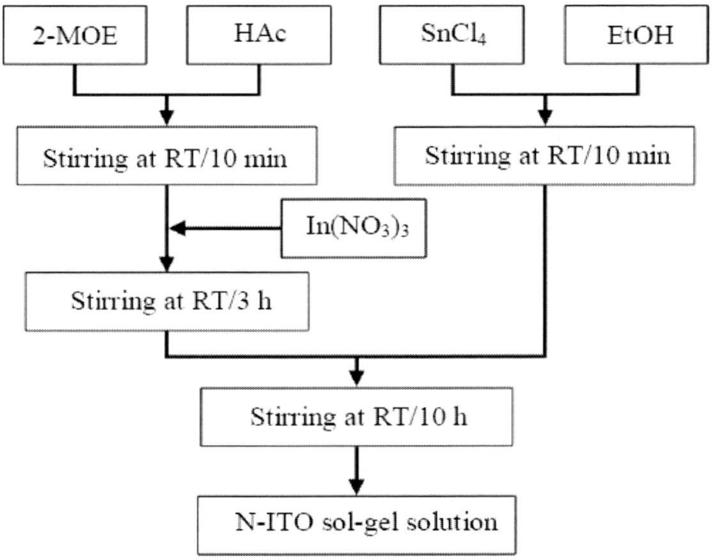

Figure 2. Schematic flowchart of the synthesis of indium nitrate-derived ITO sol-gel solution.

For the synthesis of the indium nitrate-derived ITO sol-gel solution, all of the procedures, molar ratios, and solvents were the same as that of the indium chloride-derived ITO sol-gel solution but processing was executed at room temperature to avoid the occurrence of precipitation. The schematic flowchart of the synthesis of indium nitrate-derived ITO sol-gel solution is shown in Figure 2.

2.2. Fabrication of ITO Thin Films

Figure 3 shows the schematic flowchart of the fabrication of sol-gel spin-coating ITO thin films. ITO sol-gel solutions were spin-coated on glass substrates (Corning, Eagle 2000). The as-deposited sol-gel films were first dried at 100 °C for 10 min and pyrolyzed under air atmosphere at 400 °C for 10 min at a heating rate of 12 °C /min. The procedures from coating to pyrolyzing were repeated eight times until the thickness of the pyrolyzed film was approximately 360 nm. Finally, thin films were further annealed at temperatures ranging from 400 to 700 °C for 1 h at a heating rate of 5 °C /min in air atmosphere. The average film thickness measured from three samples with the same coating condition was 360±10 nm.

2.3. Measurement of Physical Properties

The X-ray diffraction (XRD) measurement of thin films was performed by an X-ray diffractometer (Shimadzu XRD 6000). The thermal decomposition of the N-ITO and Cl-ITO xerogel powders fabricated by scraping the as-dried sol-gel films was studied by thermogravimetric analysis (TGA, TA Instrument 5100) and differential scanning calorimetry (DSC, Netzsch Instrument HT-DSC404) recorded from 30 to 1000 °C with a heating rate of 3 °C/min in air flow. X-ray photoelectron spectroscopy (XPS) was recorded by an AXIS Ultra DLD (Kratos) spectrometer equipped with Mg kα (1253.6 eV) radiation to investigate the chemical environment of elements.

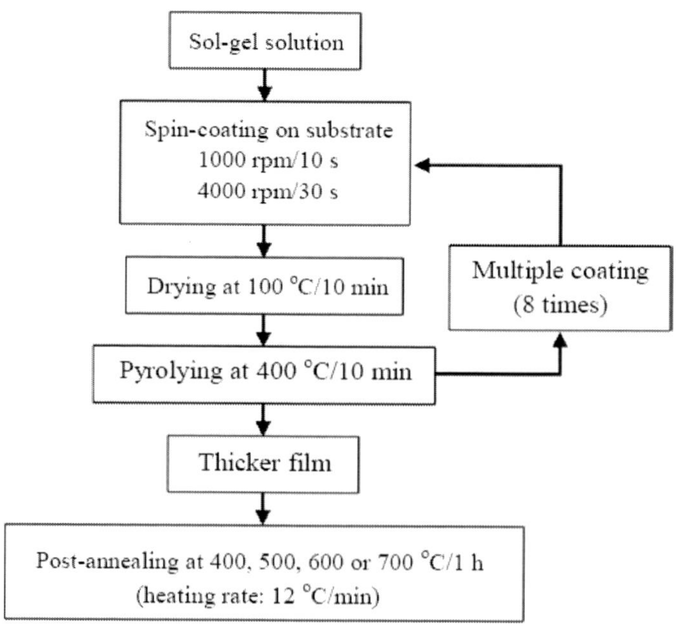

Figure 3. Schematic flowchart of the fabrication of sol-gel spin-coating ITO thin films.

The background of XPS peak was subtracted with Shirely function, and the XPS signals of C 1s and Sn 3d were fitted with mixed Lorentzian–Gaussian curves. Binding energies were calibrated with respect to the C 1s peak at 285.0 eV. Scanning electron microscope (Hitachi, S4800-I) was used for the microstructural examination. The thickness of thin films was measured by the α-step profile meter (KLA-Tencor, Alpha-Step IQ). Transmission spectra were determined on a Shimadzu UV-2100 spectrophotometer. The resistivity, Hall mobility, and carrier concentration were measured at room temperature with the four-point van der Pauw method in a magnetic field of 0.55 T (ECOPiA, HMS-3000).

3. Structure, Morphology, and Optoelectrical Properties of ITO Thin Films

3.1. TGA and DSC Analysis of ITO Xerogel Powders

Figures 4 and 5 shows the TGA and DSC curves of the N-ITO and Cl-ITO xerogel powders fabricated from two kinds of different indium precursors. The TGA curves show a significant weight loss in the temperature range from room temperature (RT) to 564 and 466 °C for N-ITO and Cl-ITO xerogel powders, respectively. The weight loss at RT-330 °C corresponds to the evaporation of physically adsorbed water and residual organic solvents. The combustion and decomposition of organic components accompanied by the crystallization process of ITO powders are at the temperature range of 330–564 °C and 330–466 °C for N-ITO and Cl-ITO xerogel powders, respectively. No weight loss was observed for the N-ITO xerogel powders heated over 564 °C and Cl-ITO xerogel powders heated over 466 °C, which implies that the thermally stabilized and well-crystallized ITO powders have formed. In addition, it can be seen that an obvious endothermal peak at 329 °C for N-ITO xerogel powders could be related to the decomposition of organic components but a very weak exothermal peak at 493 °C for Cl-ITO xerogel powders could be ascribed to the crystallization of ITO powders in DSC curves. These different TGA and DSC results will significantly induce the different structural, optical, and electrical properties between N-ITO and Cl-ITO thin films.

It is interesting to note that a certain amount of organic residues still present in N-ITO and Cl-ITO xerogel powders at 400 °C as shown in Figs. 4 and 5 but it does not means that organic residues still exist in the N-ITO and Cl-ITO thin films annealed at 400 °C for 1 h. N-ITO and Cl-ITO thin films are fabricated by spin coating method. The procedures from coating to pyrolyzing are repeated eight times until the thickness of pyrolyzed film is approximately 360 nm, and each layer is only ~45 nm thick. The contact area of the single-layered xerogel film exposing to annealing atmosphere is larger than that of aggregated xerogel powders. Therefore, the combustion rate of organic species for single-layered xerogel film is quicker than that for aggregated xerogel powders. Although the TGA curves (measurement under a heating rate of 3 °C/min) show a certain amount of organic residues in N-ITO and Cl-ITO xerogel powders at 400 °C, we believe that organic residues in N-ITO and Cl-ITO xerogel films could be eliminated after 400 °C annealing for 1 h.

Figure 6 shows the C 1s XPS peaks of N-ITO thin films annealed at 400 and 700 °C for 1 h. The C 1s spectrum can be resolved into three components with peaks located at 285 (C–C or C–H; C1), 286.3 (C–O; C2), and 289.1 (C=O; C3) eV [16, 17]. Obviously, the shape and position of C1, C2, and C3 peaks in the N-ITO thin film annealed at 400 °C for 1 h are very similar to that in the N-ITO thin film annealed at 700 °C for 1 h. These peaks are due to the presence of adventitious carbon, which is unavoidable on air-exposed materials [18]. This indicates that organic residues in the N-ITO and Cl-ITO thin films annealed at 400 °C/1 h are eliminated. Additionally, based on the TGA curves of xerogel powders, it is reasonable to deduce that the decomposition temperature of N-ITO xerogel thin films could be still higher than that of Cl-ITO xerogel thin films although the TGA curves of xerogel thin films cannot be measured.

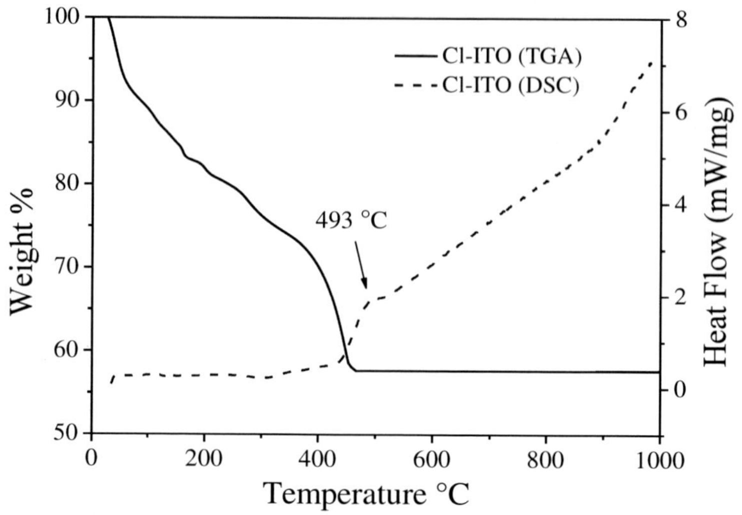

Figure 4. TGA and DSC curves of Cl-ITO xerogel powders.

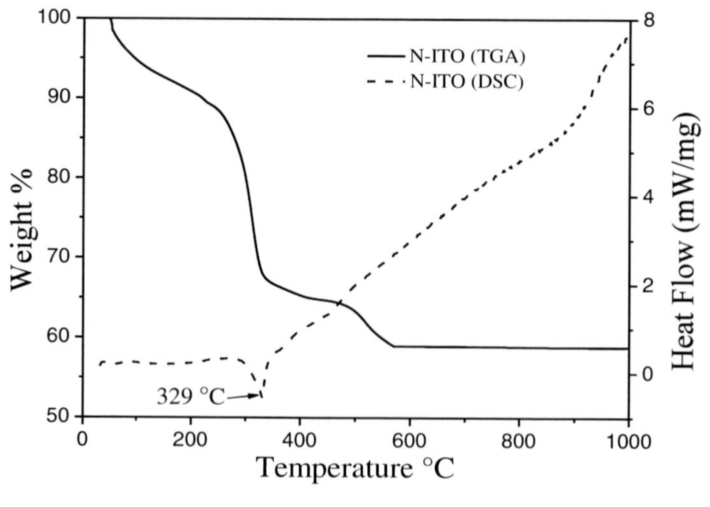

Figure 5. TGA and DSC curves of N-ITO xerogel powders.

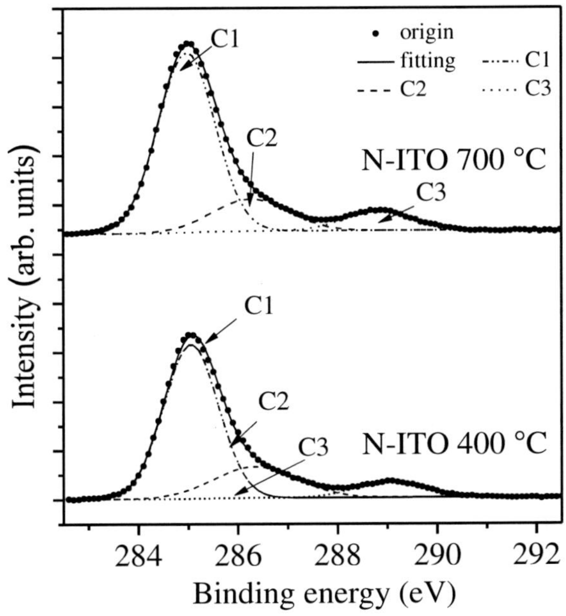

Figure 6. C 1s XPS spectra of the N-ITO thin films annealed at 400 and 700 °C for 1 h.

3.2. Crystal Structure

Figure 7 shows the XRD patterns of the N-ITO and Cl-ITO thin films annealed at 400 and 700 °C for 1 h. All samples possess well-crystallized bixbyite phase (JCPDS, Card No. 06-0416).

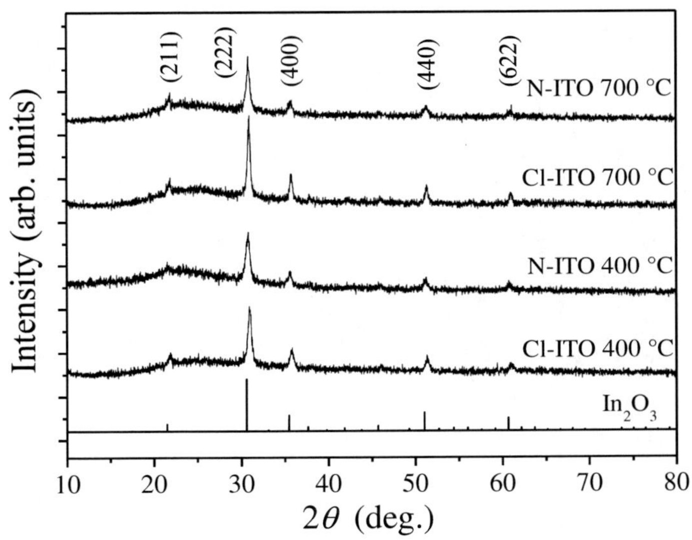

Figure 7. XRD patterns of the N-ITO and Cl-ITO thin films annealed at 400 and 700 °C for 1 h.

Because all samples have the similar thickness of 360±10 nm, the XRD peak intensity can reveal the crystallinity of thin films. High temperature annealing can increase the crystallinity of N-ITO and Cl-ITO thin films. On the other hand, the XRD peak intensity of the Cl-ITO thin films annealed at 400 and 700 °C is obviously larger than that of the N-ITO thin films annealed at 400 and 700 °C for 1 h, respectively. This means that Cl-ITO thin films have higher crystallinity, which can be attributed to the larger grain size of Cl-ITO thin films as shown in the latter scanning electron microscopy (SEM) photographs.

3.3. Film Morphology

Figures 8 and 9 illustrate the top-view and cross-sectional SEM imagines of the N-ITO and Cl-ITO thin films annealed at 400 and 700 °C for 1 h. The average grain sizes of the N-ITO and Cl-ITO thin films annealed from 400 to 700 °C for 1 h slightly increase from ~10 to ~19 nm and ~25 to ~35 nm, respectively.

Apparently, the surface morphologies of N-ITO thin films are more compact and smoother; however, Cl-ITO thin films possess larger grains and many large pores. It is noted that the cross-sectional SEM images of N-ITO thin films show the multi-layered structure but Cl-ITO thin films exhibit only single-layered structure composed of aggregated particles with semi-spherical form.

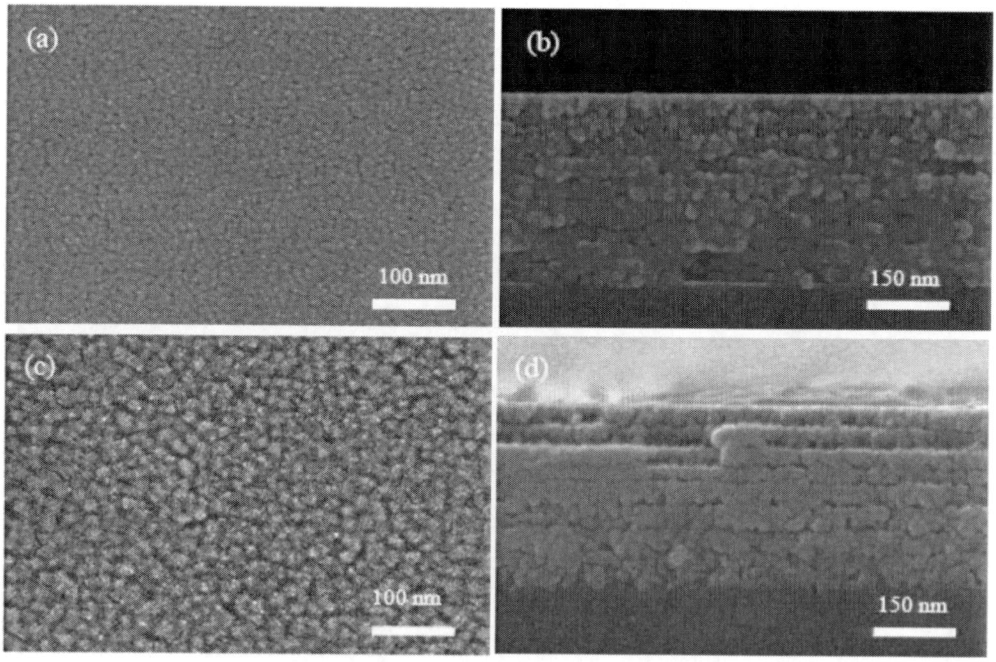

Figure 8. (a) top-view and (b) cross-sectional SEM imagines of the N-ITO thin films annealed at 400 °C for 1 h. (c) top-view and (d) cross-sectional SEM imagines of the N-ITO thin films annealed at 700 °C for 1 h.

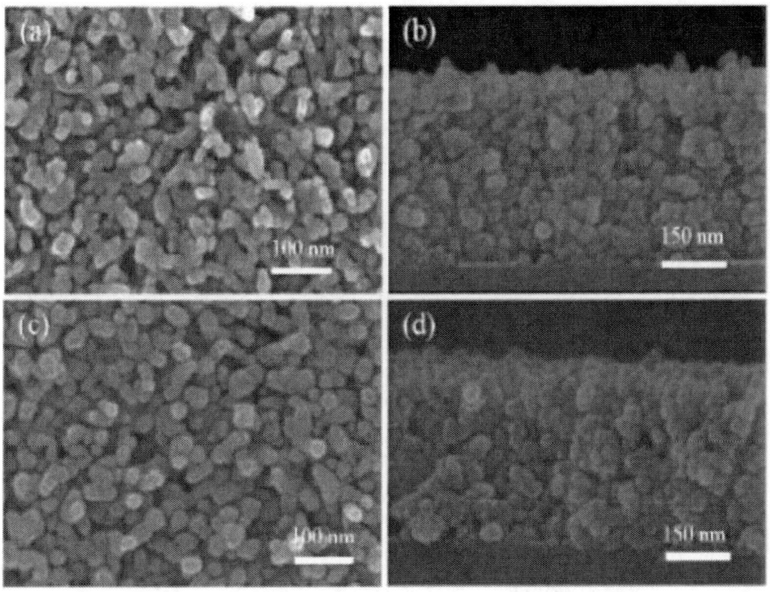

Figure 9. (a) top-view and (b) cross-sectional SEM imagines of the Cl-ITO thin films annealed at 400 °C for 1 h. (c) top-view and (d) cross-sectional SEM imagines of the Cl-ITO thin films annealed at 700 °C for 1 h.

3.4. Grown Mechanisms of ITO Thin Films

It is well known that the densification and coarsening mechanisms are usually in competition during annealing [19]. The lower decomposition temperature of Cl-ITO xerogel powders means that the elimination of organic species simultaneously associated with the breaking and reforming chemical bonds in crystallization process of ITO can occur at lower annealing temperature. This effect could make coarsening mechanism dominate, which results in the acceleration of crystal growth and the formation of larger pores. Figure 10 shows the densification and coarsening mechanisms of ITO thin films.

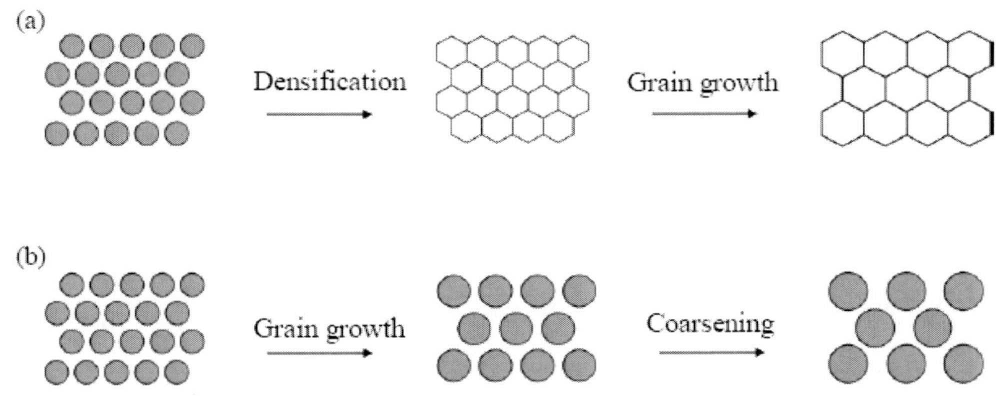

Figure 10. (a) densification and (b) coarsening mechanisms of ITO thin films.

3.4. Optoelectrical Properties

3.4.1. Electrical Properties

Figures 11 and 12 show the resistivity, mobility, and carrier concentration of N-ITO and Cl-ITO thin films as a function of different annealing temperatures (400-700 °C) for 1 h. For both N-ITO and Cl-ITO thin films, the carrier concentrations and resistivity decrease with the increase of annealing temperatures but the mobility increases with the increase of annealing temperatures.

It is well known that the conductive mechanism of ITO thin film is due to the formation of free charge carriers created by Sn^{4+} ions doping (Eq. 1) and oxygen vacancies in In_2O_3 lattice (Eq. 2) [20,21].

$$O_o^x \rightarrow V_o^{\bullet\bullet} + \frac{1}{2}O_{2(g)} + 2e' \qquad (1)$$

$$2SnO_{2(s)} \xrightarrow{In_2O_3} 2Sn_{In}^{\bullet} + 3O_0^x + \frac{1}{2}O_{2(g)} + 2e' \qquad (2)$$

Han *et al.* demonstrated that by annealing in air, the oxygen vacancies of ITO sample are filled by oxygen, resulting in the reduction of free charge carriers [22,23].

In addition, some literatures revealed that electrically active Sn dornors (Sn^{4+}) and electrically inactive Sn defects clusters (Sn^{2+}) in ITO also play an important role to influence the physical properties of ITO thin films [24-28]. If the electrically inactive Sn defects clusters (Sn^{2+}) are reduced by the activation of Sn dopants (Sn^{4+}) after thermal annealing, the carrier concentrations can be increased [24].

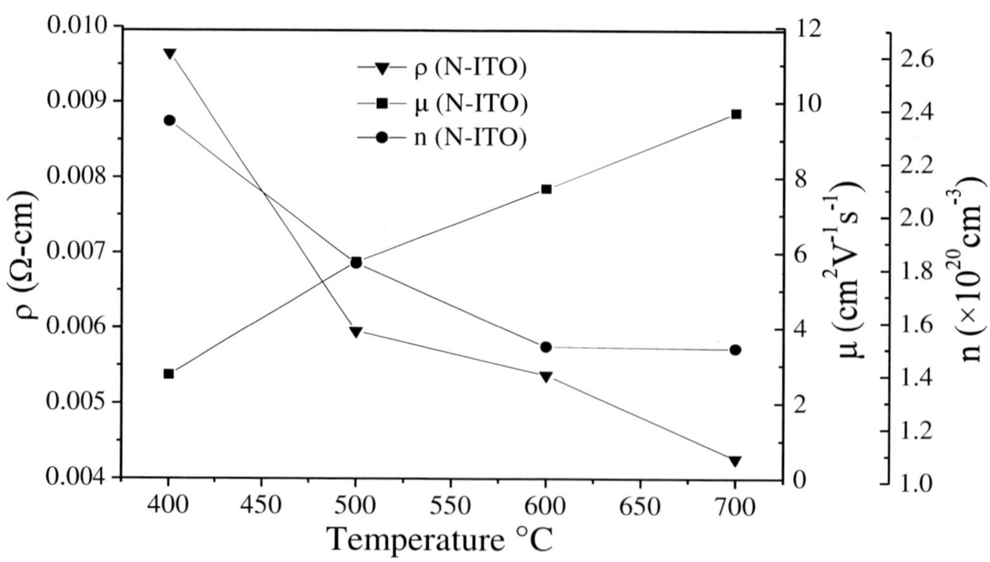

Figure 11. Resistivities, mobilities, and carrier concentrations of N-ITO thin films as a function of different annealing temperatures (400-700 °C) for 1 h.

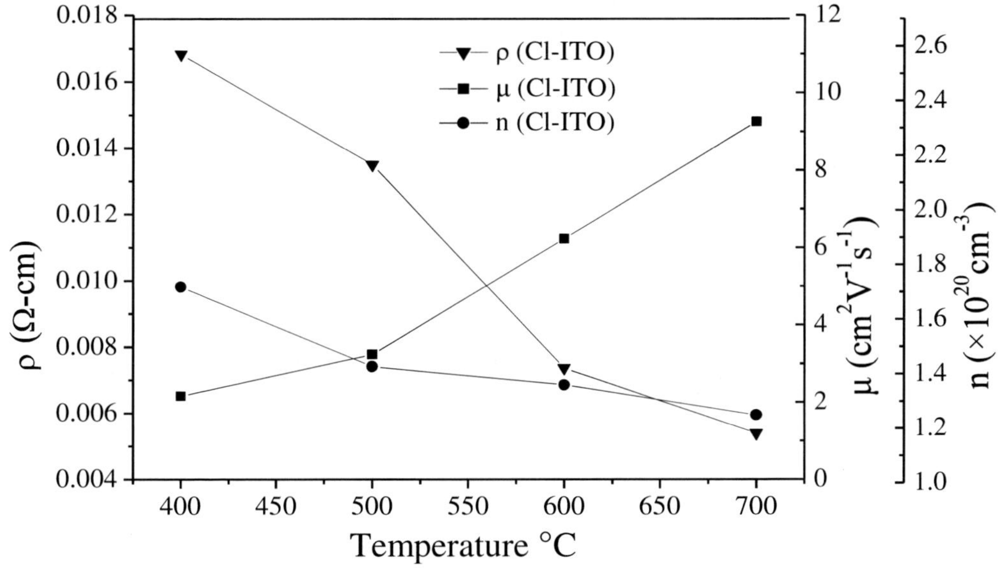

Figure 12. Resistivities, mobilities, and carrier concentrations of Cl-ITO thin films as a function of different annealing temperatures (400-700 °C) for 1 h.

Figures 13 and 14 show the Sn 3d XPS spectra of N-ITO and Cl-ITO thin films annealed at 400 and 700 °C for 1 h. The spin-orbit splittings of Sn 3d level are located at binding energies of ~486.5 (Sn 3d5/2) and ~494.9 eV (Sn 3d3/2). The deconvoluted Sn 3d5/2 spectra can be composed of two components with peaks at ~487 eV and ~486 eV which are attributed to the Sn4+ and Sn2+ bonding state, respectively.

Similarly, the deconvoluted Sn 3d 3/2 spectra can also be composed of two components. The peak positions and relative strengths of the deconvoluted Sn 3d 5/2 and Sn 3d 3/2 peaks are summarized in Table I. However, in our system there is no obvious change in the relative strengths of Sn2+/Sn4+ for N-ITO and Cl-ITO thin films annealed from 400 to 700 °C. This indicates that the number of electrically active Sn4+ ions in N-ITO thin films is similar to that in Cl-ITO thin films, and the electrically active Sn4+ ions cannot increase with the decrease of electrically inactive Sn2+ ions after thermal annealing. Therefore, the evolution of carrier concentration for both N-ITO and Cl-ITO thin films annealed at 400-700 °C is mainly related to the annihilation of oxygen vacancies.

After high temperature annealing, the crystallization and densification of N-ITO and Cl-ITO thin films are improved but oxygen vacancies are annihilated, which leads to the increase of mobility and the reduction of free charge carriers and resistivity. In addition, the surface morphology of N-ITO thin films is denser than that of Cl-ITO thin films. The mobility of N-ITO thin films annealed at 400-700 °C are always higher than that of Cl-ITO thin films, and the resistivity of N-ITO thin films annealed at 400-700 °C are always lower than that of Cl-ITO thin films.

The effect of the reduced resistivity and enhanced mobility with the increase of densification, crystallization, and crystal size had been discussed on some reported literatures [29-31]. On the other hand, the evolution of carrier concentrations for both N-ITO and Cl-ITO thin films annealed at 400-700 °C will influence the variation of optical band gaps, and this part will be discussed later.

Table I. Sn 3*d* XPS peaks of the N-ITO and Cl-ITO thin films annealed at 400 and 700 °C for 1 h

	Sn $3d_{5/2}$ (Sn^{4+})		Sn $3d_{5/2}$ (Sn^{2+})		Sn $3d_{3/2}$ (Sn^{4+})		Sn $3d_{3/2}$ (Sn^{2+})	
	Position (eV)	Relative strength(%)	Position (eV)	Relative strength (%)	Position (eV)	Relative strength (%)	Position (eV)	Relative strength(%)
N-ITO 400 °C	487.15	62.43	486.46	37.57	495.6	62.49	494.8	37.43
N-ITO 700 °C	486.87	61.59	486.16	38.41	495.4	62.38	494.6	37.62
Cl-ITO 400 °C	486.93	62.49	486.22	37.51	495.31	62.39	494.61	37.61
Cl-ITO 700 °C	486.85	62.17	486.19	37.83	495.2	62.30	494.51	37.70

Table II. Resistivities, carrier concentrations, and mobilities of ITO thin films prepared by various methods

Fabrication method and temperature	film thickness	ρ (Ω-cm)	n (cm^{-3})	μ (cm^2/vs)	reference
e-beam evaporation; 200 °C	40 nm	4.9×10^{-4}	1.1×10^{21}	11.4	Wang et al. [5]
e-beam evaporation; 150 °C	---	4.9×10^{-3}	1.17×10^{20}	11	Han et al.[23]
rf magnetron sputtering ; 350 °C	70 nm	1.15×10^{-4}	1.46×10^{21}	32.7	Tuna et al. [32]
sputtering; 400 °C	100 nm	2.3×10^{-4}	1×10^{21}	23	Guillén and Herrero [8]
sol–gel dip-coating; 550 °C	130 nm	5×10^{-3}	---	---	Beaurain et al.[30]
sol–gel spin-coating; 700 °C	360 nm	4.27×10^{-3}	1.50×10^{20}	9.7	this paper ; N-ITO
sol–gel spin-coating; 700 °C	360 nm	5.5×10^{-3}	1.25×10^{20}	9.2	this paper ; Cl-ITO

Table II summarizes the resistivities, carrier concentrations, and mobilities of the ITO thin films prepared by various methods. Our data are also shown for comparison. Wang *et al.* [11] demonstrated that the resistivity, carrier concentrations, and mobility of the ITO thin film fabricated by e-beam evaporation at room temperature and annealed at 200°C/30 min are 4.9×10^{-4} $\Omega^{-1}cm^{-1}$, 1.1×10^{21} cm^{-3}, and 11.4 $cm^2(vs)^{-1}$, respectively. Tuna *et al.* [12] reported that ITO film fabricated by magnetron sputtering at 350 °C, the resistivity, carrier concentrations, and mobility are 1.29×10^{-4} Ω-cm, 1.46×10^{21} cm^{-3}, and 32.7 $cm^2(vs)^{-1}$, respectively. Obviously, ITO thin films fabricated by vacuum processes such as thermal evaporation and magnetron sputtering exhibit the lowest resistivity (~10^{-4} Ω-cm). On the other hand, the resistivity (~10^{-3} Ω-cm) of the ITO thin films fabricated by non-vacuum dip-coating processes is at least an order of magnitude higher than that of films fabricated by vacuum processes. In our system, the resistivity of the optimal spin-coating-growth ITO film is close to that of the dip-coating-derived ITO films but is still lower than that of the sputtered ITO films. However, the spin-coating-growth method is still a good approach for the fabrication of low-cost, non-vacuum-grown ITO films.

3.4.2. Optical Transmittance Spectra

Figures 15 and 16 shows the optical transmittance spectra of the N-ITO and Cl-ITO thin films annealed at 400 and 700 °C for 1 h. The slight decrease of transmittance for high-temperature annealed thin films could be related to the thin films with larger grain size and higher surface roughness. It is interesting to note that for both N-ITO and Cl-ITO thin films, Figures 15 and 16 clearly shows the red-shift in the fundamental absorption edge with the increase of annealing temperatures.

3.4.3. Optical Band Gap

The optical band gap (E_g) of the ITO thin film which is a direct-transition-type semiconductor can be related to absorption coefficient (α) by [32]

$$\alpha h\nu = \text{const.}\ (h\nu - E_g)^{1/2} \tag{3}$$

Here we assume the absorption coefficient $\alpha = (1/d)\ \ln(1/T)$, where T is the transmittance and d is the film thickness. Figure 17 and 18 plots the relationship of $(\alpha h\nu)^2$ versus photon energy (E) of the N-ITO and Cl-ITO thin films fabricated under 400 and 700 °C/1 h annealing, and the extrapolated optical band gaps of ITO thin films are determined. The Eg of the N-ITO and Cl-ITO thin films annealed from 400 to 700 °C decrease from 3.95 to 3.87 eV and 3.71 to 3.69 eV, respectively.

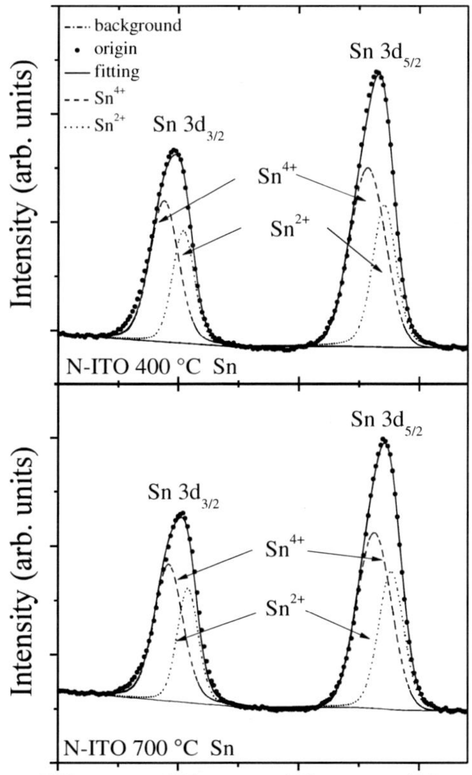

Figure 13. Sn 3d XPS spectra of the N-ITO thin films annealed at 400 and 700 °C for 1 h.

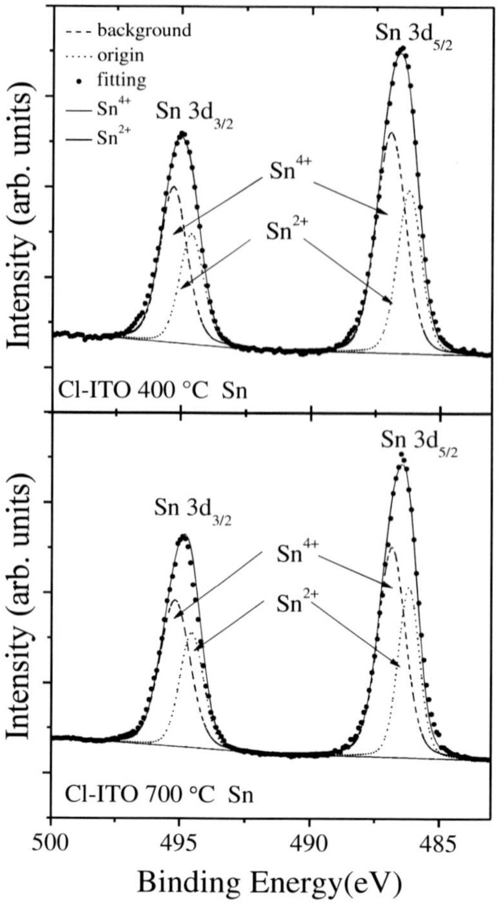

Figure 14. Sn 3d XPS spectra of the Cl-ITO thin films annealed at 400 and 700 °C for 1 h.

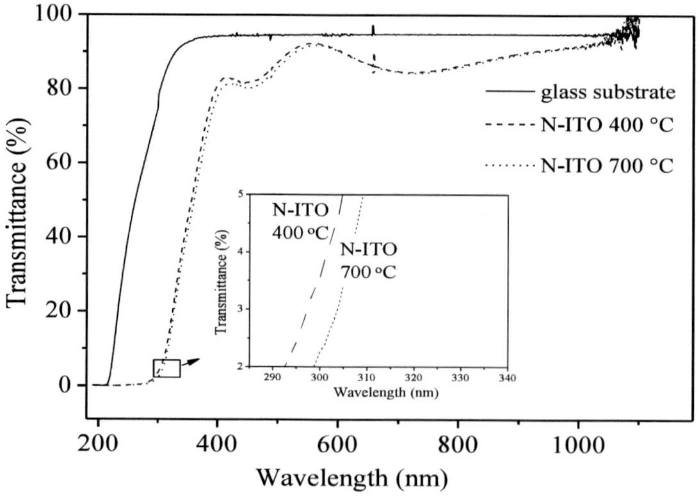

Figure 15. Optical transmittance spectra of the N-ITO thin films annealed at 400 and 700 °C for 1 h.

The generation of one oxygen vacancy can create two free charge carriers [36]. These free charge carriers can block in the lowest states of conduction band to enlarge the Eg of ITO thin films, known as Burnstein-Moss (B-M) effect [37, 38]. In our case, however, the Eg of N-ITO and Cl-ITO thin films reduce after high temperature annealing. Although the crystallization and densification of N-ITO and Cl-ITO thin films are improved after heat treatment, oxygen vacancies are annihilated.

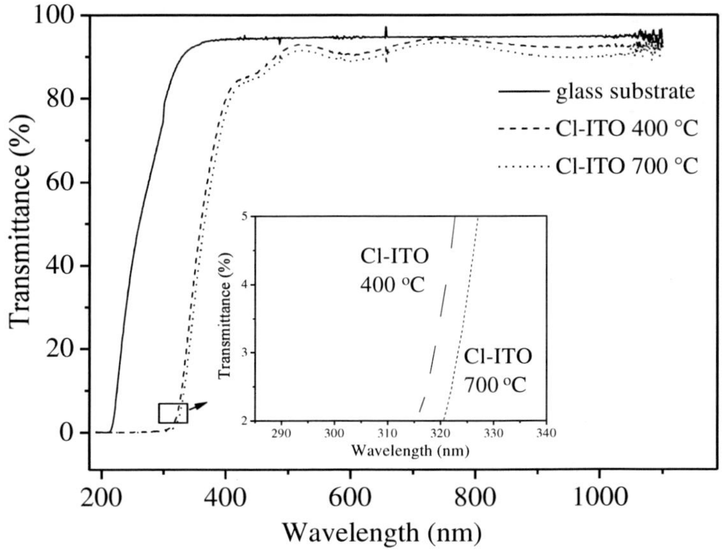

Figure 16. Optical transmittance spectra of the Cl-ITO thin films annealed at 400 and 700 °C for 1 h.

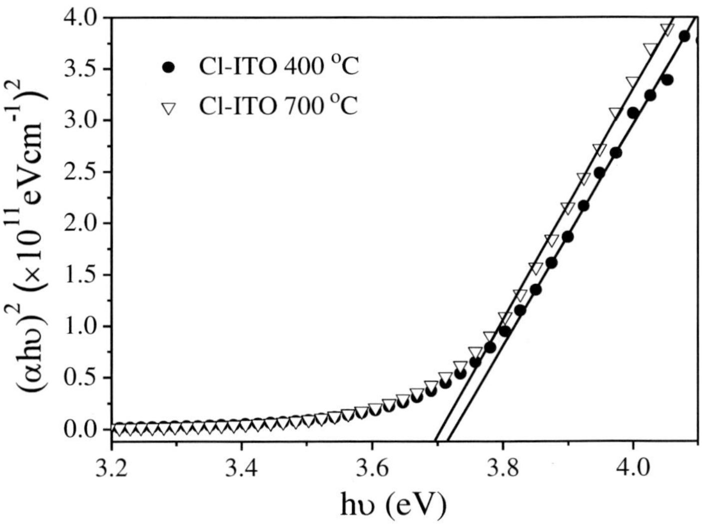

Figure 17. $(\alpha hv)^2$ as a function of photon energy (E) of the N-ITO thin films annealed at 400 and 700 °C for 1 h.

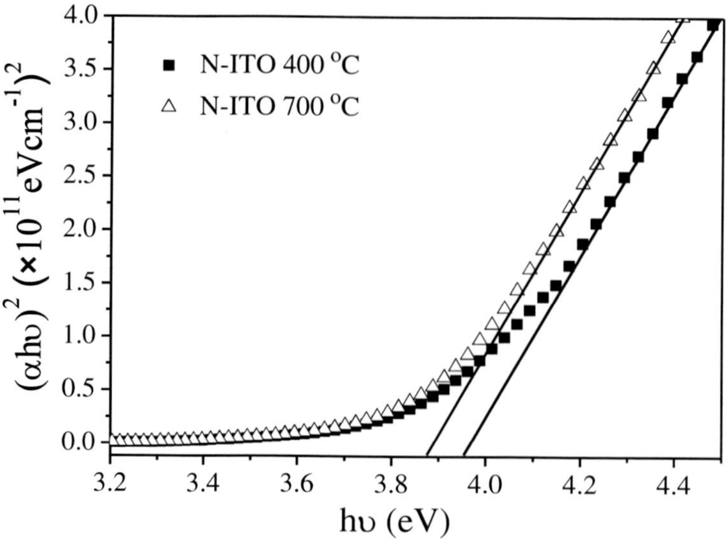

Figure 18. $(\alpha hv)^2$ as a function of photon energy (E) of the Cl-ITO thin films annealed at 400 and 700 °C for 1 h.

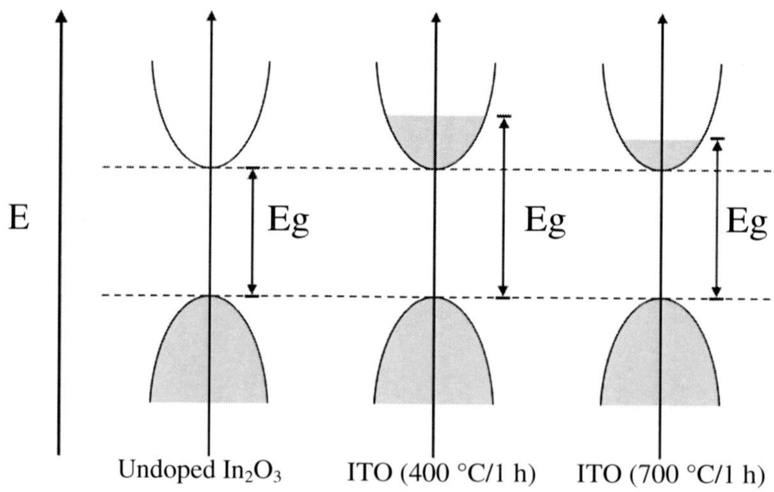

Figure 19. Schematic band structure of (a) undoped In_2O_3, (b) ITO annealed at 400 °C for 1 h, (c) ITO annealed at 700 °C for 1 h.

This phenomenon leads to the reduction of free charge carriers and Eg. Figure 19 shows the schematic band structure of undoped In_2O_3 and ITO annealed at 400 or 700 °C /1 h. On the other hand, the Eg of the Cl-ITO thin films annealed at 400 and 700 °C are lower than that of N-ITO thin films. Because Cl-ITO thin films have larger grain size and higher crystallinity implying the formation of less oxygen vacancies, they possess less carrier concentration and reduced Eg (B-M effect).

3.4.4. Refractive Index and Packing Density

Figure 20 and 21 shows that the wavelength dependence of refractive index (n) for the N-ITO and Cl-ITO thin films annealed at 400 and 700 °C for 1 h. The refractive indices are derived from the transmittance spectra by using Swanepoel's method [33].

The increase of n with the increase of annealing temperatures can be attributed to the increase in crystallinity for both of N-ITO and Cl-ITO thin films. In addition, the refractive indices of N-ITO thin films are always higher than that of Cl-ITO thin films under the same annealing condition.

Based on the SEM photographs, the surface morphologies of N-ITO thin films exhibit the densification and smaller grains. Therefore, the less porosity results in the N-ITO thin films with higher refractive index.

The packing density (P) was evaluated by using the following Bragg–Pippard formula [34]:

$$n_f^{\ 2} = \frac{(1-P)n_v^{\ 4} + (1+P)n_p^{\ 2}n_b^{\ 2}}{(1+P)n_v^{\ 2} + (1-P)n_b^{\ 2}} \qquad (4)$$

where n_f, n_v, and n_b are the refractive indices of porous films, voids (n_v=1 for empty voids), and bulk materials (n_b= 2.19 for bulk ITO [35]), respectively. After calculation, the packing densities of the N-ITO and Cl-ITO thin films annealed from 400 to 700 °C increase from 0.82 to 0.83 and 0.68 to 0.71, respectively. The packing density increases with the increase of annealing temperature and refractive index, which can be attributed to the significant reduction in the porosity and increase in the crystallinity [supporting SEM photographs, Figure 3, and XRD pattern, Figure 2].

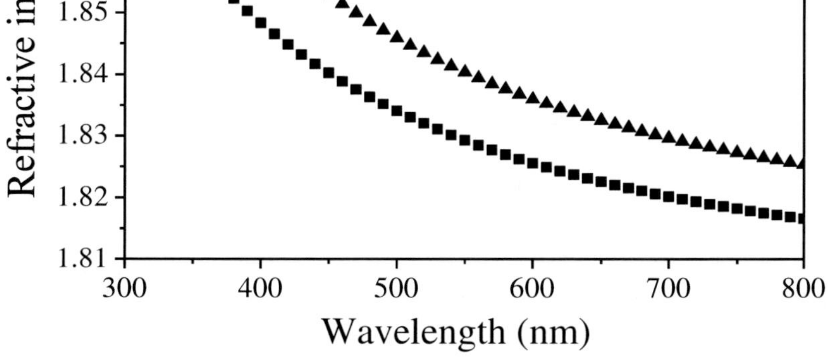

Figure 20. Wavelength dependence of refractive index for the N-ITO thin films annealed at 400 and 700 °C for 1 h.

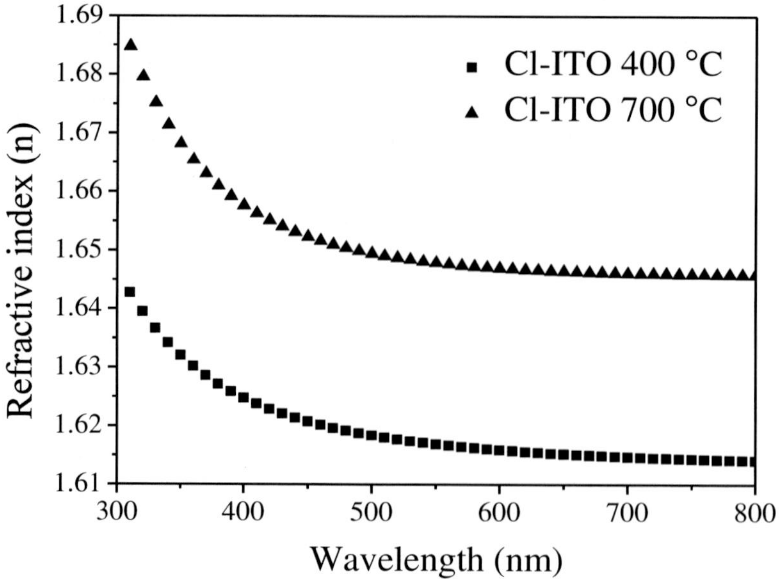

Figure 21. Wavelength dependence of refractive index for the Cl-ITO thin films annealed at 400 and 700 °C for 1 h.

Conclusion

The different sol-gel solutions synthesized from different indium precursors [$In(NO_3)_3$ or $InCl_3$] were used to fabricate ITO thin films by the spin-coating method, which obviously leads to the resulting ITO thin films with different structural, optical, and electrical properties. The refractive index (1.84 at $\lambda = 550$ nm), packing density (0.83), optical band gap (3.95 eV), and resistivity (4.27×10^{-3} Ω-cm) of the ITO thin films synthesized from $In(NO_3)_3$ are higher than that of the ITO thin films synthesized from $InCl_3$.

After high temperature annealing, the increase of crystallization/ densification and the annihilation of oxygen vacancies lead to the increased mobility and the reduced carrier concentration/resistivity for both N-ITO and Cl-ITO thin films. However, the reduced resistivity and enlarged carrier concentration/optical band gap of N-ITO thin films can be attributed to the smaller grain size, higher packing density, lower crystallinity, and more oxygen vacancies (Burnstein-Moss effect).

Acknowledgement

The authors would like to thank the National Science Council of the Republic of China for financially supporting this research under Contract No. NSC 95-2218-E-194-014.

REFERENCES

[1] S.M. Huang, Y. Yao, C. Jin, Z. Sun, Z.J. Dong, *Displays* 29 (2008) 254.
[2] A. Abderrahmen, F.F. Romdhane, H.B. Ouada, A. Gharbi, *Mater. Sci. Eng.* C 26 (2006) 538.
[3] S. Naseem, T.J. Coutts, *J. Appl. Phys.* 58 (1985) 4463.
[4] M.J. Alam, D.C. Cameron, *Thin Solid Films* 377 (2000) 455.
[5] R.X. Wang, C.D. Beling, S. Fung, A.B. Djurišić, C.C. Ling, S. Li, *J. Appl. Phys.* 97 (2005) 033504.
[6] X.D. Liu, E.Y. Jiang, D.X. Zhang, *J. Appl. Phys.* 104 (2008) 073711.
[7] K. Sreenivas, T.S. Rao, A. Mansingh, S. Chandra, *J. Appl. Phys.* 57 (1985) 384.
[8] C. Guillén, J. Herrero, *J. Appl. Phys.* 101 (2007) 073514.
[9] Q. Wei, H. Zheng, Y. Huang, *Sol. Energy Mater. Sol. Cells* 68 (2001) 383.
[10] M.K. Jeon, M. Kang, *Mater. Lett.* 62 (2008) 676.
[11] M.S. Tsai, C.L. Wang, M.H. Hon, *Surf. Coat. Tech.* 172 (2003) 95.
[12] S. Kundu, P.K. Biswas, *Chem. Phys. Lett.* 414 (2005) 107.
[13] M.A.F. Mendoza, R.C. Perez, G.T. Delgado, J.M. Marín, O.Z. Angel, *Thin Solid Films* 517 (2008) 681.
[14] T.F. Stoica, V.S. Teodorescu, M.G. Blanchin, T.A. Stoica, M. Gartner, M. Losurdo, M. Zaharescu, *Mater. Sci. Eng. B* 101 (2003) 222.
[15] T.F. Stoica, M. Gartner, M. Losurdo, V. Teodorescu, M. Blanchin, T. Stoica, M. Zaharescu, *Thin Solid Films* 455–456 (2004) 509.
[16] T.P. Nguyen, J. Ip, C. Renaud, C.H. Huang, C. Guillen, J. Herrero, *Appl. Surf. Sci.* 252 (2006) 8388.
[17] T.G. Andersson, M. Andreasson, U. Klement, C.B. Lee, B. Pujilaksono, A. Uddin, *Mater. Sci. Eng. B* 145 (2007) 48.
[18] T.L. Barr, S. Seal, *J. Vac. Sci. Technol. A* 13 (1995) 1239.
[19] M.W. Barsoum, *Fundamentals of Ceramics*, McGraw-Hill, New York, 2003, p. 304.
[20] S. Ohno, Y. Kawaguchi, A. Miyamura, Y. Sato, P.K. Song, M. Yoshikawa, P. Frach, Y. Shigesato, *Sci. Technol. Adv. Mater.*7(2006)56.
[21] J.C.C. Fan, J.B. Goodenough, *J. Appl. Phys.* 48 (1977) 3524.
[22] H. Han, J.W. Mayer. T.L. Alford, *J. Appl. Phys.* 99 (2006) 123711.
[23] H. Han, J.W. Mayer. T.L. Alford, *J. Appl. Phys.* 100 (2006) 083715.
[24] H. Han, Y. Zoo, S.K. Bhagat, J.S. Lewis, T.L. Alford, *J. Appl. Phys.* 102 (2007) 063710.
[25] H.Y. Yu, D. Feng, D. Grozea, Z.H. Lu, R.N.S. Sodhi, A.M. Hor, H. Aziz, *Appl. Phys. Lett.* 78 (2001) 2595.
[26] S. Archambeau, I. Seguy, P. Jolinat, J. Farenc, P. Destruel, T.P. Nguyen, H. Bock, E. Grelet, *Appl. Surf. Sci.* 253 (2006) 2078.
[27] J.H. Wi, J.C. Woo, D.S. Um, J. Kim, C.I. Kim, *Thin Solid Films* 518 (2010) 6228.
[28] D. Maestre, A. Cremades, L. Gregoratti, J. Piqueras, *J. Phys. Chem. C* 114 (2010) 3411.
[29] Y. Takahashi, S. Okada, R.B.H. Tahar, K. Nakano, T. Ban, Y. Ohya, *J. Non-Cryst. Solids* 218 (1997) 129.
[30] A. Beaurain, D. Luxembourg, C. Dufour, V. Koncar, B. Capoen, M. Bouazaoui, *Thin Solid Films* 516 (2008) 4102.

[31] H. Tomonaga, T. Morimoto, *Thin Solid Films* 392 (2001) 243.

[32] O. Tuna, Y. Selamet, G. Aygun, L. Ozyuzer, *J. Phys. D: Appl. Phys.* 43 (2010) 055402.

[33] R. Swanepoel, *J. Phys. E: Sci. Instrum.* 16 (1983) 1214.

[34] W.L. Bragg, A.B. Pippard, *Acta Crystallogr.* 6 (1953) 865.

[35] J.K. Kim, S. Chhajed, M.F. Schubert, E.F. Schubert, A.J. Fischer, M.H. Crawford, J. Cho, H. Kim, C. Sone, *Adv. Mater.* 20 (2008) 801.

[36] M.W. Barsoum, *Fundamentals of Ceramics*, McGraw-Hill, New York, 2003, p. 150.

[37] E. Burstein, *Phys. Rev.* 93 (1954) 632.

[38] T.S. Moss, *Proc. Phys. Soc. B* 67 (1954) 755.

In: Indium
Editors: Hsaio G. Woo and Huang Tsai Choi

ISBN: 978-1-62257-696-8

Chapter 11

THE INFLUENCE RARE EARTH IONS DOPING ON THE PHOTOLUMINESCENT PROPERTIES OF INDIUM HYDROXIDE

F. V. Motta[1*], A. P. A. Marques[2], C. A. Paskocimas[1], M. R. D. Bomio[1], R. P. Souza[3], M. F. C. Abreu[4], M. S. Li[5], J. A. Varela[4] and E. Longo[4]

[1]Laboratório de Síntese Química, Departamento de Engenharia de Materiais, Universidade Federal do Rio Grande do Norte, Natal, RN, Brazil
[2]Departamento de Ciências Exatas e da Terra, Universidade Federal de São Paulo, Diadema, SP, Brazil
[3]DEP, Universidade Federal do Rio Grande do Norte, Natal, RN, Brazil
[4]Laboratório Interdisciplinar de Eletroquímica e Cerâmica, Instituto de Química, Universidade Estadual Paulista, Araraquara, SP, Brazil
[5]Instituto de Física de São Carlos, Universidade de São Paulo, São Carlos, SP, Brazil

ABSTRACT

Crystalline rare earth ions -doped indium hydroxide ($In(OH)_3$:RE^{3+}) nanostructures were prepared by a rapid and efficient Microwave-Assisted Hydrothermal (MAH) method. The influence of doping rare earth (RE) ions (Eu, Tb and Tm) on the luminescence of $In(OH)_3$ was studied. Nanostructures were obtained at a low temperature. FE-SEM images confirm that these samples are composed of 3D nanostructures. XRD, optical diffuse reflectance and photoluminescence (PL) measurements were used to characterize the products. Emission spectra of RE-doped indium hydroxide samples under excitation (350.7 nm) presented broad band emission referent to the indium hydroxide ($In(OH)_3$) matrix. Relative intensities of the RE^{3+} emissions increased as the concentration of this ion increased from 0, 1, 2, 4 and 8 mol %, of Eu^{3+}, but the luminescence is drastically quenched for the $In(OH)_3$ matrix.

Keywords: Hydroxide, indium, microwave, hydrothermal, nanostructures

1. INTRODUCTION

Self-assembled micro-, meso- and nanostructures with specific morphologies, novel properties and great application potential in many fields are of great interest to chemists and material scientists because of their fundamental significance for addressing some basic issues of the quantum confinement effect and space-confined transport phenomena as well as their potential applications as advanced materials in functional nanodevices [1].

Recently, indium and indium oxide have received considerable attention because of their semiconductor and optical properties [2, 3]. The indium hydroxide ($In(OH)_3$) is important n-type semiconductor material with a wide band gap (Eg) estimated to be 5.15 eV. Nanostructural $In(OH)_3$ have show photocatalytic behavior to TiO_2 for degradation of aromatic hydrocarbons under UV irradiation. Recently, Shi et al. [4] reported a $In(OH)_3$ truncated polyhedral microcrystal formation via conventional hydrothermal treatment at 180°C for 16 h and 18 h that showed PL spectrum with two emission peaks at 496.6 and 419.2 nm using an excitation wavelength (380 nm at room temperature). Yan et al. [3] showed a broad band PL blue-green emission peak of $In(OH)_3$ nanocubes centered at 480 nm.

Rare earth ions such as Eu^{3+} can be added in semiconductors to increase PL. Europium in its trivalent state is one of the most studied among the rare earth ions by luminescence spectroscopy as a consequence of the simplicity of its spectra and its wide application as a red phosphor in color TV screens. This ion has also attracted significant research attention due to its potential application as biological sensors, phosphors, electroluminescent devices, optical amplifiers or lasers when it is used as doping in a variety of materials [5, 6]. Dutta et al. [7] showed Eu^{3+} and Dy^{3+} doped indium oxide nanoparticles synthesized by a sonochemical technique. The transitions of Eu^{3+} were observed after excitation at 235 nm which is possibly due to the highly strained and distorted environment around the Eu^{3+} ions in the In_2O_3 lattice matrix. Gaponenko et al. [8] showed Er^{3+} and Tb^{3+} doped In_2O_3 films fabricated on porous silicon synthesized by Sol-Gel. The compositions coating porous silicon are considered as promising host materials for erbium and terbium optical activation, allowing to detect both visible and infrared luminescence from samples annealed at low temperatures.

Indium hydroxides with different morphologies such as spheres, flowers [9, 10], cubes [11-17], rods [18, 19], etc., have been synthesized by several synthesis methods. The physical and chemical properties of these powders can be influenced by shape, size and size distribution of the particles which depend on the synthesis method. A conventional hydrothermal synthesis has been used to obtain structures of $In(OH)_3$ using a simple, practical and environmentally friendly method [3, 8, 13].

Komarneni et al. [20] introduced the use of microwave radiation in the hydrothermal system by promoting the development of a new technique offering reaction kinetic enhancement, formation of materials with different morphologies, low synthesis temperature and reduced processing times. Recently, the MAH method was successfully employed to obtain $In(OH)_3$ [16] and other oxides such as ZnO [21], $BaTiO_3$ [22], CuO [23], $SrMoO_4$ [24], $BaMoO_4$ [25].

Considering that the MAH method has been successful in preparing oxide powders, this chapter reports the preparation, characterization and photoluminescent properties of $In(OH)_3$:xRE^{3+} (Eu and Tb) nanostructures com concentrantion of x= 0 and 1 mol% of RE^{3+}.

2. Experimental

$In(OH)_3$ powders were obtained by microwave-assisted hydrothermal (MAH) method.

Indium hydroxide was formed by the dissolution of $InCl_3$, indium (III) chloride, in 80mL of deionized water under constant stirring for 30 min. The pH of the solution was adjusted to 10 by adding ammonium hydroxide. The reaction system was heated at 140°C for 30 min. Europium nitrate and terbium nitrate were dissolved in 80 mL of deionized water under constant stirring. The ion RE^{3+} was added with 1 mol %. The heating rate in this system was fixed at 25°C/min and the pressure into the autoclave was stabilized at 294 kPa. The MAH (Brazilian patent 2008-PI0801233- 4) one hole was drilled into the wall of a domestic oven (2.45 GHz/800W) for a polytetrafluoroethylene (PTFE) vessel in the oven top. The magnetron was monitored to a very efficient proportional-integral-derivative (PID) external controller. Intuitive software enables the user to control everything from process conditions (temperature, time and heating rate). The user simply defines the desired temperature–time profile, and the MAH will automatically carry out the programmed cycle [9, 10]. After MAH treatment, the white precipitate was washed with distilled water and dried at room temperature.

The structural evolution and the unit cell volume of the $In(OH)_3$ powders were determined by XRD using a Rigaku diffract meter (Model D/max-2500/PC) with Cu K_α radiation.

The microstructure and surface morphology of the nanocrystalline powders of the $In(OH)_3$ heat treated at 140°C were observed by field emission scanning electron microscopy (FE-SEM; Jeol JSM 6330F) images. UV–vis reflectance spectra of $In(OH)_3$ powders were taken using Cary5Gequipment.

The photoluminescence spectra of the $In(OH)_3$ powders were collected with a Thermal Jarrel-Ash Monospec 27 monochromator and a Hamamatsu R446 photomultiplier. The 350.7nm line of a krypton ion laser (Coherent Innova 90 K) was used as the excitation source with out- put power of the lasers kept at 200mW. All characterizations were taken at room temperature.

3. Results and Discussion

X-ray patterns of the $In(OH)_3$ powders synthesized with different percentages of Eu and Tb ion at 140°C are presented in Figure 1 and 2. Figure 1a and 2a shows that all diffraction peaks could be indexed to the cubic lattice [space group *Im*$_3$(204)] to pure $In(OH)_3$ (JCPDS card No. 16-0161). It is noted in Figures 1 and 2 that all difratograms of samples nanostructures present the relative intensity of the (200) face higher compared with the standard value in the JCPDS card N° 16-0161, suggesting a preferential growth of the indium hydroxide nanostructures along the (200) plane, which coincides well with literature [26-27]. With europium percentage increments (Figure 1b-e), the phases obtained by MAH treatment were basically $In(OH)_3$ with a cubic structure and a second phase, the indium oxyhydroxide (InOOH) with an orthorhombic structure [space group *P21nm*(31)] (JCPDS card No.71-2283). With terbium percentage increments (Figure 2b-e), the phases obtained by MAH treatment remained $In(OH)_3$ with a cubic structure, which indicates the substitution of Tb^{3+} in the In^{3+} sites. No other phases, such as InOOH or In_2O_3, were verified. The absence of deleterious phases indicates the purity of the samples.

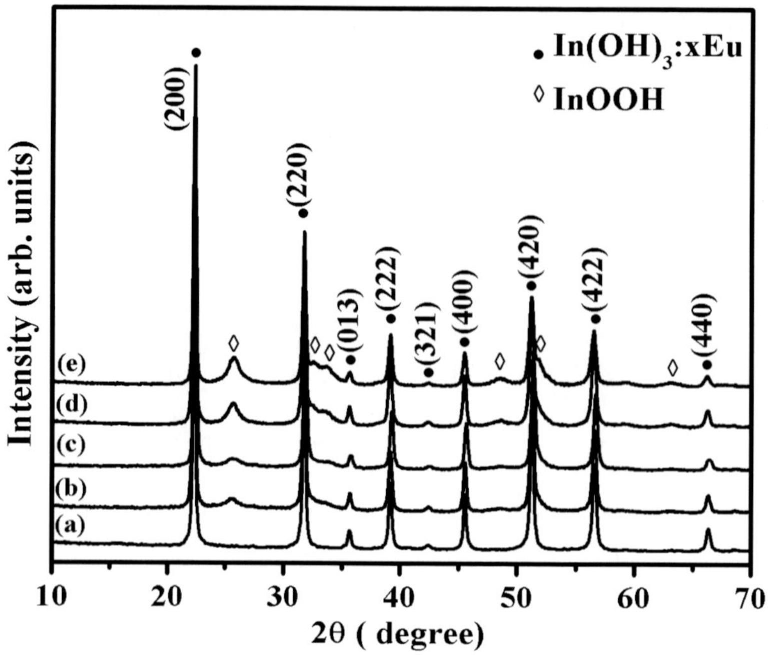

Figure 1. XRD patterns of $In(OH)_3$:xEu^{3+} powders obtained by the MAH method at 140°C for 30 min: (a) x = 0, (b) x = 1, (c) x = 2, (d) x = 4 and (e) x = 8 mol %.

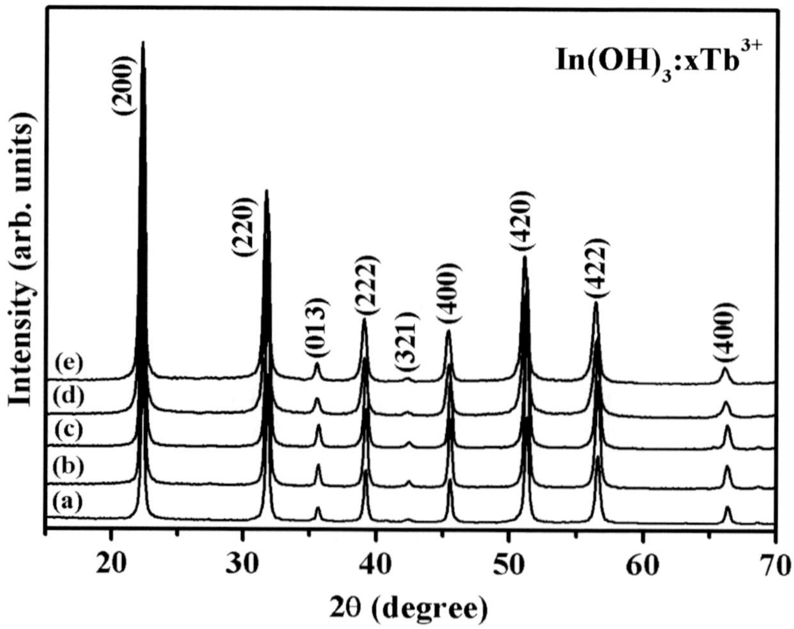

Figure 2. XRD patterns of $In(OH)_3$:xTb^{3+} powders obtained by the MAH method at 140°C for 30 min: (a) x = 0, (b) x = 1, (c) x = 2, (d) x = 4 and (e) x = 8 mol %.

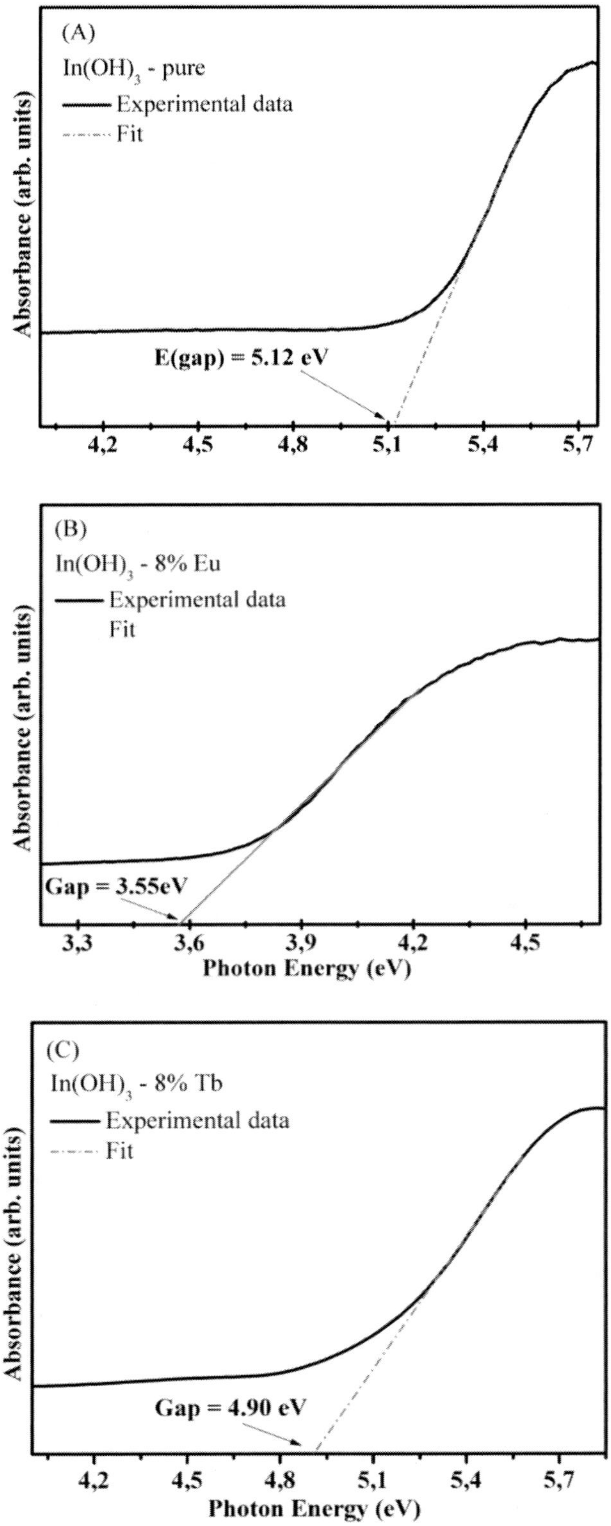

Figure 3. UV–Vis absorbance spectra for the $In(OH)_3$:RE^{3+} powders processed by the MAH method at 140°C for 30 min: (A) pure; (B) x = 8 mol% Eu^{3+}; (C) x = 8 mol% Tb^{3+}.

The synthesis of $In(OH)_3$ and InOOH nanostructures is based on reactions (1 and 2) [28]. Chen et al. [29] reported that the InOOH phase and the morphology synthesized by a solvothermal route under the assistance of ultrasound can be controlled by the choice of the alkaline media and also the solvent. The factors reported for the formation of InOOH hollow spheres synthesized by a conventional hydrothermal method were temperature treatment, the time of the reaction, reagents and the pH of the solution [30, 31]. In this chapter, the pH value of the solutions was controlled by addition of NH_4OH (pH = 10), therefore, the pH value no affect the formation phase. As well as the temperature treatment, the time of the reaction and the reagents used were the same in all experimental procedures. We speculate that the formation of InOOH is related to the influence of the Eu ion. With the addition and the increase of the content of europium in the structure was noted the presence of InOOH phase simultaneously with the $In(OH)_3$ phase and an increase in the intensity of the peaks for InOOH phase.

$$In^{3+} + 3OH^{-} \rightarrow In(OH)_3 \quad (1)$$

$$In(OH)_3 \rightarrow InOOH + H_2O \quad (2)$$

To our knowledge, the optical properties of $In(OH)_3:RE^{3+}$ have rarely been studied. Figures 3 (A-C) show absorbance of $In(OH)_3:xRE^{3+}$ structures at room temperature. Optical diffuse reflectance measurements have been carried out on $In(OH)_3$:xRE particles obtained under MAH conditions. An estimation of the optical band gap "E_{gap}" was obtained using the Wood and Tauc method [32]. Table 1 presents the calculated values of E_{gap} for the samples treated at 140°C for 30 min, pure and with 0, 1, 2, 4 and 8 mol % RE^{3+}.

Figure 3 shows that the increase of the percentage of ion RE^{3+} leads to a reduction in the E_{gap}. The Egap of the pure indium hydroxide cube-like (Egap = 5.12 eV) is like the $In(OH)_3$:xRE (Egap = 3.55 eV - 8%Eu and Egap = 4.90 eV - 8%Tb). Pure $In(OH)_3$ or $In(OH)_3$:xRE belongs to n-type semi-conductors where oxygen vacancies can induce the formation of new energy levels in the band gap. Thus, during the MAH process, the displacement of oxygen related to the In or RE modulates different species of trapped holes ($V_O^{\bullet}$, and species) around $[InO_5.V_O^{x}]$ clusters and gives rise to complex cluster vacancies [21]. Before donor excitation, a hole in the acceptor and an electron in a donor are created according to the following equations, where a Kröger-Vink notation is used for complex clusters [33]:

$$[InO_6]^{x} + [InO_5.V_O^{x}]\ [InO_6]' + [InO_5.V_O^{\bullet}], \quad (01)$$

$$[InO_6]^{x} + [InO_5.V_O^{\bullet}]\ [InO_6]' + [InO_5.V_O^{\bullet\bullet}], \quad (02)$$

$$[InO_6]' + [REO_6]^{\bullet} \rightarrow [InO_6]^{x} + [REO_6]^{x}, \quad (03)$$

$$[InO_5.V_O^{\bullet\bullet}] + [REO_6]^{x}\ [InO_5.V_O^{\bullet}] + [REO_6]^{\bullet}. \quad (04)$$

where $[InO_6]'$ and $[REO_6]^x$ are donors; $[InO_5.V_O^{\bullet}]$ are donors/acceptors and $[InO_5.V_O^{\bullet\bullet}]$ and $[REO_6]^{\bullet}$ are acceptors. RE = Eu^{3+} or Tb^{3+}.

Efficient red and green emission can be observed when this matrix is doped with Eu^{3+} and Tb^{3+} respectively, showing a potential application for the nanoarchitectures to serve as efficient red and green phosphors in luminescent nanodevices. Much interest has been dedicated to the study of PL in nanostructural materials since the first visible PL at room temperature was observed in porous silicon for the first time. Rare ions are used to dope oxides not only as a probe to investigate local centers and energy [34-36], but also to provoke changes in the optical behavior.

Figure 4 depicts room temperature emission spectra of $In(OH)_3$:xRE^{3+} powders prepared by the MAH method, heat-treated at 140°C for 30 min, λ_{EX} 350.7nm. Assuming the PL emission peaks can be described with a Gaussian function, the emission envelope in Figure 4(i) can be fitted to multiple peaks. In the emission envelope from pure $In(OH)_3$, each color represents a different type of electronic transition and can be linked to a specific structural arrangement [37]. The pure $In(OH)_3$ matrix PL curves for all samples are composed of four PL components with the following nm maxima: blue = 450 nm, green = 512 nm, yellow = 570 nm and orange = 623 nm. Green maximum percentage area was observed for the pure $In(OH)_3$. This study of $In(OH)_3$:xTb^{3+} powder emission shows that increases in terbium concentration promote changes in the $In(OH)_3$:xTb^{3+} emission properties related to $[TbO_8]$ clusters (Figure 3ii). These materials produced the predominant band at around 550 nm (the $^5D_4 \rightarrow {}^7F_5$ transition of Tb^{3+}) which is responsible for its green emission. In the emission spectra of samples, another weaker emission at around 495 nm were assigned to the $^5D_4 \rightarrow {}^7F_6$ transition; the emission at around 590 nm was attributed to the $^5D_4 \rightarrow {}^7F_4$ transition, and the group emission at around 627 nm were assigned to the $^5D_4 \rightarrow {}^7F_3$ transition of the Tb^{3+} ions [4]. The relative intensities of the Tb^{3+} emissions increase as the concentration of this ion increases from 1 to 8 mol %; however the broad emission band of the $In(OH)_3$ cube-like clusters was not observed in $In(OH)_3$:xTb^{3+} powders. When the Tb^{3+} concentration is higher, the mechanism which prevails is the efficient energy transfer from the $In(OH)_3$ cube-like clusters to [TbO8] clusters in these materials. This energy migration process is responsible for the increase in Tb^{3+} luminescence [38].

Emission spectra of the $In(OH)_3$:Eu structures exhibit one efficient red PL emission peak centered at around 618 nm, corresponding to the forced electric dipole transition $^5D_0 \rightarrow {}^7F_2$ which is allowed in this case because the europium does not occupy a symmetry center in the indium. Another weaker emission at $^5D_0 \rightarrow {}^7F_0$, $^5D_0 \rightarrow {}^7F_1$, $^5D_0 \rightarrow {}^7F_3$ and $^5D_0 \rightarrow {}^7F_4$ referent to the Eu^{3+} transitions can be observed at around 582, 596, 653 and 701 nm, respectively [39]. The relative intensities of the Eu^{3+} emissions (Figure iii) increase as the concentration of this ion increases from 1 to 8 mol %; however the luminescent emission of the $In(OH)_3$ matrix is drastically quenched. When the RE^{3+} concentration is higher, the mechanism which prevails is the energy transfer from the $In(OH)_3$ matrix to one RE^{3+}. This energy migration process is responsible for the increase in RE^{3+} luminescence [39, 40].

Figure 5 depicts FE-SEM images of the $In(OH)_3$:xRE nanostructures obtained at 140°C for 30 min. Figure 5a shows cubic and irregularly-shaped structures of pure $In(OH)_3$ with a smooth and uniform surface where the angle between the adjacent edges is relatively close to 90°; however, a difference in the morphology of the $In(OH)_3$ to $In(OH)_3$:RE^{3+} was apparent.

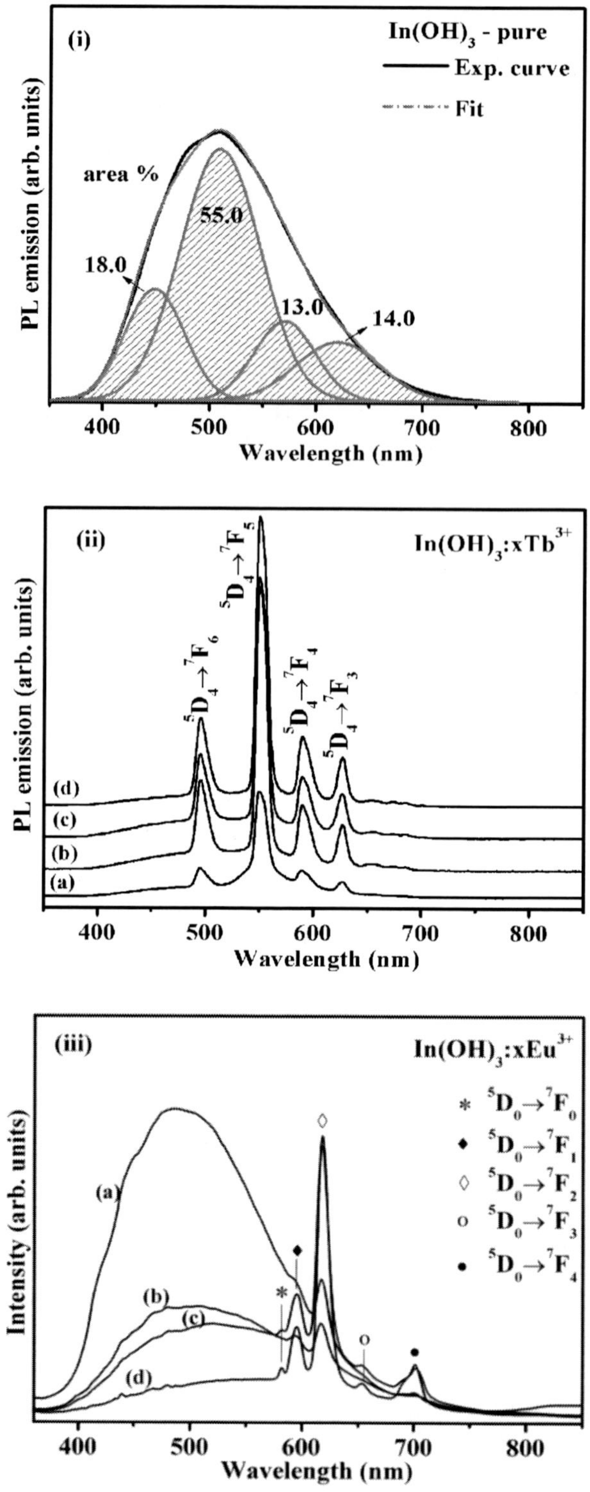

Figure 4. PL spectra of $In(OH)_3$:xRE structures obtained by the MAH method: (i) x = 0 (deconvolution of a PL curve fitted) and (ii) (a) x = 1; (b) x = 2; (c) x = 4; (d) x = 8 mol % Tb^{3+} and (iii) (a) x = 1; (b) x = 2; (c) x = 4; (d) x = 8 mol % Eu^{3+}.

Table I. The calculated values of E_{gap} for the samples treated at 140°C for 30 min, pure and with 0, 1, 2, 4 and 8 mol % RE^{3+}

mol %	In(OH):xEu E_{gap} (eV)	In(OH):xTb E_{gap} (eV)
x = 0	5.12	5.12
x = 1	4.88	5.06
x = 2	3.63	5.12
x = 4	3.56	4.77
x = 8	3.55	4.90

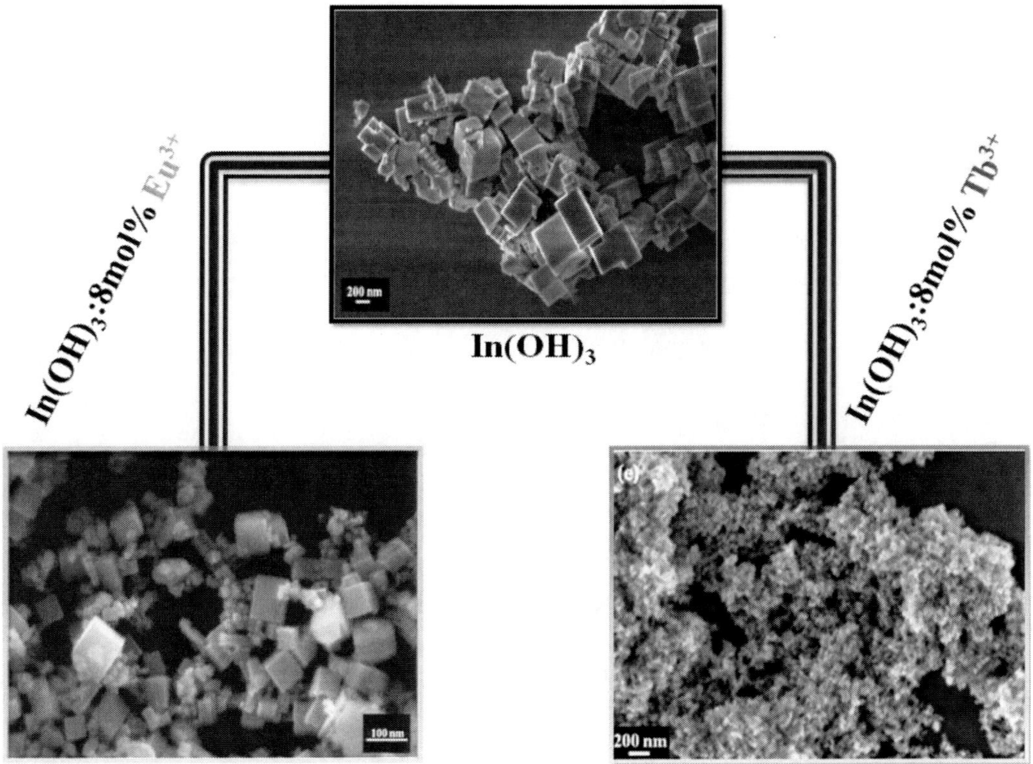

Figure 5. FE-SEM images of $In(OH)_3:xRE^{3+}$ structures obtained by the MAH method at 140 ◦C for 30 min: (a) x = 0, (b) x = 8 mol % Eu, (c) x = 8 mol % Tb.

In the samples with 8 mol % Eu (Figure 5b) and Tb (Figure 5c), the formation of a very defined cube-like structure and a medium-sized decrease for europium and totally desapear when the amount of 8mol % of terbium was used during the synthesis.

Based on the FE-SEM micrograph, the synthesis of $In(OH)_3:xRE^{3+}$ in this work resulted in changes in the formation and morphology of the materials due to an intense and continuous nucleation-dissolution-recrystallization mechanism promoted by microwave hydrothermal processing, [41, 42].

However, when it is added RE^{3+} ions, can be seen the principle the nucleation-dissolution-recrystallization mechanism, where cubic structures of $In(OH)_3$ are dissolved and recrystallized in a different morphology due to presence of RE^{3+} inside of the structure.

As the concentration of RE^{3+} is increases, notes that the cubic morphology of the $In(OH)_3$ starts to disappear and result in anisotropic growth and the formation of irregular morphology. When the quantities of Tb^{3+} achieves 8%, at this point, the high concentration of Tb^{+3}, the lattice parameters in $In(OH)_3$ matrix expand as a function of expansion of dopant ionic radii Tb; therefore an increase of percentage molar of Tb^{3+} ion contained in the arrangement can cause one more distortion of the lattice and consequently provoke one variation in morphology observed.

On the other hand, when the same concentration used for Eu^{3+}, it is possible to observe that lattice of $In(OH)_3$ is less affect, consequently, the cubic morphology still persists in the material, indicanting that in the present system, the behavior observed in figure 5 shows that nucleation-dissolution-recrystallization mechanism has a stronge influence of the RE^{3+} used in the preparation of $In(OH)_3$:xRE.

Conclusion

$In(OH)_3:RE^{3+}$ powders were efficiently obtained in their crystalline phase using the MAH method at 140°C for 30 min. The relative intensities of the RE^{3+} ion emissions increase as the concentration of this ion increases from 1 to 8 mol % while in the $In(OH)_3$ matrix sample the luminescence intensity was drastically quenched. The mechanism which prevails in this case is the energy transfer from the $In(OH)_3$ matrix to one RE ion. As shown in the FE-SEM images, increasing the percentage of earth rare in the structure of $In(OH)_3$ intensily influence on crystal size and morphology.

These optical properties exhibited by $In(OH)_3$: RE^{3+} suggest that this material is a highly promising candidate for photoluminescent applications.

Acknowledgments

The authors thank the financial support of the Brazilian research financing institutions: RECAM (Rede de Catalisadores Ambientais), CNPq and FAPESP-CEPID 98/14324-8.

References

[1] A. M. Cao, J. S. Hu, H. P. Liang, L. J. Wan, *Angew. Chem.-Int. Edit.* 44 (2005) 4391.

[2] S. Avivi, Y. Mastai, A. Gedanken, *Chem. Mater.* 12 (2000) 1229.

[3] T. J. Yan, X. X. Wang, J. L. Long, P. Liu, X. L. Fu, G. Y. Zhang, X. Z. Fu, *J. Colloid Interface Sci.* 325 (2008) 425.

[4] Z. Shi, W. Wang, Z. K. Zhang, *Mater. Lett.* 62 (2008) 4293.

[5] C. C. Silva, F. P. Filho, A. S. B. Sombra, I. L. V. Rosa, E. R. Leite, E. Longo, J. A. Varela, *J. Fluoresc.* 18 (2008) 253.

[6] E. A. Morais, L. V. A. Scalvi, A. Tabata, J. B. B. De Oliveira, S. J. L. Ribeiro, *J. Mater. Sci.* 43 (2008) 345.

[7] D. P. Dutta, V. Sudarsan, P. Srinivasu, A. Vinu, A. K. Tyagi, *J. Phys. Chem. C* 112 (2008) 6781.
[8] N. V. Gaponenko, A. V. Mudryi, O. V. Sergeev, V. E. Borisenko, J. C. Pivin, A. S. Baran, E. A. Stepanova, A. I. Ratko, J. F. McGilp, *Phys. Status Sol.* (A) 165 (1998) 131.
[9] H. Zhu, X. L. Wang, Z. J. Wang, C. Yang, F. Yang, X. R. Yang, *J. Phys. Chem. C* 112 (2008) 15285.
[10] H. Zhu, X. L. Wang, F. Yang, X. R. Yang, *Cryst. Growth Des.* 8 (2008) 950.
[11] F. V. Motta, R. C. Lima, A. P. A. Marques, E. R. Leite, J. A. Varela, E. Longo, *Mater. Res. Bull.* 45 (2010) 1703.
[12] F. V. Motta, R. C. Lima, A. P. A. Marques, M. S. Li, E. R. Leite, J. A. Varela, E. Longo, *J. Alloy. Compd.* 497 (2010) L25.
[13] H. L. Zhu, N. Y. Wang, L. Wang, K. H. Yao, X. F. Shen, *Inorg. Mater.* 41 (2005) 609.
[14] Q. Tang, W. J. Zhou, W. Zhang, S. M. Ou, K. Jiang, W. C. Yu, Y. T. Qian, *Cryst. Growth Des.* 5 (2005) 147.
[15] H. L. Zhu, Y. Wang, N. Y. Wang, Y. Li, J. Yang, *Mater. Lett.* 58 (2004) 2631.
[16] N. Koga, T. Kimizu, *J. Am. Ceram. Soc.* 91 (2008) 4052.
[17] L. Y. Chen, Y. G. Zhang, W. Z. Wang, Z. D. Zhang, *Eur. J. Inorg. Chem.* (2008) 1445.
[18] L. Wang, L. A. Pérez-Maqueda, E. Matijevic, *Colloid Polym. Sci.* 276 (1998) 847.
[19] L. A. Pérez-Maqueda, L. Wang, E. Matijevic, *Langmuir* 14 (1998) 4397.
[20] S. Komarneni, R. Roy, Q. H. Li, *Mater. Res. Bull.* 27 (1992) 1393.
[21] R. C. Lima, L. R. Macario, J. W. M. Espinosa, V. M. Longo, R. Erlo, N. L. Marana, J. R. Sambrano, M. L. d. Santos, A. P. Moura, P. S. Pizani, J. Andrés, E. Longo, J. A. Varela, *J. Phys. Chem. A* 112 (2008) 8970.
[22] M. L. Moreira, G. P. Mambrini, D. P. Volanti, E. R. Leite, M. O. Orlandi, P. S. Pizani, V. R. Mastelaro, C. O. Paiva-Santos, E. Longo, J. A. Varela, *Chem. Mater.* 20 (2008) 5381.
[23] D. P. Volanti, D. Keyson, L. S. Cavalcante, A. Z. Simoes, M. R. Joya, E. Longo, J. A. Varela, P. S. Pizani, A. G. Souza, *J. Alloy. Compd.* 459 (2008) 537.
[24] J. C. Sczancoski, L. S. Cavalcante, M. R. Joya, J. A. Varela, P. S. Pizani, E. Longo, *Chem. Eng. J.* 140 (2008) 632.
[25] J. C. Sczancoski, L. S. Cavalcante, N. L. Marana, R. O. da Silva, R. L. Tranquilin, M. R. Joya, P. S. Pizani, J. A. Varela, J. R. Sambrano, M. Siu Li, E. Longo, J. Andrés, *Curr. Appl. Phys.* 10 (2010) 614.
[26] C. Li, S. Lian, Y. Liu, S. Liu, Z. Kang, *Mater. Res. Bull.* 45 (2010) 109.
[27] J. Huang, L. Gao, *Cryst. Growth Des.* 6 (2006) 1528.
[28] X. X. Xu, X. Wang, *Inorg. Chem.* 48 (2009) 3890.
[29] L. Chen, X. Ma, Y. Liu, Y. g. Zhang, W. Wang, Y. Liang, Z. Zhang, *Eur. J. Inorg. Chem.* (2007) 4508–4513.
[30] H. L. Zhu, K. H. Yao, H. Zhang, D. R. Yang, *J. Phys. Chem. B* 109 (2005) 20676.
[31] S. G. Chen, Y. F. Huang, Y. Cheng, Q. Xia, H. W. Liao, C. G. Long, *Mater. Lett.* 62 (2008) 1634.
[32] D. L. Wood, J. S. Tauc, *Phys. Rev. B* 5 (1972) 3144.
[33] F. A. Kröger, H. J. Vink, I. F. Seitz, D. Turnbull, Sol. State Phys., Academic Press, New York, 1956.
[34] H. M. Xiong, D. P. Liu, Y. Y. Xia, J. S. Chen, *Chem. Mater.* 17 (2005) 3062.

[35] S. Fujihara, Y. Ogawa, A. Kasai, *Chem. Mater.* 16 (2004) 2965.
[36] E. W. Draeger, J. C. Grossman, A. J. Williamson, G. Galli, *J. Chem. Phys.* 120 (2004) 10807.
[37] F.V. Motta, A.P.A. Marques, J.W.M. Espinosa, P.S. Pizani, E. Longo, J. A. Varela, *Curr. Appl. Phys.* 10 (2010) 16.
[38] Z. Li, J. Zeng, C. Chen, Y. Li, *J. Cryst. Growth* 286 (2006) 487.
[39] I. L. V. Rosa, A. P. A. Marques, M. T. S. Tanaka, F. V. Motta, J. A. Varela, E. R. Leite, E. Longo, *J. Fluoresc.* 19 (2009) 495.
[40] Y. Yu, D. Chen, Y. Wang, P. Huang, F. Weng, M. Niu, *Phys. Chem. Chem. Phys.* 11 (2009) 8774.
[41] Z. Luo, H. Li, H. Shu, K. Wang, J. Xia, Y. Yan, *Cryst. Growth Des.* 8 (2008) 2275.
[42] A. Cameirão, R. David, F. Espitalier, F. Gruy, *J. Cryst. Growth* 310 (2008) 4152.

In: Indium
Editors: Hsaio G. Woo and Huang Tsai Choi
ISBN: 978-1-62257-696-8

Chapter 12

HYDROTHERMAL SYNTHESIS OF INDIUM TIN OXIDE NANOPARTICLES WITHOUT CHLORINE CONTAMINATION

Jianrong Zhang[1,2,*], Ling Ling[1], Guodong Xu[1] and Lian Gao[2]
[1]Department of Chemistry, Key Laboratory for Advanced Materials of Ministry of Education, East China University of Science and Technology, Shanghai, P. R. China
[2]State Key Lab of High Performance Ceramics and Superfine Microstructure, Shanghai Institute of Ceramics, Chinese Academy of Sciences, Shanghai, P. R. China

ABSTRACT

Indium tin oxide ($In_2Sn_{1-x}O_{5-y}$) nanoparticles were synthesized by hydrothermal method from stable indium tin acetylacetone complexes and postannealed at 600 °C. The absence of chlorine ions shortens the synthesis process, decreases the particle agglomeration and improves the particle purity. The introduced complexing ligand acetylacetone decreases the obtained nanoparticle size. The improved powder properties accelerates the sintering of the $In_2Sn_{1-x}O_{5-y}$ nanoparticles and reaches a relative density of 96.4% of theoretical density (TD) when pressureless sintered at 1400 °C.

1. INTRODUCTION

Indium tin oxide (ITO) is the most widely used transparent conductive oxide (TCO) owing to its combination of high conductivity and high optical transparency in visible region. Most of the transparent electrodes (e. g. in displays, OLEDs, and solar cells) are made of ITO films. ITO also reflects infrared radiation and can be used as heat mirrors to reduce energy

* Corresponding author: Jianrong Zhang. Tel.: +86 21 64253294; fax: +86 21 51685620. E-mail: zhangjr@ecust.edu.cn.

consumption [1-5]. Almost all the ITO materials are applied in form of films and only a small amount is used in the form of particles added in the polymer to increase electrical conductivity [6]. Among the various techniques to deposit ITO films, chemical vapor deposition, vacuum evaporation, spray pyrolysis, and the newly developed printing methods, magnetron sputtering is the most preferred because it is advantageous in controllability of deposition conditions and adaptability to large area [1,2]. Large area deposition of ITO films has been the trend from view of market and cost. It is well known that sputtering efficiency and properties of the ITO films are strongly dependent on characteristics of ITO target and the most important parameter is target density [1,3,4,5,7]. Study for higher ITO target density, more microstructural homogeneity and lower fabrication cost has been devoted too much works. Obtaining monodispersed indium tin oxide particles with high sinterability and high purity is of the first importance for ITO materials.

Coprecipitation is a simple and economic method to obtain ITO nanoparticles and almost all ITO powders appeared in the market are produced by this method [8,9]. But coprecipitation method has several drawbacks. The chemical homogeneity is difficult to control and the indium tin hydroxide precursors require calcination to be transformed to oxide, which leads to particle aggregation. Previous study showed that the resistivity of SnO_2 phase is several orders of magnitude higher than that of ITO matrix. So, a less homogeneous ITO target results in nodule formation and in turn decreases the target utility and film deposition efficiency [3]. The state of powder aggregation strongly affects the sintering behavior. The increase of aggregate content decreases green density, increases sintering temperatures and decreases microstructural homogeneity [10]. In order to decrease the particle aggregation, emulsion techniques were developed to synthesize ITO nanoparticles [11,12]. Buhler et al. obtained conductive ITO nanocrystals by microwave-assisted coprecipitation in ionic liquids (ILs) [2]. But these methods need comparatively large amounts of organic compounds, which makes the hydroxide washing step very tedious and increases cost sharply. No suppliers acknowledge that their ITO powders are produced by emulsion methods. To overcome the drawbacks of coprecipitation, hydrothermal method is introduced to synthesize oxide powders and also applied to ITO particles. The main advantages of hydrothermal synthesis are homogeneous nucleation, which leads to chemical homogeneity; avoidance of calcination step which decreases degree of aggregation [13]. Yanagisawa et al. hydrothermally treated the indium tin hydroxides obtained by coprecipitation, followed by calcination at 500-1000 °C and obtained $In_2Sn_{1-x}O_{5-y}$ nanoparticles. They pressurelessly sintered the powder compact at 1450 °C and obtained a relative density of 93% which is the highest value in the literature [5,14,15]. This result revealed that the hydrothermally obtained indium tin oxide powder showed high sinterability. In order to decrease the hydrothermal pressure, Xu et al. changed the mineralizer NH_3 to NaOH and obtained the indium tin oxide nanoparticles after heat-treated at 500 °C [16]. But the introduced Na^+ ions increase the impurity content and deteriorate the ITO film conductivity when these powders were used to fabricate ITO sputtering target. Yanagisawa and Xu used indium tin hydroxides as the hydrothermally starting materials and the hydroxides underwent dissolution-precipitation transformation during the hydrothermal process which needed a high temperature and a long period of time. For example, Yanagisawa et al. used 300 °C [14,15] and Xu et al. decreased the hydrothermal temperature to 250 °C [16]. Fang et al. tried to synthesize ITO nanoparticles in sub- and supercritical water from the mixture solution of $SnCl_4$ and $InCl_3$ [17]. They only obtained a mixture of SnO_2 and

In_2O_3 nanoparticles, which may be ascribed to the strong acidic synthesis condition. The ITO powders produced by the traditional wet chemical methods usually need an additional reducing treatment to increase the particle conductivity. Choi et al. changed the hydrothermal solvent water to ethylene glycol (solvothermal) and obtained bluish conductive cubic ITO nanopowders, directly [13]. Indium nitrate or chloride and tin chlorides are commonly used as the starting materials because they are inexpensive and easy to control the synthesis parameters. But the chlorine ions are very difficult to be rinsed from the colloidal precipitates even after ten times of washing and such a procedure is very inefficient. It is well known that the presence of chlorine ions in oxide powders forms hard agglomeration between nanoparticles, hinders powder densification and decreases particle purity. Kim et al. [8] found that the introduced chlorine ions changed the morphology of the obtained ITO nanoparticles from spherical to needle-like, which is disadvantageous for ITO densification. The presence of high concentrated chlorine ions in autoclave is very corrosive under the high temperature and high water pressure working condition. Indium and tin alkoxides are good choice to avoid the contamination of chlorine ions, but they are expensive, water-sensitive and hard to control [18].

In the present study, we synthesized indium tin oxide nanoparticles by a novel hydrothermal method. We firstly synthesized stable indium tin complexing solution without introducing chlorine ions. Then hydrothermally treated the solution at 200 °C for 10 h. The obtained particles were calcined at 600 °C and the yellowish indium tin oxide nanopowders were produced. We sintered the powders in air and reached a relative density of 96.4% of theoretical density at 1400 °C.

2. EXPERIMENTAL PROCEDURE

The indium tin oxide nanoparticles were synthesized by a novel hydrothermal method. Metal indium (5N purity) was dissolved in concentrated HNO_3. Metal tin (5N purity) was dissolved in 8 $molL^{-1}$ HNO_3 in ice water bath and then mixed these two solutions. Calculated amount of acetylacetone(acac) (molar ratio of acac / ([In]+ [Sn]) was set at 2:1) was added to the above solution. Aqueous ammonia (25 wt%) was added dropwise into the solution to adjust pH to ~7. Then this solution was transferred into a sealed stainless Telfon-lined autoclave with a capacity of 100-mL and maintained at 200 °C for 10 h. After air-cooled to room temperature, the derived black product was washed with distilled water and absolute alcohol, repeatedly. The yellowish indium tin oxide nanoparticles were obtained after heat-treated in muffle furnace at 600 °C for 3 h.

The indium tin oxide nanopowders (SnO_2 content was 10 wt%) were formed into pellets of 10 mm in diameter and 1.5 mm thick by uniaxial pressing, followed by cold isostatic pressing (CIP) at a pressure of 200 MPa. Then the compacts were sintered at different temperatures for 5 h in air at a constant heating rate of 5 ^{o}C/min and obtained green to black ITO ceramics.

The phase structure of the indium tin oxide nanoparticles and sintered ceramics was characterized by powder X-ray diffraction (XRD, D/max 2550V, Rigaku, Tokyo, Japan) (Cu K_{α}, λ=1.5418 Å). Weight variation upon thermal treatment of the hydrothermally obtained particles was characterized by simultaneous thermogravimetry and differential scanning calorimetry (TG-DSC, Netasch STA 449 analyser, Germany). The specific surface area (A) of the indium tin oxide particles was measured by multipoint Brunauer, Emmett and Teller adsorption (BET, Micrometritics ASAP 2010, London, British) The morphology of the indium tin oxide nanoparticles was examined by transmission electron microscopy (TEM, JEOL-2100F, Tokyo, Japan). The microstructures of the polished and thermally etched ITO ceramics were observed by scanning electron microscope (SEM, JSM 6700F, Tokyo, Japan). The density of the ITO ceramics was determined using Archimedes' method in water.

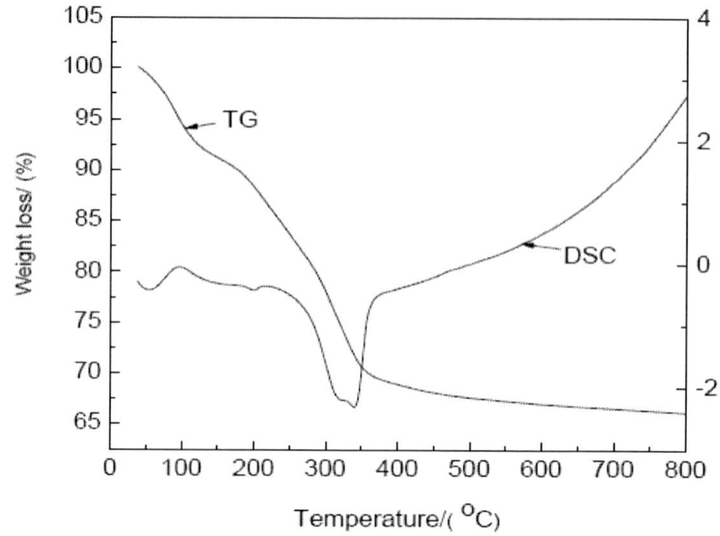

Figure 1. TG-DSC curves of the hydrothermally obtained powders.

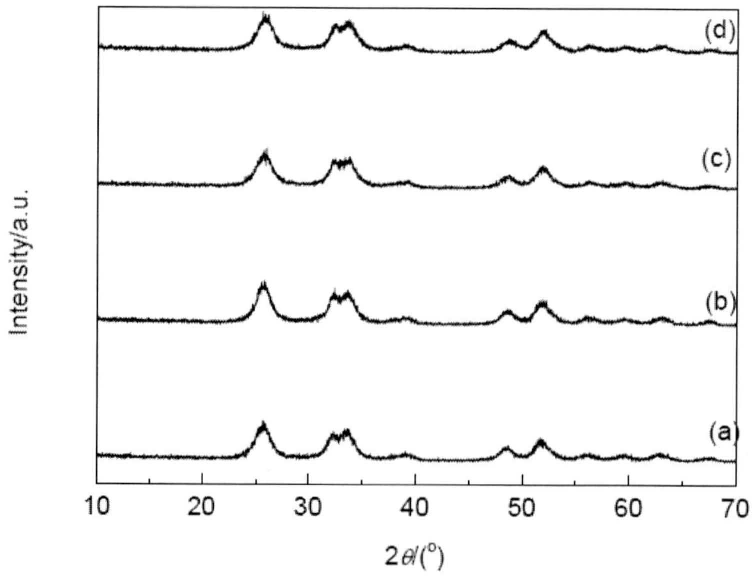

Figure 2. XRD patterns of the hydrothermally obtained InOOH nanoparticles with different tin/indium molar ratios (a) 3 at%; (b) 6 at%; (c) 9 at%; (d) 12 at%.

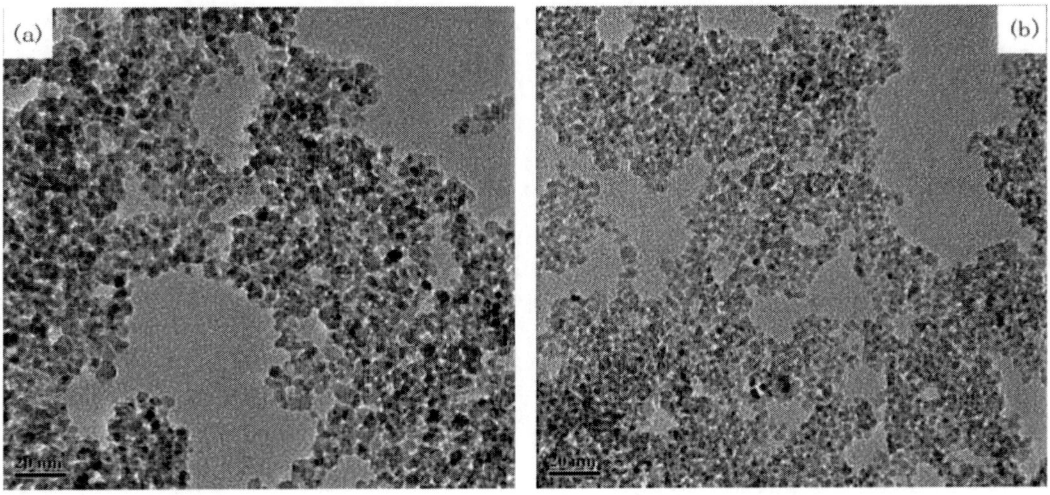

Figure 3. TEM photographs of the hydrothermally obtained InOOH nanoparticles with tin/indium molar ratios at (a) 3 at%; (b) 9 at%.

3. Results and Discussion

Figure 1 presents the typical TG-DSC curves of the hydrothermally obtained powders (molar ratio of Sn/In at 9 at.%). The first weight loss of 6.5% (from room temperature to 130 °C) with an endothermic feature is attributed to the evaporation of the adsorbed water and residual EtOH. The second weight loss of 23% (ca. 400 °C) with a heavy exothermal feature at 340 °C originates from the oxidation of the residual carbon (the black appearance of the hydrothermally obtained particles revealed that acac was decomposed under the hydrothermal condition). Above 400 °C, the weight remains constant and has no endo- or exothermal character, these results reveal that the particles transformed into stable phase before 400 °C.

Figure 2 shows XRD patterns of the hydrothermally synthesized particles with different tin doping content. All peak positions correspond well with the reflections of orthorhombic InOOH (JCPDF 17-0549). This result agrees with the particles obtained by the hydrothermal method from the indium tin hydroxide precursors but we decreased the hydrothermal temperature from 250 °C or 240 °C to 200 °C [5,15,16]. In our experiment, the free indium and tin ions were released from the stable complexing ions steadily under the hydrothermal conditions and then nucleated and grew into the InOOH particles. In the hydroxide precipitates hydrothermal process, the hydroxides $In(OH)_3$ and $Sn(OH)_4$ underwent a dissolution-precipitation process and transformed to InOOH phase. This comparison revealed that hydrothermal treatment is essential for the partly dehydrated InOOH formation. Kim et al. [19] found that precipitated $In(NO_3)_3$ and $SnCl_4$ at a temperature of 5 °C also obtained the InOOH phase, which revealed that a relatively low temperature accelerated the $In(OH)_3$ dehydration. The XRD peak position shifted steadily toward lower angle side with the increase of tin doping level, indicating an increase in the lattice parameter. We failed to calculate the lattice parameters by least-squares refinement because of the relatively low particle crystallinity. The particle crystallinity decreased with the increase of tin content, in agreement with what Yanagisawa et al. [14] observed. The average crystallite size calculated by the Scherrer equation based on the width of the (100) reflection plane shows that the crystallte size is about 6.3 nm for the 3 at% doping level and decreases to 4.5 nm when doping content was 12 at%. Figure 3 shows TEM images of the 3 at% and 12 at% tin doped InOOH nanoparticles. The particle size ranges from 5 to 9 nm for the 3 at% doped sample and decreases to 3 to 6 nm for the 12 at% InOOH particles. The particles are weakly connected with one another ascribed to the high surface energy of the nanoparticles and the adhesive decomposed acac residuals. Yanagisawa et al. obtained InOOH particles in the size of 15 nm with heavy degree of agglomeration. Yin et al. [20] suggested that in hydrothermal process, the complex ions coordinated to the metal ions, making the nucleation completed at the early stage of hydrothermal process and inhibiting crystal growth.

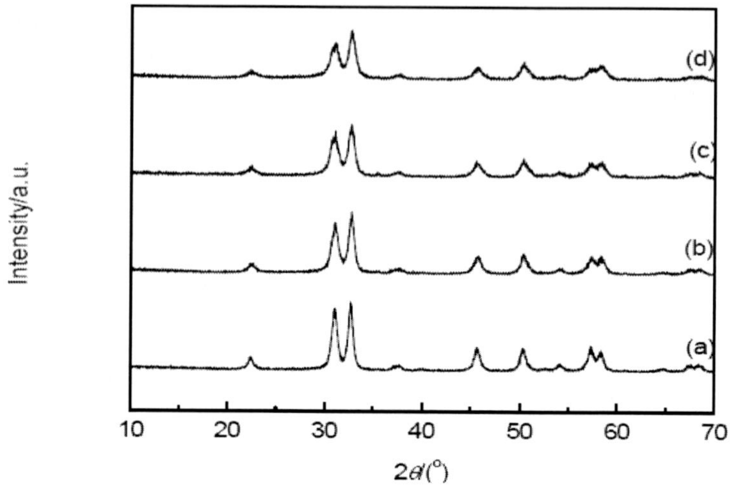

Figure 4. XRD patterns of the $In_2Sn_{1-x}O_{5-y}$ nanoparticles with different tin/indium molar ratios (a) 3 at%; (b) 6 at%; (c) 9 at%; (d) 12 at%.

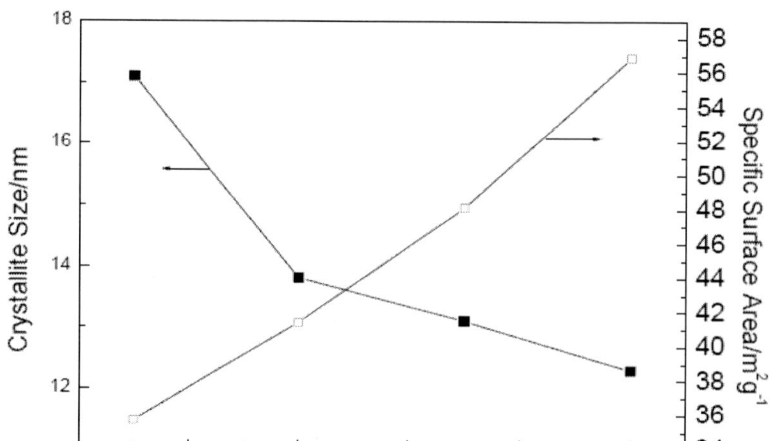

Figure 5. Crystallite size and specific surface area of the $In_2Sn_{1-x}O_{5-y}$ nanoparticles as a function of tin doping content.

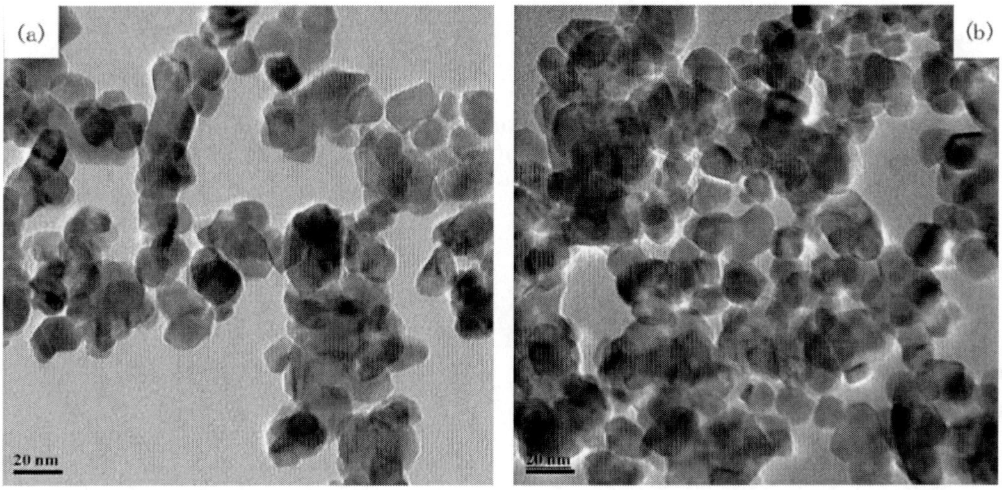

Figure 6. TEM photographs of the $In_2Sn_{1-x}O_{5-y}$ nanoparticles with tin/indium molar ratios at (a) 3 at%; (b) 9 at%.

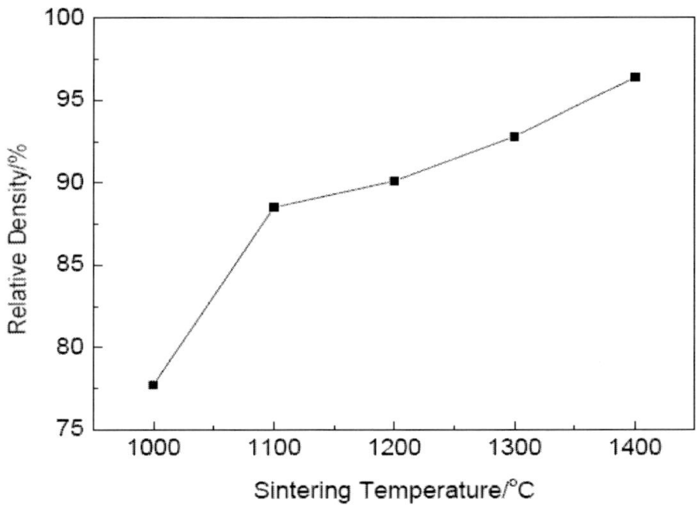

Figure 7. Relative density of the fabricated ITO ceramics as a function of sintering temperature.

The hydrothermally obtained tin doped InOOH nanoparticles were calcinated at 600 °C to be transformed to indium tin oxide particles. Figure 4 shows XRD patterns of the particles. The InOOH phase has been dehydrated completely to the substitutional vacancy type interstitial solid solution of $In_2Sn_{1-x}O_{5-y}$ agreed with what the previously obtained.[5,14,15] The crystallite size calculated from the plane 2θ at 31° and the specific surface area of the particles with different tin content are presented in Figure 5. The crystallite size decreases from 17.1 nm gradually to 12.3 nm with the increase of tin content from 3 at % to 12 at%, while the specific area increases from 35.7 m^2g^{-1} to 56.9 m^2g^{-1}. TEM images provided in Figure 6 show that the 3 at% doped $In_2Sn_{1-x}O_{5-y}$ nanoparticles have a size of 15 to 23 nm and size of 10 to 17 nm for the 9 at% tin doped sample, growing 2 to 3 times larger than the just hydrothermally obtained InOOH particles. The particles are well-faceted and show high degree of crystallinity and this is advantageous for particle sintering.

Figure 7 shows the relative density of the fabricated ITO ceramics at different sintering temperatures. A relative density of 77.8% is obtained at 1000 °C and this is the highest value for the ITO particles sintered at this temperature in the literature. Yanagisawa et al. sintered the $In_2Sn_{1-x}O_{5-y}$ particles with different tin content at 1250 °C and all the relative density values were below 60%.[14] Sawada et al. used ITO particles with size of 2.97 μm and no densification was observed when sintered up to 1500 °C [21]. This comparison indicates that our $In_2Sn_{1-x}O_{5-y}$ nanoparticles exhibit high sinteribility, which is ascribed to the small particle size, low degree of particle agglomeration and free from chlorine contamination. As is well known, the driving force for densification is lowering the surface free energy by the elimination of solid-vapor interface.

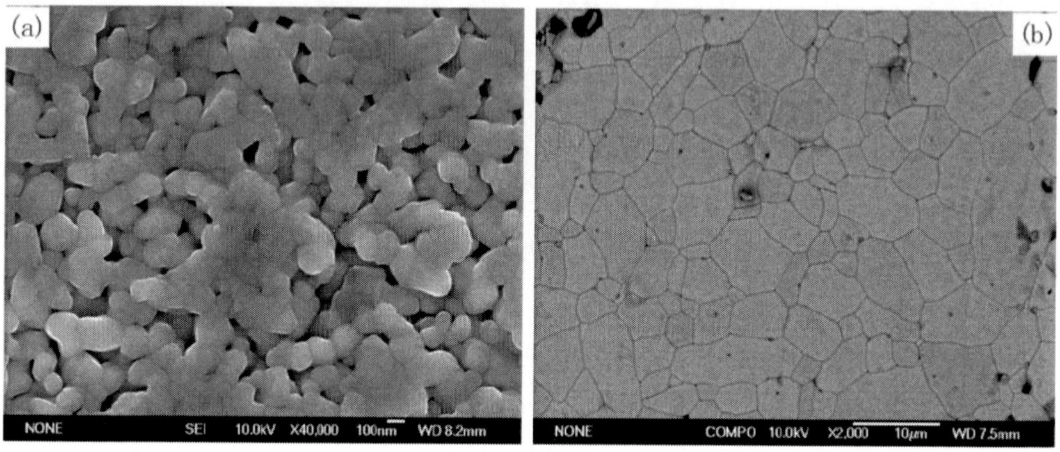

Figure 8. SEM micrographs of the ITO ceramics sintered at (a) 1000 °C and (b) 1400 °C.

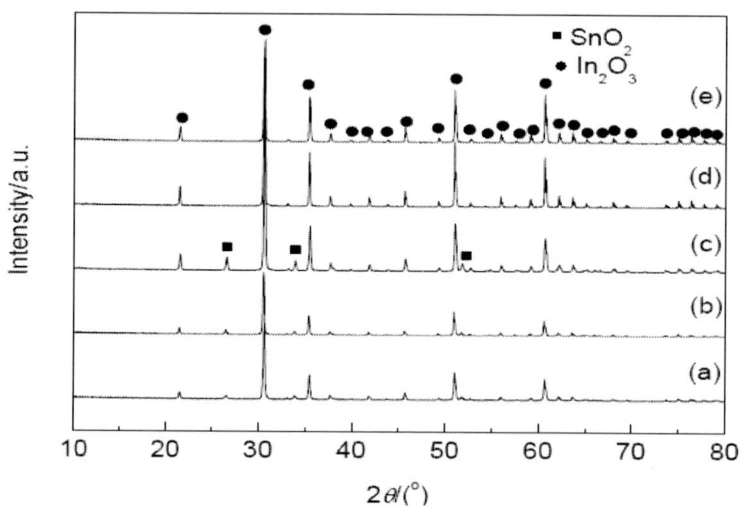

Figure 9. XRD patterns of the ITO ceramics sintered at different temperatures (a) 1000 °C; (b) 1100 °C; (c) 1200 °C; (d) 1300 °C; (e) 1400 °C.

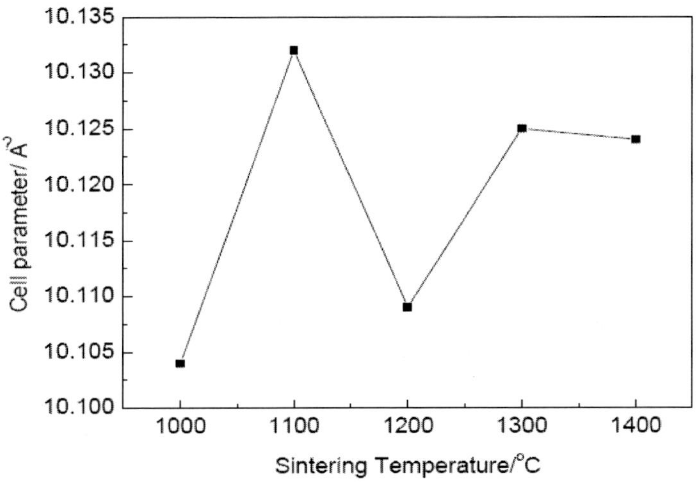

Figure 10. Lattice parameter of the ITO ceramics as a function of sintering temperature.

The smaller the particle size, the higher the surface free energy, the easier of surface atom diffusion between neighbour particles. The densification rate, dρ/dt, is dependent on the grain size (d), as predicted by several well-known sintering models: $d\rho/dt \propto (1/d)^n$, where the grain size exponent n has a valve of 3 or 4, depending on the dominant diffusion path. A decrease in the grain size by a factor of 2 will increase the densification rate by approximately one order of magnitude [22]. So, the synthesis of nanoparticles is of the first importance in fabrication ceramics having improved properties. But nanoparticles are not always beneficial for sintering, the agglomeration which is usually present in the nanoparticles due to the high particle surface free energy retards the compaction of the particles, decreases the sintering rates and leads to inhomogeneous microstructure in the ceramics [23]. Figure 8 (a) is SEM photograph of the ITO pellet sintered at 1000 °C. The particles sintered by neck formation and the particle size grew to about 150 nm. Further raising the sintering temperature increases the relative density of the ITO ceramics gradually and the highest value 96.4% was reached when sintered at 1400 °C. Figure 8 (b) shows that the ceramics has a dense surface with a few small pores. The grain size is about 5.9 μm and is about forty times larger than the 1000 °C sintered pellet. The grain growth is much faster in the final stage of densification when the density is above 90% TD. To achieve densification without grain growth, the grain boundary diffusion needs to be maintained while grain boundary migration should be suppressed. Solute drag is an effective strategy for controlling rapid grain growth in the final stage sintering [25]. Densification of In_2O_3 ceramics is very difficult by the conventional sintering methods due to the rapid vaporization of In_2O_3 at temperatures above 1200 ℃ [4]. In order to suppress evaporation, the fabrication of ITO ceramics is usually sintered in oxygen atmosphere [9].

Figure 9 shows XRD patterns of the sintered ITO ceramics. Cubic In_2O_3 is the major phase in all patterns which indicates that high temperature sintering accelerates $In_2Sn_{1-x}O_{5-y}$ phase transforming to cubic In_2O_3 (JCPDF 44-1087). Small amount of SnO_2 (JCPDF 41-1445) is detected in the ITO ceramics sintered up to 1200 °C and is not observed in the 1300 °C, 1400 °C sintered pellets. This result reveals that the tin ions precipitate in the form of SnO_2 when its content is above the solubility in In_2O_3 (the tin content is 10 wt% in our sintered pellets)[8,24]. Previous studies indicated that SnO_2 showed a high evaporation rate for temperatures higher than 1200 °C due to the peritectic decomposition according to the reaction [22]:

$$SnO_2\,(s) \rightarrow SnO\,(g) + 1/2\,O_2\,(g) \quad (1)$$

So the disappearance of SnO_2 phase in the 1300 °C, 1400 °C sintered pellets may be ascribed to the evaporation of the precipitated SnO_2. The sublimation of SnO_2 in ITO is accelerated when temperature is above 1200 °C has also been observed directly by Palmer et al [26]. The XRD patterns also show the peak sharpening increases with the increase of sintering temperature, which witnesses the growth of the ITO grains.

Figure 10 shows lattice parameters of the ITO ceramics determined by a least-squares method based on the XRD patterns. The lattice parameter a shows random change with the sintering temperature. The a is 10.104 Å for the 1000 °C sintered ITO ceramics, which is little smaller than that of bulk pure In_2O_3 (a = 10.118 Å, JCPDF 44-1087). The lattice parameter shows an abrupt increase to 10.132 Å for the 1100 °C sintered pellet and decreases to 10.109 Å for 1200 °C sintered ceramics, increases again to 10.125 Å and 10.124 Å for the 1300 °C and 1400 °C sintered pellets, respectively. Previous studies showed that only Sn^{4+} is present in ITO samples. So the lattice parameter should decrease when Sn^{4+} substitutes for In^{3+} since the radius of Sn^{4+} (0.71 Å) is smaller than that of In^{3+} (0.81 Å). Similar results have been detected by other researchers. This result reveals that the lattice parameter of In_2O_3 is governed by several reasons other than the substitution mechanism when doped with tin ions [24].

CONCLUSION

Indium tin oxide ($In_2Sn_{1-x}O_{5-y}$) nanoparticles with particle size of about 15 nm were successfully synthesized by hydrothermal and post heat treatment. We obtained the stable indium tin acetylacetone complexes to avoid the contamination of the chlorine ions without using metal alkoxides. The hydrothermal treatment of stable indium tin acetylacetone complexes free from chlorine ion contamination brings several good results: greatly shorten the particle washing steps, improve the powder purity, decrease the nanoparticle size, decrease the particle agglomeration and increase the particle sinterability.

REFERENCES

[1] K. Utsumi, H. Ligusa, R. Tokumaru, P. K. Song, Y. Shigesato, *Thin Solid Films* 445 (2003) 229-34.

[2] G. Buhler, D. Tholmann, C. Feldmann, *Adv. Mater.* 19 (2007) 2224-227.

[3] K. Nakashima, Y. Kumahara, *Vacuum* 66 (2002) 221-26.

[4] N. Nadaud, M. Nanot, and P. Boch, *J. Am. Ceram. Soc.* 77 (1994) 843-46.

[5] C. P. Udawatte, K. Yanagisawa, *J. Am. Ceram. Soc.* 84 (2001) 251-53.

[6] Y. S. Cho, G. R. Yi, J. J. Hong, S. H. Jang, S. M. Yang, *Thin Solid Films* 515 (2006) 1864-71.

[7] T. Omata, M. Kita, H. Okada, S. O. Y. Matsuo, N. Ono, H. Ikawa, *Thin Solid Films* 503 (2006) 22-8.
[8] S. M. Kim, K. H. Seo, J. H. Lee, J. J. Kim, H. Y. Lee, J. S. Lee, *J. Euro. Ceram. Soc.* 26 (2006) 73-80.
[9] B. C. Kim, J. H. Lee, J. J. Kim, H. Y. Lee, J. S. Lee, *Mater. Res. Bull.* 40 (2005) 395-404.
[10] E. A. Barringer, H. K. Bowen, *J. Am. Ceram. Soc.* 65 (1982) C199-C201.
[11] D. W. Kim, S. G. Oh, J. D. Lee, *Langmuir* 15 (1999) 1599-603.
[12] D. W. Kim, S. G. Oh, S. C. Yi, S. Y. Bae, S. K. Moon, *Chem. Mater.* 12 (2000) 996-1002.
[13] J. S. Lee, S. C. Choi, *J. Euro. Ceram. Soc.* 25 (2005) 3307-314.
[14] C. P. Udawatte, K. Yanagisawa, *J. Solid State Chem.* 154 (2000) 444-50.
[15] K. Yanagisawa, C. P. Udawatte, *J. Mater. Res.* 15 (2000) 1404-408.
[16] H. R. Xu, G. S. Zhu, H. Y. Zhou, A. B. Yu, *J. Am. Ceram. Soc.* 88 (2005) 986-88.
[17] Z. Fang, H. Assaaoudi, R. I. L. Guthrie, J. A. Kozinski, *J. Am. Ceram. Soc.* 90 (2007) 2367-371.
[18] M. Toki, M. Aizawa, *J. Sol-Gel Sci. Tech.* 8 (1997) 717-20.
[19] K. H. Seo, J. H. Lee, J. J. Kim, *J. Am. Ceram. Soc.* 89 (2006) 3431-46.
[20] H. B. Yin, Y. Wada, T. Kitamura, T. Sumida, Y. Hasegawa, S. Yanagida, *J. Mater. Chem.* 12 (2002) 378-83.
[21] M. Suzuki, M. Muraoka, Y. Sawada, J. Matsushita, *Mater. Sci. Engin. B* 54 (1998) 46-50.
[22] E. R. Leite, J. A. Cerri, E. Longo, J. A. Varela, C. A. Paskocima, *J. Euro. Ceram. Soc.* 21 (2001) 669-75.
[23] F. W. Dynys, J. W. Halloran, *J. Am. Ceram. Soc.* 67 (1984) 596-601.
[24] G. B. Gonzalez, Northwest University, Evanston, Ellinois, U. S. A. 2003.
[25] X. H. Wang, P. L. Chen, I. W. Chen, *J. Am. Ceram. Soc.* 89 (2006) 431-37.
[26] G. B. Palmer, K. R. Poeppelmeir, T. O. Mason, *Chem. Mater,* 9 (1997) 3 121-3126.

INDEX

B

C

D

E

F

I

J

K

L

M

N

Q

R

S